D1203803

DEVELOPMENT 1992 SUPPLEMENT

GASTRULATION

EDITED BY
CLAUDIO D. STERN AND PHIL W. INGHAM

Papers presented at a meeting of the
British Society for Development Biology
at the University of Sussex
April 1992

THE COMPANY OF BIOLOGISTS LIMITED
CAMBRIDGE

Typeset, Printed and Published by
The Company of Biologists Limited
Department of Zoology, University of Cambridge, Downing Street,
Cambridge CB2 3EJ

Cover designed by Raphael Whittle and Adam Throup

Contents

1992 Supplement

Preface

Gastrulation is the developmental process, involving extensive cell reorganizations, which results in the formation of the mesoderm and gut endoderm of the embryo. The prefix gastr- in fact means 'stomach', which is perhaps both a reference to the shape of the gastrula stage of miolecithal eggs (e.g. sea urchin, *Amphioxus*) and to the fact that the endoderm, which lines the digestive tube, arises during gastrulation. Its importance was eulogised by Lewis Wolpert's famous statement that '*it is not birth, marriage or death but gastrulation that is truly the most important time in your life*'. Not only does the embryo become trilaminar, but it is also during gastrulation that the basic body plan is laid down, the three axes of the embryo become established and many cells receive the signals that lead them to acquire developmental fates and positional information.

We have attempted to collate a volume that reflects the current emphasis of research into gastrulation. Thus, four main themes recur: (a) investigations of the patterns of movement of embryonic cells, (b) analysis of cell fate and developmental potential, (c) the role of induction in cell fate allocation, and (d) the acquisition of more refined regional pattern ('regionalization') beyond differentiation into the major classes of cell types. But we have also tried to put gastrulation into a broader context, both in terms of its evolution and of the history of its study.

The last 5-10 years have seen an explosion in the number of publications concerned with gastrulation and the early development of the mesoderm. From this, one might be tempted to assume that we now understand much more about gastrulation than we did 20 years ago. But a deeper analysis might reveal that our understanding has not increased greatly; perhaps the only thing that we now understand about gastrulation is that the process is far more complex and sophisticated than our predecessors thought. To the existing descriptions of cell movements, made mainly from histological observations at the turn of the century and from vital dye mapping in the 1930s and 1940s, we are now adding more descriptions of morphogenetic movements, and constantly re-discovering the accuracy of the earlier observations. We are also adding descriptions of the patterns of gene expression, particularly of genes suspected of playing a role in gastrulation. Yet unravelling the functional significance of these patterns continues to present us with a major challenge. In Drosophila, of course, such genes have for the most part been discovered through their mutant phenotypes, so that the characterisation of their expression patterns has gone hand in hand with the analysis of their functions. This mutational approach has provided us with some fascinating insights into the cellular basis of the morphogenetic movements that underlie the gastrulation process. Many now hope that a similar approach can be applied to at least one vertebrate, the zebrafish, with similarly fruitful results. Even in Drosophila, however, it is clear that the genetics cannot yield all the answers, and in this and all the other systems that we study we have to face the prospect of characterising large numbers of spatially restricted genes of unknown function. Our goal then will be to make the link between gene expression and morphogenesis, a familiar problem but none the less exciting for that!

For a reader who wants to become familiar with gastrulation, we hope that this Supplement will represent an introduction to this fascinating process and encouragement to make a contribution to the field. We hope to help dispel the view that a knowledge of the genes expressed during gastrulation will suffice to understand the process itself, just as much as we wish to discourage the equally extreme view that genes are irrelevant to this largely epigenetic process.

To a reader already familiar with the embryology of gastrulation, the volume may seem to concentrate unduly on mesoderm, at the expense of endoderm. This is a reflection of the direction of current research, and the paucity of good endodermal markers may be partly responsible. We feel that the importance of the endoderm both during and shortly after gastrulation has been neglected. Perhaps we should not be surprised if the *Development* Supplement volume for the year 2002, containing papers presented at the SDB meeting in, say, Berkeley, were entitled 'The Endoderm'.

> *"What has been achieved is but the first step; we still stand in the presence of riddles, but not without hope of solving them. And riddles with the hope of solution? What more can a scientist deserve?"*. (Hans Spemann, 1927)

C.D.S./P.W.I., Oxford, Summer of 1992

Editors' note:

When we planned this meeting, we were anxious to introduce young students of gastrulation to some of the colourful history of the subject. Johannes Holtfreter would have been an ideal choice, but his health precluded him from being able to accept. We were very pleased when Salome Gluecksohn-Waelsch, who was also a graduate student in Hans Spemann's laboratory when some of the 'organizer' experiments were being conducted, did accept our invitation to give a plenary after-dinner talk. She was to reminisce on the atmosphere in Spemann's lab and on her other colleagues during her distinguished career. A few weeks before the meeting, she was forced to decline because of pressure of grant applications and teaching. Our disappointment at her decision diminished only when she very kindly agreed to our inclusion of the text of this unpublished lecture in this volume. The lecture was given at a Conference on 'Embryonic origins and control of neoplasia' in Dubrovnik, October 13-16, 1986.

Development 1992 Supplement, 1-5 (1992)
Printed in Great Britain © The Company of Biologists Limited 1992

The causal analysis of development in the past half century: a personal history

SALOME G. WAELSCH

Albert Einstein College of Medicine, 1100 Morris Park Avenue, Bronx, NY 10461, USA

I feel greatly privileged having been asked to talk to you here and I want to begin by thanking the organizers for this invitation.

My task of preparing this talk has caused me considerable worry. Obviously, I shall not be able to present here a sound and objective history of embryology over the past 50 years. If nothing else, my great admiration for my close friend Jane Oppenheimer would keep me from being bold enough to step onto her territory, and there have been other serious attempts of an analytical evaluation of embryology during the past half century, e.g. the Nottingham symposium in 1983, published in 1986. What I intend to present here are my personal reflections based on reminiscences over the years during which I had the good fortune of seeing our science develop and of getting to know personally many of the scientists actively involved in the causal analysis of development. Because I am convinced that scientists do not operate intellectually or experimentally in a vacuum totally divorced from personal, social and political phenomena in their environment, I have always paid attention to these extraneous factors, and I am glad to have an opportunity to share here some personal and perhaps unorthodox impressions which I have gained over the years. To me, at least, they add to the total historical image of our branch of science.

I came onto the scene of experimental embryology, as it was called at that time, in 1928. I had completed several semesters studying Zoology and Chemistry in Königsberg, my home town in East Prussia, and in Berlin, and in a course in Embryology I had learned of Spemann's work. In Germany at that time students were still encouraged to migrate as much as possible, and so I decided to go and see if Spemann would accept me as a graduate student. Our first meeting made it quite clear that we were not meant for each other, but I suppose Spemann did not have quite enough courage to turn me down outright. Actually, I had come there at the same time as a fellow student, a young man who became an object of Spemann's love at first sight and who remained his favourite pupil. So Spemann decided that I could be of some help by carrying out a rather boring descriptive study of limb development for my own Ph.D. dissertation which would provide the essential basis for the young man's quite exciting experimental problem of the respective roles of ectoderm and mesoderm in limb pattern formation that Spemann had proposed for his doctoral research. There was no doubt about Spemann's prejudice against women, and my own case serves as an illustration.

Nevertheless, the years in Spemann's Institute were stimulating to the utmost. As a graduate student I shared a lab that housed six pre- and post-doctoral students, among them Hall, an American, Schmidt, a Russian, Sato, a Japanese, and Oscar Schotté of hopelessly mixed nationality. Spemann came to our lab every day, stopped at every desk and discussions between him and the rest of us were extremely instructive. Spemann had high standards of perfection, and we had to learn to live up to them. During the 'season', i.e. that of Amphibian egg laying, all of us worked day and night and we shared results, interpretations, etc. Spemann's eminence in his field of experimental embryology attracted visitors from all over the world, and I remember meeting Walter Vogt who introduced Nile blue as a vital stain of Amphibian embryos, Richard Goldschmidt, Paul Weiss and others there. Spemann was very outspoken in his criticisms, and I never forgot the scathing words with which he demolished Paul Weiss after a seminar the latter gave. In contrast to this, Spemann's admiration for Boveri, who had established the individuality of the chromosomes and their role in heredity, knew no limits and to this day I feel the strong influence this had on me. I never met Boveri, but knew his widow who taught Biology at a girls' college in New Haven, and I had the good fortune of being asked some years ago to translate Boveri's 1902 paper into English. While working on that, I experienced some of the excitement Boveri must have felt in the course of interpreting his experimental results with polyspermy in sea urchins and the development of the chromosome theory of inheritance.

In spite of our great admiration for Spemann, some of us in the institute were surprised by his obvious limitations. I was particularly conscious of his total lack of recognizing and taking into account any genetic considerations in the interpretations of his experimental results, in spite of his being a student of Boveri. When Oscar Schotté in xenoplastic transplantation experiments accomplished the demonstration that anuran tissues when transplanted into urodeles were subject to embryonic induction by urodele, i.e. host, tissues but that the developing structures expressed donor, i.e. anuran, specific traits, Spemann used a military analogy in his interpretations: he compared the response of the induced cells to that of a German soldier to the order of 'salute' after having been put into the French army. The soldier understood the order but in response he would salute the German way. In using this analogy, Spemann totally left out of consideration the possible role of the genetic makeup of the induced tissues. The narrowness of mind expressed in this incident, both on the intellectual and the scientific level, surprised me greatly.

I would like to say a few words about Spemann's political attitudes without analyzing them in detail here. He was a strong German nationalist, full of mistrust towards other nationalities and sharing prejudices of his fellow nationals. I already mentioned his prejudice against women expressed also in his dealings with Hilde Proescholt[1], the discoverer of the 'organizer,' who is reported not to have appreciated it when Spemann added his name to the publication of her thesis while other male students were permitted to publish their work alone. Many years later, the story of the 'dead' organizer provides another example: it was a female graduate student, Else Wehmeier, who first observed embryonic induction with a Bouin-fixed piece of upper blastopore. I was around at that time. Her name did not even appear on the first publication reporting this exciting result!

However, it is not my intention to give you a totally negative picture of Spemann here. I think of him as a very influential and productive figure in 20th century experimental embryology - but I do not worship him as a hero, and in many ways he is a true representative of the German academic atmosphere at that time, which was full of prejudices of various kinds, including anti-semitism. Some of the narrowness expressed in such prejudices extended also into Spemann's science. Conceptually, the 'dead' organizer did not really appeal to him since he was essentially a vitalist whose mysterious concepts actually even served to prevent bold experiments, and he was full of mistrust towards people like Waddington who wanted to explore and identify the chemical nature of the organizer. Waddington came to Freiburg as a young graduate of Cambridge and we became very close friends. He opened our eyes to the biochemical and molecular problems inherent in inductive interactions between cells. He also fully recognized the involvement of genetic mechanisms in developmental phenomena. Waddington remained one of my closest friends until the time of his death.

Among those who influenced me strongly in my scientific development is Viktor Hamburger. He was Spemann's so-called 'assistant' when I became a graduate student in Freiburg. Spemann turned over to him the close supervision of my dissertation research, and Viktor saw to it that I remained on a straight path. At that time, Hamburger's work was concerned with the analysis of the role of innervation in limb development. His approach to science was broad, and he was the only one who provided us students with some introduction to the principles of genetics. He also engaged us in theoretical discussions, and was most likely responsible for arranging the joint seminars with the Department of Philosophy and Heidegger, the phenomenologist, which, though memorable, did not accomplish much. Hamburger also had to leave Germany in 1933, and of course you know of him as one of the founders and leaders of modern Neuroembryology.

In the United States, I had the good fortune of meeting Ross G. Harrison soon after my arrival there in 1933 as a refugee from Hitler. Harrison's dynamic approach to problems of development and differentiation was in many ways no less impressive than that of Spemann, and it seems that he would have deserved to share the Nobel Prize with Spemann. I enjoyed my contacts with Harrison immensely, and I remember particulary his great sense of humour. His early experiment, in which he proved that the nerve fiber arose as an outgrowth of a single neuronal cell, is most impressive. For the purpose of this experiment, he developed a totally new method, i.e. that of tissue culture; however, unlike many scientific technologists, he did not continue to pursue this exciting method for its own sake but only in the service of significant problems.

My own scientific career was decisively influenced by my association with L. C. Dunn at Columbia University who had recognized early the significance of genes in processes of development and differentiation. Furthermore, Dunn appreciated the potential of mutations affecting mammalian development for purposes of identification and analysis of the corresponding normal developmental mechanisms and their genetic control. In particular, Dunn focused his attention on the T-locus and its mutational effects on mouse development. He realized that some knowledge of experimental embryology would be helpful in these studies, and in 1935 he invited me to join his laboratory because of my previous experience. The environment at Columbia University differed greatly from that which I had left behind in Freiburg. Politically, Dunn was extremely progressive. He helped Nazi refugees, fought Fascism and participated in activities of the 1930s, e.g. those of supporting the Spanish Loyalists and the American-Soviet Friendship Committee. In my early years at Columbia, I even met E. B. Wilson who came to the Department on crutches and only rarely, but I still appreciate these meetings. I consider Wilson and his book among the most important milestones in the history of developmental biology.

During the middle and late 1930s and the 1940s, I witnessed the expression of a strong liaison between embryology and genetics. I know that this view contrasts with that held by others, and I believe it may be due to a large extent to my own close contacts with particular people such as Dunn, Goldschmidt, Waddington and Ephrussi. Of course, E. B. Wilson's book also lent strong support to the early and close relationship between the two fields. I got to know Richard Goldschmidt well, after he came to the USA, and I came to admire his breadth of vision, his courage, his imaginative thinking and his knowledge. His well-known and feared arrogance did not bother me, and to this day I appreciate his writings, be it in the book he created for his children, *Ascaris*, his *Physiological Genetics*, his book on 'the material basis of evolution,' or his *Theoretical Genetics*.

Waddington continued his contributions to Developmental Genetics more on the theoretical than the experimental level. His books *Organizers and Genes*, *New patterns in Genetics and Development*, and *The Strategy of the Genes*

[1]Hilde Proescholt is perhaps better known by her married name, Hilde Mangold, co-author with Spemann of their famous 1924 paper. (editors' note)

provided much interpretation and speculation concerning development and its genetic control. As quoted by Needham not long ago, Waddington predicted that for the analysis of development, people would have to come back to experimental embryology since concentration on the genetic code and genetic engineering would never reveal all the secrets of embryogenesis. Waddington's relatively early death prevented him from seeing how this is actually happening now. I would have loved to discuss the current state of affairs with him!

Ephrussi's significant contributions to the causal analysis of development are unfortunately not sufficiently known, e.g. that to the analysis of T-locus effects in 1935 where he used explantation techniques to exclude the possibility that the T mutation acted as a cell lethal. He was a brilliant scientist who perhaps paid too much attention to what he himself called 'l'ordre du jour', and thus changed experimental systems too frequently, instead of pursuing problems in greater detail and depth.

If we return now once more to the question of the Amphibian 'organizer' and ask ourselves about real progress in its analysis since the time of Spemann, i.e. more than 50 years later, answers are restricted largely to the negative category. Many chemical substances have been excluded, and we know to some extent what the organizer is **not**, but we still do not know precisely what it actually represents. I remember a long talk between Spemann, Rudolf Schoenheimer and myself in 1932 in which we tried to design a joint project in order to identify the biochemical basis of the organizer. The project never materialized because of the Nazis. But in the final analysis even now the point has not been reached where the chemical nature of the neural inducing factor, i.e. the organizer, has been determined. Much important work in the years since, on cell and tissue interactions e.g. in the development of the kidney and various glands as studied under in vitro conditions, is in a way an outcome of the pursuit of the question of inductive phenomena during early embryogenesis. I am thinking particularly of the contributions by Edgar Zwilling, a developmental geneticist who worked with Walter Landauer at the University of Connecticut in Storrs, and who made use of methods similar to those developed by Grobstein. Zwilling was a good friend who died quite young.

Among the concepts which have undergone radical changes in the course of the past 50 years is that of the constancy of the genome during development. Whereas the dynamic state of the developing and differentiating cell came to be accepted in the course of transplantation and explantation experiments, the stability of the genes and the constancy and integrity of the DNA during embryogenesis attained the state of a dogma that lasted until not very long ago. Today this dogma is being reevaluated in the light of dynamic progress of modern developmental genetics. In this connection, I remember discussions with Barbara McClintock more than 40 years ago. In my personal history of developmental analysis, these talks with her play an important role, even though Barbara and I disagreed about possible solutions for problems of women in science! I also remember with pleasure discussions with Briggs and King, who presented the first evidence for possible irreversible modifications of the nuclear genome during differentiation. At a meeting many years ago, it was Leroy Hood who first called my attention to the immunoglobulin system and its role in the formulation of the concept of genomic plasticity during development. Of course today the elegant experiments of nuclear transplantation, e.g. those performed by Davor Solter, give the strongest promise of eventually unravelling some of the puzzles of genomic modifications during differentiation.

It is essential at least to make reference to several additional systems of developmental analysis. My personal memories include a meeting with Andrej Tarkovsky, the creator of the chimaera system for studies of development. His first paper on this was published in 1968! Since that time, many variations of experimental chimaera production have been developed, and at this time sophisticated methods for chimaera production are being used by many people, e.g. Jaenisch who injects single transfected cells into blastocysts. These approaches appear most promising in developmental studies and will hopefully be as successful as Jaenisch's earlier experimental design of using insertional mutagenesis for the identification of developmental genes.

In my talk here, I have so far focused exclusively on vertebrate and particularly mammalian embryos, which is appropriate because they are closest to my own scientific activities. In this personal account of the history of causal analysis of development, I must however transcend the mammalian organism and include the recent most exciting work on *Drosophila* by E. B. Lewis of whose fan club I have been a strong member for many years. Lewis's work focuses on the so-called bithorax gene complex and its mutations which disrupt the characteristic segmentation of the fly's body and cause the transformation of particular segments and their specific organs into body segments of different types and with different specific organs, i.e. so-called homeotic mutations. Lewis attempted to correlate these mutational changes of morphogenesis with underlying changes in the relevant genes and their organization. His own classical approach - that of a developmental geneticist who works his way from the mutant phenotype back to the genome - received strong complementation by the interest and active collaboration of molecular biologists such as David Hogness and Welcome Bender and their direct attacks on the DNA level itself with sophisticated molecular genetic techniques. In this way, they succeeded in mapping a considerable stretch of the bithorax complex and locating mutational changes of its DNA and the eventual discovery of homeoboxes. The problems of homeoboxes in various organisms and their developmental effects, are far from solved. But the analysis of the bithorax complex presents a striking example of progress in developmental genetics made possible by the study of the genetic and developmental details of a system with the best tools available and with the imaginative foresight of a classical geneticist. Thus, Lewis prepared the stage for an approach with techniques of molecular genetics that has the potential of promoting the general understanding of the molecular basis of the genetic control of development and differentiation.

I also knew Hadorn well, who actually was the first to make use of homeotic mutations in *Drosophila* in his analysis of the genetic control of cell determination. Hadorn was remarkable in combining a strict and straightforward attitude to life, to science and to all his activities with an unusual degree of imagination in experimental design and interpretation. I remember with great amusement a small meeting in London where Hadorn and Waddington, protagonists of two diametrically opposed Weltanschauungen, clashed repeatedly on issues including science as well as life style.

I am fully aware of many omissions in this account. To mention just one particularly glaring one, it is that of teratocarcinomas and stem cell lines, which offer such promising and potentially valuable systems of analysis.

Obviously one of the most exciting areas, in which much progress has been achieved in recent years, is that of growth factors and the analysis of their role in development and differentiation. It all started with the Nerve Growth Factor and Rita Levi-Montalcini. I remember meeting Rita at a Growth Symposium in the USA in the mid 1950s and hearing the exciting story of NGF, the first growth factor to be discovered. We have remained close friends ever since, and I was delighted to hear of the recent Lasker Award that she received. Of course, throughout the years, the number of growth factors has grown astronomically, and it appears significant that the nucleotide sequences of the genes encoding them and their receptors show some homology to those of oncogenes. The prospects of causal analysis of growth factor action are underlined by the finding of DNA sequence homology between the gene encoding EGF and the *Notch* gene in Drosophila. *Notch* is instrumental in the regulation of epidermal versus nerve tissue differentiation of ectoderm. The gene's deletion, analyzed so beautifully years ago by Poulson who failed to receive the well deserved recognition for his discovery, results in overgrowth of nerve tissue at the expense of epidermal tissue. This adds fascination to the sequence homology!

To return once more to a vertebrate species, I want to mention the zebra fish that George Streisinger only a few years ago put on the map in such a brilliant fashion as a potentially very productive system of developmental genetic analysis. I knew George when he used to visit the Zoology Department at Columbia University as an eager young student at the famous Bronx High School of Science in the 1940s. In this connection, I must also mention Joshua Lederberg who as an undergraduate hung around the Zoology Department at Columbia, and whom I supplied with discarded mice even though the departmental chairman had strictly forbidden this. Lederberg used the mice for experiments of parabiosis where he wanted to study the question of whether liver regeneration might be promoted by a substance circulating in the blood. Lederberg later (1966) rewarded me with statements, e.g. "embryology should be studied with embryos" and that "the mouse should be the central material in developmental biology."

If I appear amiss for not even mentioning the elegant developmental analysis of *Caenorhabditis elegans*, it is because I never had the good fortune of personal experience with this worm as an analytical system. The only exception is my acquaintance with nematode cultures which we grew in Spemann's laboratory on potatoes for the purpose of feeding our Salamander larvae. But seriously, there is no doubt that the intriguing combination of apparently rather strictly determinate cell lineage with the discoveries of an ever increasing amount of cellular interactions and regulative phenomena in this nematode, harbours an enormous potential for the identification of general principles of developmental genetics. This potential is underscored by the limited size and cell number of the worm which make possible a complete analysis.

Finally, I must add my personal view of immediate and future concerns of causal developmental analysis. Our ultimate problem remains that of finding out how the zygote, which contains one set of maternal and one set of paternal genes, manages to differentiate into the multicellular organism with its multitude of cells derived by mitosis and therefore presumably of identical genotypes but totally heterogeneous phenotypes. The expression of particular genes in certain cell types, their failure to be expressed in others etc. etc. is the result of strict regulation of gene activity during differentiation. It is the identification of *cis* and *trans* acting regulatory genes, their gene products and the mechanisms by which they affect various cell type specific structural genes that is among important tasks today.

I cannot resist the temptation of showing here at least one slide which demonstrates a system of analysis that we are exploiting in the search for regulatory genes and their mode of action in mammalian development. With the help of a series of overlapping chromosomal deletions, we have identified a chromosomal region that includes gene sequences instrumental in the trans regulation of expression of a cluster of unlinked structural genes encoding liver specific traits. I shall try to summarize the crucial details of this system in the following figure.

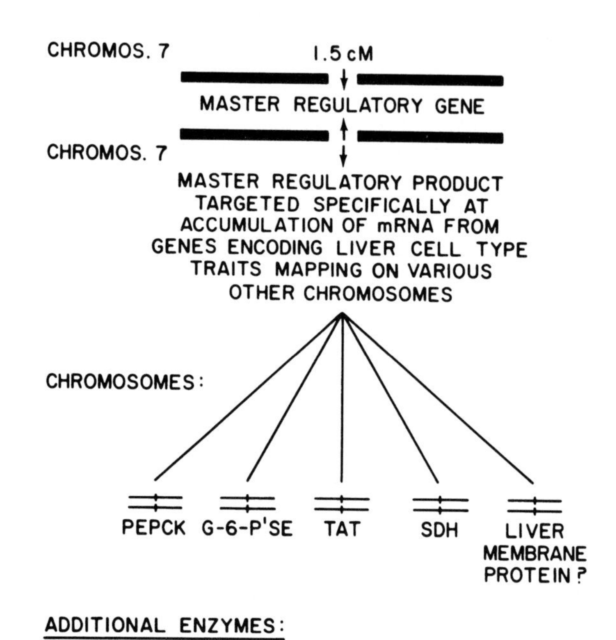

The following questions are in the centre of our interest at this time:

(1) The level of regulation. Inducible but not constitutive expression of the structural gene targets seems to be affected by the deletion. We postulate that this is due to the absence of a *trans* acting regulatory factor essential for the induction of the relevant DNA sequences by inducing stimuli, be they metabolic or hormonal.

(2) Identification of the regulatory factor.

(3) Analysis of the molecular defect.

(4) Cloning of the regulatory gene(s). (Schütz et al. in Heidelberg).

Having reached the present, a look into the future seems appropriate. I personally am increasingly impressed with the degree to which molecular developmental biology and molecular genetics are merging into one science. While giving thought to this lecture and the period it covers, I came to realize the extent to which Boveri's concepts of the role of chromosomes in development and differentiation had prepared the ground for the concepts of molecular geneticists in 1986. The list of contents of a recent symposium entitled 'Molecular Developmental Biology' and held in 1985 in the USA includes four headings, all of them dealing with 'Gene Expression in Development'. The unifying concept of the vast majority of molecular approaches in developmental biology is that of gene expression and its regulation by *cis* as well as *trans* acting regulatory factors. Nonetheless, there is no doubt that problems of development extend beyond the level of gene expression and that molecular developmental biology will have to pay attention to mechanisms of differentiation on these additional levels. I have in mind phenomena such as the role of cell surface molecules in cell interactions, the formation of membranous structures, the role of phosphorylation in differentiation, the general problem of form in all its ramifications - I doubt that the molecular analysis of gene expression by itself can provide answers to these and similar questions. I therefore would expect increasing awareness of the need to focus on molecular and biochemical problems of development beyond the level of gene expression, i.e. that of epigenesis.

To quote and paraphrase Ephrussi and Lederberg once more: embryologists may still have to look at embryos in the pursuit of their analytical studies of development and differentiation.

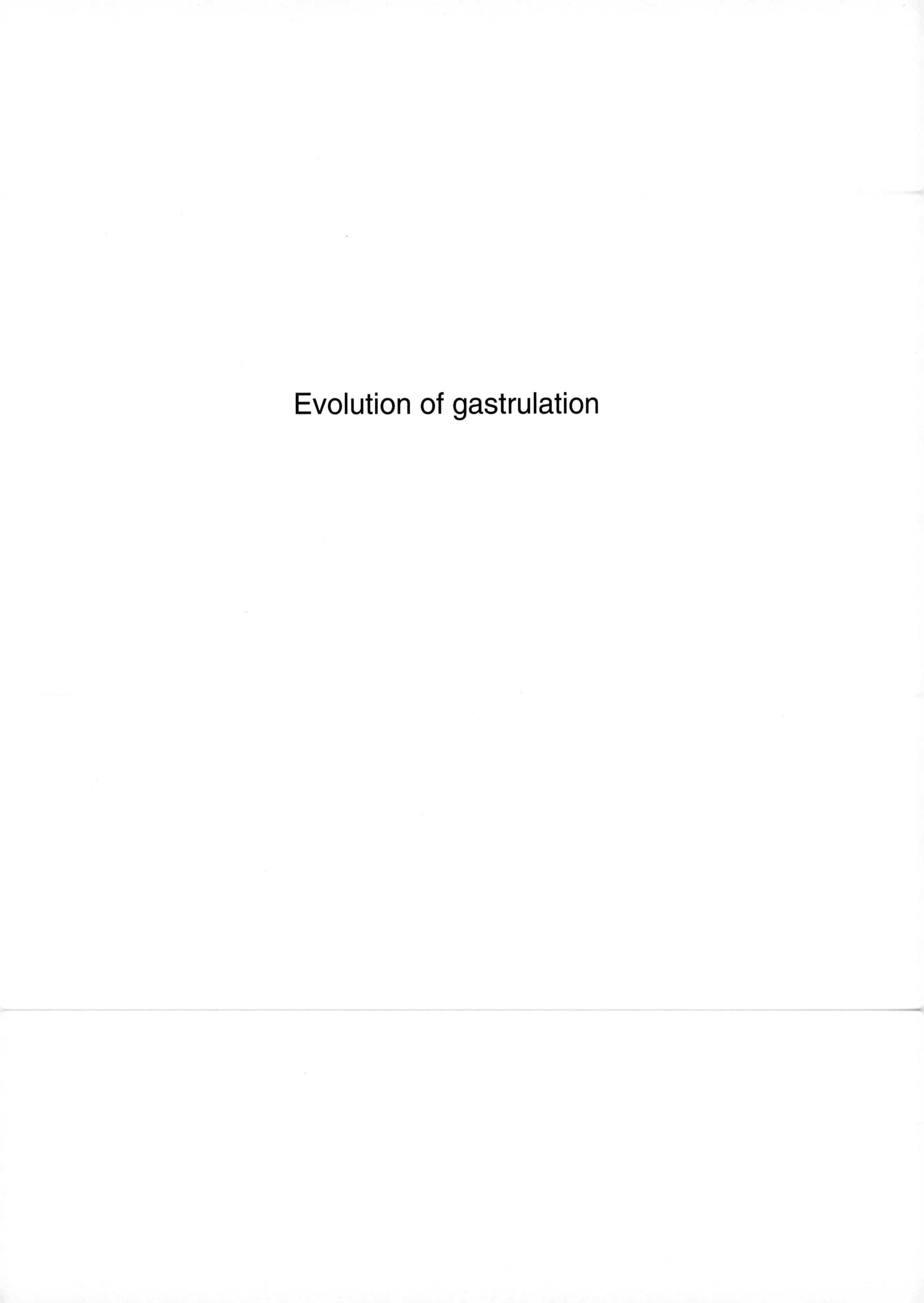

Evolution of gastrulation

Development 1992 Supplement, 7-13 (1992)
Printed in Great Britain © The Company of Biologists Limited 1992

Gastrulation and the evolution of development

L. WOLPERT

Department of Anatomy and Developmental Biology, University College and *Middlesex School of Medicine, London, UK*

Summary

The original eukaryotic cell may have possessed the key processes necessary for metazoan development - cell differentiation, patterning and motility - and these are present in the cell cycle. Protozoa also possess key patterning processes. It remains a problem as to why there should be two main modes of development - one based on asymmetric cell division and the other on cellular interactions. The latter may be related to asexual reproduction.

The morphogenetic movements of gastrulation - as distinct from specifying the body plan - are highly conserved in a wide variety of organisms. This may reflect the requirement for patterning being specified in two dimensions, sheets of cells, and a third dimension being created by cell infolding.

The origin of the gastrula can be accounted for in terms of Haeckel's gastrea theory - an early metazoan resembling the gastrula. Gastrulation in Cnidaria may resemble the primitive condition but there is nevertheless considerable diversity. While this may reflect, for example, yolkiness, it seems that there is little selection on developmental processes other than for reliability.

Thus it is possible that the embryo is privileged with respect to selection and this may help account for the evolution of novel processes like the origin of the neural crest.

Reliability is the key demand made on development. This may be provided by apparent redundancy. Since many developmental processes involve switches and spatial patterning reliability is provided by parallel buffering mechanisms and not by negative feedback.

Key words: gastrulation, evolution, pattern formation, selection, neural crest.

Introduction

By the evolution of development, I mean the evolution of development itself: the way the various developmental mechanisms and embryonic stages evolved and the selective pressures that were acting. Thus one wants to understand the origin of the key process in development - patterning, morphogenesis and cell differentiation, the origin of embryonic stages like gastrulation, and how developmental novelty, like, for example, the neural crest could have arisen. One also needs to understand variants in developmental processes, like different modes of gastrulation and patterns of cleavage, radial and spiral. In considering selective pressures, it is also necessary to try and understand redundancy. We may never know the true answers but that is no reason not to attempt to try and answer such questions, for they lie at the heart of the whole of evolution. In a sense all evolution of multicellular animals is the brilliant result of altering developmental programmes.

If we consider the three basic processes in development - differentiation, spatial patterning and change in form - these are already, it can be argued, well developed in the eukaryotic cell (Wolpert, 1990). One may assume that the original eukaryotic cells had all the basic structures that characterise all cells - membranes, nucleus, mitochondria and so on, and in fact it is remarkable how similar the basic components of all cells are. Given these characteristics little more was required to generate multicellular animals, and so development.

The cell cycle can be taken as a paradigm for several of the key processes in development. Every cell cycle involves a temporal programme of gene activity, a spatial differentiation that assigns the results of growth to the daughter cells, and cell motility, which ensures that the chromosomes are distributed to daughter cells and in animal cells brings about cell cleavage by active constriction. The evidence for a temporal programme of gene activity is well documented (Alberts et al., 1989) and it does not seem unreasonable to consider G_1, S, and G_2, as being homologous with different differentiated cell states.

Two further features of the cell cycle merit attention. The first involves the decision whether or not to enter the cycle following mitosis. This decision is closely linked to growth and the nuclear/cytoplasmic ratio. It also involves some sort of threshold event. In more general terms, it involves intracellular signalling and a switch of a kind that is very similar to many developmental processes.

The other feature meriting attention is mitosis and cleavage. Here there is highly organized spatial patterning within the cell which is also linked to motility. The development of the plane of cleavage at right angles to the spindle foreshadows the orthogonality that is fundamental to embryonic development. In addition, mitosis provides the opportunity for unequal distribution of components to daughter cells that could be the basis for the origin of divergence in cell fate and differentiation. This autonomous generation of differences is clearly present in the division of yeast cells in relation to mating type (Wolpert, 1989a).

It requires little imagination to derive other properties from the primitive cell for development. Cell adhesion and cell-to-cell signalling may be novel but require little modification of a system in which there is flow of material to the membrane with an external coat. Even the provision of an extracellular matrix would seem to present little difficulty given an endoplasmic reticulum and Golgi. There is, perhaps, one cellular process that may require a novel evolution and that is cellular memory. The inheritance of the differentiated state through a cell cycle might require new methods for controlling gene action.

The single cell protozoa show complex patterning with very precise spatial location of organelles (Frankel, 1991). Many of the mechanisms required for the evolution of metazoan development are present in the protozoa (Goodwin, 1989), their presence suggesting that they represent fundamental biological processes possibly linked to the cytoskeleton that arise, as it were, easily and naturally in evolution. Protozoa clearly show polarity in form. The rows of cilia suggest that a mechanism for generating spacing patterns - stripes - is well developed. In the ciliated protozoa, fission involves considerable reorganization of existing structures. For example, in *Tetrahymena,* a new oral apparatus, which is normally located at one end of the animal, begins to develop near the zone of fission - in fact two new cell surface organizations are formed within what was originally one, the two now being in tandem alignment along the anteroposterior axis. This reorganization is essentially complete before fission occurs.

Thus it is clear that single cell organisms have many of the mechanisms required for multicellular development. But these mechanisms must now be used for multicellular patterning. With a mechanism for spacing patterns and positional fields, many patterns are possible (Wolpert, 1989b).

The evolution of early differentiation and patterning

The simplest conceivable multicellular organism would consist of two cells that differ from one another (Wolpert, 1990). This unlikely ancestor would have manifested two key processes of multicellular development, diversity in cell state, or cell differentiation, and spatial heterogeneity; for it was essential that the two cells be different. From the beginning, there needed to be a mechanism to ensure this difference and either cytoplasmic differences or a signal could be the basis (Fig. 1). Both became firmly established. Almost all eggs - mammals providing a major exception - have a well-defined polarity due to cytoplasmic differences, but in most embryos at least some later differences arise not because of cytoplasmic localization but because of cell-to-cell interactions.

It is a major problem in the evolution of development to understand why some animals have very well-defined cell lineages, the asymmetric behaviour of daughter cells being a key feature of their development, whereas others rely largely on cell-to-cell interactions. What does this diversity reflect? One possible scenario for thinking about the difference is to relate it to how cells are specified. In the former, cell fate is specified on a cell by cell basis, whereas, in the latter, groups of cells are specified, and this may be related to asexual reproduction.

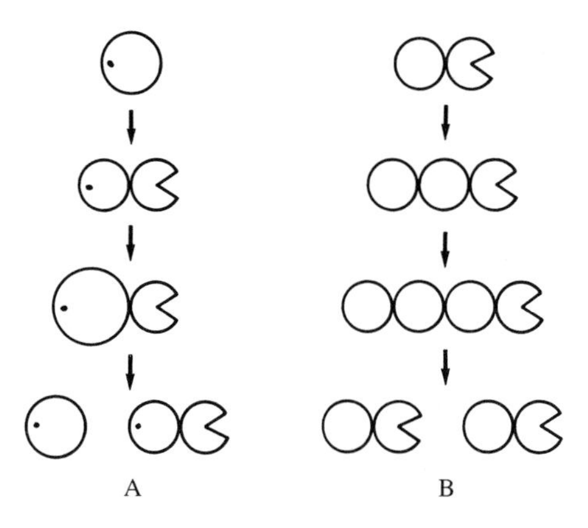

Fig. 1. Possible modes of development and reproduction in primitive metazoa, made up of just two cell types, one of which can be thought of being specialized for feeding. In A, development and reproduction is based on lineage and a stem cell. The dot represents some determinant that maintains the non-feeding cell as a stem cell. In B, development and reproduction is based on cell-to-cell interactions. One of the cells divides and the organism then reproduces by fission, a new feeding cell developing in just one of the two cells that have split off. This requires both cell signalling and polarity.

Again, what underlies the differences between spiral and radial cleavage? Metazoans arose from protozoans that were almost certainly ciliated. All known protozoa are asymmetrical with no known examples of bilateral symmetry. This may provide the clue to spiral cleavage. For the asymmetric structure of protozoa may have led inevitably to spiral cleavage, the essence of which is cytoplasmic asymmetry. It also seems reasonable to think of some primitive organisms relying on this asymmetry to generate differences between cells by asymmetric distribution of some component. Once, however, cellular interactions are involved then asymmetrical divisions are no longer required. This provides a plausible explanation for the rather good correlation between spiralean development and cytoplasmic localization and its loss in systems based on cell interactions. While this in no way implies that in spirally cleaving embryos there are no cellular interactions, it is quite impressive that in vertebrates and insects, which are not spirally cleaving, there are no good examples of autonomous asymmetric cell divisions.

Gastrulation: conservation and dimensionality

In the evolution of gastrulation and its conservation in different animals, there are at least two separate processes that need to be considered - the setting up of the main body plan and the morphogenetic movements. Curiously, it seems that it is the latter as distinct from the former that has been most widely conserved. There is nothing equivalent to the action of the organizer of vertebrates or the micromeres of sea urchins in spirally cleaving embryos. By contrast, in a very wide variety of animals as pointed out a long time ago by Haeckel, the formation of the endomesoderm involves invagination or at least a related process (Willmer, 1990). As far as is known, there is no case in which a metazoan develops its three germ layers without undergoing a substantial rearrangement of the cells, usually of the type characterized by invagination. Indeed the similarity of invagination in *Drosophila* and sea urchins is remarkable, a picture of one could easily be mistaken for the other (compare Leptin and Grunewald (1990) with Gustafson and Wolpert, 1963).

A possible explanation for the remarkable conservation of gastrulation movements throughout the animal kingdom is that it is a fundamental requirement for specifying pattern in just two dimensions in the early embryo. Patterning occurs in cell layers, often only one cell thick. Given this requirement, then gastrulation movements are the inevitable result. And in a way the 'layer' requirement makes biological sense for it reduces the problem of specifying pattern in three dimensions to specification in two dimensions, which is a considerable simplification. In this way, the third dimension is acquired or specified by cell movement.

The origin of the mesoderm and endoderm in a very wide variety of animals is rather similar, namely the movement of cells from an outer sheet into the hollow interior. Often variants reflect the yolkiness of the egg as in birds. The timing of such events - that is whether the endoderm or mesoderm moves in first, and when they enter accounts for much variation but the same basic mechanism is involved. However, a major variant is seen when the inner layers form by delamination, that is tangential planes of cleavage giving rise to inner cells from the outer layer. Many of these variants can be seen in cnidarian gastrulation (see below).

There have been attempts to classify animals into two major super-phyla on the basis of gastrulation - the Protostomea and the Deuterostomea. In prostostomes the blastopore is supposed to become the mouth, whereas in deuterostomes the mouth is supposed to form secondarily, at some distance from the blastopore which may become the anus or close up. However, classification on this basis has been questioned, since in almost every phylum there are animals that form their blastopores into anus when their classification requires they should be forming mouths there (Willmer, 1990). It seems rather that the relationship between mouth and anus to the blastopore should be seen as a continuum.

The origin of the gastrula

The origin of gastrulation is linked to the origin of the metazoa. Haeckel's views on metazoan origins have had a dominant influence: he supposed that some protozoan became colonial and gave rise to the 'Blastea', a hollow sphere of cells, the ancestor of all metazoa. Further development of the Blastea gave rise to the 'Gastrea' which was a two-layered structure, the inner layer entering through invagination at the posterior pole. A somewhat different explanation was offered by Metschnikoff in 1886 (see Jaegerstern, 1956) and has been promoted by Hyman (1942). Invagination is seen as a secondary process, multiple ingression of cells being primary; in essence the idea is that phagocytosis would be carried out by outer cells and digestion by the inner cells, while in Haeckel's 'Gastrea' the invagination provides a primitive gut with a mouth. As Jaegerstern points out, there is much to support the view of the origin of gastrulation as being linked to the evolution of an animal - the gastrea - whose form corresponds to a very large number of animals at early stages of development. In support of the Gastrea theory, Jaegerstern has provided quite a detailed scenario for the transformation of the Blastea into a Gastrea based on its settling on the ocean bottom and ingesting food particles from the bottom (Fig. 2). In his view the primitive intestine retained the cilia and that these may have assisted feeding. He also proposes that this bottom dweller had dorsoventrality and was bilaterally symmetrical.

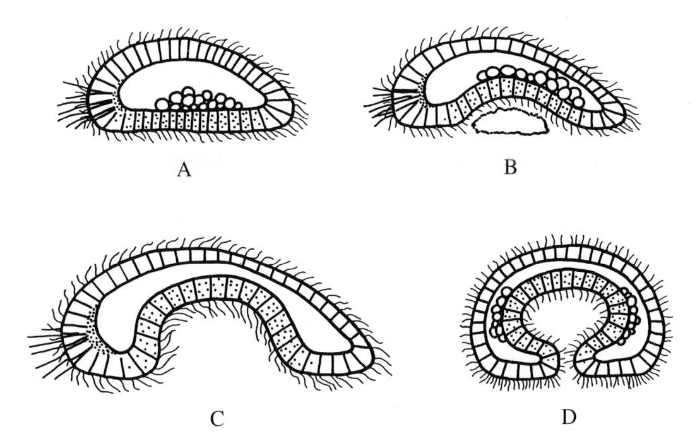

Fig. 2. A possible scenario for the development of the gastrula based on Jaegerstern (1956). In A, the 'Blastea' may have sexual cells inside. Phagocytosis may have become limited to the ventral side and this could have led to a small invagination to assist feeding and decomposing larger food particles. (B) Further development led to the development of a primitive gut with ciliary activity aiding the movement of food particles (C,D).

The Cnidaria provide a valuable phylum in which to try and understand the evolution of early stages of development (Campbell, 1974). The phylum present some of the simplest animal structures which are made up of only two germ layers but which develop into a variety of forms characteristic of other metazoa phyla. There are clear signs of segmental and spacing patterns and moderately complex spatial differentiation. Lacking a mesoblast they are less complex internally than other phyla. Most Cnidaria give rise to a ciliated larva known as a planula. This is essentially a two-layered hollow cylinder. An extraordinary feature of Cnidarian development is the apparent diversity of pathways from fertilization to this rather simple structure (Fig. 3). Cleavage, which is somewhat variable depending often on the yolkiness of the egg, is usually radial to begin with, the first few cleavages being at right angles to one another. In some cases, by contrast, cleavage is anarchic. Cleavage leads to two common blastula types - the stereoblastula, which is a solid mass of cells, and the coeloblastula in which there is a sheet of cells surrounding a hollow

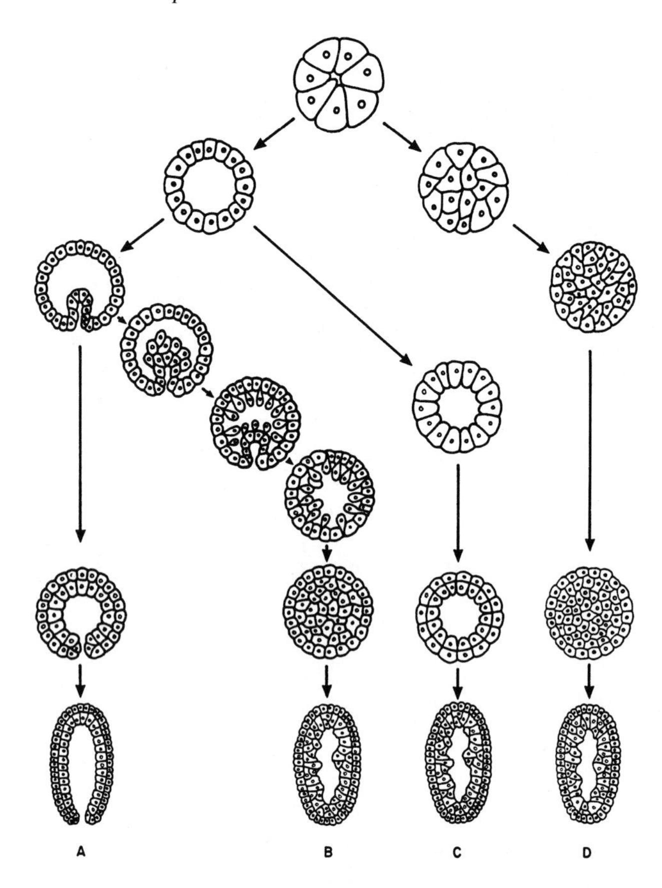

Fig. 3. Some different modes of development leading to the formation of the two-layered planula in Cnidaria (after Tardent, 1978). This may occur by gastrulation of the cells as a sheet (A) or by multipolar ingresssion (B). A quite different mode is by delamination (C). A yet further variant starts from a solid, rather than a hollow blastula, which then develops a cavity (D).

interior. There is, at first sight, no obvious correlation between these two kinds of blastulae and either the yolkiness or size of the egg (see Table I, Campbell, 1974) but it may correlate with relative cell size (see below).

Gastrulation, the formation of a two-layered structure, comprising ectoderm and endoderm, is no less diverse. Even within one class there is great diversity. Gastrulation may occur by invagination of the sheet from one pole, or by ingression of cells from one pole, or by ingression from multiple sites, or by delamination by tangential cleavage. This last mechanism seems to be quite different from the others, being based on a specific plane of cleavage such that one cell remains at the embryo's surface and the other is effectively placed in the interior.

The other modes of gastrulation might be thought of as having a common theme. Studies on sea urchin gastrulation have shown that cell movement and change in cell adhesion are key processes (Gustafson and Wolpert, 1967; Hardin, 1990). Initially the mesenchyme cells enter at the vegetal pole and this is followed by invagination of the gut:ingression and invagination are rather similar. Since most the different forms of gastrulation in Cnidaria involve movement of cells from the surface layer towards the centre, the differences between them may merely reflect spatial differences as to where this occurs and differences in the adhesion of the cells to one another. Multipolar ingression is not hard to understand, though there is the problem of specifying which cells will go in; unipolar ingression localizes the process which involves motility and loss of adhesion. Invagination will occur if loss of adhesion in unipolar ingression is delayed, a feature that can be directly observed in sea urchin morphogenesis when, in some embryos, the entry of the primary mesenchyme cells is delayed and a small invagination occurs. It is but a small step, no loss of adhesion, that might lead to invagination proper. A quite different theory about the origin of the gastrula based on competition between cell lineages has been proposed by Buss (1987) and critically discussed by Wolpert (1990).

So many aspects of the variability can be seen as variations on a basic theme. But the problem then, is, why should there be this variability? What, if any, is the selective pressure? And why should delamination sometimes occur?

Berrill (1949) has analysed development of scyphomedusae and claims a correlation of egg size and type of gastrulation. In general, he claims, the smaller eggs gastrulate by unipolar ingression and the largest by invagination alone, the intermediate sizes combining both methods. He rightly draws attention to the relationship of blastula wall thickness and blastula size. The most relevant correlation may lie here - polar ingression and invagination correlate with a thin wall and many cells, and multipolar with thick wall and few cells. Thus differences in gastrulation may relate to the mechanics of gastrulation, which are determined by the number of cleavages prior to gastrulation. This, I would suggest, might be related to selection for the appropriate amount of yolk to support development (see Elinson, 1987, for vertebrates).

From his consideration of the embryology of the annelids and arthropods, Anderson (1973) has concluded that "The blastula, or its equivalent stages of embryonic development, has a greater stability of functional configuration than any stage that precedes or follows it, no doubt because this is the stage at which the fundamental framework of bodily organization is established". Many of the differences may again be accounted for by changes in the amount of yolk and the manner in which it is used.

Nevertheless it is far from clear that the divergent modes of gastrulation give a functional significance. It simply may not matter to the embryo which way it gastrulates provided the end result is appropriate. There may thus be no selection for 'improving' gastrulation providing that it is reliable. Variants may reflect unimportant differences in developmental terms.

Selection on developmental processes: the privileged embryo

Selection on development will act to ensure reliable generation of reproductive organisms. Reliability will be at a premium. But that selective process is not so much on development but on the organism itself. For the evolution of development the more interesting question is whether there is selection on the developmental processes themselves. Is economy of energy of one developmental process over another to be preferred? Will irrelevant development pathways be lost? Is there a tendency to shorten a developmental process? These are not only questions of interest in their own right, but they relate directly to the variable pathways embryos use to achieve similar results. Are all these variants merely drift or the result of selection for something else, like yolkiness or rapid development, with a knock-on effect on development (Sander, 1983; Dickinson, 1988).

It is possible that the embryo is evolutionarily privileged; that is, the main selective pressure is for reliable development. When one compares an embryo to an adult, this privileged status is evident: the embryo need not seek food, avoid predators, mate and reproduce, all functions for which there is such strong selective pressure. There is not even any evidence that conservation of energy is important and cell death, which is clearly wasteful, accompanies many developmental processes. If this is the case, it may facilitate the evolution of novelty in development.

Evolution of developmental novelty: the neural crest

A feature of developmental systems is that there are examples of apparently different pathways leading to a similar structure. Gastrulation in cnidaria has already been described; another example is the different ways of forming the neural tube in vertebrates - folding of the neural plate and hollowing out of a solid rod. Both occur in mice and chicks, the former in anterior regions and the latter in posterior ones. It is possible that there is no selective advantage of one mechanism over the other and there are merely variants that have arisen by chance. And indeed examination of the timing of the stages of gastrulation in sea urchins shows a considerable variation within embryos from the same species. Such variations may be neutral since the embryo is privileged.

If morphological variants that do not affect the later stages in development are not subject to selection, then they will persist and can provide the basis for the evolution of novelty. Over long periods, new variants will arise and this greatly increases the possibility that some of these will alter later development in a positive way. The variants, because of their neutrality, offer the opportunity for further diversity. They provide, in a sense, an opportunity to explore new developmental modes, some of which will be deleterious and so selected against, but others will be advantageous. These ideas can be illustrated with respect to the origin of the neural crest.

The origin of the neural crest is an important and unexplained event in the evolution of vertebrates. I propose the following scenario that naturally links it to the folding of the neural tube. The development of the neural tube by folding, involves two main processes. In the first process, the cells in an epithelial sheet change shape so that a sharp curvature results and the two ridges come together in the midline. In the second process, the cells in the sheet fuse across the midline and the outer sheet, the ectoderm, separates from the inner sheet, the neural tube. This second process involves rearrangement of cell contacts. It does not seem unreasonable that during this process, in the primitive condition at least, there were errors in the fusion process and some cells were left in the space between the ectoderm and neural tube. These cells would have provided for the origin of the neural crest.

These primitive neural crest cells would not have been subject to selection since they would be neutral to the development of the animal. They would have persisted for many generations and during this period new variants could have arisen to give rise to Schwann or pigment or sensory cells. The key point is that the primitive population persisted long enough to, as it were, explore new possibilities.

This view of the evolution of development makes it much easier to understand how novelties could have arisen in development. Mutants, for example, that caused local invagination or ingression of cells may have been treated as neutral and so further variants generated until one provided the basis for a new structure and became subject to positive selection.

Selection on developmental processes acts primarily on reliability and this requires consideration of buffering and redundancy in developmental processes.

Buffering, redundancy and precision

The idea that redundancy may be quite common in cell and developmental biology probably has its modern origin in Spemann's (1938) idea of double assurance, a term taken from engineering "The cautious engineer makes a construction so strong that it will be able to stand a load which, in practice, it will never have to bear." Spemann's observations on lens induction led him to this view. In *Rana esculenta,* a lens will form without being induced by the eyecup even though the eyecup can induce a lens. Two different processes "are working together, either of which would be sufficient to do the same alone." Thus redundancy could provide a mechanism for ensuring reliability and precision.

More recently there have appeared an increased number of reports of redundancy in both cells and developmental systems (Wolpert and Stein, in preparation). To many workers' surprise, the complete loss of certain specific genes, like some actins, has had remarkably little, if any, effect on the phenotype of the cell or embryo. Another example in development is provided by the specification of the spatial pattern of expression of the gap gene *Krüppel* in early *Drosophila* development. This pattern is controlled both by *bicoid* and *hunchback* (Hülskamp and Tautz, 1991).

A number of different types of redundancy can be imagined; an example would be two very similar genes coding for two very similar proteins which fulfil the same function. Here I will focus on the case where there are two different genes controlling the same process but in different ways, as in the case of *Krüppel* or the three cyclin genes in budding yeast, any two of which seem to be dispensable. By buffering, is meant the control of some process in the face of variations like ambient temperature, pH and other molecules. If the process is the production of a defined concentration of some biochemical species then negative feedback is the control mechanism that is almost invariably used. In such systems, there is no evidence for redundancy. And it is hard to see how a redundant system would be better and so have a selective advantage.

By contrast in those many systems in cells and embryos where a particular molecular species is constrained to vary in space and time to give a defined spatiotemporal profile, negative feedback cannot be the mechanism for buffering, precisely because things are not kept constant. Moreover such spatiotemporal profiles are usually themselves the result and cause of on/off switches, like gene activity. Such switching mechanisms for gene activity in eukaryotes, at least, almost never involve negative feedback but positive feedback is often present to keep a gene switched on by autocatalysis. In many systems, the amount of protein made is directly related to the number of gene copies as in amplification of chorion genes or when additional copies have been involved.

How can buffering be achieved in a system with spatiotemporal switching? An answer would seem to lie in providing multiple parallel mechanisms with differing - appositely directed - responses to environmental variations. In this way, buffering and hence precision could be achieved. Such a mechanism has been referred to as canalization when applied to developmental systems.

As an example, consider the specification of the activity of the *Krüppel* in the early *Drosophila* development. *Krüppel* is a gap gene, which has a broad band of expression about half way along the early embryo. Its pattern of expression is controlled by both activation and repression by proteins which have graded concentration. Activation is by low levels of both *bicoid* and *hunchback* and repression by high levels of these same two genes, and there are indeed separate binding sites adjacent to the *Krüppel* gene, both of which specify its expression in the same region. Two further genes *tailless* and *giant* control its final expression. There is apparent redundancy since the pattern of expression of *Krüppel* is very similar in the absence of either *bicoid* or *hunchback*. So why should specification require this apparent redundancy? We propose that this is required for precision in the expression of *Krüppel*. In general terms, precision will be improved if there are multiple parallel mechanisms for the specification, and this will be particularly so if the effect of environmental perturbation, like temperature, move the region of activation in opposite directions. We have, however, no evidence for this. On the other hand, it seems very unlikely that two or more behaviourly identical parallel processes, such as would be provided by two identically redundant genes, would improve buffering or precision.

The same arguments would apply to other systems where it is claimed that redundancy is present. We assert that in all such cases the redundancy is only apparent and the presence of more than one system in parallel has a selective advantage ensuring the precise buffering of the process. Where absence of a gene has no clear phenotype effect, we believe this merely reflects that the system has not been tested in the appropriate environment.

We do however have to recognise that some cases of redundancy may reflect evolutionary relics, since genes coding for processes that no longer have a selective advantage but for which there is no selective disadvantage, may persist for millions of years.

Precision in specifying spatiotemporal patterns is a major problem for the developing embryo. Waddington's theory of canalization was to account for this (Falconer, 1964). He suggested that selection would operate against the causes of deviation from optimal shapes, such as, for example, that of the insect wing. Apparent redundancy thus provides the mechanism for such a process.

Conclusion

Gastrulation can be thought of as one of the most important processes in early development. Its evolution probably did not require any significant mechanisms not already present in the eukaryotic cell. The striking conservation of the formation of the endo-mesoderm by ingression may reflect the fact that it enables patterning to occur in cell sheets - that is two dimensions - and the three-dimensional structure to be created by gastrulation movements.

Because of its importance, it is necessary for the end result to be reliable. However, the precise pathway by which gastrulation takes place may be quite variable. Reliability and precision probably require apparently redundant processes.

Gastrulation may represent a very primitive metazoan whose development has been elaborated and extended in evolution. It is one of the few examples, possibly, of the discredited theory of recapitulation.

References

Alberts, B., Bray, D., Lewis, J., Roberts, K. and Watson, D. (1989). *Molecular Biology of the Cell.* New York: Garland

Anderson, D. T. (1973). *Embryology and Phylogeny in Annelids and Arthropods.* Oxford: Pergamon

Balinsky, B. I. (1965). *An Introduction to Embryology.* Philadelphia: Saunders

Berill, N. J. (1949). Developmental analysis of scyophomedusae. *Biol.Revs.* **24**, 393-410

Buss, L. W. (1987). *The Evolution of Individuality.* Princeton: Princeton University

Campbell, R. D. (1974). Cnidaria. In *Reproduction of Marine Invertebrates'*, **1**, 133-199. New York: Academic Press

Dickinson, W. J. (1988). On the architecture of regulatory systems: evolutionary insights and implications. *BioEssays* **8**, 204-208

Elinson, R. P. (1987). Change in developmental patterns: embryos of amphibians with large egg. In *Development as an Evolutionary Process* (eds. R.A. Raff and E.C. Raff), pp.1-21. New York: Liss

Falconer, D. S. (1964). *Introduction to Quantitative Genetics.* Edinburgh: Oliver and Boyd

Frankel, J. (1991). The patterning of ciliates. *J.Protozool.* **38**, 519-525

Goodwin, B. C. (1989). Unicellular morphogenesis. In *Cell Shape: Determination, Regulation and Regulatory Role'* (eds. W.D. Stein and F. Bronner). London: Academic Press

Gustafson, T. and Wolpert, L. (1963). The cellular basis of sea urchin morphogenesis. *Int.Rev.Cytol.* **15**, 139-213

Gustafson, T. and Wolpert, L. (1967). Cellular movement and contact in sea urchin morphogenesis. *Biol.Revs.* **42**, 442-498

Hardin, J. (1990). Contact-sensitive cell behaviour during gastrulation. Seminars *Dev.Biol.***1**, 335-345

Hülskamp, M. and Tautz, D. (1991). Gap genes and gradients - the logic behind the gaps. *BioEssays* **13**, 261-268

Hyman, L. (1942). *The Invertebrates*, Vol.II. New York: McGraw-Hill

Jaegerstern, G. (1956). The early phylogeny of the metazoa. The bilaterogastrea theory. *Zool.Bidrag.(Uppsala)* **30**, 321-354

Leptin, M. and Grunewald, B. (1990). Cell shape change during gastrulation in *Drosophila. Development* **110**, 73-84

Sander, K. (1983). The evolution of patterning mechanisms: gleanings from insect embryogenesis and spermatogenesis. In *Development and Evolution*' (eds.B.C.Goodwin, N.Holder and C.C.Wylie), pp.137-159. Cambridge: Cambridge University Press

Spemann, H. (1938). *Embryonic Development and Induction.* New Haven: Yale University Press

Willmer, P. (1990). *Invertebrate Relationships.* Cambridge: Cambridge University Press

Wolpert, L. (1989a). Stem cells: a problem in asymmetry. *J.Cell Sci.*, **10 Suppl.** 1-9

Wolpert, L. (1989b). Positional information revisited. *Development* **Suppl.**, 3-12

Wolpert, L. (1990). The evolution of development. *Biol.J.Linn.Soc.*,**39**, 109-124

Wolpert, L. and Stein, W. D. (1984). Positional information and pattern formation. In *Pattern Formation'* (eds. G.M. Malacinski and S.V. Bryant), pp.2-21. New York: Macmillan.

Development 1992 Supplement, 15-22 (1992)
Printed in Great Britain © The Company of Biologists Limited 1992

Evolution of developmental decisions and morphogenesis: the view from two camps

RUDOLF A. RAFF

Institute for Molecular and Cellular Biology, and Department of Biology, Indiana University, Bloomington, IN 47405 USA

Summary

Modern developmental biology largely ignores evolution and instead focuses on use of standard model organisms to reveal general mechanisms of development. Evolutionary biologists more widely hold developmental biology to be of major consequence in providing potential insights into evolution. Evolutionary insights can enlighten our views of developmental mechanisms as much as developmental data offer clearer views of mechanisms which underlie evolutionary change. However, insights have been limited by the long-term disengagement of the two fields dating to the rise of experimental embryology in the 1890s. Molecular genetics now provides a powerful tool to probe both gene function and evolutionary relationships, and a greater connection has become possible. The expansion of experimental organisms beyond the standard model animals used in most studies of development allows us to ask deeper questions about the interaction of development and evolution. This paper presents an analysis of the complementary uses of the resulting data in the two fields as they grope for accommodation. Analysis of the radical changes in early develement seen in closely related sea urchins with alternate modes of development illustrate the complementarity of developmental and evolutionary data. These studies show that what have been thought to be constrained mechanisms of axial determination, cell lineage patterning, and gastrulation in fact evolve readily and provide the means for the rapid evolution of development.

Key words: gene function, evolution, sea urchin.

In the Tower of Babel

In 1981, John Tyler Bonner organized a major cross-disciplinary conference to explore effective ways of recombining the study of evolution and development (Bonner, 1982). I was at that stimulating meeting in Dahlem, and came away with an interesting education in just how difficult it really is to combine two fields that for nearly a century have had different goals, approaches and vocabularies. The feeling that it ought to be done has persisted, and has motivated subsequent meetings which have revealed that the rapprochement is still slow in coming. What is different now from a decade ago is that some experimenters have taken steps to make a practicable fusion and have framed problems that can be addressed with appropriate experimental systems.

The separation of developmental biology from evolutionary biology occurred at about the turn of the century, when a focus on mechanistic controls of development replaced the search for phylogenetic relationships through embryonic resemblances as a central concern of the field. Embryology became an experimental discipline with a very different paradigm for exploring biology from that developed by evolutionary biology during the same period (Mayr, 1982). Table 1 summarizes the distinctions that presently characterize developmental biology and evolutionary biology as disciplines. For evolutionary biologists, the critical issue is to explain the diversity of life. Genes provide the raw material for the generation of diversity, and thus population genetics has been developed as the major tool for relating the behavior of genes in populations to evolutionary events. Another important focus of evolutionary biology has been the definition of phylogenetic relationships using both morphological and molecular tools. On the other hand a very different set of principles are important to developmental biologists. Questions of phylogenetic relationship are seldom part of mainstream developmental biology. Instead, several organisms have been selected from various taxa purely for their convenience as experimental systems in the study of various developmental processes. Mechanistic universality in developmental processes is considered to exist despite the diversity of organisms: thus common mechanisms for gastrulation are sought. Genes are regarded as important not as sources of variation, but as controllers and executers of developmental processes.

The consequence of these differences in disciplinary histories is that when developmental and evolutionary biologists are brought together there is some difficulty in deciding just what people from that other discipline are talking about. Yet, a boundary discipline exists, and its investiga-

Table 1. *Disciplinary digressions: evolution versus development*

EVOLUTION	DEVELOPMENT
Exploration of diversity	Mechanistic universality
Genes as sources of variation	Functional role of genes
Phylogenetic relationships	Standard model organisms

tion can yield important complementary insights in the two fields. This paper attempts to show how that can be done.

Common ground and experimental systems

The most crucial issue to evolutionary biologists interested in development has been one of how morphology evolves. Both deBeer (1958) and Gould (1977) laid out their considerations of the interface between ontogeny and phylogeny in terms of a unifying developmental mechanism that would explain the relatively easy transformation of form suggested by evolutionary histories. That mechanism was heterochrony, the concept that events in development can shift in timing relative to each other to produce new ontogenies. For example, humans resemble young apes more closely than adult ones, suggesting that human evolution might have involved developmental changes resulting in a more juvenilized morphology in sexually mature adults. [In Aldous Huxley's novel, *After Many a Summer Dies the Swan*, that concept is followed out to a bizzare denouement in which tampering to prolong life has predictable if unseemly consequences.] In the hands of its proponents, heterochrony has continued to provide a major explanatory concept for interpreting evolutionary change in fossil as well as living organisms. The insistence that heterochrony is the dominant mechanism for evolutionary changes is overdone, but the goal of providing a simple unifying mechanism around which data can be ordered is clear.

The study of the role of development in evolution must shift from a focus on theoretical considerations to the framing of experimental questions that can reveal mechanisms by which developmental processes influence evolutionary change. Subsidiary questions about the evolution of development grow from other themes in evolutionary biology. Thus, the demonstration of developmental innovations becomes a major part of understanding the origins of novel features that have led to origins of new groups of animals. Comparative studies become important to trace the directions of evolutionary changes in development (Wray and Raff, 1991; Raff, 1992a), the number of times they might have occurred independently (Emlet, 1990), and even their potential reversibility (Raff et al., 1993). Evolutionary data also reveal that some aspects or stages of development evolve readily and others do not, providing clues that different mechanisms might govern various stages of development (Raff et al., 1991; Raff, 1992b). The basic issue is whether natural selection can elicit responses in any direction from the mutations that appear in organisms, or if evolution is somehow constrained by existing genetic and developmental systems. If so, a major principle of evolution exists beyond an all-powerful selection working on randomly generated variation, and the inner workings of genetic regulatory systems and developmental processes as well as their histories must be considered as key elements of evolution (Alberch, 1982; Jacob, 1977; Müller, 1990; Thomson, 1991).

Because of the ability to clone regulatory genes, purely developmental questions have quickly become issues in evolutionary biology. The finding that homeobox-containing genes are major regulators of axial specification in both insects and vertebrates suggests that very ancient developmental-genetic regulatory mechanisms are shared by widely diverged animal groups. These genes have been detected even in the most primitive metazoans, the cnidarians (jelly fish, anemones, etc.), as well as many other phyla (Holland, 1991; Murtha et al., 1991; Schiewater et al., 1991). The phylogenetic distribution of homeobox genes, and the additional roles that can be shown for them in aspects of vertebrate development indicates a pattern of gene duplication, divergence, and co-option for new functions (eg. Holland 1991; Hunt et al., 1991). Other regulatory gene families, such as the steroid receptor family (Evans, 1988; Amero et al., 1992), show analogous patterns of evolutionary expansion and co-option to provide genetic raw material for regulatory innovations in the evolution of development.

Old pathways and evolutionary innovations

It is perhaps not so easy to define precisely what constitutes an evolutionary innovation. Certainly a novel feature must differ in some important qualitative way from the ancestral feature. Discussions of innovation (Mayr, 1960; Cracraft, 1990; Liem, 1990; Müller, 1990; Raff et al., 1990) visualize novelty in two ways. First, novel features are considered to be significant if they permit assumption of a new function or provide a key element upon which new evolutionary directions can be taken by an evolving lineage; key innovations are thus recognized in evolutionary retrospect. Because multiple features emerge during an evolutionary history, there is considerable disagreement on the existence and phylogenetic role of key innovations (eg. Cracraft, 1990 versus Liem, 1990). The second aspect of novelty is that it represents a departure from an existing pattern of development; thus, a new feature can be novel even if no new group arises as a result (Raff et al., 1990).

Responses to selection may be limited by the properties of the ancestral developmental program: hypothetically adaptive responses to selection may not be attainable because the starting genetic control networks cannot reach certain states; interactions between developmental processes may be too complex to modify; developmental programs may follow epigenetic rules that exclude certain outcomes. Some descendant states indeed may be favored by features of the existing developmental program that bias the response to selection in particular directions. On the other hand, some selective pressures might push developing systems over a threshold to yield a new state or interactions (Müller, 1990). The descendant states ultimately achieved only can be those which are both adaptive and attainable by some evolutionary trajectory not blocked by developmental constraints.

These considerations suggest that we might need to look

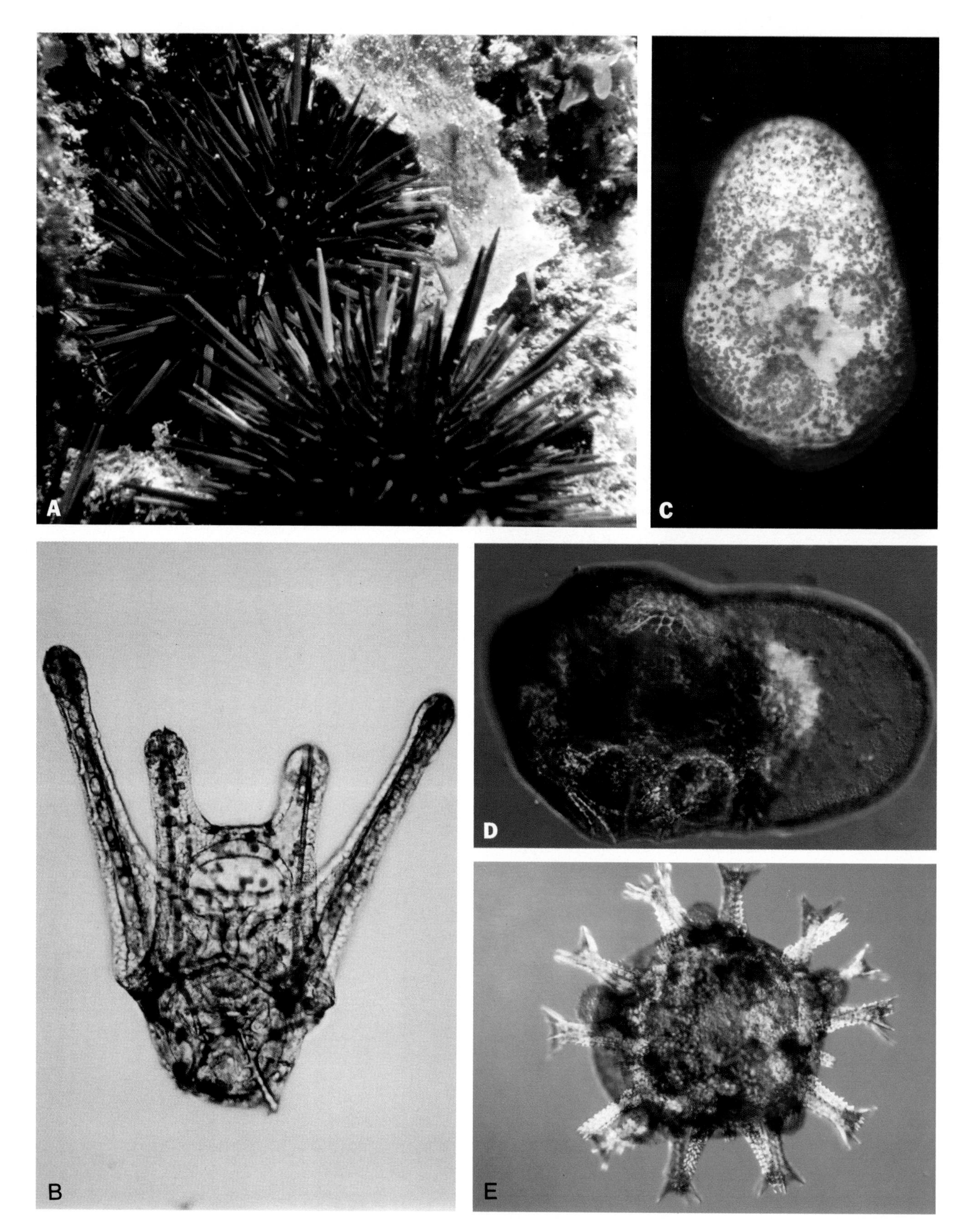

Fig. 1. Divergent development in two closely related sea urchins. A, foreground, adult of *H. tuberculata*. Note the longer, spatulate spins as compared to *H. erythrogramma* in the background. The two species live in the same shallow intertidal habitats on the East coast of Australia. B, pluteus larva of *H. tuberculata*. Typical structures of the 4-arm pluteus include pigment cells, spicular skeleton, ciliated larval arms, and functioning gut. Length from vertex to arm tips 300 µm. C, oral side of *H. erythrogramma* "larva" about half way through development. Note that the only larval features are the pigment cells and the ciliary band at one end; the five primary adult tube feet show prominantly. There is no spicular skeleton or larval gut. Length 550 µm. D, polarizing optic view of cleared *H. erythrogramma* larva from the side. Tube feet are visible on the oral side. Two kinds of adult skeletal structures are visible in polarizing light; juvenile adult spines facing orally, and fenestrated juvenile adult test plates. E, cleared newly metamorphosed juvenile *H. erythrogramma* under polarized light.

for key innovations as initial genetic events that make possible a series of subsequent evolutionary modifications in ontogeny. One's first reaction would be to seek initial additions to, striking modifications of, or duplications and subsequent divergence of pre-existing features. However, many developmental regulatory genes act by repressing the function of other genes. The evolutionary result of the insertion of such repressors would be to remove some aspect of expression of a primitive feature. For example, a major aspect of the function of homoeotic genes in insect development is to prevent features of more anterior or more posterior segments from being expressed in the "wrong" segments. The result is tagmosis (Raff and Kaufman, 1983). The unitary head has been evolved from a primitive condition in arthropods in which there were several separate limb-bearing anterior segments (Della Cave and Simonetta, 1991). Mutations in the *Drosophila* homoeotic gene *fork head* transform the head into separate segments that express features of limb-bearing segments. The gene thus normally acts in development to prevent segmentation in the head (Jürgens and Weigel, 1988; Finkelstein and Perrimon, 1991). The initial crucial step in an evolutionary innovation thus may be a suppression of part of an older program. The suppression event then resets the stage for additions to the modified feature, such as addition of head specializations to the fused segments. The suppression of older genetic controls can be a key innovation, because it provides the basis for subsequent genetic modifications and additions.

Direct development and alternative ontogenies

In many cases larval development is extraordinarily conservative in evolution over long periods, with similar larvae conserved over 500 Myr of evolutionary time (Müller, K. J. and Walossek, D., 1986). The sea urchin pluteus has been conserved for at least 250 Myr (Wray, 1992). Yet, radically different modes of early development exist among direct developing species in many taxa. These forms offer important experimental systems. They represent alternate ways to achieve the same developmental end as achieved by their relatives with feeding larvae, and provide natural experiments in developmental genetics. Although indirect development is widely conserved, it has been often replaced by direct development in such diverse groups as corals, polychaete annelids, starfish, sea urchins, ascidians, and frogs (Raff and Kaufman 1983; del Pino, 1989; Emlet, 1990; Jeffery and Swalla, 1992). In many instances, indirect- and direct-developing species occur within the same genus, and can even produce hybrids, indicating rapid evolutionary changes in developmental mode (Levin et al., 1991; Jeffery and Swalla, 1992; Raff, 1992a). The replacement of complex patterns of larval development by direct-development offers an instance of creation of an evolutionary novelty by an initial suppression of an old pattern.

This proposition is illustrated by two congeneric sea urchin species, *Heliocidaris tuberculata* and *H. erythrogramma* (Fig. 1). The two species of *Heliocidaris* are similar as adults but differ greatly in development. *H. tuberculata* develops from a small egg via a typical feeding pluteus, whereas *H. erythrogramma* has omitted the pluteus larva and undergoes direct development in which a large egg develops rapidly into a small sea urchin without feeding. The two species diverged about 10 million years ago, as estimated from single copy and mitochondrial DNA distances (Smith et al., 1990; McMillan et al., 1992). The initial innovation setting the stage for subsequent remodeling of early development in direct-developing species probably lies in oogenesis. Egg sizes in sea urchins fall into two classes. Indirect developers which produce feeding larvae generally have small eggs of about 100 μm diameter. Direct developers feature eggs of 300 μm to 1500 μm, and produce non-feeding larvae (Wray and Raff, 1991). In between are a very few species with eggs of about 300 μm that produce feeding plutei. In one case, it has been demonstrated that the large pluteus can metamorphose even if not fed; it is facultative (Emlet, 1986). Thus, there is a threshold. Species that pass over it via a facultative feeding larva can afford to suppress pluteus features and assemble a new complex of features that result in rapid, direct development of the juvenile.

Evolution of gastrulation and origins of a novel morphogenesis

The larva of *H. erythrogramma* looks quite different from the pluteus of *H. tuberculata* (Fig. 1). A superficial view suggests that the main evolutionary change has been a loss of feeding structures in the larva and an acceleration of adult development, relatively simple heterochronies. In fact, the underlying changes are pervasive and have reorganized gametogenesis and development. In *H. erythrogramma*, the eggs are 100 times the volume of *H. tuberculata* eggs, the sperm heads are longer and narrower, the nuclear genome is about 30 percent larger, storage proteins are different (Raff, 1992a). Maternal localization and important aspects of axial determination have been modified (Henry and Raff, 1990; Henry et al., 1990). As development begins, cleavage is radial, but no micromeres are produced as in indirect developing sea urchins. The result is that the cellular precursors and cell fates have been changed from those of indirect development (Wray and Raff, 1990). Because no larval skeleton is produced, and because adult structures start to form at the end of gastrulation, the morphogenetic processes that shape the post-gastrula larva have been highly modified. No pluteus with its elaborate feeding structures is made; all morphogenetic processes are directed at rapid production of the juvenile sea urchin. The overarching changes in early development of *H. erythrogramma* have been recently reviewed (Raff, 1992a; Raff et al., 1992), and do not need to be detailed here. I instead focus on gastrulation to show how its evolutionary modification makes possible the accelerated development of the adult.

Gastrulation is the most fundamental morphogenetic movement in animal development, and establishes the primary germ layers and their topological relationships for subsequent inductive interactions. Gastrulation varies between classes, but would be expected to be conserved between closely related species, because these would share many inductive interactions and downstream morphogenetic processes that would constrain evolutionary changes in gastrulation. It is evident from experimental manipulations of indirect-developing sea urchins that abundant inductive interactions occur prior to and during gastrulation (Ettensohn and McClay, 1988; Henry et al., 1989; Hardin and McClay, 1991). However, the timing and topological differences in morphogenesis in *H. erythrogramma* versus *H. tuberculata* indicate that cell-cell communication has been substantially modified despite the clear role for such inductive interactions. Thus, the expected constraints on gastrulation might not actually exist.

In the indirect-developing sea urchins that have been studied, gastrulation has two phases. In the first, there is an active and symmetrical movement of cells from the vegetal region of the blastula to form a short, wide tube (Burke et al., 1991). This initial archenteron precursor is transformed during the second phase of gastrulation into a long slender thin walled tube (Ettensohn, 1985; Hardin and Cheng, 1986). No involution occurs during the second phase; instead, the extension is driven by changes in cell shape and position (Ettensohn, 1984; Ettensohn, 1985; Hardin and Cheng, 1986; Hardin, 1989). It is important to note that no addition of cells into the archenteron occurs during the second phase of archenteron elongation in this ancestral mode of sea urchin gastrulation.

As in typical sea urchin gastrulation, in *H. erythrogramma*, there is an initial symmetric phase of invagination (Fig. 2). However, the cell rearrangements characteristic of the second phase of the ancestral archenteron elongation process have been replaced by a novel asymmetric involution. This change meets both definitions of evolutionary innovations in that the ancestral elongation process is not merely modified but is replaced by a new process that has novel consequences. In *H. erythrogramma* (Fig. 3), gastrulation is immediately followed by generation from the tip of the archenteron of a very large coelom that provides the extensive hydrocoel mesoderm which interacts with the ingressing ectodermal vestibule. Together these cell layers form much of the developing juvenile sea urchin. These same coelomic-vestibular interactions occur in indirect development as well. However, they occur much later and involve a different strategy of cellular recruitment. Only a few (10-12 cells) from the archenteron contribute to each left and right coelomic pouch precursor (Cameron et al., 1991). Subsequent increases in coelomic cells are produced by cell growth and division in the feeding and growing pluteus larva (Pehrson and Cohen, 1986). In *H. erythrogramma*, the origins of coelomic cells are different. There is no growth. The juvenile adult is formed directly from cells of the embryo. Thus, whereas in the fate map of an indirect developer only a part of two cells of the 32-cell embryo give rise to vestibule, a full eight cells of the 32-cell *H. erythrogramma* embryo produce the vestibule (Wray and Raff, 1990).

In order to allow large scale elaboration of the coelom by *H. erythrogramma*, cells must be added to the archenteron. Isovolumetric cell-cell rearrangements alone would be inadequate. Thus, the second phase of the ancestral mode of gastrulation has been abandoned and involution continues instead of terminating early. There are two phases of involution in *H. erythrogramma*. The first is symmetric, and evidently corresponds to the initial involution of indirect development. The second phase of involution is asymmetric, and involves the ingression of ventral ectoderm. This asymmetric ingression combined with the involution of a full quarter of the cellular volume of the embryo to form the vestibule, results in the apposition of vestibule and coelomic cells sufficient for morphogenesis of the ventral portion of the juvenile adult sea urchin within a few hours after gastrulation (Fig. 3). Finally, since gastrulation is the topic of this symposium it is important to note that evolution shows that parts of the gastrulation process are dissociable from each other. Some features of *H. erythrogramma* gastrulation have been conserved, and others have changed in substantial ways (Wray and Raff, 1991). Thus, the position of gastrulation initiation, invagination by involution, and timing of primary mesenchyme cell ingression have been conserved. The number of primary mesenchyme cells, the origins of cells contributing to the archenteron, the mechanism of archenteron elongation, the symmetry of cellular movements, and the origin of coelomic cells have all been substantially changed. In total the evolution of the novel ontogeny of *H. erythrogramma* has resulted from a suite of changes encompassing changes in timing of developmental events, cell fates, cell-cell interactions, and gastrulation movements.

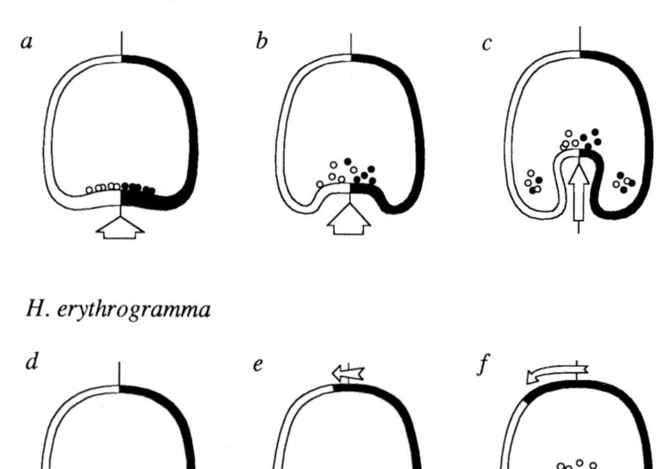

Fig. 2. Cell movements in the two modes of sea urchin gastrulation. A, In indirect development, primary mesenchyme cells enter the blastocoel prior to the start of archenteron ingression. B, Initial ingression of cells of the vegetal plate. C, Elongation of the archenteron involves only cell rearrangements and cell shape without involution. In *H. erythrogramma*, D and E, primary mesenchyme cells enter the blastocoel as ingression progresses. Ingression begins similarly to that of indirect development, but rapidly shows a ventral bias in the involution of the vegetal plate. F, As gastrulation continues, archenteron extension is by asymmetric involution of cells from the ventral side. Reproduced from Wray and Raff (1991), with permission of *Evolution*.

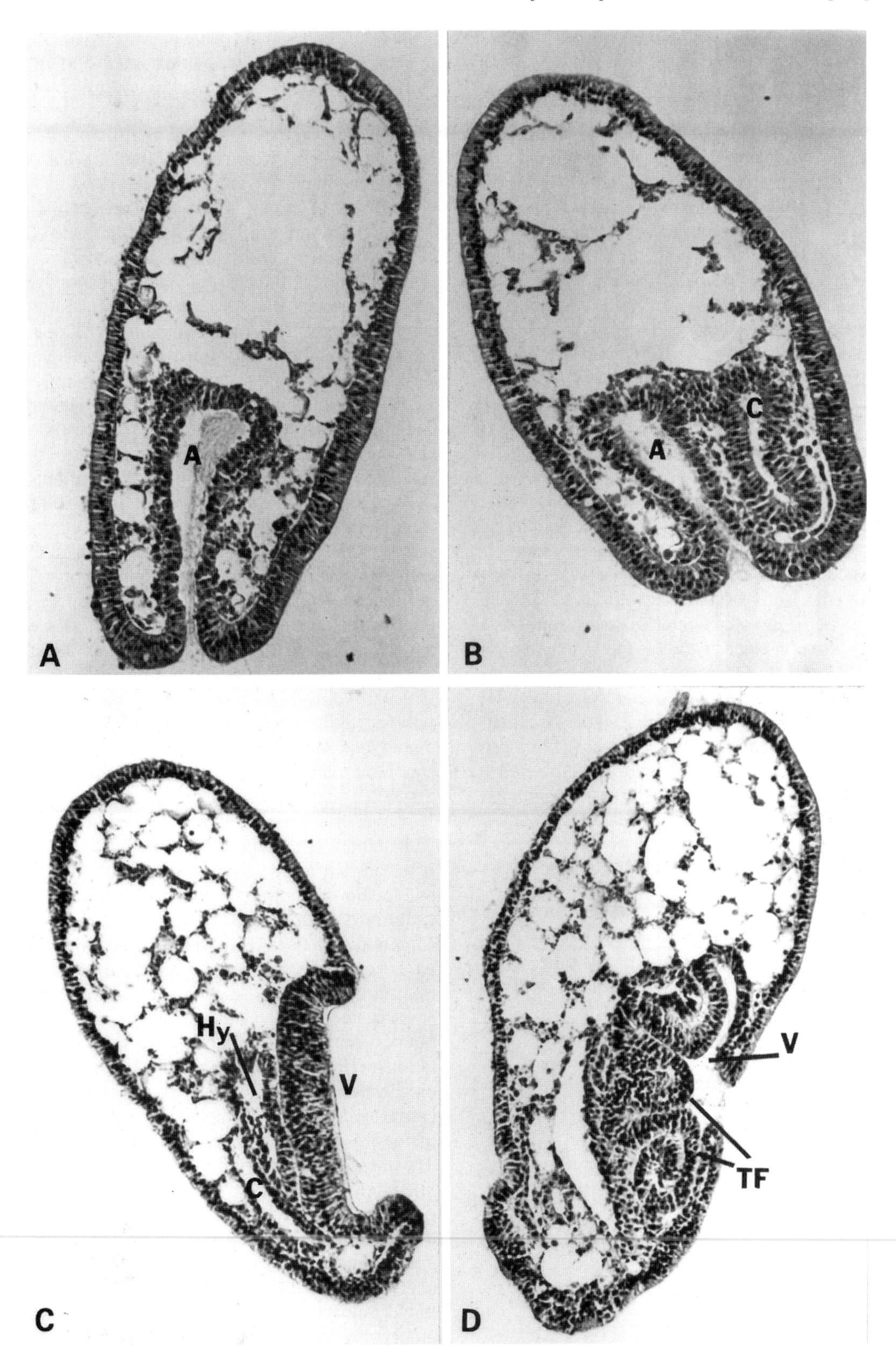

Fig. 3. Gastrulation and subsequent coelom formation in *H. erythrogramma*. Embryos are oriented with animal poles up and ventral side to the right. A, Full extension of the archenteron (A) in the 18 hour mid-gastrula. Note the initial bulging of the archenteron tip as coelomic pouches begin to form. B, Late (22 hour) gastrula with a large coelomic sac (C). The ectoderm overlying the coelom has begun to thicken. C, Early (30 hour) larva with vestibular (V) invagination of ventral ectoderm. The hydrocoel (Hy), which is a derivative of the coelom has formed and is closely associated with the vestibule. D, Mid larval development (34 hour) showing rapid development of tube feet (TF) from the interaction of vestibular ectoderm and hydrocoel mesoderm.

How two disciplines read the story

The usual ending to chapters on development/evolution is to show how the research under discussion advances the fusion of the two fields or provides data on one of the questions posed in the boundary field. That is obviously the objective motivating our research as well. However, at this point in history it is also of interest to look at features important in an evolutionary embryology context and dissect out what is significant about each separately to developmental and evolutionary biology. Table 2 lists several such features, each of which has quite different meanings in the currency of the respective fields.

The most notable aspect of *H. erythrogramma* is its radical reorganization of early development. From a developmental viewpoint, *H. erythrogramma* reveals that early development is governed by independent mechanisms which function in an integrated manner, but are dissociable from each other in evolution. Relationships between mechanisms can be changed in ways that produce other viable ontogenies. That this is so may seem surprising. It is commonly expected that early development must be rigidly constrained because early events must serve to specify the broad foundations of later development. Indeed, much current work in developmental genetics supports a view that early development includes the refinement of maternal information by zygotic gene action to produce the initial patterning of the embryo (Nüsslein-Volhard, 1991). The dissociability revealed by comparative studies is a manifestation of what developmental biologists have long recognized as the regulative abilities of embryos subjected to experimental perturbations. The phenomenon of regulation is still not understood mechanistically, but evolution gives us an insight into its potentially great importance.

Table 2. *What we learn from Development/Evolution*

FEATURE	MEANING TO DEVELOPMENT	MEANING TO EVOLUTION
Stages of Development Exhibit Different Amounts of Evolution	Different Mechanisms or Degrees of Integration Between Stages	Existence of Phylotypic Stages and Stable Bauplans
Convergent Evolution of Development	"Standard" Morphogenetic Processes	Convergent Features Confound Phylogenetic Information
Radical Reorganization of Early Development	Multiple Genetic Mechanisms Determine Early Development	Flexible Responsiveness of Development to Selection
Cell Lineage Reorganization	Readily Recognizable Common Features may not be Regulatory Controls	Developmental Basis of Homology Unreliable
Common Regulatory Genes	Homologous Genes may not Play same Roles	Gene Co-option Frequent
Genes that Repress Pathway	Prevent Incorrect Program in Portion of Body	Create Novel Features that Provide Basis for Further Evolution

For evolutionary biology, the finding that early development can be reorganized means that development may respond more flexibly to selection than generally appreciated. Just how this occurs, and under what developmental constraints is clearly an important matter for understanding the origins of evolutionary novelties, evolutionary trends, and evolution of life histories. Evolutionary discussions of ontogeny are generally about later stages of development. Yet early development is as able to respond to selection as are later stages, and evolutionary innovations can result from changes in early development. Tropical frogs, for example, exhibit nearly thirty distinct adaptations in early development ranging from gastric brooding of tadpoles to direct development on land (del Pino, 1989; Tyler, 1983). Such experiments in early development should remind us that the key innovation to the vertebrate conquest of the land was probably the evolution of the amniotic egg.

Evolution of specific processes is also highly revealing in having quite different implications for the two disciplines. One of the most striking features of *H. erythrogramma* development is the reorganization of cell fates and cell lineages. The mechanistic consequences to one subsequent developmental process, gastrulation, were mentioned above, but there are more profound implications as well. The focus of developmental biologists on model organisms is usually done so that the species that is most experimentally accessible for the study of a particular process is employed. Without an evolutionary dimension to such studies, a pattern of development that is consistent within a species may be interpreted as mechanistically required. However, such a conclusion is weak. For example, in the well studied sea urchin *Strongylocentrotus purpuratus* there is a constant relationship between the first cleavage plane and the dorso-ventral axis (Cameron et al., 1987). This relationship is reflected in cell placement, cell lineage precursors, and cell fates in morphogenesis. When seen from an evolutionary perspective, constancy breaks down (Henry et al., 1992). Different species do exhibit constant relationships between first cleavage plane and dorso-ventral axis, but several different relationships exist between the orientation of first cleavage and the dorso-ventral axis among closely related species. This is a lesson in what evolution can provide to development; it lets us sort out real mechanisms from the idiosyncrasies of any one model organism. The message for evolutionary biology is a quite different one. Cell types and structures that appear to be homologous arise from quite different cell lineage precursors and different developmental processes. The temptation to seek a developmental basis for homology is always strong. However, early as well as late development proves to be an unreliable guide to homology, a point well realized by earlier embryologists (Wilson, 1895; deBeer, 1971). If a developmental basis for homology is to be defined, it must lie in the genetic regulatory systems that underly particular developmental features.

I thank my collaborators for making the exploration of *Heliocidaris* as enjoyable and stimulating as it has been. I also thank Elizabeth Raff, Jessie Kissinger and Eric Haag for critically reading the manuscript. Brent Bisgrove, Louis Herlands, and Chris McDonald provided the photographs used in Fig. 1, and Annette Parks those used in Fig. 3. This work was supported by NIH grant HD 21337.

References

Alberch, P. (1982). Developmental constraints in evolutionary processes. In *Evolution and Development.* (ed. J. T. Bonner), pp. 313-332. Berlin: Springer-Verlag.

Amero, S. A., Kretsinger, R. H., Moncrief, N. D., Yamamoto, K. R., and Pearson, W. R. (1992). The origin of nuclear receptor proteins: A single precursor distinct from other transcription factors. *Mol. Endocrinol.* **6**, 3-7.

Bonner, J. T. (ed.) (1982). *Evolution and Development.* Berlin: Springer-Verlag.

Burke, R. D., Myers, R. L., Sexton, T. L., and Jackson, C. (1991). Cell movements during the initial phase of gastrulation in the sea urchin embryo. *Dev. Biol.* **146**, 542-557.

Cameron, R. A., Hough-Evans, B. R., Britten, R. J., and Davidson, E. H. (1987). Lineage and fate of each blastomere of the sea urchin embryo. *Genes Dev.* **1**, 75-84.

Cameron, R. A., Fraser, S. E., Britten, R. J., and Davidson, E. H. (1991). Macromere cell fates during sea urchin development. *Development* **113**, 1085-1091.

Cracraft, J. (1990). The origin of evolutionary novelties: Pattern and process at different hierarchical levels. In *Evolutionary Innovations.* (ed. M. Nitecki), pp. 21-44. Chicago: University of Chicago Press.

deBeer, G. (1958). *Embryos and Ancestors.* 3rd. ed. Oxford: Oxford University Press.

deBeer, G. (1971). Homology, an unsolved problem. *Oxford Biology Readers No. 11.* (ed. J. J. Head and O. E. Lowenstein). London: Oxford University Press.

del Pino, E. M. (1989). Modifications of oogenesis and development in marsupial frogs. *Development* **107**, 169-187.

Della Cave, L. and Simonetta, A. M. (1991). Early Paleozoic arthropods and problems of arthropod phylogeny; with some notes on taxa of doubtful affinities. In *The Early Evolution of Metazoa and the Significance of Problematic Taxa.* (ed. A. M. Simonetta and S. Conway Morris), pp. 189-244. Cambridge: Cambridge University Press.

Emlet, R. B. (1986). Facultative planktotrophy in the tropical echinoid *Clyeaster rosaceus* (Linnaeus) and a comparison with obligate planktotrophy in *Clypeaster subdepressus* (Gray) (Clypeasteroidea: Echinoidea). *Jour. Exp. Mar. Biol. Ecol.* **95**, 183-202.

Emlet, R. B. (1990). World patterns of developmental mode in echinoid echinoderms. In *Advances in Invertebrate Reproduction 5.* (ed. M. Hoshi and O. Yamashita). pp. 329-335. Amsterdam, Elsevier Science Publishers.

Ettensohn, C. A. (1984). Primary invagination of the vegetal plate during sea urchin gastrulation. *Amer. Zool.* **24**, 571-588.

Ettensohn, C. A. (1985). Gastrulation in the sea urchin embryo is accompanied by the rearrangement of invaginating epithelial cells. *Dev. Biol.* **112**, 383-390.

Ettensohn, C. A. and McClay, D. R. (1988). Cell lineage conversion in the sea urchin embryo. *Dev. Biol.* **125**, 396-409.

Evans, R. M. (1988). The steroid and thyroid hormone receptor superfamily. *Science* **240**, 889-895.

Finkelstein, R. and Perrimon, N. (1991). The molecular genetics of head development in *Drosophila melanogaster. Development* **112**, 899-912.

Gould, S. J. (1977). *Ontogeny and Phylogeny.* Cambridge: Harvard University Press.

Hardin, J. (1989). Local shifts in position and polarized motility drive cell rearrangement during sea urchin gastrulation. *Dev. Biol.* **136**, 430-445.

Hardin, J. and Cheng, L. Y. (1986). The mechanisms and mechanics of archenteron elongation during sea urchin gastrulation. *Dev. Biol.* **115**, 490-501.

Hardin, J., and McClay (1991). Target recognition by the archenteron during sea urchin gastrulation. *Dev. Biol.* **142**, 86-102.

Henry, J. J. and Raff, R. A. (1990). Evolutionary change in the process of dorsoventral axis determination in the direct developing sea urchin, *Heliocidaris erythrogramma. Dev. Biol.* **141**, 155-169.

Henry, J. J., Amemiya, S., Wray, G. A., and Raff, R. A. (1989). Early inductive interactions are involved in restricting cell fates of mesomeres in sea urchin embryos. *Dev. Biol.* **136**, 140-153.

Henry, J. J., Wray, G. A., and Raff, R. A. (1990). The dorsoventral axis is specified prior to first cleavage in the direct developing sea urchin *Heliocidaris erythrogramma. Development* **110**, 875-884.

Henry, J. J., Klueg, K. M. and Raff, R. A. (1992). Evolutionary dissociation between cleavage, cell lineage and embryonic axes in sea urchin embryos. *Development* **114**, 931-938.

Holland, P. W. H. (1991). Cloning and evolutionary analysis of *msh*-like homeobox genes from mouse, zebrafish and ascidian. *Gene* **98**, 253-257.

Hunt, P., Whiting, J., Muchamore, I., Marshall, H., and Krumlauf, R. (1991). Homeobox genes and models for patterning the hindbrain and branchial arches. *Development 1991 Suppl.* **1**, 187-196.

Huxley, A. (1955). *After Many a Summer Dies the Swan.* London: Penguin Books.

Jacob, F. (1977). Evolution and tinkering. *Science* **196**, 1161-1166.

Jeffery, W. R. and Swalla, B. J. (1992). Evolution of alternate modes of development in ascidians. *BioEssays* **14**, 219-226.

Jürgens, G. and Weigel, D. (1988). Terminal versus segmental development in the *Drosophila* embryo: the role of the homoeotic gene *fork head. Roux's Arch. Dev. Biol.* **197**, 345-354.

Levin, L. A., Zhu, J., and Creed, E. (1991). The genetic basis of life-history characters in a polychaete exhibiting planktotrophy and lecithotrophy. *Evolution* **45**, 380-397.

Liem, K. F. (1990). Key evolutionary innovations, differential diversity, and symecomorphies. In *Evolutionary Innovations.* (ed. M. Nitecki), pp. 147-170. Chicago: Univ. Chicago Press.

Mayr, E. (1960). The emergence of evolutionary novelties. In *Evolution After Darwin. Vol. I. The Emergence of Life.* (ed. S. Tax), pp. 349-380. Chicago: University of Chicago Press.

Mayr, E. (1982). *The Growth of Biological Thought. Diversity, Evolution,and Inheritance.* Harvard University Press. Cambridge.

McMillan, W. O., Raff, R. A., and Palumbi, S. R. (1992). Population genetic consequences of developmental evolution and reduced dispersal in sea urchins, (Genus *Heliocidaris*). *Evolution*, in press.

Müller, G. B. (1990). Developmental mechanisms at the origin of morphological novelty: A side-effect hypothesis. In *Evolutionary Innovations.* (ed. M. Nitecki), pp. 99-130. Chicago: Univ. Chicago Press.

Müller, K. J. and Walossek, D. (1986). Arthropod larvae from the Upper Cambrian of Sweden. *Trans. Roy. Soc. Edinburgh: Earth Sciences.* **77**, 157-179.

Murtha, M. T., Leckman, J. F., and Ruddle, F. H. (1991). Detection of homeobox genes in development and evolution. *Proc. Nat. Acad. Sci. USA* **88**, 10711-10715.

Nüsslein-Volhard, C. (1991). Determination of the embryonic axes of *Drosophila. Development 1991 Suppl.* **1**, 1-10.

Pehrson, J. and Cohen, L. (1986). The fate of the small micromeres in sea urchin development. *Dev. Biol.* **113**, 522-526.

Raff, R. A. (1992a). Direct-developing sea urchins and the evolutionary reorganization of early development. *BioEssays* **14**, 211-218.

Raff, R. A. (1992b). Developmental mechanisms in the evolution of animal form: Origins and evolvability of body plans. In *Early Life on Earth.* (ed. S. Bengtson). New York: Columbia University Press, in press.

Raff, R. A. and Kaufman, T. C. (1983). *Embryos, Genes, and Evolution.* New York: MacMillan.

Raff, R. A., Parr, B. A., Parks, A. L., and Wray, G. A. (1990). Heterochrony and other mechanisms of radical evolutionary change in early development. In *Evolutionary Innovations* (ed. M. Nitecki). pp. 71-98. Chicago: University of Chicago Press.

Raff, R. A., Wray, G. A., and Henry, J. J. (1991). Implications of radical evolutionary changes in early development for concepts of developmental constraint. In *New Perspectives on Evolution.* (ed. L. Warren and H. Koprowski). pp. 189-207. New York: Wiley-Liss.

Raff, R. A., Henry, J. J., and Wray, G. A. (1992). Rapid evolution of early development: Reorganization of early morphogenetic processes in a direct-developing sea urchin. In *Gastrulation, Movements, Patterns, and Molecules.* (ed. R. Keller, W. H. Clark, Jr., and F. Griffin). pp. 251-280. New York: Plenum Press.

Raff, R. A., Marshall, C. R. and Raff, E. C. (1993). Dollo's Law and the death and resurrection of genes. Submitted.

Schierwater, B., Murtha, M., Dick, M., Ruddle, F. H., and Buss, L. W. (1991). Homeoboxes in cnidarians. *J. Exp. Zool.* **260**, 413-416.

Smith, M. J., Boom, J. D. G., and Raff, R. A. (1990). Single copy DNA distance beteween two congeneric sea urchin species exhibiting radically different modes of development. *Mol. Biol. Evol.* **7**, 315-326.
Tyler, M. J. (ed.), *The Gastric Brooding Frog*. London: Croon Helm.
Thomson, K. S. (1991). Parallelism and convergence in the horse limb: The internal-external dichotomy. In *New Perspectives on Evolution*. (ed. L. Warren and H. Koprowski). pp. 101-122. New York: Wiley-Liss.
Wilson, E. B. (1985). The embryological criterion of homology. *Biological Lectures at the Marine Biological Laboratory, Woods Hole, Mass.* pp. 21-42. Boston: Ginn and Company.
Wray, G. A. (1992). The evolution of larval morphology during the post-Paleozoic radiation of echinoids. *Paleo. Biol.*, in press.
Wray, G. A., and Raff, R. A. (1990). Novel origins of lineage founder cells in the direct-developing sea urchin Heliocidaris erythrogramma. *Dev. Biol.* B, 41-54.
Wray, G. A. and Raff, R. A. (1991). The evolution of developmental strategy in marine invertebrates. *Trends Ecol. Evol.* **6**, 45-50.
Wray, G. A. and Raff, R. A. (1991). Rapid evolution of gastrulation mechanisms in a sea urchin with lecithotrophic larvae. *Evolution* **45**, 1741-1750.

Fate maps and the morphogenetic movements of gastrulation

Development 1992 Supplement, 23-31 (1992)
Printed in Great Britain © The Company of Biologists Limited 1992

Mechanisms of early *Drosophila* mesoderm formation

MARIA LEPTIN, JOSÉ CASAL, BARBARA GRUNEWALD and ROLF REUTER

Max Planck Institut für Entwicklungsbiologie, Spemannstrasse 35,7400 Tübingen, Germany

Summary

Several morphogenetic processes occur simultaneously during *Drosophila* gastrulation, including ventral furrow invagination to form the mesoderm, anterior and posterior midgut invagination to create the endoderm, and germ band extension. Mutations changing the behaviour of different parts of the embryo can be used to test the roles of different cell populations in gastrulation. Posterior midgut morphogenesis and germ band extension are partly independent, and neither depends on mesoderm formation, nor mesoderm formation on them. The invagination of the ventral furrow is caused by forces from within the prospective mesoderm (i. e. the invaginating cells) without any necessary contribution from other parts of the embryo. The events that lead to the cell shape changes mediating ventral furrow formation require the transcription of zygotic genes under the control of *twist* and *snail*. Such genes can be isolated by molecular and genetic screens.

Key words: *Drosophila*, mesoderm formation, ventral furrow formation, germ band extension, *twist*, snail.

Introduction

The basic rules for some developmental processes - like pattern formation in the *Drosophila* embryo, or differentiation - are now nearly understood. However, we still know little about the mechanisms and the molecules involved in morphogenesis. How do cells, once they have been assigned their fates and their positions in the developing organism, build ordered structures and organs? Morphogenetic mechanisms include cell proliferation and growth, cell migration, shape changes of individual cells and of groups of cells, for instance epithelia. The most vigorous period of morphogenesis during development is gastrulation, when the spatial relationships of cells within the embryo are continuously changing until the basic body plan is established. All of the morphogenetic mechanisms listed above occur during *Drosophila* gastrulation, although the initial and most dramatic events are mediated only by shape changes of epithelia and by cell intercalation within epithelia. Our own work, and this review, concentrates on an example of epithelial invagination, the formation of the ventral furrow.

Ventral furrow formation is the beginning of mesoderm development. It is the first morphogenetic event of *Drosophila* gastrulation and a particularly clear example of epithelial folding. It has the advantage over the classically investigated cases, such as amphibian gastrulation or neurulation, of being quick and uncomplicated (no cell division or growth, only a single homogeneous cell layer), and of being amenable to genetic analysis. We already know the genes that determine the fates of the cells involved (reviewed in Anderson, 1987 and Stein and Stevens, 1991), and the ventral furrow forms less than an hour after these genes begin to be transcribed. Therefore, the interval between cell fate determination and morphogenetic activity is very short, and hopefully the genetic regulatory cascade, beginning downstream of the known fate determining genes and leading to change in cell behaviour, will be correspondingly simple.

Materials and methods

Staining, in situ hybridisation and sectioning of embryos

Embryos were collected, fixed and processed for in situ hybridisation and antibody staining as described previously (Leptin and Grunewald, 1990).

Stocks

The $twist^{RY50}$ stock was obtained from P. Simpson, Strasbourg. Lethal mutations that had accumulated on the $twist^{RY50}$-carrying chromosome were separated from the $twist^{RY50}$ mutation by recombination. All other stocks were from the Tübingen stock collection (Tearle and Nüsslein-Volhard, 1987). We used the following mutant alleles: sna^{IIG}, Df(2R)twi^{S60} (Simpson 1983), tor^{XR1}, scw^{N5}, fog^{4a6}, $kni^{IID}hb^{7M}$, $Toll^{9QRE}$, $Toll^{r444}$ and $Toll^{10b}$. The *dpp* embryos were transheterozygous for dpp^{Hin37} and Df(2L)DTD2.

cDNA subtraction

RNA was prepared from embryos (blastoderm to early gastrulation) derived from mutant mothers ($Toll^{9QRE}/Toll^{r444}$ and $Toll^{10b}/+$) (details to be published elsewhere), reverse transcribed, digested with *Alu*I and *Rsa*I, linkered, amplified by PCR and enriched through several cycles of subtractive hybridisation as described by Wang and Brown, 1991.

Germ line clones

Germ lines mutant for the gene *flightlessI* were generated as described by Wieschaus and Noell, 1986. The *flightless* allele was *fli*^{WC2} (see Perrimon et al., 1989).

Results and discussion

Drosophila *gastrulation*

Several morphogenetic processes occur in parallel during *Drosophila* gastrulation (Fig. 1). Three invaginations create the germ layers. First, the presumptive mesoderm begins to invaginate as a broad band of cells on the ventral side of the embryo. Then, while mesoderm invagination continues, the endoderm is made from two invaginations, one at the posterior pole (the posterior midgut invagination), and one on the ventral side of the head region (the anterior midgut invagination). During this time, a process called germ band extension moves the posterior end of the embryo onto the dorsal side.

We will describe the cellular events and possible mechanisms that bring about ventral furrow formation and show to what extent mesoderm formation depends on other cell populations and events during gastrulation (an extensive review of these and other aspects of gastrulation has recently been published by Costa et al., 1992). We will conclude with an outlook on how we plan to define the genetic pathway that leads to mesoderm morphogenesis.

Ventral furrow formation and mesoderm invagination

Before gastrulation begins (Fig. 1A), all 5000 cells in the blastoderm epithelium look morphologically identical (except that cellularisation is not completely finished on the dorsal side). However, cells in different regions are already distinguished by their gene expression patterns. The genes *twist* and *snail* are expressed in ventral cells, including all future mesoderm cells, and some endodermal and ectodermal cells (Thisse et al., 1988; Leptin and Grunewald, 1990). The mesoderm is made from the *snail*-expressing cells between ~10% and ~70% egg length (measured from the posterior end of the embryo). All of these cells also express *twist*, but unlike *snail*, *twist* expression extends slightly beyond the lateral edge of the prospective mesoderm (Kosman et al., 1991; Leptin, 1991).

As soon as the cells on the ventral side of the embryo have formed, they begin to invaginate. The apical sides (facing the outside of the embryo) of the prospective mesodermal cells first flatten and then a subpopulation, a central band of cells approximately 8-10 cells wide, constrict apically and their nuclei migrate towards their basal ends (Fig. 1C. Kam et al., 1991; Leptin and Grunewald, 1990; Sweeton et al., 1991). The cells are thereby turned from a nearly columnar to a wedge shape. The band of cells that undergo these changes will be called the 'central population'. A 4- to 5-cell-wide band of prospective mesoderm cells on each side of the central population, the peripheral population, does not constrict apically. Within about 10 minutes, the changes in the central population result in the appearance of an identation in the ventral surface of the embryo (the ventral furrow), which deepens and invaginates into the interior of the embryo, followed by the peripheral cells on each side which have not constricted apically (Fig. 1B,D). This first phase of mesoderm invagination, which lasts about 15 minutes, is characterized by the absence of individual cell movements (either within the plane of the epithelium, or out of the epithelium), and of cell division or growth. The epithelium remains intact as it invaginates.

It seems likely that the driving force for this phase comes from the cell shape changes in the central population. However, the causal relationships between the events associated with cell shape changes are not clear. Like in other epithelial invaginations (Burnside, 1973), the actin cytoskeleton probably plays a major role in apical constriction. This is supported by the finding that myosin, concentrated at the base of the cell before ventral furrow formation, becomes localized to the apical side as constriction occurs (Young et al., 1991 and Fig. 2). However, apical constriction cannot alone be responsible for the movement of nuclei and shape changes. In *twist* mutant embryos (see below) no apical constriction occurs, but nuclei nevertheless move away from the apical end of the cells and the cells change their shape sufficiently to make a small transient furrow (Leptin and Grunewald, 1990; Sweeton et al., 1991). Surprisingly, apical flattening and nuclear movement occur even in mutants in which the blastoderm is not properly cellularized (Fig. 3; Kristina Straub and M. L., in preparation). All

Fig. 1. Whole mounts and transverse sections of embryos at various stages of gastrulation, stained with antibodies against the *twist* gene product, a nuclear protein expressed in ventral cells. (A) Cellular blastoderm. The nuclei on the ventral side are beginning to move basally; the posterior midgut primordium with the pole cells is shifting dorsally. (B) Early germ band extension. The mesoderm has invaginated on the ventral side. The posterior midgut has begun to invaginate and move onto the dorsal side, carrying the pole cells with it. (C-E) Transverse sections to show mesoderm invagination. The embryo in C is a few minutes older than that in A, the one in D the same age as that in B. (F-H) Drawings of the sections shown in C-E to illustrate the subpopulations of prospective mesoderm cells with different behaviours. Twist protein (yellow) is expressed in future mesoderm cells (blue), but initially a gradient of twist protein extends through the mesectodermal cells (red) into the extoderm (Kosman et al., 1991; Leptin 1991). All cells between the mesectodermal cells will invaginate to form the mesoderm, but in wild-type embryos only the central cells (dark blue) undergo the typical shape changes that we believe to bring about mesoderm invagination, while the peripheral cells (light blue) appear to follow into the furrow passively. These two populations are no longer distinguishable at the stage shown in E and H. The colour differences in this panel are therefore based entirely on conjecture. (I-L) Comparison of mesoderm formation in wild-type (I,K) and mutant (J,L) embryos. (I) Ventral view of a wild-type embryo at the same age as that in B. (J) Ventral view of a *snail* mutant embryo at the same age. The mesodermal region has not invaginated, but has formed irregular small folds. (K) Wild-type embryo at the extended germ band stage. The mesoderm, still expressing *twist*, has spread out as a single cell layer on the inside of the ectoderm. (L) *twist* mutant embryo at the extended germ band stage. No mesodermal cell layer has formed.

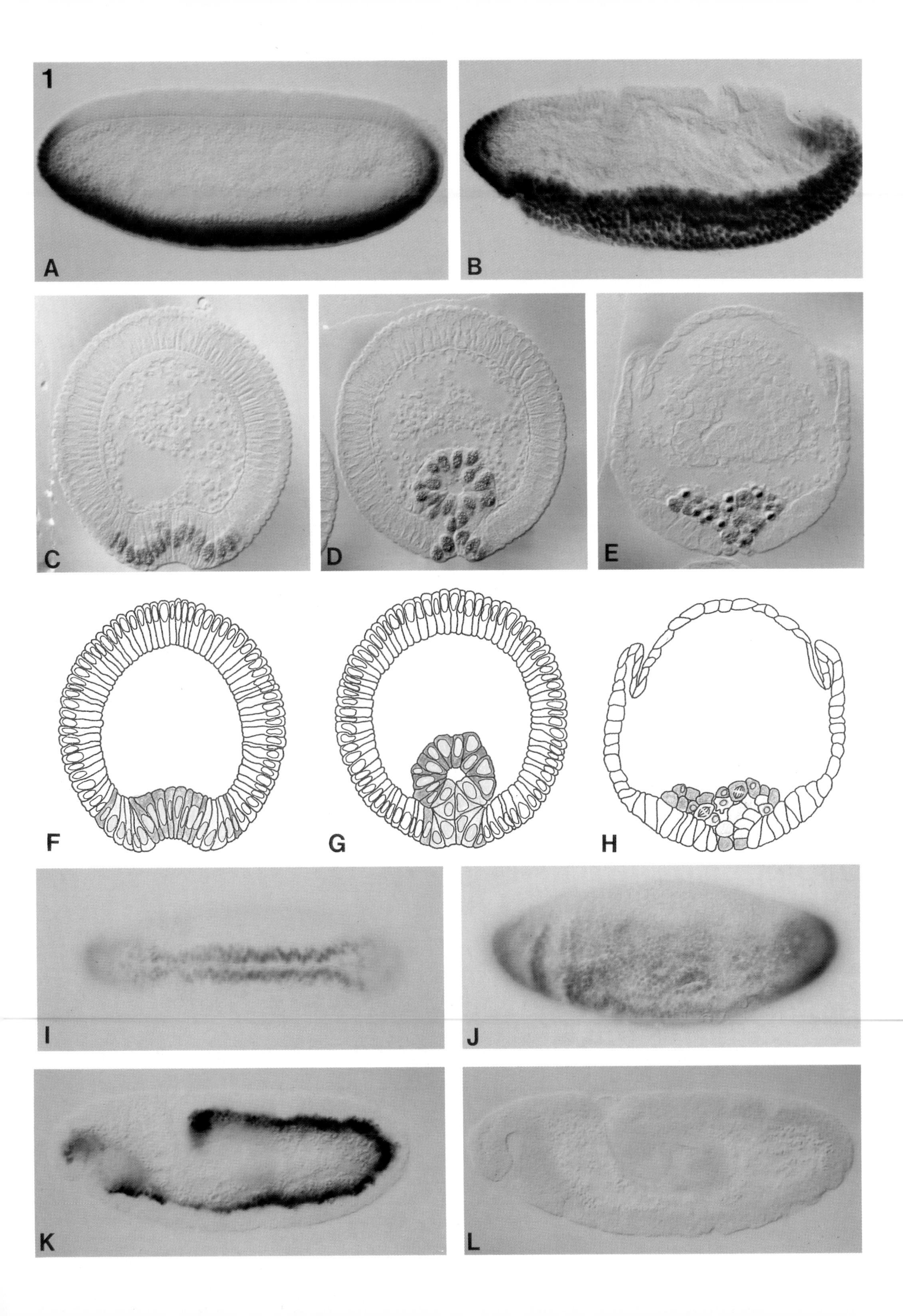
1
A
B
C
D
E
F
G
H
I
J
K
L

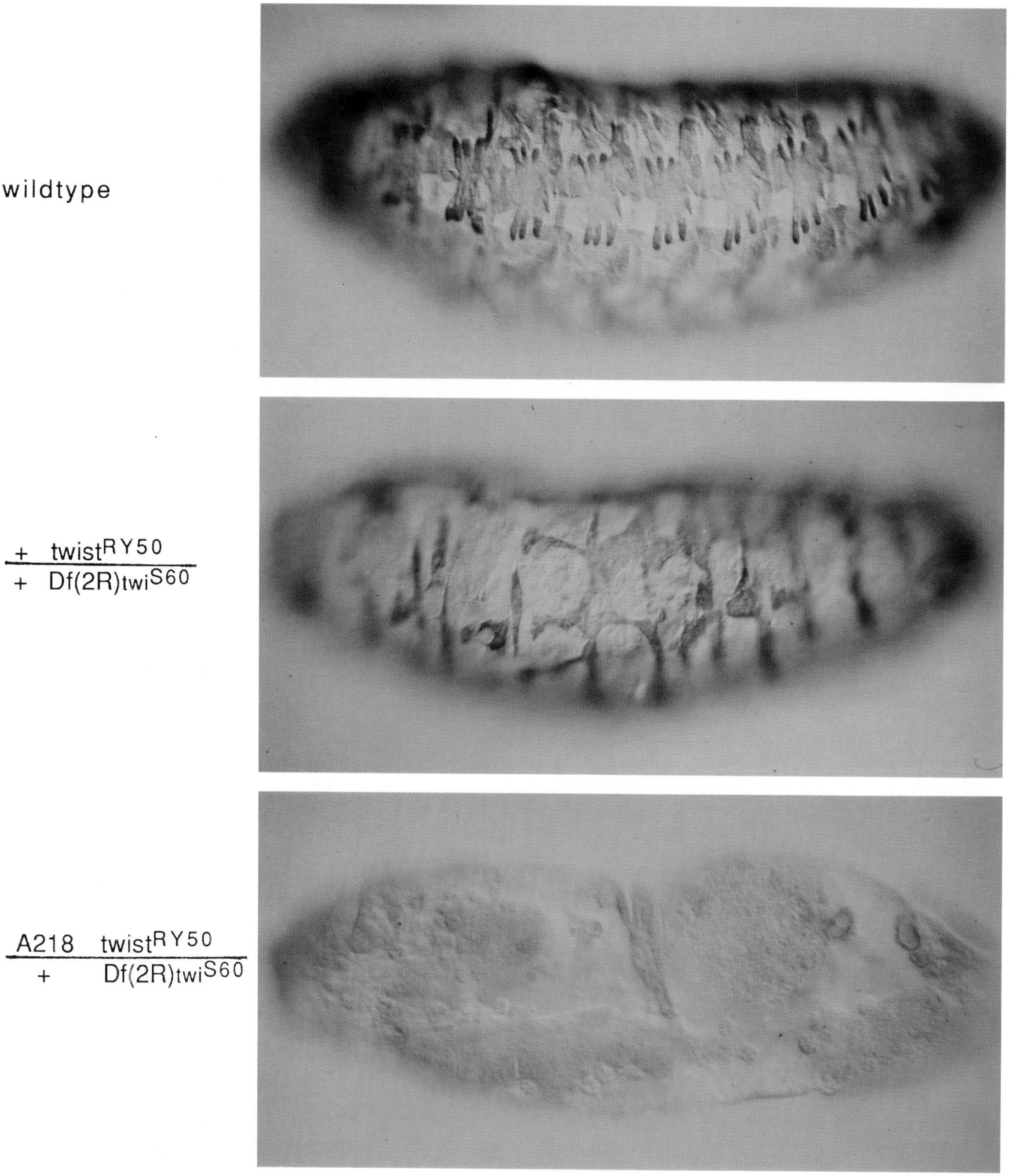

wildtype

$\frac{+ \quad twist^{RY50}}{+ \quad Df(2R)twi^{S60}}$

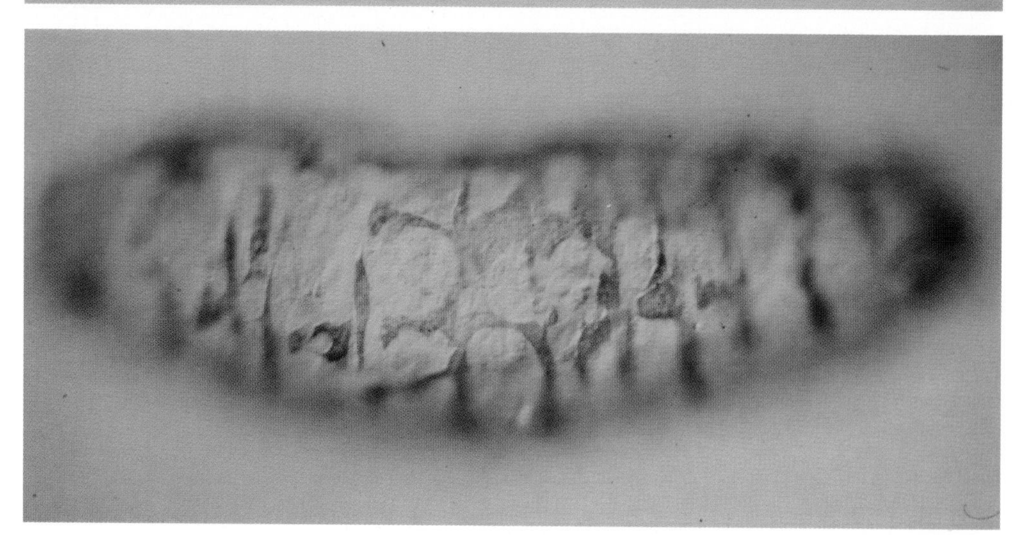

$\frac{A218 \quad twist^{RY50}}{+ \quad Df(2R)twi^{S60}}$

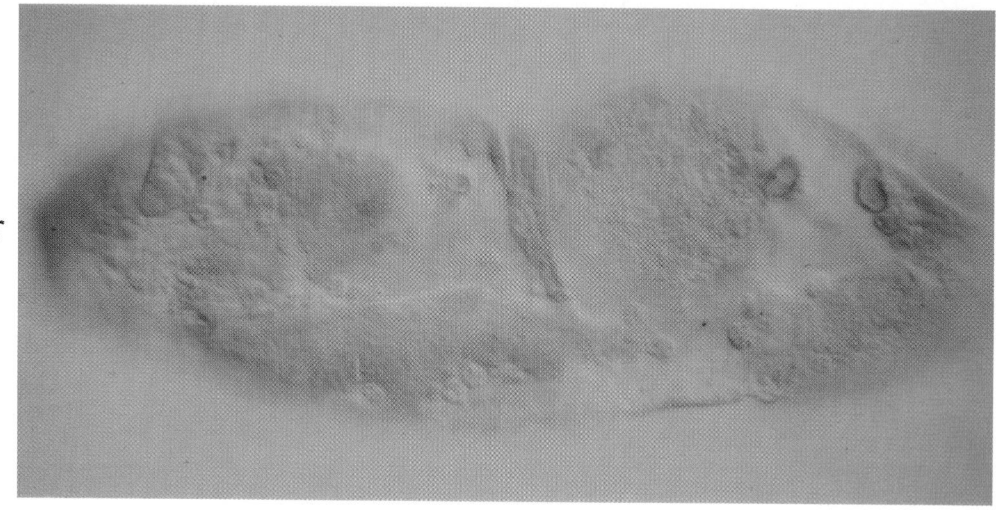

Fig. 7. Muscle development in mutant embryos. Late embryos were stained with antibodies against muscle myosin to visualize muscles. The typical wild-type muscle pattern (top) is disrupted in embryos carrying a weak *twist* allele over a *twist* deficiency (middle). If the embryos are also heterozygous for a mutation in another gene (*twist*-enhancer A218), muscles fail to develop altogether.

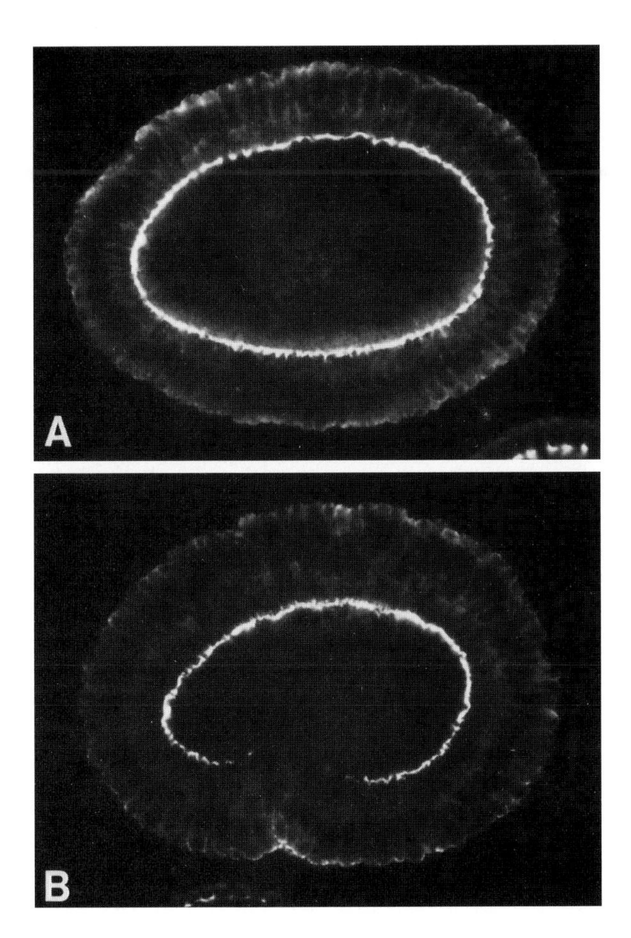

Fig. 2. Transverse sections of embryos before gastrulation begins (A), and when the ventral furrow is forming (B), stained with antibodies against cytoplasmic myosin. Myosin is initially concentrated at the basal end of each cell, but relocates to the apical sides of the ventral cells as they constrict.

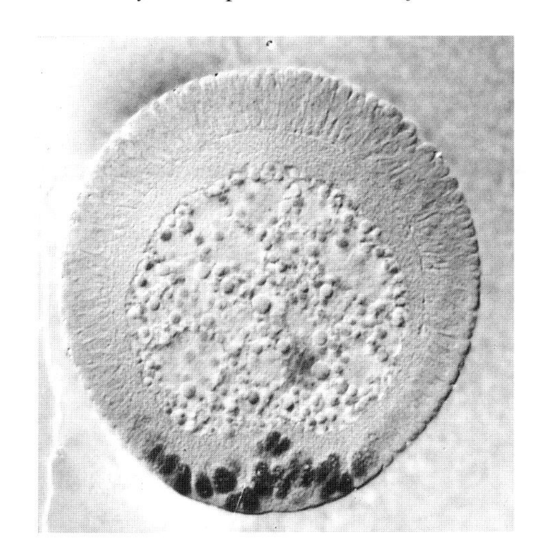

Fig. 3. Transverse section of an embryo derived from a germ line mutant for the gene *flightlessI.* Although no proper cell membranes are formed in ventral cells, apical flattening and nuclear migration still take place.

of these findings are consistent with the notion that the movement of nuclei is passive, i. e. that nuclei are merely being released from the apical end of the cell. This would mean that the role of nuclear movement is to permit the cell shape changes rather than to cause them.

The actual mesodermal germ layer is formed in the next phase of mesoderm invagination. *snail* expression is now lost from the mesoderm, but *twist* continues to be expressed (Thisse et al., 1988; Alberga et al., 1991; Leptin 1991; Kosman et al., 1991). The tube of prospective mesoderm created by the invagination of the ventral furrow loses its epithelial structure and disperses into single cells (Fig. 1F). These divide, migrate out on the ectoderm to form a single cell layer (Fig. 1H) and then divide again. The cell divisions during this phase have no morphogenetic role, as shown by the finding that the mesoderm disperses and spreads out normally in mutant embryos in which these divisions fail to occur (Leptin and Grunewald, 1990).

Relationship between ventral furrow formation and other morphogenetic events in the embryo

Mutations that affect different parts of the embryo can be used to assess the roles of different cell populations in gastrulation. Mutations in both maternal-effect and zygotic genes have been useful for this purpose. Most of these are mutations that delete or change certain cell fates in the developing embryo either along the anterior-posterior axis, or along the dorsoventral axis.

Mutations in maternal-effect genes that set up pattern along the dorsoventral axis affect the fates of most cells along the dorsoventral axis (Roth et al., 1989; Ray et al., 1991). Ventralizing mutations lead to the expression of ventral-specific genes in all cells of the embryo, while dorsalizing mutations produce embryos in which no ventral genes are expressed. In contrast, mutations in zygotic genes have more restricted effects. Usually only the fates of the cells that normally express the gene are changed, while other cells in the embryo develop according to their normal programme. Mutations in the genes *twist* and *snail* abolish or change the fates of the cells that normally give rise to the mesoderm, and mutant embryos fail to make a mesoderm (Fig. 1J,L; Simpson, 1983). Another group of genes is responsible for the development of dorsal fates (Ferguson and Anderson, 1992; Arora and Nüsslein-Volhard, 1992), and embryos mutant for these genes fail to differentiate the amnioserosa and varying proportions of the dorsal ectoderm.

Observations in mutants with changed ventral fates:regional and cellular autonomy of ventral furrow formation

Maternally dorsalized embryos do not form a ventral furrow, nor do any of their cells undergo the changes usually seen in ventral cells. Their apical surfaces remain rounded and the nuclei stay in the apical parts of the cells (Leptin and Grunewald, 1990). In contrast, in completely ventralized embryos, the apical surfaces of all cells flatten, and nuclei at all positions along the dorsoventral axis migrate basally, behaviour characteristic of the most ventral cells in wild-type embryos. This indicates that this behaviour is part of the autonomous morphogenetic program of cells expressing ventral genes, and not simply a mechanical response to activities from neighbouring non-ventral cells.

A similar argument can be made for the formation of the ventral furrow. Some ventralized embryos still have residual dorsoventral polarity such that they still form a ventral invagination although *all* cells express ventral genes. Since these embryos contain no cells with lateral or dorsal fates, one can conclude that the activities of dorsal and lateral cells are not required for ventral furrow formation. That their activities are also not sufficient can be seen in *twist* and *snail* mutant embryos. These mutants make no ventral furrow, although dorsal and lateral cells develop normally. These findings suggest that all information for furrow formation resides within the cells that make the furrow, and that the necessary forces are generated within this region (Leptin and Grunewald, 1990).

These conclusions are further supported by several findings that suggest a high degree of cellular autonomy of the processes that cause the ventral furrow to form. First, in certain maternally dorsalized embryos, ventral cell fates can be induced by injection of wild-type cytoplasm from wild-type embryos (Anderson et al., 1985), and a furrow forms at the site of injection (Siegfried Roth and M. L., in preparation). The shape, direction and dimension of the furrow depends on the shape of induced ventral gene expression. This indicates that the shape and site of the furrow do not depend on the geometry of the egg, but are determined only by the patch of cells expressing ventral genes. Second, the cells of the central population in normal embryos begin their shape changes nearly, but not completely simultaneously. There is no particular order in which they constrict (Kam et al., 1991; Sweeton et al., 1991), suggesting that each cell begins its shape change independent of its neighbours. Certainly cells at all positions within the prospective mesoderm have the capacity to constrict, even when they are surrounded by cells that are genetically unable to constrict (M. L. and Siegfried Roth). Therefore, no large scale coordination by cell interactions appears to be necessary.

These conclusions apply to the *initiation* of ventral furrow formation and the mechanisms of cell shape changes. The speed and efficiency of the process may be coordinated by communication between cells. This view is based on the mutant phenotypes of two genes, *concertina* and *folded gastrulation,* (Sweeton et al., 1991) and the predicted structure of one of their products (Parks and Wieschaus, 1991). Ventral cells in mutant embryos have the capacity to constrict apically and begin these activities at the appropriate time, but the furrow then forms too slowly and with an irregular appearance. Since *concertina* codes for a G-protein homolog, these findings suggest the involvement of signal transduction mechanisms, and therefore possibly cell communication in the process of furrow invagination (Parks and Wieschaus, 1991; Costa et al., 1992).

Timing of ventral furrow formation

If all cells that make the ventral furrow initiate their cells shape changes independently of each other, how is the starting point for these changes determined? Since the changes begin as soon as cellularisation of the blastoderm is completed ventrally, cellularisation itself might provide the signal. However, this cannot be the case, since apical flattening and nuclear movement occur even in the absence of proper cellularisation. It seems much more likely that the accumulation of one or more crucial zygotic gene products above a critical level initiates ventral furrow formation. This notion is based on the observation that embryos that have only one functional copy of the zygotically active gene *twist* (and therefore probably only half the amount of this transcription factor) begin ventral furrow formation several minutes later than embryos with two copies (unpublished observation). Thus, the product of one or more genes transcribed under the control of *twist* must be limiting for furrow formation.

Several aspects of ventral furrow formation are easily explained by this interpretation. The cells begin to change their shapes approximately at the same time, because they all transcribe their genes at approximately the same rate. The apparently random initiation of cell shape changes is due to slight differences in the time when the critical level of zygotic gene products is reached in individual cells. The cells nearest the midline have a higher chance of constricting early compared to more lateral cells because the earliest *twist* (and *snail*) expression is restricted to a narrow ventral band corresponding in width to the central population of prospective mesoderm cells (Leptin, 1991) and, within this band, *twist* levels are highest near the midline. Finally, the early expression of *twist* and *snail* in this region might be the only genetic difference between the central and the peripheral population of prospective mesoderm cells. The peripheral cells would then in principle also have the capacity to constrict, but only accumulate enough of the critical gene product by the time that most of the central cells have invaginated and begun to pull the peripheral cells into the deepening furrow. This view is supported by mosaics in which patches of wild-type cells in the region of the peripheral population do constrict if they are in an environment of mutant, non-constricting central cells (M. L. and Siegfried Roth, in preparation).

Observations on mutations affecting other fates: relationships between germ band extension, endoderm formation and mesoderm formation

Endoderm formation

The posterior midgut invaginates by the same cell shape changes as the mesoderm (Sweeton et al., 1991). The cell surfaces flatten and then constrict apically while cell nuclei move basally. Shortly afterwards, the posterior midgut forms an indentation which invaginates, drawing neighbouring non-constricting cells along. It appears that posterior midgut formation is more sensitive to subtle interference with these shape changes since the severity of the phenotypes of *concertina* and *folded gastrulation* differs in ventral furrow and posterior midgut although the progression through cell shape changes is affected in both (Sweeton et al., 1991).

The anterior midgut is formed by different mechanisms (for review see Costa et al., 1992) and will not concern us further here.

Germ band extension

Germ band extension begins as soon as the ventrolateral

furrow has begun to invaginate and the posterior midgut cells are beginning to change their shapes. The germ band consists of the invaginated mesoderm and the overlying ectoderm, situated between the head fold and the hindgut primordium. This region begins to lengthen, but since the embryo is enclosed in membranes and cannot change its overall dimensions, this results in the posterior midgut primordium being displaced dorsally. The dorsal epithelium of the embryo, the future amnioserosa, does not lengthen but becomes thin and folds up as the advancing posterior midgut invagination is pushed dorsally. Germ band extension is later reversed by germ band retraction so that the original anterior-posterior order is re-established. The function of germ band extension and retraction are not understood. They may have more to do with anterior-posterior pattern formation within segments than with morphogenesis (Wieschaus et al., 1991).

Mutations interfere with germ band extension in two ways. They can abolish the mechanisms that provide the driving force (the active mechanisms) or they can interfere with processes that allow the gastrulating embryo to respond to these forces by proper morphogenetic movements.

The driving force for germ band extension appears to be cell intercalation in the ventral ectoderm (Wieschaus et al., 1991). This process is disrupted by mutations that abolish positional values along the anterior-posterior axis (mutations in maternal axis determining genes and segmentation genes; Fig. 4). In the most extreme cases, no germ band extension movements occur at all (Wieschaus et al., 1991). However, the mesoderm and the anterior and posterior midgut invaginate normally in these embroyos.

The second aspect of germ band extension affected by mutations is the movement of the posterior midgut primordium. For the germ band to extend properly, the posterior midgut has to move dorsally. In mutants that do not form a posterior midgut (e.g. *folded gastrulation*, or *torso*), the posterior pole of the embryo cannot be pushed onto the dorsal side. Instead, the germ band stops extending or buckles into folds on the ventral side of the embryo.

The movement of the posterior midgut towards the head is made possible by the dorsal epithelium (the amnioserosa) becoming very thin and folding up between head and posterior midgut. In mutants whose amnioserosa does not form, the posterior midgut begins to be pushed dorsally, but then stalls and eventually sinks into the inside of the embryo, underneath the mutant amnioserosa, which has not folded up (Fig. 4).

Independence of gastrulation movements

From the phenotypes of the mutants described so far, it is already clear that proper germ band extension depends on proper posterior midgut development, while the posterior midgut develops independently of germ band extension. Neither germ band extension nor posterior midgut formation depend on mesoderm development, since they occur normally in *twist* and *snail* mutants, which do not make any mesoderm (Simpson, 1983; Leptin and Grunewald, 1990; Sweeton et al., 1991).

How is mesoderm development affected by disruptions of germ band extension or posterior midgut formation? The ventral furrow forms and invaginates normally even if the posterior midgut is absent, for example in *torso* embryos. This is also the case if the germ band does not extend, either because the driving force for extension is abolished by mutations in anterior-posterior patterning genes, or if posterior midgut movement is inhibited by the failure of the amnioserosa to form due to mutations in dorsoventral patterning genes (Fig. 4). However, the spreading of the invaginated mesoderm on the ectoderm appears abnormal in these mutant embryos. Especially in *dpp* mutant embryos, which lack all dorsal fates, the mesoderm migrates all the way to the dorsal midline (Fig. 4). This is seen to a lesser extent in embryos mutant for the other genes of this group. This behaviour has two causes. One is the reduction of germ band extension. In the absence of germ band extension, the germ band does not fold over the back of the embryo, and the path is free for the mesoderm to migrate much further dorsally than normal. However, in mutants that fail to extend their germ band due to anterior-posterior patterning defects, the mesoderm migrates less far dorsally than in *dpp* mutants (B. G. and M. L., in preparation). Therefore the fate of the underlying ectoderm probably also plays a role in determining the extent of mesoderm migration.

Genetic regulation of ventral cell behaviour

We know the genes responsible for determining the cells that invaginate to form the mesoderm (the maternal dorsal group genes) and we know what the activities of ventral cells are that bring about the first steps of the invagination. We do not know how these two are linked. What is the cascade of gene activities that leads from fate determination to morphogenetic activity?

The first manifestation of ventral identity under the control of the maternal dorsoventral morphogen gradient is the expression of *twist* and *snail*. These genes code for transcription factors that are found in the nuclei of prospective mesoderm (and a few other) cells (Alberga et al., 1991; Boulay et al., 1987; Kosman et al., 1991; Leptin, 1991; Leptin and Grunewald, 1990; Thisse et al., 1987). They are the only known genes that are essential for all aspects of mesoderm differentiation and morphogenesis (Fig. 1). In *twist* and *snail* mutants, no mesoderm develops and the failure is visible already at the time when the ventral cells should begin their characteristic shape changes. Ventral cells in *twist* mutants flatten but do not constrict their apical sides (Leptin and Grunewald, 1990; Costa et al., 1991). They become tall and thin and then often form a small, transient furrow. Ventral cells in *snail* mutants do not flatten. They become short and wide, so that the ventral epithelium turns into a thin sheet which buckles into irregular folds (Fig. 1). In embryos mutant for both genes, ventral cells undergo none of these changes. Thus, *twist* and *snail* regulate separate aspects of early mesoderm formation.

Since *twist* and *snail* code for transcription factors, their roles can be shown more directly by assaying the expression of their potential target genes in mutant embryos (Leptin, 1991). Several genes are known that are expressed early in the mesoderm, or excluded from the mesoderm, or expressed only at the boundary between mesoderm and ectoderm. The function of *twist* is to activate genes in the

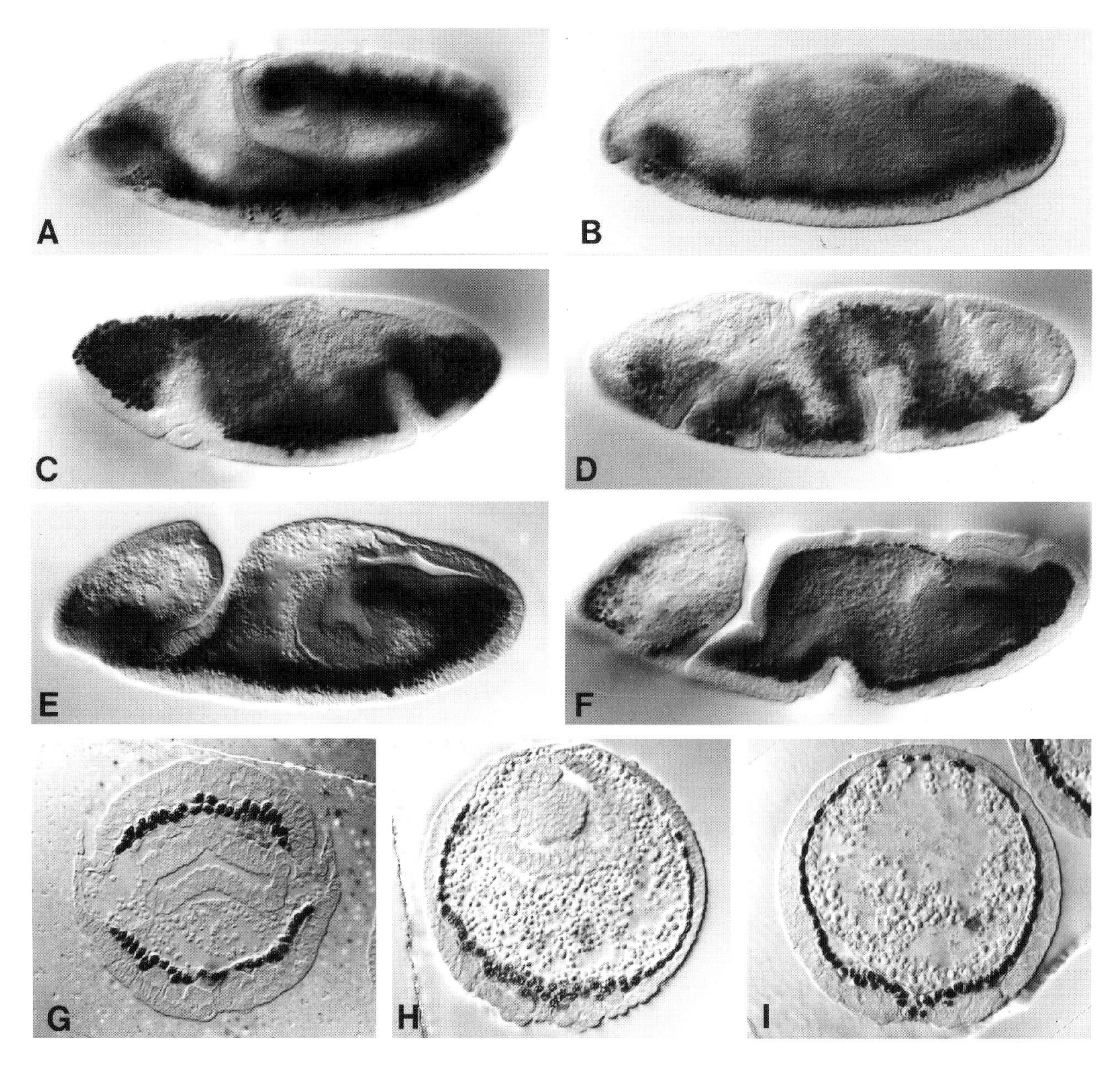

Fig. 4. Germ band extension and mesoderm spreading in wild-type and mutant embryos. All embryos are stained with *twist* antibodies and are at the stage corresponding to the extended germ band stage of the wild-type embryo shown in A. (A) Wild-type extended germ band. (B) Embryo that has failed to undergo proper germ band extension due to pattern formation defects along the anterior-posterior axis (*hunchback knirps* double mutant embryo). (C,D) Mutant embryos without posterior midgut invagination: *torso* embryo (C; derived from homozyogus *torso* mutant mother), and *folded gastrulation* embryo (D). (E,F) Embryos with defects in dorsal fate determination in which the amnioserosa fails to form. *screw* embryo (E) and *dpp* embryo (F). (G-I) Transverse sections through embryos similar to those in A, E and F. G, wild-type; H, *screw*; I, *dpp*.

region of the prospective mesoderm, but it has no effect on genes transcribed in the ectoderm. In contrast, *snail* is not required directly for the activation of mesodermal genes (at least those examined so far), but in the mesodermal region it represses genes destined to be active only outside the mesoderm. Of course *snail* can thereby be indirectly required for the activation of mesodermal genes if one of its targets normally represses mesodermal genes. *twist* and *snail* together define the mesoderm/ectoderm border and the expression of genes expressed only at this border (see also paper by Levine in this issue).

We have shown that the ventral cellular activities that cause the ventral furrow to form do not occur properly in *twist* and *snail* mutants, and that *twist* and *snail* act by regulating the expression of other genes in the ventral region of the embryo. To understand the pathway from fate determination to morphogenesis, one has to identify these genes. They are unlikely to be easily recognisable by their mutant phenotypes. Neither saturation screens for embryonic lethal mutations, nor a specific screen of the whole genome for early acting zygotic genes have identified any genes whose mutant phenotypes resemble that of *twist* or *snail*. There-

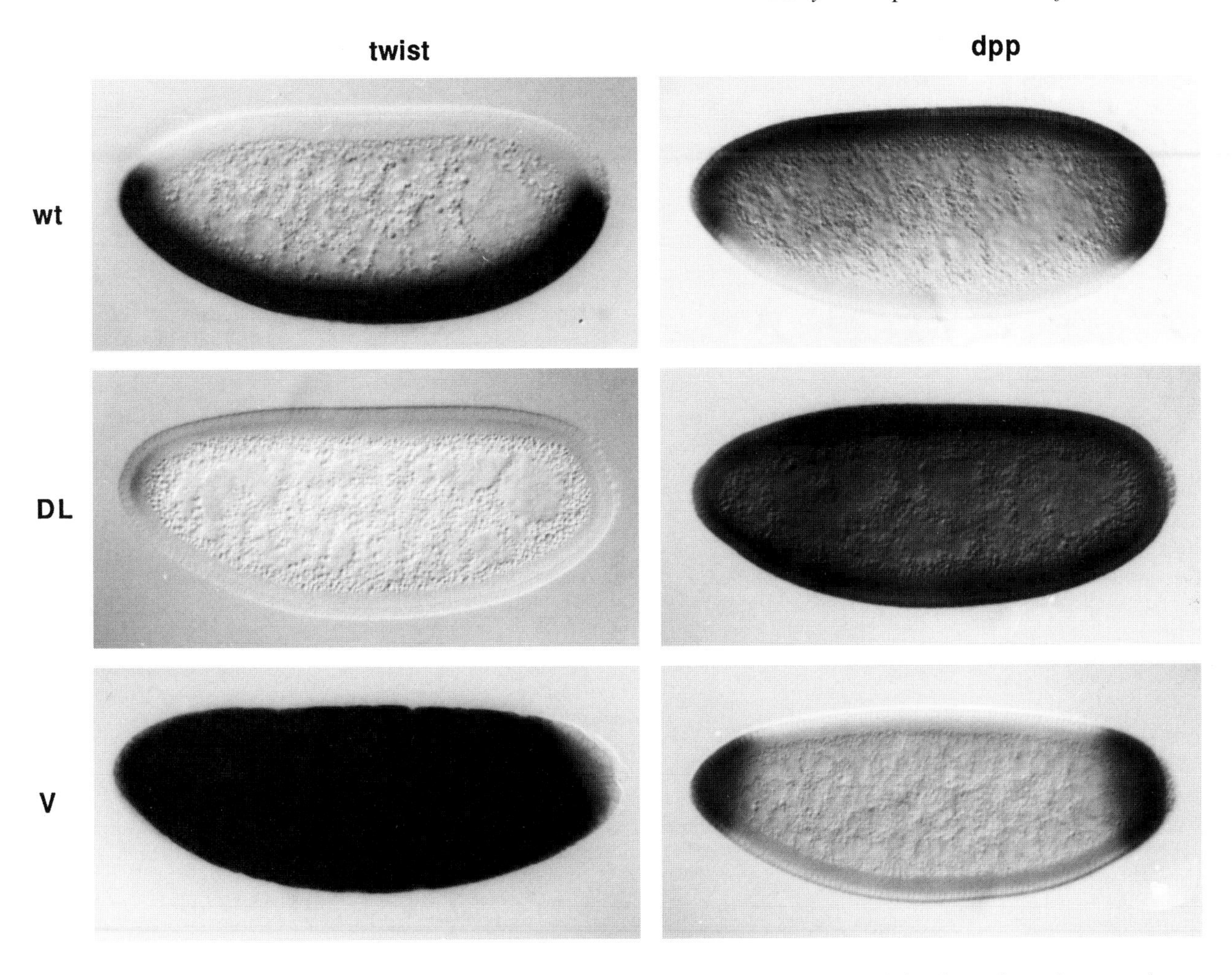

Fig. 5. Dorsal and ventral gene expression in maternally ventralized and lateralized embryos. The left column shows the expression pattern of *twist,* the right column the expression of *dpp* in wild-type (wt), 'dorsolateral' (DL) and 'ventralized' embryos. The DL embryos were derived from mothers transheterozyogous for the *Toll* alleles $Toll^{9QRE}/Toll^{r444}$, the V embryos from mothers heterozygous for the dominant ventralizing Toll allele $Toll^{10b}$ (Anderson et al., 1985).

fore, mutations in *twist* and *snail* target genes involved in furrow formation may give very subtle phenotypes. Indeed, one gene that was identified only because its mutation causes defects in posterior midgut formation turned out upon further analysis to show defects in ventral furrow formation as well (Sweeton et al., 1991). If the desired genes cannot be found in a conventional screen for visible phenotypes, other methods have to be designed to identify them.

Search for new genes active in mesoderm morphogenesis

We have made use of two aspects of ventral furrow formation to conduct molecular and genetic screens. A genetic screen is based on the dosage sensitivity of the ventral-fate-determining system, and a molecular screen on the knowledge that at least some of the genes that we are searching for have to be expressed ventrally.

The molecular screen

Genes that are expressed ventrally in wild-type embryos are not expressed in maternally dorsalized embryos, but are expressed in all cells of maternally ventralized embryos (Roth et al., 1990 and Fig. 5). Therefore ventral genes can be identified by using subtractive hybridisation to isolate all those genes expressed only in ventralized and not in dorsalized embryos. Theoretically one could also find ventrally expressed genes by isolating all genes expressed in wild-type embryos but not in *twist* mutant embryos. However, this is more difficult, because *twist* mutant embryos constitute only one quarter of the progeny of a cross and would have to be hand-selected at the appropriate time of development. In contrast, **all** embryos from a mother carrying a maternal effect mutation express the phenotype, and large numbers of appropriately staged embryos can be collected. Fig. 6 shows cDNA from ventralized and dorsalized embryos before and after several cycles of subtractive hybridisation. cDNA from the dorsally expressed gene *dpp* becomes enriched in dorsalized cDNA, while *twist* becomes enriched in ventralized cDNA. At the same time, RNA from ubiquitously expressed genes like tubulin disappear from both populations. A library constructed from the enriched ventral cDNA should be a good source for new ventrally expressed genes, as has indeed been confirmed by the analysis of the first 72 clones isolated.

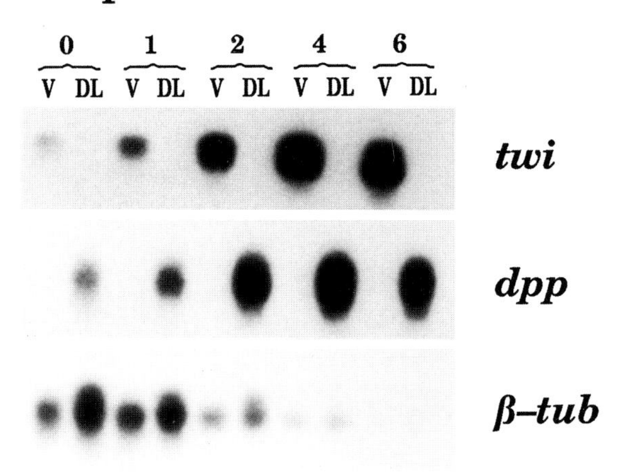

Fig. 6. Subtraction of 'ventral' (V) versus 'dorsolateral' (DL) cDNA. Amplified cDNA populations isolated from mutant embryos as those shown in Fig. 5 were subtracted against each other in successive cycles as described by Wang and Brown, 1991. Southern blots of the cDNAs were probed for a ventrally expressed gene (*twist*), a dorsally expressed gene (*dpp*) and a ubiquitously expressed gene (β-tubulin).

The genetic screen

As described above, embryos heterozygous for *twist* mutations gene begin ventral furrow formation several minutes later than embryos carrying two intact copies of *twist*. If the embryo also has only half the normal amount of maternal *dorsal* product (a condition that normally has no effect on development), it cannot survive at 27°C (Simpson, 1983). Similarly, embryos carrying a particular combination of *twist* alleles survive at 22°C, but die at raised temperatures (Thisse et al., 1987), or, more importantly, when the dose of *snail* is reduced by half. In both cases, the product of a gene (or genes) acting in parallel or downstream of *twist* is probably reduced below a critical level required for viability. We based a screen on the assumption that halving the level of such a gene product by inducing a mutation in it should also cause lethality in embryos carrying the above combination of *twist* alleles. The effect of one of the mutations found in this screen is shown in Fig. 7. Embryos carrying the *twist* alleles mentioned above show the characterstic pattern of larval muscles when stained with antibodies against muscle myosin. The introduction of an additional mutation in a new, unrelated gene that was isolated because it causes lethality leads to the loss of these muscles (Fig. 7). Thus, rather than causing a non-specific enhancement of lethality, the new mutation does indeed interfere with mesoderm formation and therefore probably represents a gene whose product normally interacts with *twist* or is controlled by *twist*.

The genetic and molecular approach together should identify genes acting in the pathway from fate determination to the expression of ventral fate by ventral-specific cell behaviour. It will also be important to investigate directly the cellular mechanisms involved in ventral cell shape changes. Determining the role of the actin and tubulin cytoskeleton and finding the cytoskeleton-associated proteins that mediate the rearrangement of cellular components during shape changes will be a main task towards this goal. By working down from the genes directly controlled by the fate-determining genes, and up from the genes whose products control the state of the cytoskeleton in ventral cells, we hope to be able to fill in the pathway from cell fate determination to morphogenesis.

We thank Mike Costa for comments on the manuscript, Dan Kiehart for antibodies against cytoplasmic and muscle myosin, Siegfried Roth for *twist* antibodies, Donald Brown and Zhou Wang for hospitality and for help with the cDNA subtraction, and Heike Schauerte for staining some of the embryos shown here.

References

Alberga, A., Boulay, J.-L., Kempe, E., Dennefeld, C. and Haenlin, M. (1991). The snail gene required for mesoderm formation is expressed dynamically in derivatives of all three germ layers. *Development* **111,** 983-992

Anderson, K. (1987). Dorsal-ventral embryonic pattern genes of *Drosophila. Trends Gen.* **3,** 91-97

Anderson, K. V., Bokla, L. and Nüsslein-Volhard, C. (1985). Establishment of dorsal-ventral polarity in the *Drosophila* embryo: the induction of polarity by the *Toll* gene product. *Cell* **42,** 791-798

Arora, K. and Nüsslein-Volhard, C. (1992). Altered mitotic domains reveal fate map changes in *Drosophila* embryos mutant for zygotic dorsoventral patterning genes. *Development* **114** ,1003-1024

Boulay, J. L., Dennefeld, C. and Alberga, A. (1987). The Drosophila developmental gene *snail* encodes a protein with nucleic acid binding fingers. *Nature* **330,** 395-398

Burnside, B. (1973). Microtubules and Microfilaments in Amphibian Neurulation. *Amer. Zool.* **13,** 989-1006

Costa, M., Sweeton, D. and Wieschaus, E. (1992). Gastrulation in Drosophila: Cellular Mechanisms of Morphogenetic Movements. In *The Development of* Drosophila, (ed. M. Bate and A. Martinez-Arias) New York: CSH Laboratory Press.

Ferguson, E. L. and Anderson, K. V. (1992). Localized enhancement and repression of the activity of the TGF-β family member, *decapentaplegic*, is necessary for dorso-ventral pattern formation in the *Drosophila* embryo. *Development* **114,** 583-597

Kam, Z., Minden, J. S., Agard, D. A., Sedat, J. W. and Leptin, M. (1991). *Drosophila* gastrulation: Analysis of cell behaviour in living embryos by three-dimensional fluorescence microscopy. *Development* **112,** 365-370

Kosman, D., Ip, Y. T., Levine, M. and Arora, K. (1991). The establishment of the mesoderm-neuroectoderm boundary in the *Drosophila* embryo. *Science* **254,** 118-122

Leptin, M. (1991). *twist* and *snail* as positive and negative regulators of during *Drosophila* mesoderm development. *Genes Dev.* **5,** 1568-1576

Leptin, M. and Grunewald, B. (1990). Cell shape changes during gastrulation in *Drosophila*. *Development* **110,** 73-84

Parks, S. and Wieschaus, E. (1991). The Drosophila Gastrulation Gene *concertina* encodes a Gα-like Protein. *Cell* **64,** 447-458

Perrimon, N., Smouse, D. and Miklos, G. L. G. (1989). Developmental Genetics of Loci at the Base of the X Chromosome of *Drosophila melanogaster*. *Genetics* **121,** 313-331

Ray, R. P., Arora, K., Nüsslein-Volhard, C. and Gelbart, W. M. (1991). The control of cell fate along the dorsal-ventral axis of the *Drosophila* embryo. *Development* **113,** 35-54

Roth, S., Stein, D. and Nüsslein-Volhard, C. (1989). A gradient of nuclear localization of the dorsal protein determines dorsoventral pattern in the *Drosophila* embryo. *Cell* **59,** 1189-1202

Simpson, P. (1983). Maternal-zygotic gene interactions during formation of the dorsoventral pattern in *Drosophila* embryos. *Genetics* **105,** 615-632

Stein, D. S. and Stevens, L. M. (1991). Establishment of dorsal-ventral and terminal pattern in the Drosophila embryo. *Current Opinion in Genetics and Development* **1,** 247-254

Sweeton, D., Parks, S., Costa, M. and Wieschaus, E. (1991). Gastrulation in *Drosophila*: the formation of the ventral furrow and posterior midgut invaginations. *Development* **112,** 775-789

Tearle, R. and Nüsslein-Volhard, C. (1987). Tübingen mutants and stocklist. *Dros. Inf. Serv.* **66,** 209-269

Thisse, B., Messal, M. E. and Perrin-Schmitt, F. (1987). The twist gene: isolation of a *Drosophila* zygotic gene necessary for the establishment of dorsoventral pattern. *Nucleic Acids Res.***15,** 3439-53

Thisse, B., Stoetzel, C., Gorostiza, T. C. and Perrin-Schmitt, F. (1988). Sequence of the twist gene and nuclear localization of its protein in endomesodermal cells of early *Drosophila* embryos. *EMBO J.***7,** 2175-2183

Wang, Z. and Brown, D. D. (1991). A gene expression screen. *Proc. Natl. Acad. Sci. USA* **88,** 11505-11509

Wieschaus, E. and Noell, E. (1986). Specificity of embryonic lethal mutations in *Drosophila* analyzed in germ line clones. *Roux's Arch. Dev. Biol.* **195,** 63-73

Wieschaus, E., Sweeton, D. and Costa, M. (1991). Convergence and extension during germ band elongation in *Drosophila* embryos. In *Gastrulation: Movements, Patterns and Molecules*, (ed. R. Keller, W. H. Clark Jr and F. Griffin), 213-224. New York: Plenum Press

Young, P. E., Pesacreta, T. C. and Kiehart, D. P. (1991). Dynamic changes in the distribution of cytoplasmic myosin during *Drosophila* embryogenesis. *Development* **111,** 1-14

Development 1992 Supplement, 33-41 (1992)

Pattern formation during gastrulation in the sea urchin embryo

DAVID R. McCLAY[1,*], NORRIS A. ARMSTRONG[1] and JEFF HARDIN[2]

[1]*Department of Zoology, Duke University, Durham, NC 27706, USA*
[2]*Department of Zoology, University of Wisconsin, 1117 W. Johnson St. Madison, WI 53706, USA*

[*]Author for correspondence

Summary

The sea urchin embryo follows a relatively simple cell behavioral sequence in its gastrulation movements. To form the mesoderm, primary mesenchyme cells ingress from the vegetal plate and then migrate along the basal lamina lining the blastocoel. The presumptive secondary mesenchyme and endoderm then invaginate from the vegetal pole of the embryo. The archenteron elongates and extends across the blastocoel until the tip of the archenteron touches and attaches to the opposite side of the blastocoel. Secondary mesenchyme cells, originally at the tip of the archenteron, differentiate to form a variety of structures including coelomic pouches, esophageal muscles, pigment cells and other cell types. After migration of the secondary mesenchyme cells from their original position at the tip of the archenteron, the endoderm fuses with an invagination of the ventral ectoderm (the stomodaem), to form the mouth and complete the process of gastrulation. A larval skeleton is made by primary mesenchyme cells during the time of archenteron and mouth formation.

A number of experiments have established that these morphogenetic movements involve a number of cell autonomous behaviors plus a series of cell interactions that provide spatial, temporal and scalar information to cells of the mesoderm and endoderm. The cell autonomous behaviors can be demonstrated by the ability of micromeres or endoderm to perform their morphogenetic functions if either is isolated and grown in culture. The requirement for cell interactions has been demonstrated by manipulative experiments where it has been shown that axial information, temporal information, spatial information and scalar information is obtained by mesoderm and endoderm from other embryonic cells. This information governs the cell autonomous behavior and places the cells in the correct embryonic context.

Key words: sea urchin, pattern formation, gastrulation, endoderm, mesoderm.

Introduction

Each of the deceptively simple morphogenetic events of gastrulation is driven by a number of interactions that are complex at both the molecular and cellular levels. For example, the mesodermal and endodermal lineages are partitioned early by a precise spatiotemporal sequence of cleavage divisions that divide the egg along the animal-vegetal axis. At the 16-cell stage, micromeres at the vegetal pole contain information that allows them to form the skeleton, even if these micromeres are isolated and grown in culture (Okazaki, 1975). In a beautiful series of manipulations, Horstadius (1939) showed that cell interactions are crucial for maintaining the correct distribution of cell fates along the animal-vegetal axis. If cells along that axis are removed, cells toward the animal pole alter their normal cell fates and replace structures formed by the missing vegetal cells. Horstadius even found that if he placed micromeres at the animal pole they would induce a secondary axis (Horstadius, 1939).

Recently, cell marking experiments have shown that the dorsal-ventral axis is also specified very early during cleavage (Cameron et al., 1989), although, like the animal-vegetal axis, the dorsal-ventral axis can be modified experimentally until some time during gastrulation (see below). To the embryo these axial asymmetries are important because they provide blastomeres with spatial information necessary for organized cellular rearrangements. This information is somehow supplied to each cell and then used to perform a function appropriate to that cell's particular position in the embryo.

Recent progress, especially in the *Drosophila* embryo (Ferguson and Anderson, 1991; St Johnston and Nusslein-Volhard, 1992; Wharton and Struhl, 1991; Wang and Lehmann, 1991; Nusslein-Volhard, 1991) and in the frog embryo (Cooke, 1991; Cho et al., 1991; Green and Smith, 1991) have begun to identify genes and growth factors that appear to be involved in specification of the embryonic axes. Presumably then, in those embryos as well as in the sea urchin, the cells, upon receiving the appropriate axial information, go on to perform the embryonic function appropriate to that position. If specification were that

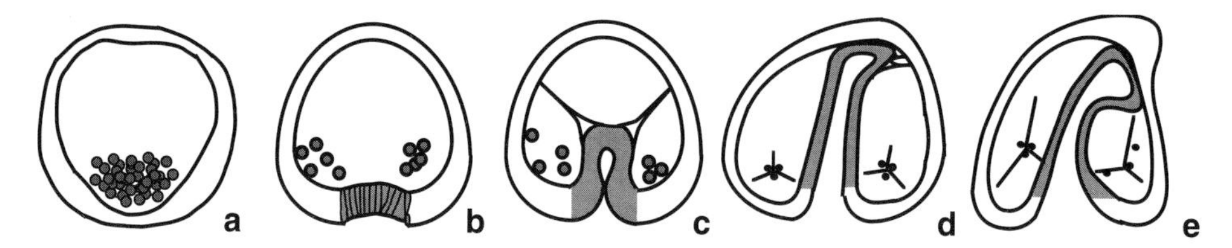

Fig. 1. Gastrulation in the sea urchin embryo. Primary mesenchyme cells have ingressed from the vegetal plate, through the basal lamina lining the blastocoel, and begin migrating along the blastocoelar wall (a). The thickened vegetal plate then bends inward to begin the process of invagination (b). During invagination endoderm cells converge and extend to elongate the archenteron. The filopodial extensions of secondary mesenchyme cells are not necessary for this phase of elongation (c). The archenteron reaches the wall of the blastocoel using filopodial contraction to complete the last one third of the elongation (d) (Hardin, 1988). In the final stage of gastrulation, the tip of the archenteron makes contact with a stomodael invagination and will soon form the mouth (e).

simple, it should be possible to isolate a cell after it has been specified and (assuming the cell is healthy), it should carry out its eventual fate. This rarely happens, however, probably because, although axial information is very important, many other interactions influence the developmental repertoire of a given cell. Here we examine primary mesenchyme cells, secondary mesenchyme cells and endoderm of the sea urchin embryo to ask what combination of cell autonomous behaviors and cell interactions are necessary to form and position the archenteron, and to form the embryonic skeleton. How simple or how complicated is the information necessary for morphogenesis? As our experimental story has developed it has become increasingly clear that these cells utilize a mixture of cell autonomous activities with simultaneous input through a number of cell interactions to regulate pattern.

Invagination of the archenteron

Invagination of the archenteron occurs in three recognizable stages (Fig. 1). First, a thickened plate of the cells at the vegetal pole of the embryo bends inward (Burke et al., 1991). Recent experiments have suggested that movements lateral and more animal to the site of inward bending may provide the driving force for this initial shape change in the vegetal plate. Burke et al. (1991) have shown that, just prior to the first signs of invagination, cells along the sides of the embryo move toward the vegetal plate suggesting that forces from all directions bearing down on the vegetal plate could provide the force that initiates inbending of the vegetal plate. It is hypothesized that somehow the forces generated by the movements of the blastular cells are focused on the vegetal plate so that inward bending is initiated at the center of the plate. Other experiments indicate that there are no additional movements of cells from lateral to the vegetal plate once the initial inbending is started. Marking experiments originally performed by Horstadius (reviewed by Horstadius, 1973) and repeated by Ettensohn (1984) have shown that only the vegetal plate contributes cells to the archenteron. It has also been shown that isolated vegetal plates support invagination (Moore and Burt, 1939; Ettensohn, 1984). Thus forces in the immediate vicinity of the invagination are all that are necessary to drive further invagination (Moore and Burt, 1939; Ettensohn, 1984). Because these movements, once started, are confined to the vegetal plate, they can be said to be cell autonomous (Ettensohn, 1985b).

After the initial inward bending of the vegetal plate, the archenteron begins to elongate. During the first stage of elongation, endoderm cells rearrange by convergent-extension movements in which the cells of the endodermal sheet change location in an organized fashion that elongates the primitive gut tube. This conclusion is based on three sorts of studies. First, an ultrastructural study deduced that cell rearrangements must occur (Ettensohn, 1985a). Second, it was shown that the cell rearrangments in the archenteron do not require participation of secondary mesenchyme cells (Hardin and Cheng, 1986; Hardin, 1988). Finally, labeling studies directly showed the localized shifts in position of endodermal cells in a convergent-extension pattern of movement that elongates the archenteron (Fig. 2) (Hardin, 1989). Although these experiments include time-lapse films that directly visualize the cellular rearrangements, several

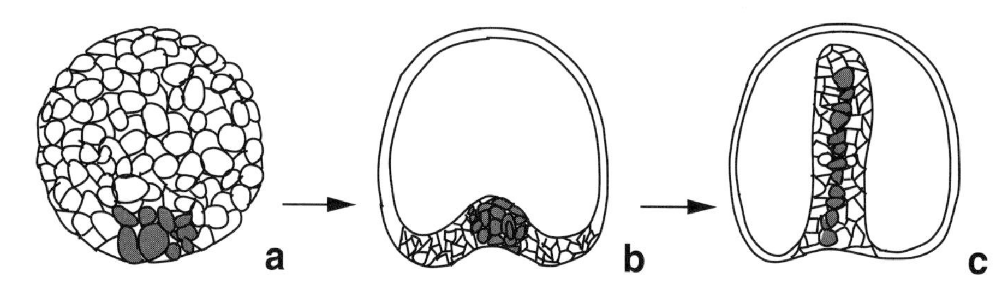

Fig. 2. The archenteron elongates by convergence-extension movements of the endoderm (from Hardin, 1989). Single labeled macromere derivatives were incorporated into embryos at the morula stage, and these cells divided to form a patch of labeled cells in the blastula (a). When the patch was located at the vegetal plate, it was included in the indentation movements of the archenteron (b), and by convergent-extension rearrangements the cells changed neighbors to form the elongated archenteron (c).

crucial properties of invagination are not known. These properties include the molecular basis of the motility involved, the importance of possible cell adhesion changes as guidance cues, and the mechanism that imparts directionality to the movements.

The second stage of elongation of the archenteron utilizes secondary mesenchyme cells located at the tip of the archenteron. These cells extend filopodia throughout gastrulation (Gustafson and Kinnander, 1956; Gustafson and Wolpert, 1967) and, though it was once thought that the filopodial extension behavior was important for all of the archenteron elongation movements, it is now appreciated that the filopodial movements are necessary only for about the last third of the elongation, at least in the species that have been examined experimentally (Hardin and Cheng, 1986; Hardin, 1988). Several experimental approaches support this conclusion, including a laser ablation study in which the filopodial extensions were destroyed without affecting archenteron elongation until it reached about two thirds of its final length (Hardin, 1988). In other species, it remains possible that filopodial extension contributes to the mechanics of early archenteron elongation. For example, in species where the archenteron invaginates close to the dorsal or to the ventral wall of the blastocoel (Hardin and McClay, 1990), filopodia may be used. In *Lytechinus*, the genus examined in our studies, the archenteron invaginates into the center of the blastocoel and, while filopodial extension could be used, experiments show filopodia are not necessary mechanistically to provide an elongation force until the final third of archenteron extension.

The archenteron tip ultimately makes contact with a specific region of the blastocoel wall. Experiments have shown that this region is a specific target that is recognized, selectively, by the filopodial attachments of the secondary mesenchyme cells (Hardin and McClay, 1990). Throughout gastrulation, filopodia are extended from the secondary mesenchyme cells in all directions. When the filopodia make contact with the wall of the blastocoel, they adhere for up to ten minutes. When they finally reach the target region, the filopodia adhere, are not released, and the behavior of the secondary mesenchyme cells changes. Only contact with the specific target region will cause the cell behavior change. This change can be brought about precociously if this region is pushed into close proximity to the archenteron tip, or the behavior change can be greatly delayed by contorting the embryo to prevent the filopodia from reaching the target region, or by removing the target region from the embryo (Fig. 3) (Hardin and McClay, 1990). These

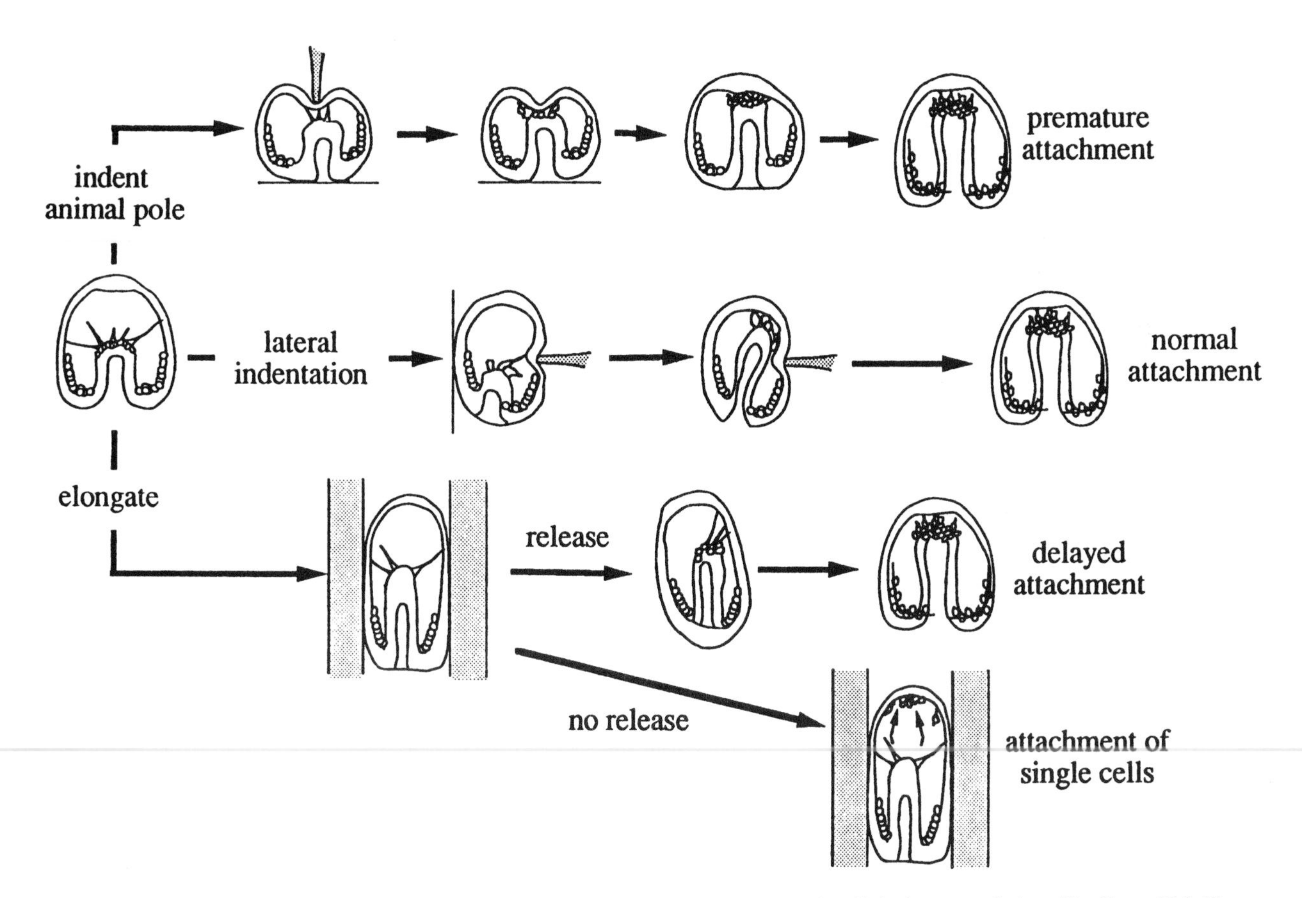

Fig. 3. Secondary mesenchyme cells attach to a specific target region of the blastocoel wall during gastrulation (Hardin and McClay, 1990). Lateral indentations show that although contact can be made to any part of the blastocoelar wall, only contact with a target region ends filopodial extension behavior and halts the movement of the archenteron. By indenting the animal pole one can establish contact with the target region and change the filopodial behavior precociously. By elongating the embryo one can delay contact with this region for several hours, and the filopodial behavior change is delayed as well. If released from the elongation, attachment occurs and the embryo develops normally. If the embryos remain elongated, the secondary mesenchyme cells eventually migrate away from the archenteron and collect at the target region at the animal pole.

experiments show the existence of an anatomical target for the extension and attachment of the archenteron to the blastocoel roof. Contact with the target promotes adhesion and a change in behavior of the secondary mesenchyme cells (Hardin and McClay, 1990). Coincident with this contact-stimulated behavioral change, secondary mesenchyme cells lose their ability to change fate as shown experimentally by a loss of an ability to become skeleton-producing cells (Ettensohn and McClay, 1988; Ettensohn, 1990a). The secondary mesenchyme cells then enter a complicated migratory and differentiation pathway to give rise to the coelomic pouches, esophageal muscle, pigment cells and blastocoelic cells of unknown fate (Tamboline and Burke, 1989, 1992). It is not known whether contact with the target region provides information for these differentiation events but it is of interest to know whether these programs of differentiation can be stimulated to occur early or be delayed due to time of contact with the target.

Control over the invagination process appears to be exercised at the beginning of invagination and at the completion of the target contact period. Several inhibitors block gastrulation with the most sensitive window being just prior to the beginning of invagination (Wessel and McClay, 1987; Schneider et al., 1978). Once invagination begins both the presumptive endoderm and the secondary mesenchyme cells appear to act cell autonomously. The presumptive endoderm cells aggregate and sort out to the interior of an aggregate, if mixed with ectoderm (Bernacki and McClay, 1989). The endoderm further hollows out to form an epithelial layer that looks like gut tissue. From the aggregation and sorting data, it would appear that endoderm cells recognize one another (McClay et al., 1977), and will form an epithelial layer either by invagination or by cavitation. Antigen expression data suggest that archenteron morphogenesis is more complicated than a relatively simple cell autonomous series of behaviors. As the gut forms, regional differences in antigen expression become apparent (McClay et al., 1983). Monoclonal antibodies that recognize different regions of the gut appear in localized regions before invagination is completed and before there are anatomical landmarks to delineate regions of the gut (McClay et al., 1983). Thus, while cells of the gut behave cell autonomously when the vegetal plate is isolated, cell interactions between the cells of the gut must convey directional information regarding position. In response to that information, the cells produce molecules appropriate to their location in the gut.

Pattern formation of the skeleton

In most species, there is an invariant number of primary mesenchyme cells that do not divide as the skeleton forms. In *Lytechinus*, 64 primary mesenchyme cells ingress from the vegetal pole, migrate within the blastocoel for a period of time, then associate with one another to form a syncytial ring around the base of the invaginating archenteron. In two ventrolateral areas of this ring, clusters of primary mesenchyme cells synthesize triradiate $CaCO_3$ crystals that initiate spiculogenesis. The arms of these spicules elongate, branch, and eventually form a skeleton that has a morphology with species-specific characteristics (Fig. 4). This section examines the interactions required for skeletal pattern formation.

Two kinds of studies demonstrate that skeletogenesis is largely a cell autonomous property of the primary mesenchyme cells. First, Okazaki's studies (Okazaki, 1975) demonstrated that micromeres could be isolated at the 16-cell stage, and grown in culture (with only the addition of horse serum), where they grew spicules. The spicule rods were not precisely of the pattern produced in vivo, but they had some of the pattern properties that are normally displayed in vivo.

Cell autonomy has been demonstrated recently in a different way. Primary mesenchyme cells from an embryo of one species were transplanted to the blastocoel of a primary mesenchyme-depleted embryo of a second species (Armstrong and McClay, in preparation). The skeleton produced was of the donor phenotype, again demonstrating cell autonomy of pattern (Fig. 4). Examination of single, or small numbers of cells, transplanted into a second species revealed that cell autonomy was present. In some cases, single donor primary mesenchyme cell made a portion of the skeleton appropriate to the region of the host but of the donor pattern. Thus, while cell autonomy exists in the expression of skeletal morphology, the inserted cell is provided with information regarding its position within the embryo so that the appropriate piece of skeleton is made.

What kind of information is given to the primary mesenchyme cells that allows them to make a skeleton in the correct pattern? The experiments below show that primary mesenchyme cells make the piece of skeleton appropriate to their position in the embryo, at the appropriate time, of the appropriate size, and in coordination with other primary mesenchyme cells (the skeleton is a single crystal of $CaCO_3$). Earlier studies suggested that mesenchyme cells do receive external information: in Okazaki's experiments (Okazaki, 1975), for example, although skeletons were made in culture, they did not fully resemble the final patterns that are seen in vivo, but those experiments did not reveal what types of cues were being received by the primary mesenchyme cells. The following series of experiments show the diversity of cues affecting skeletal pattern.

Timing of skeletogenesis

Micromeres, if grown in vitro, divide, form small clusters, migrate, and finally associate with one another and begin to form spicules, a behavior that parallels the behavior of these cells in vivo, but the timing is somewhat different relative to the behavior in vivo. (Fink and McClay, 1985). If primary mesenchyme cells are transferred from one embryo to another, one can ask whether the timing of skeletogenesis is intrinsic to the primary mesenchyme cells, or whether there are interactions with the surrounding embryo that somehow coordinate the timing of skeletogenesis. If older donor primary mesenchyme cells were transplanted into the blastocoel of younger host embryos, the skeleton was produced on the host's timetable. When the older cells were transplanted into blastulae prior to host primary mesenchyme cell ingression, the donor cells did not begin to

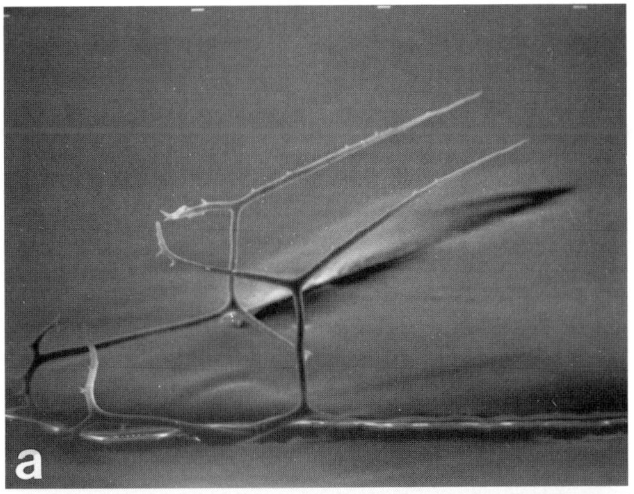

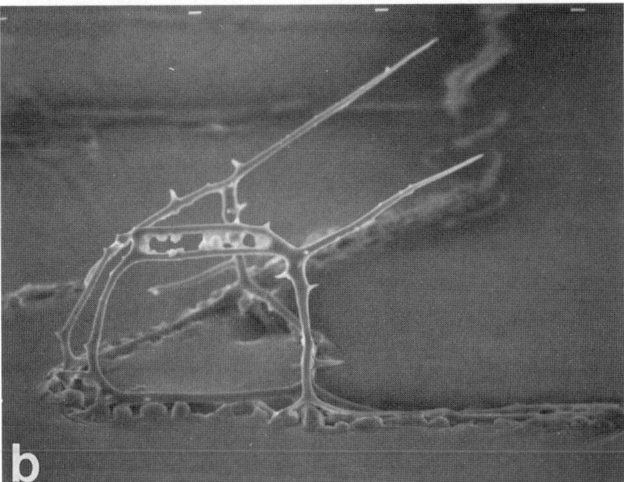

Fig. 4. Spicule morphology. Scanning electron micrographs of skeletons from pluteus larvae. At the top is a skeleton of *Lytechinus variagatus.* At the bottom is a skeleton of *Tripneustes esculentus.* The embryo in the middle contained *Lytechinus* ectoderm and endoderm, but had *Tripneustes* primary mesenchyme cells substituted for its own primary mesenchyme cells. Note that the dorsal portion of the skeleton (to the left) is more like the donor pattern (bottom) than the host pattern (top).

produce a skeleton until the host primary mesenchyme cells had ingressed and started skeletogenesis (Ettensohn and McClay, 1986). Thus the timing of skeletogenesis is controlled, at least in this experiment, by the host environment. (Ettensohn and McClay, 1986). In experiments where cells of one species are transplanted into the blastocoel of a second species, this observation is reinforced. The two species used, *Lytechinus* and *Tripneustes*, make spicules at different rates, yet, when together in the same embryo, the primary mesenchyme cells are able to coordinate (on the timetable of the host) to make a skeleton that appears normal, albeit with a modified pattern that fits the pattern input of each species (Armstrong and McClay, in preparation). In each of these transplantation experiments, the skeleton grows at a rate appropriate to the host embryo. The timetable used by micromeres grown in vitro may therefore be some sort of default timing pathway. Indeed, in vitro spiculogenesis follows in same behavior sequence but with a timing delay relative to the in vivo pattern (Fink and McClay, 1985).

The size and scale of the skeleton

Experiments by Horstadius (Horstadius and Wolsky, 1936), and by others since, have demonstrated that each of the cells from the 4-cell stage, when isolated, will make a whole embryo, though the embryo is one quarter the size of the normal embryo. Similarly, blastomeres separated at the 2-cell stage give rise to perfect half-sized embryos (Fig. 5). In each case, the organs of the embryo are scaled appropriately. The scale regulation could be an intrinsic property of the cells of each tissue, or there could be a regulation imposed by one or more tissues to coordinate the size of the others. A hint that the skeleton's size is regulated by other tissues was seen in experiments by Ettensohn (1990b). When extra primary mesenchyme cells were transplanted into the blastocoel of host embryos, the skeleton that resulted was no bigger than that seen in control embryos. Even when more than twice the normal number of primary mesenchyme cells were added to the blastocoel, the skeleton that grew was of a normal size and mass. Examination of the cells and skeletons indicated that the donor cells joined into the syncitium and appeared to participate in skeletogenesis but the increase in number of cells had no effect on size.

Components of skeletal size regulation can be seen in experiments contrasting growth in vitro with the size of skeletons in vivo, and in transplantation experiments where primary mesenchyme cells are inserted into embryos of different size (Fig. 5). If primary mesenchyme cells are cultured they have the ability to produce skeletons that are more than five times as long as those produced in vivo (Armstrong and McClay, in preparation). Thus, in vitro there appears to be a lack of constraint on the size of the spicules that can grow. In half- and quarter-sized embryos there normally are 32 or 16 primary mesenchyme cells respectively. These produce half- or quarter-sized skeletons. If one experimentally supplements the number of primary mesenchyme cells in the half- or quarter-sized embryos, the half- and quarter-sized embryos with extra primary mes-

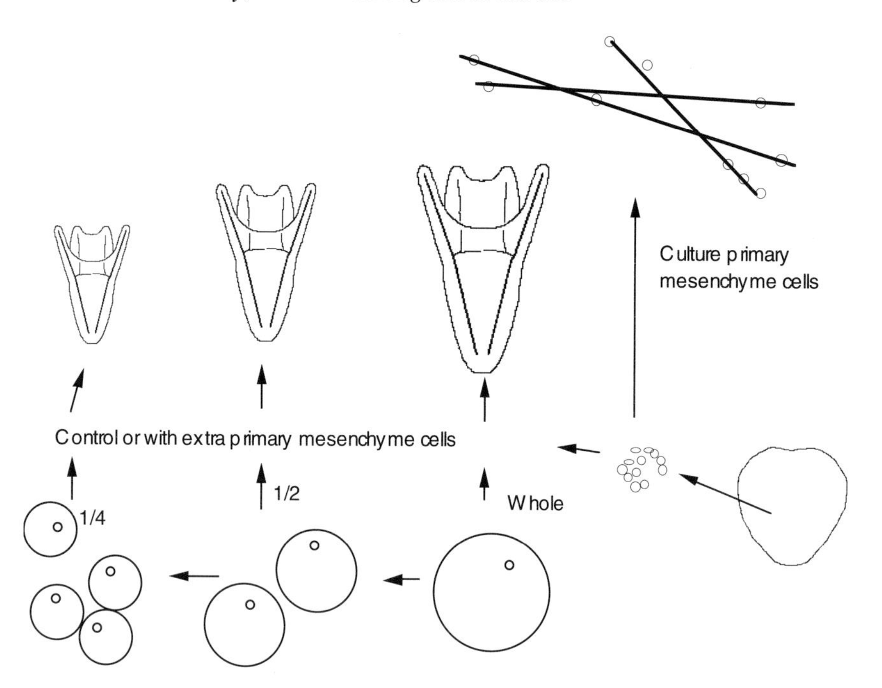

Fig. 5. Perfect half-, or quarter-size embryos result from single blastomeres isolated at the 2- or 4-cell stage. If one adds extra primary mesenchyme cells to the bastulae of quarter-, half-, or whole embryos, the skeletons continue to scale appropriately. This is in spite of the fact that the same primary mesenchyme cells, if cultured in vitro, would make spicules several times the length of those grown in vivo.

enchyme cells still produce half- or quarter-sized skeletons (Fig. 5) (Armstrong and McClay, in preparation). Scale, therefore, is somehow imposed on the skeleton-producing machinery. The timing of the regulation must occur at the time of skeletogenesis because cells transplanted from normal-sized embryos (and therefore in a full-sized embryo from fertilization until the time of transplantation), produce skeletons appropriate to the size of the host. Thus somehow information on scale is delivered to primary mesenchyme cells as they produce a skeleton of size appropriate to the host. Experiments to be described below will show how the primary mesenchyme cells receive spatial information telling them where to make particular skeletal structures. It is possible, indeed likely, that 'scale' may simply be a function of the dorsal-ventral and animal-vegetal coordinate information received by primary mesenchyme cells. That is, somehow the cells receive the correct positional information and it is the positional information that is scaled appropriately. What is unusual is that not only is the pattern of the skeleton positionally correct but also the mass of the skeleton is appropriate to the size of the embryo. Thus primary mesenchyme cells receive positional information that is scaled correctly, and they also receive information that somehow governs the amount of spicule that can be produced at any given spot.

Spatial information for pattern formation

Skeleton formation begins with two triradiate spicules that grow bilaterally from ventrolateral clusters of primary mesenchyme cells. The skeletal rods elongate from each triradiate spicule and primary mesenchyme cells join the syncitium along the way to make two half-skeletons that are the mirror image of one another (Fig. 6b).

The next series of experiments show the kinds of information received by the primary mesenchyme cells. The ability to produce a skeleton may be cell autonomous, but the spatial organization of a correct pattern requires several sorts of cues from the surrounding embryo.

If primary mesenchyme cells are inserted into the blastocoel, the cells move toward the vegetal plate (Ettensohn and McClay, 1986). Only primary mesenchyme cells respond in this way. When other cell types are injected into the blastocoel, they either do not move or they wander about without direction (Ettensohn and McClay, 1986). Therefore, the first cue that ingressed primary mesenchyme cells apparently recognize is a unique site near the vegetal plate.

Although each of the primary mesenchyme cells is equivalent in its pattern-forming potential (Ettensohn, 1990b), the two triradiate spicule primordia arise in the embryo from a ventrolateral cluster of cells. The normal initiation of only two spicule primordia appears to be a restriction of the full potential of the primary mesenchyme cells because when these cells are cultured in vitro they will initiate a number of spicules, often with a spicule arising from a cluster of only two cells (Armstrong and McClay, in preparation). If embryos are grown in $NiCl_2$ from midblastula until mesenchyme blastula stage, many more than two spicule primordia arise (Figs. 6,7) (Hardin et al., 1992; Armstrong et al., in preparation). Nickel has been found to disrupt the dorsal-ventral axis if administered during the late blastula stage (Hardin et al., in preparation). As a result, ventral marker proteins are expressed all around the embryo and dorsal markers are severely underexpressed. In nickel-treated embryos, there are an average of six spicule primordia. Do the additional primordia arise from a loss of embryonic control over the primary mesenchyme cells (so they could make more spicules as was seen in vitro)? Or, did the additional spicules arise because ventrolateral-like information was now present all around the floor of the blastocoel in the radialized embryos? To distinguish between these possibilities, extra primary mesenchyme cells were inserted into the blastocoel (Fig. 8). If there were a

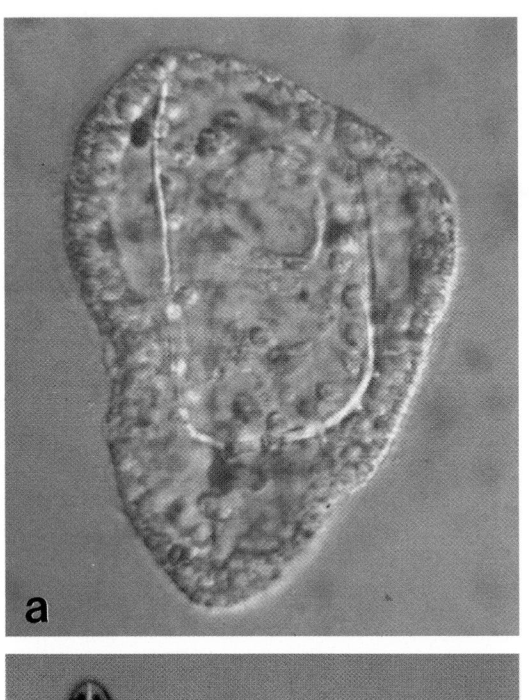

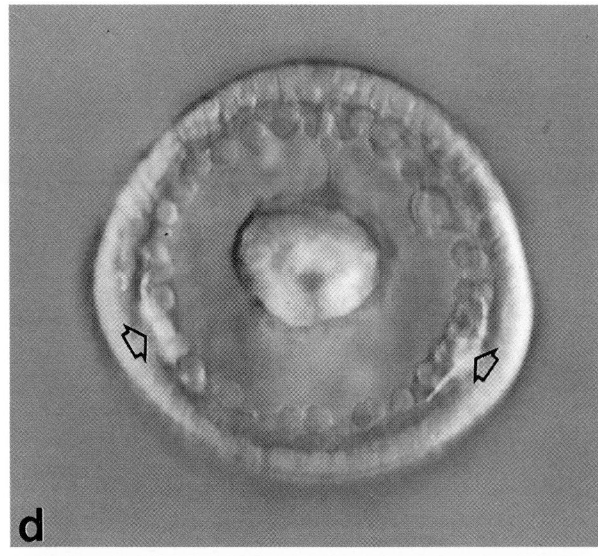

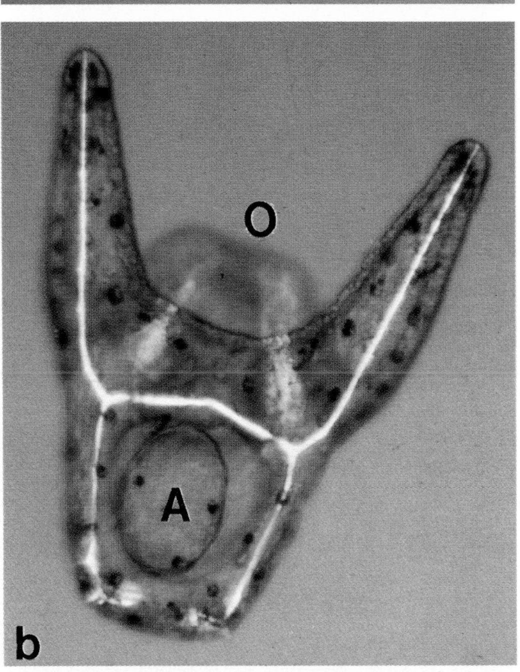

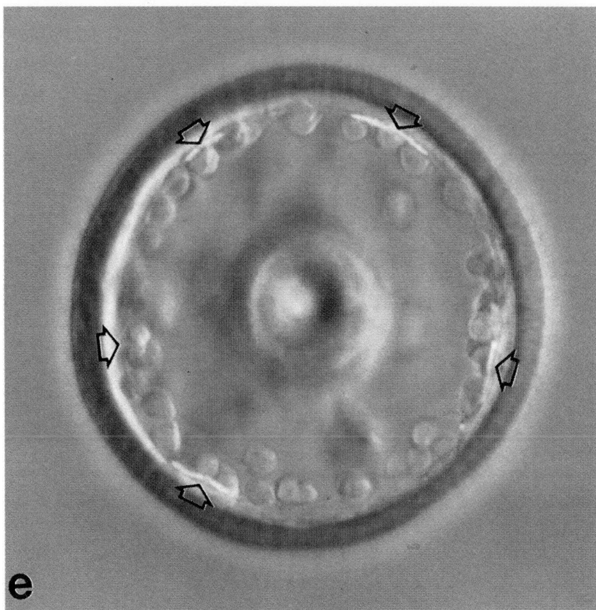

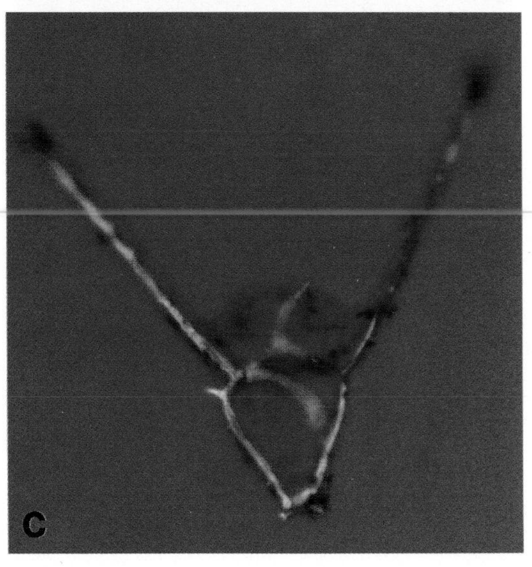

Fig. 6. The skeletal pattern requires cues from the ectoderm along the animal-vegetal axis and along the dorsal-ventral axis. The normal skeleton is shown in b, with the oral hood (O) out of focus and the anus in the center of the aboral region (A). If the embryo is cut at the equator (see Fig. 9), the two halves round up. If primary mesenchyme cells are inserted into the animal half embryos, a skeleton of the morphology of the oral hood results (a) (note the mouth in the center of this half embryo). If one cultures the vegetal half embryo, a skeleton appropriate to the vegetal portion of the embryo results (c). In experiments examining the dorso-ventral axis contribution, two spicule primordia arise from ventro-lateral clusters of primary mesenchyme cells (d). (This view looks up from the vegetal plate). If one alters the dorsal-ventral axis and ventralizes the embryos, more spicule primordia arise (e).

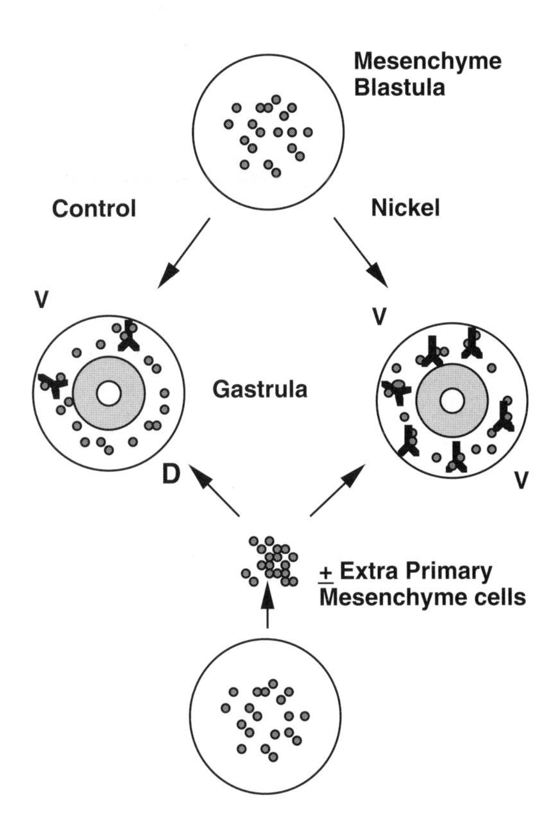

Fig. 7. Experiments with nickel-treated embryos. Treatment with nickel causes the embryo to synthesize supernumerary spicules. Other experiments suggest the extra spicules result from an alteration of the dorso-ventral axis. Addition of extra primary mesenchyme cells to the nickel-treated embryos has no augmenting effect on the nickel-induced pattern, just as additional primary mesenchyme cells has no effect on the two primordia produced in normal embryos.

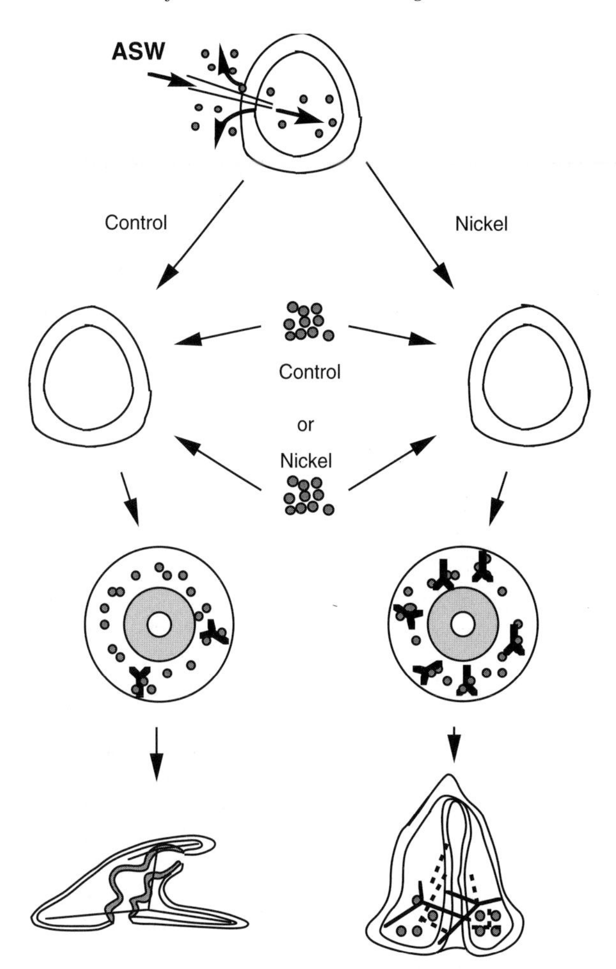

Fig. 8. The number of spicule primordia is, at least in part, a function of the ectoderm. If ectoderm is treated with nickel and control primary mesenchyme cells added, extra primordia are observed. If one adds nickel-treated primary mesenchyme cells to untreated ectoderm, the normal number of primordia results.

loss of mesenchymal regulation, the experimental expectation would be to see an increase in primordia as more and more primary mesenchyme cells were added to the $NiCl_2$-treated embryos. It was found, however, that when additional primary mesenchyme cells were added to nickel-treated embryos, the average number of spicule primordia remained at six. Therefore there still was some kind of regulation limiting the number of primordia. Was that regulation a function of the ectoderm or of the mesoderm? To address this question, control and nickel-treated embryos were depleted of primary mesenchyme cells and donor primary mesenchyme cells were added back to the embryos (Fig. 8). This allowed us to add nickel-treated primary mesenchyme cells to control ectoderm, or control primary mesenchyme cells to nickel-treated ectoderm. The multiple spicule primordia turned out to be a function of nickel influence on the ectoderm (Fig. 8). As a result, instead of the usual two ventrolateral regions, there were additional territories capable of supporting growth of spicule primordia. Those ectodermal territories nevertheless maintained a tight regulation over the spicule-producing capacity of the primary mesenchyme cells when extra cells were added to the blastocoel.

The number of ectodermal territories that stimulate primary mesenchyme cells appears to be a function of the size of the embryo. If one examines control embryos that are half-, quarter-, or super-sized (made by osmotic swelling of the blastocoel through inclusion of sucrose), the number of spicule primordia seen is two in each case. After nickel treatment, the number of spicule primordia is proportional to the size of the embryo (Armstrong et al., in preparation). Therefore, the skeletal regulating capacity of the ectoderm sheet appears, itself, to be sensitive to size. The molecular basis of this property is not known.

The experiments with nickel show that spicule pattern is influenced by the dorsal-ventral axis. Is the pattern also influenced by the animal-vegetal axis? To address this question embryos were cut in half at the equator at various times between the mesenchyme blastula stage and early gastrula stages (Hardin and Armstrong, in preparation) (Figs. 6, 9). The half embryos seal themselves into spheres. After the hemispheres rounded up, primary mesenchyme cells were placed in the animal half (the vegetal half already had primary mesenchyme cells, the animal halves required transplantation of cells). Both halves gave rise to skeletal elements (Fig. 9). The animal halves synthesized elements with the morphology of spicules of the oral hood and the

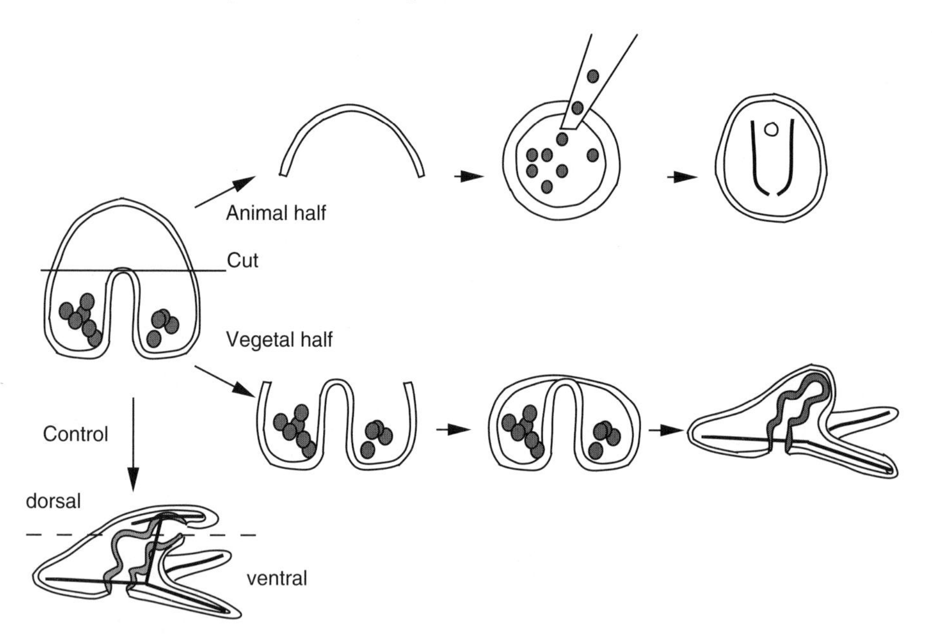

Fig. 9. Animal-vegetal half experiments. Embryos are cut in half at the equator. The half embryos round up again in culture. Primary mesenchyme cells are added to the animal halves and vegetal half embryos are also cultured. The skeletons that are produced are of a pattern appropriate to the half embryo in which the primary mesenchyme cells are contained (see also Fig. 6). A mouth is formed only in the animal half embryos and the vegetal half embryos have a gut but no mouth.

vegetal halves synthesized aboral elements and lacked oral structures. These results show that primary mesenchyme cells synthesize the portion of the skeleton appropriate to the ectodermal region surrounding these spicule-producing cells. Apparently, therefore, the ectoderm contains all, or much of the axial information necessary for spicule pattern formation and transfers the correct information to the primary mesenchyme cells. The degree to which primary mesenchyme cells might also influence the pattern of their immediate neighbors is not known.

The interactions described above represent only a few of the signals that an embryo must use during embryogenesis. Elsewhere in this volume (Ettensohn, 1992), primary mesenchyme cell interactions with secondary mesenchyme cells are shown to regulate cell lineage decisions. Here it is shown that the primary mesenchyme cells receive spatial, temporal and scalar information necessary for pattern formation. Secondary mesenchyme cells interact with a specific region to regulate the correct positioning of the archenteron. In each of these cases, information is transferred from one cell to another. This information can be fairly complex because, for example, individual primary mesenchyme cells will produce a skeletal element that is spatially correct. These cells must receive the correct three-dimensional coordinates for that to happen. In other cases, primary mesenchyme cells actively prevent secondary mesenchyme cells from expressing primary mesenchyme cell fates, an interaction that is sensitive to the number of primary mesenchyme cells (Ettensohn and McClay, 1988). For each of these cases, there appears to be a default mode. For example, if one removes the animal pole recognition region altogether, the secondary mesenchyme cells eventually stop displaying filopodial behavior and make many of their normal derivatives. Primary mesenchyme cells, in the absence of ectodermal cues, make skeletons, though the spicules that form are made on a different schedule, without the correct pattern, and without the correct scale. Our impression of embryogenesis, given these results, is that cells require large numbers of interactions for the correct pattern to form. There are many cell autonomous morphogenetic properties in the embryos but those autonomous activities appear to receive frequent inputs of information from other cells for correct patterns to emerge.

In each case that we have examined, there appears to be a complex mixture of cues. Endodermal cells appear to be programmed to rearrange. In order to coordinate that movement they require directional information. Secondary mesenchyme cells display stereotypical behavior by extending filopodia on a certain schedule. This movement is terminated following an interaction with a specific region of the ectoderm. Primary mesenchyme cells appear to be programmed (are cell autonomous) in their ability to make spicules. In order to make the correct pattern they require input of several kinds of information from the surrounding ectoderm. The signals are highly varied. Each activity must have a beginning and an ending and all the activities are coordinated with developmental events elsewhere in the embryo. The mixture of cues described here appears to be part of a large network that coordinates morphogenesis.

Given these data, it is important that we now learn how the cells of the embryo store structural information and how the cues are transferred between cells. In other systems analyses of this sort are already underway, though mostly at the genetic level. In the formation of the vulva of *C. elegans*, for example, genetic analyses suggest that a whole sequence of cell interactions regulates a cellular hierarchy, and that, in turn, regulates vulval pattern formation (Sternberg and Horvitz, 1989; Horvitz and Sternberg, 1991). A number of mutants disrupt the normal pattern-forming capacity during neurogenesis in *Drosophila* (Camposortega and Jan, 1991). Many of these genes appear to be involved in adhesive functions, or in information transfer between cells, or as DNA-binding proteins. During amphibian gastrulation and neurulation, there appear to be a number of genes that provide spatial cues for the correct pattern to form (Cooke, 1991). In these, and in other examples, the emerging picture is one of a hierarchy of regulatory cues governing pattern formation. Many of the cues are sent and received via cell interactions. Presumably, those cues are transduced intracellularly and the cell responds by making a portion of a structure appropriate to the mixture of cues that it receives. At the same time the cell provides information to other cells. Thus pattern formation is a highly interactive property of embryonic cells. Both the embryos that were traditionally thought of as 'mosaic' as well as the more regulative embryos appear to utilize these interactions. In the end we will begin to understand how the organism encodes the information necessary for assembling a three-dimensional structure. Clearly genes for the "bricks and mortar" of that structure are going to be few in number relative to the many interactions required to fit the structure into the correct context.

References

Bernacki, S. H. and McClay, D. R. (1989). Embryonic cellular organization: differential restriction of fates as revealed by cell aggregates and lineage markers. *J. Exp. Zool.* **251,** 203-216.

Burke, R. D., Myers, R. L., Sexton, T. L. and Jackson, C. (1991). Cell movements during the initial phase of gastrulation in the sea urchin embryo. *Dev. Biol.* **146,** 542-557.

Cameron, R. A., Fraser, S. E., Britten, R. J. and Davidson, E. H. (1989). The oral-aboral axis of a sea urchin embryo is specified by first cleavage. *Development* **106,** 641-647.

Campos-Ortega, J. A. and Jan, Y. N. (1991). Genetic and molecular bases of neurogenesis in Drosophila melanogaster. *Annu. Rev. Neurosci.* **14,** 399-420.

Cho, K. W. Y., Blumberg, B., Steinbeisser, H. and De Robertis, E. M. (1991). Molecular nature of Spemanns organizer - the role of the Xenopus homeobox gene goosecoid. *Cell* **67,** 1111-1120.

Cooke, J. (1991). Inducing factors and the mechanism of body pattern formation in vertebrate embryos. *Current Topics in Dev. Biol.* **25,** 45-75.

Ettensohn, C. A. (1984). Primary invagination of the vegetal plate during sea urchin gastrulation. *Amer. Zool.* **24,** 571-588.

Ettensohn, C. A. (1985a). Gastrulation in the sea urchin embryo is accompanied by the rearrangement of invaginating epithelial cells. *Dev. Biol.* **112,** 383-390.

Ettensohn, C. A. (1985b). Mechanisms of epithelial invagination. *Quart. Rev. Biol.* **60,** 289-307.

Ettensohn, C. A. (1990a). Cell interactions in the sea urchin embryo studied by fluorescence photoablation. *Science* **248,** 1115-1118.

Ettensohn, C. A. (1990b). The regulation of primary mesenchyme cell patterning. *Dev. Biol.* **150,** 261-271.

Ettensohn, C. A. (1992). Cell interactions and mesodermal cell fates in the sea urchin embryo. *Development 1992 Supplement, (this issue)*

Ettensohn, C. A. and McClay, D. R. (1986). The regulation of primary mesenchyme cell migration in the sea urchin embryo: transplantations of cells and latex beads. *Dev. Biol.* **117,** 380-391.

Ettensohn, C. A. and McClay, D. R. (1988). Cell lineage conversion in the sea urchin embryo. *Dev. Biol.* **125,** 396-409.

Ferguson, E. L. and Anderson, K. V. (1991). Dorsal-ventral pattern formation in the Drosophila embryo - the role of zygotically active genes. *Current Topics in Dev. Biol.* **25,** 17-43.

Fink, R. D. and McClay, D. R. (1985). Three cell recognition changes accompany the ingression of sea urchin primary mesenchyme cells. *Dev. Biol.* **107,** 66-74.

Green, J. B. A. and Smith, J. C. (1991). Growth factors as morphogens - do gradients and thresholds establish body plan. *Trends Genet* **7,** 245-250.

Gustafson, T. and Kinnander, H. (1956). Microaquaria for time-lapse cinematographic studies of morphogenesis of swimming larvae and observations on gastrulation. *Exp. Cell Res.* **11,** 36-57.

Gustafson, T. and Wolpert, L. (1967). Cellular movement and contact in sea urchin morphogenesis. *Biol. Rev.* **42,** 442-498.

Hardin, J. (1988). The role of secondary mesenchyme cells during sea urchin gastrulation studied by laser ablation. *Development* **103,** 317-324.

Hardin, J. (1989). Local shifts in position and polaried motility drive cell rearrangement during sea urchin gastrulation. *Dev. Biol.* **136,** 430-445.

Hardin, J. and McClay, D. R. (1990). Target recognition by the archenteron during sea urchin gastrulation. *Dev. Biol.* **142,** 86-102.

Hardin, J. D. and Cheng, L. Y. (1986). The mechanisms and mechanics of archenteron elongation during sea urchin gastrulation. *Dev. Biol.* **115,** 490-501.

Horstadius, S. (1939). The mechanics of sea urchin development as studied by operative methods. *Biol. Rev.* **14,** 132-179.

Horstadius, S. (1973) *Experimental Embryology of Echinoderms.* London: Oxford Press

Horstadius, S. and Wolsky, A. (1936). Studien uber die Determination der Bilateralsymmetrie des jungen Seeigelkeimes. *Wilhelm Roux' Arch. EntwMech. Org.* **135,** 69-113.

Horvitz, H. R. and Sternberg, P. W. (1991). Multiple intercellular signalling systems control the development of the Caenorhabditis elegans vulva. *Nature* **351,** 535-541.

McClay, D. R., Cannon, G. W., Wessel, G. M., Fink, R. D. and Marchase, R. B. (1983). Patterns of antigenic expression in early sea urchin development. In *Time, Space, and Pattern in Embryonic Development* (W. Jeffery and R. Raff, eds.), pp. 157-169. New York: Alan R. Liss.

McClay, D. R., Chambers, A. F. and Warren, R. G. (1977). Specificity of cell-cell interactions in sea urchin embryos. Appearance of new cell-surface determinants at gastrulation. *Dev. Biol.* **56,** 343-355.

Moore, A. R. and Burt, A. S. (1939). On the locus and nature of the forces causing gastrulation in the embryos of Dendraster excentricus. *J. Exp. Zool.* **82,** 159-171.

Nusslein-Volhard, C. (1991). Determination of the embryonic axes of *Drosophila. Development* 1991 **Supplement 1**, 1-10.

Okazaki, K. (1975). Spicule formation by isolated micromeres of the sea urchin embryo. *Amer. Zool.* **15,** 567-581.

Schneider, E. G., Nguyen, H. T. and Lennarz, W. J. (1978). The effect of tunicamycin, an inhibitor of protein glycosylation, on embryonic development in the sea urchin. *J. Biol. Chem.* **253,** 2348-2355.

Sternberg, P. W. and Horvitz, H. R. (1989). The combined action of two intercellular signalling pathways specifies three cell fates during vulval induction in C. elegans. *Cell* **58,** 679-693.

St. Johnston, D. and Nusslein-Volhard, C. (1992). The origin of pattern and polarity in the Drosophila embryo. *Cell* **68,** 201-219.

Tamboline, C. R. and Burke, R. D. (1989). Ontogeny and characterization of mesenchyme antigens of the sea urchin embryo. *Dev. Biol.* **136,** 75-86.

Tamboline, C. R. and Burke, R. D. (1992). Secondary mesenchyme of the sea urchin embryo: Ontogeny of blastocoelar cells. *J. Exp. Zool.* **262,** 51-60.

Wang, C. and Lehmann, R. (1991). Nanos is the localized posterior determinant in Drosophila. *Cell* **66,** 637-647.

Wessel, G. M. and McClay, D. R. (1987). Gastrulation in the sea urchin embryo requires the deposition of crosslinked collagen within the extracellular matrix. *Dev. Biol.* **121,** 149-165.

Wharton, R. P. and Struhl, G. (1991). RNA regulatory elements mediate control of Drosophila body pattern by the posterior morphogen nanos. *Cell* **67,** 955-967.

Development 1992 Supplement, 43-51 (1992)
Printed in Great Britain

Cell interactions and mesodermal cell fates in the sea urchin embryo

CHARLES A. ETTENSOHN

Department of Biological Sciences and Center for Light Microscope Imaging and Biotechnology, Carnegie Mellon University, 4400 5th Avenue, Pittsburgh, PA 15213, USA

Summary

Cell interactions during gastrulation play a key role in the determination of mesodermal cell fates in the sea urchin embryo. An interaction between primary and secondary mesenchyme cells (PMCs and SMCs, respectively), the two principal populations of mesodermal cells, regulates the expression of SMC fates. PMCs are committed early in cleavage to express a skeletogenic phenotype. During gastrulation, they transmit a signal that suppresses the skeletogenic potential of a subpopulation of SMCs and directs these cells into an alternative developmental pathway. This review summarizes present information concerning the cellular basis of the PMC-SMC interaction, as analyzed by cell transplantation and ablation experiments, fluorescent cell labeling methods and the use of cell type-specific molecular markers. The nature and stability of SMC fate switching, the timing of the PMC-SMC interaction and its quantitative characteristics, and the lineage, numbers and normal fate of the population of skeletogenic SMCs are discussed. Evidence is presented indicating that PMCs and SMCs come into direct filopodial contact during the late gastrula stage, when the signal is transmitted. Finally, evolutionary questions raised by these studies are briefly addressed.

Key words: cell interaction, mesoderm, cell fate, sea urchin.

Introduction

Intercellular signaling is a key mechanism of cell fate specification in all multicellular animals, even those that have traditionally been thought to develop in a highly mosaic manner. The sea urchin embryo has been an attractive experimental system for discerning how cell interactions control cell fates because of its optical clarity, the ease with which its cells can be isolated, transplanted and cultured, and the availability of a large collection of cell type-specific molecular markers (see reviews by Davidson, 1989; Ettensohn and Ingersoll, 1992). Blastomere isolation and recombination experiments have shown that cell interactions during cleavage are important in establishing embryonic patterns (Hörstadius, 1973; Henry et al., 1989; Khaner and Wilt, 1991). Critical cell interactions continue during gastrulation, and are involved both in regulating the extensive morphogenetic movements that reorganize the embryo during this stage of development and in specifying the fates of embryonic cells. We are investigating a key regulatory cell interaction during gastrulation that plays an important role in the specification of mesodermal cell fates and the process of skeletogenesis.

Mesenchyme cell populations in the sea urchin embryo

The mesoderm of the sea urchin embryo arises from the vegetal plate; a thickened epithelial placode at the vegetal pole of the blastula. In most species of commonly studied sea urchins, the mesoderm consists of two distinct populations of cells, primary and secondary mesenchyme cells (PMCs and SMCs, respectively) (Fig. 1). Both cell populations are highly motile and migrate actively within the blastocoel following their release from the vegetal epithelium. The PMCs and SMCs differ, however, with respect to their time of ingression, lineage and developmental fates. (For more detailed reviews of the development of these cell populations, see Gustafson and Wolpert (1967), Okazaki (1975a), Harkey (1983), Solursh (1986), Decker and Lennarz (1988), Wilt and Benson (1988), and Ettensohn (1991a)).

The PMCs are the sole descendants of the large micromere daughter cells, four cells that arise from the unequal cleavage of the micromeres at the fifth cleavage division. These cells undergo either 3 or 4 additional rounds of cell division depending upon the species to give rise to an average of 32 or 64 PMCs per embryo, a number that remains constant throughout gastrulation and early larval development. The PMCs ingress into the blastocoel at the start of gastrulation and migrate to specific target sites on the blastocoel wall between the vegetal pole and the equator of the embryo, where they become arranged in a characteristic ring-like pattern. As the subequatorial ring pattern forms, filopodial processes of the PMCs fuse, forming thick cable-like extensions that join the cells in a syncytial network. Within these fused filopodial cables, the PMCs deposit a branched skeletal framework of crystalline rods (spicules) composed of $CaCO_3$, $MgCO_3$, and several spicule matrix glycoproteins (Wilt and Benson, 1988). The skeleton serves as a structural framework for the distinctively angular pluteus larva.

The PMCs are restricted to a skeletogenic pathway of differentiation very early in embryogenesis, as micromeres isolated from 16-cell-stage embryos give rise only to skeletogenic cells when cultured in vitro, transplanted to ectopic positions in the embryo, or reassociated with other blastomeres in a variety of combinations (Hörstadius, 1973; Okazaki, 1975b; Livingston and Wilt, 1990; Khaner and Wilt, 1991). Associated with this cellular phenotype is the expression of a collection of PMC-specific gene products that have been identified by means of monoclonal antibodies and cDNAs (Carson et al., 1985; Wessel and McClay, 1985; Leaf et al., 1987; George et al., 1991; Katoh-Fukui et al., 1991).

Unlike the PMCs, the SMCs are a heterogeneous population of cells and express several different fates. In all species that have been carefully examined, a population of prospective pigment-forming SMCs ingresses relatively early in gastrulation (Gustafson and Wolpert, 1967; Gibson and Burke, 1985; Ettensohn and McClay, 1988). Later in gastrulation, larger numbers of cells are released from the tip of the archenteron. Some of these cells move into the blastocoel and adopt a fibroblast-like phenotype; they have been referred to as blastocoelar cells or basal cells (Cameron et al., 1991; Tamboline and Burke, 1992), although this population might well include several distinct cell types. Shortly before the completion of gastrulation, the anterior tip of the archenteron expands bilaterally to form the two coelomic pouches. The classification of the cells of the coelomic pouches as SMCs is debatable, since most of these cells remain as part of a coherent epithelium during coelom formation despite their intense filopodial activity (Gustafson and Wolpert, 1963). After the completion of gastrulation, 10-15 cells move away from each coelomic pouch and surround the foregut; these cells form the circumesophageal musculature of the pluteus larva and express several distinctive cytoskeletal proteins (Ishimoda-Takagi et al., 1984; Cox et al., 1986; Burke and Alvarez, 1988; Wessel et al., 1990).

The SMCs are considered to be descendants of the veg_2 layer of blastomeres that arises at the 64-cell stage (Hörstadius, 1973) (Fig. 1). Some cells in the coelomic pouches, however, are derived instead from the small micromeres, the siblings of the four founder cells of the PMC lineage (Endo, 1966; Pehrson and Cohen, 1986; Tanaka and Dan, 1990; Cameron et al., 1991). These cells normally undergo one additional round of division to produce eight cells that remain at the tip of the archenteron during gastrulation and later contribute to both the right and left coelomic pouches. At present, nothing is known of the mechanisms by which the descendants of the veg_2 blastomeres become progressively restricted in their fate, or when such fate restrictions occur during embryogenesis.

Evidence for PMC-SMC signaling

Fate mapping studies have shown that, in the undisturbed embryo, only PMCs express a skeletogenic fate and contribute to the larval skeleton (Hörstadius, 1973). Studies examining the development of isolated blastomeres of cleavage-stage embryos, however, demonstrate that the *potential* for skeletogenic differentiation is not restricted to cells of the micromere-PMC lineage. In particular, macromeres or their veg2 descendants, the progenitors of the SMCs and the endoderm (Fig. 1), have a significant capacity to give rise to skeletogenic cells when cultured in isolation (Hörstadius, 1973; Khaner and Wilt, 1991). Evidence that cells other than PMCs possess skeletogenic potential even at much later stages of development came from a little-known study by Fukushi (1962), who removed PMCs from embryos of *Glyptocidaris crenularis* and observed that SMCs moved to the region normally occupied by the PMCs and synthesized spicules. Langelan and Whiteley (1985) used low concentrations of SDS to inhibit micromere and PMC formation and reported that skeletal elements were nevertheless synthesized in such embryos, either by PMCs that ingressed in a delayed fashion or by SMCs. These studies suggested that in the absence of a signal transmitted by the PMCs, SMCs might alter their fate and adopt a skeletogenic phenotype. A clear demonstration of this interaction and a detailed analysis of its cellular basis has come from a combination of cell transplantation and ablation experiments, fluorescent cell marking techniques and the use of PMC-specific molecular markers, as described below. These studies show that PMCs interact with SMCs in a unidirectional manner late in gastrulation, and that the effect of this interaction is to regulate the specification of SMC fates and the process of skeletogenesis.

SMC conversion

Elimination of the entire complement of PMCs at the early gastrula stage by microsurgical or fluorescence-based methods (Figs. 2 and 8A) leads to a spectacular change in the developmental program of the SMCs. 65-75 of these cells switch fate and adopt the PMC phenotype, a process that has been termed SMC conversion (Ettensohn and McClay, 1988). During this regulative event, SMCs migrate to PMC-specific target sites in the embryo, activate a PMC-specific program of gene expression, and assemble a correctly patterned larval skeleton.

Changes in directional cell movements

The normal motile activity of PMCs and SMCs is similar in two respects; both migrate as individuals within the blastocoel by means of contractile filopodia and both have a tendency to fuse with like cells. However, these two populations of cells exhibit very different directional patterns of movement. The PMCs migrate to specific target sites on the blastocoel wall and arrange themselves in a subequatorial ring pattern, while SMCs either invade the embryonic epithelium and differentiate as pigment cells or remain scattered within the blastocoel, often in the interior of the cavity or (in the case of muscle cells) along the basal surface of the foregut. Cell transplantation experiments show that even when these two cell types are microinjected into recipient embryos of the same developmental stage and are con-

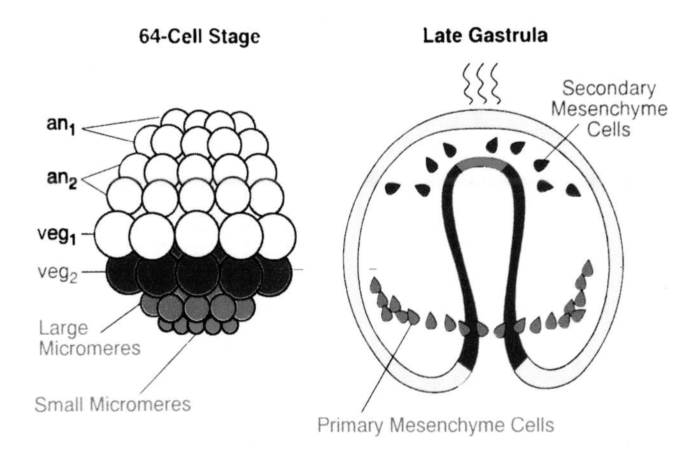

Fig. 1. Partial fate maps of the 64-cell-stage embryo and late gastrula, showing the origin of the mesoderm. The primary mesenchyme cells (PMCs) are derived from the large micromeres (red), while the secondary mesenchyme cells (SMCs) and endoderm are derived from the veg_2 layer of blastomeres (blue). The small micromeres (green), the siblings of the large micromeres, remain at the tip of the archenteron during gastrulation and later contribute to the coelomic pouches (see text for details).

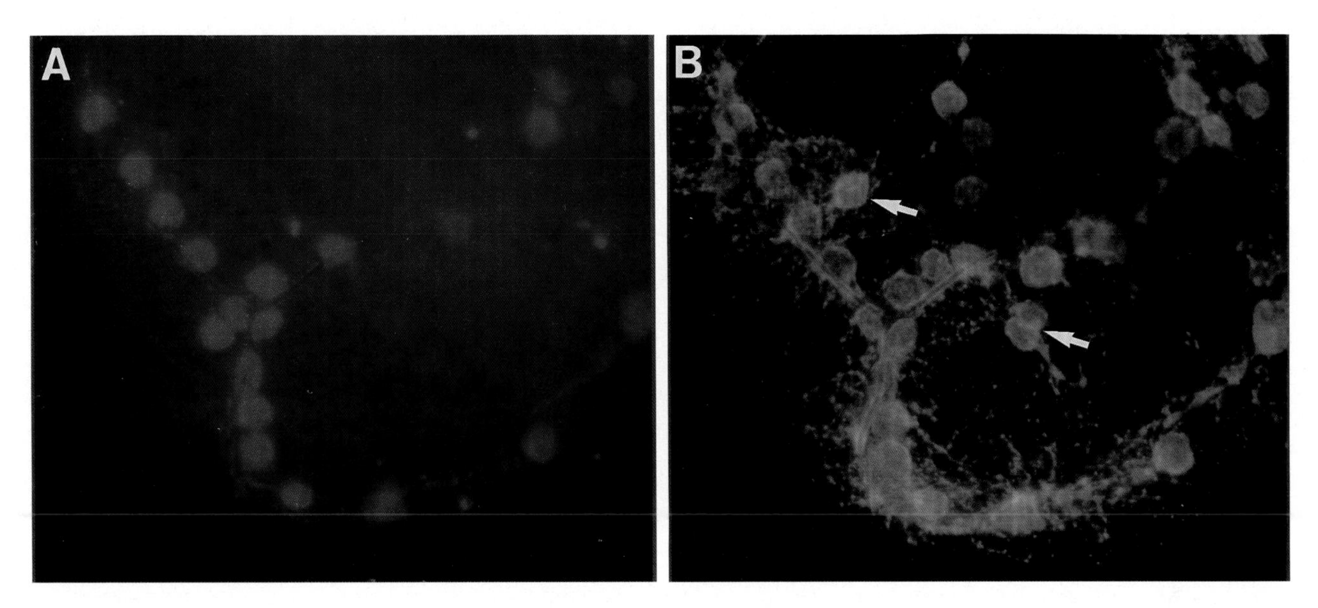

Fig. 6. SMC conversion in an embryo with an intermediate number of PMCs, as assayed by the method shown in Fig. 5. (A) Rhodamine fluorescence, showing labeled donor PMCs. (B) Immunostaining with an anti-msp protein MAb and a fluorescein-conjugated secondary antibody. All donor cells are both red and green, while converted SMCs derived from the recipient embryo (arrows) fluoresce green because they express the skeletogenic marker proteins, but are not rhodamine-labeled.

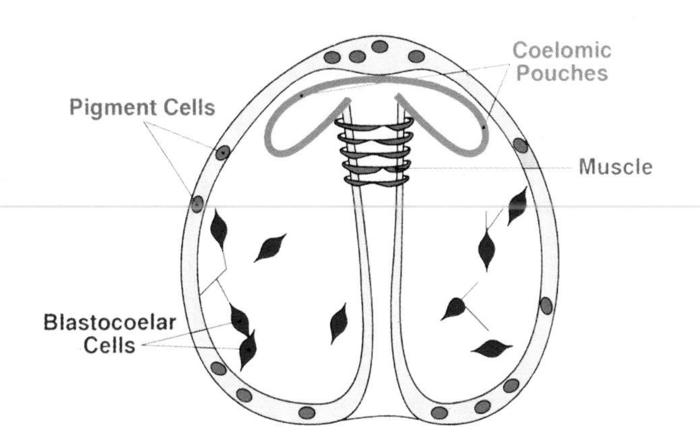

Fig. 9. SMC fates. SMCs give rise to four major derivatives. (1) Pigment cells arise from both early- and late-ingressing populations of SMCs. (2) Other SMCs (so-called basal or blastocoelar cells) that ingress late in gastrulation remain scattered in the blastocoel and have an unknown function, although they may be fibroblast-like (Tamboline and Burke, 1992). (3) Cells at the tip of the archenteron, including the small micromeres, give rise to the two coelomic pouches. (4) At the end of gastrulation, 10-15 cells migrate out of each coelomic pouch and surround the foregut, giving rise to the circumesophageal musculature (see text for details).

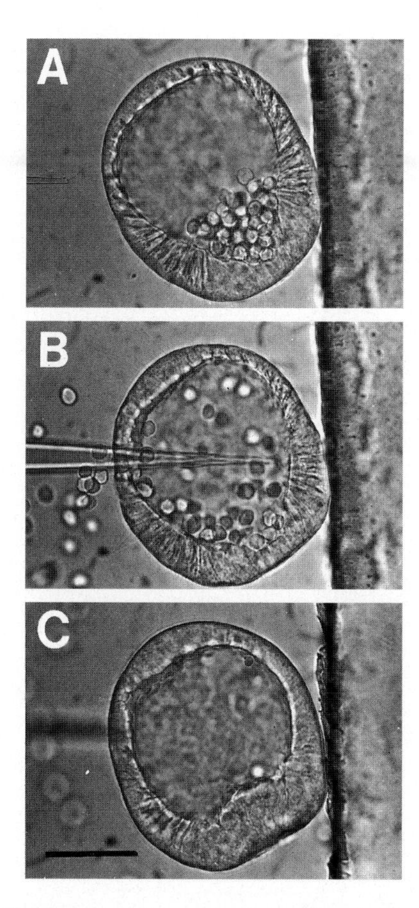

Fig. 2. PMC ablation. At the mesenchyme blastula stage, before the migration of the PMCs away from the vegetal plate, all or any fraction of the cells can be removed by directing a stream of seawater into the blastocoel from the tip of a micropipette. The PMCs flow out of the blastocoel through the wound created by the pipette. Reprinted from Ettensohn and McClay (1988), with permission from Academic Press, Inc. Scale bar = 50 μm.

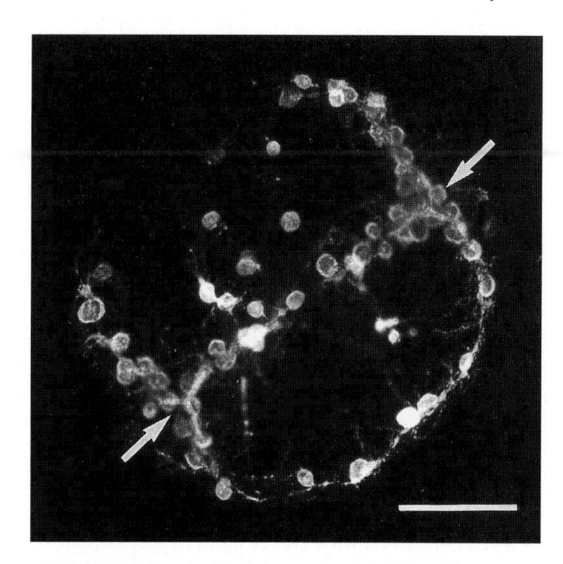

Fig. 3. Whole mount of a PMC(–) embryo after SMC fate-switching, immunostained with a monoclonal antibody (MAb) that recognizes msp proteins in order to reveal the arrangement of the converted cells. Skeletogenic SMCs form a pattern that closely resembles the subequatorial ring normally formed by PMCs. The SMC ring arises after control embryos have completed gastrulation, 8-10 hours after the PMC ring would normally form. Spicule formation begins in two ventrolateral clusters of skeletogenic cells (arrows), as in control embryos. Scale bar = 50 μm.

fronted with identical directional cues, each migrates in its characteristic pattern (Ettensohn and McClay, 1986). Therefore, there is a high degree of specificity in the ability of these cells to detect or respond to guidance signals in the blastocoel. During conversion, SMCs acquire the capacity to recognize PMC-specific directional cues and they arrange themselves in a subequatorial ring pattern remarkably similar to that formed by the PMCs (Fig. 3). Spiculogenesis is initiated within two clusters of cells that form along the ventrolateral aspects of the subequatorial ring, as in control embryos. Because the migration and patterning of the converting SMCs is delayed by several hours relative to the normal developmental schedule of the PMCs, the relevant directional cues must persist in PMC(–) embryos for at least that long.

Expression of PMC-specific molecular markers

The PMCs express a collection of unique cell surface glycoproteins that have been studied in several laboratories (Carson et al., 1985; Wessel and McClay, 1985; Leaf et al., 1987). The best characterized of these proteins is msp130, a PMC-specific, phosphatidylinositol-linked, cell surface glycoprotein that has been implicated in calcium transport (Farach-Carson et al., 1989; Parr et al., 1990). Msp130 is antigenically related to several other PMC-specific proteins (Shimizu-Nishikawa et al., 1990; Kabakoff et al., 1992). In *Lytechinus variegatus*, immunological and biochemical methods have been used to demonstrate that msp130 is one member of a family of antigenically related, PMC-specific cell surface glycoproteins ('msp proteins'). The pattern of expression of these proteins, both with respect to their overall abundance in the cell and their distribution on the cell surface, changes during the morphogenetic program of the PMCs (Fuhrman et al., unpublished observations).

The earliest indication of SMC fate-switching thus far detected is the de novo expression of the complete repertoire of msp proteins. Expression of these proteins can first be observed by indirect immunofluorescence 7-8 hours after PMC removal, at a time when many SMCs are migrating away from the tip of the archenteron but before they have accumulated in a ring pattern (Ettensohn, 1991b). Surface expression of these proteins by converted SMCs persists throughout skeletogenesis. Immunoblot analysis of msp protein expression in PMC(–) embryos shows that following conversion, skeletogenic SMCs undergo the same modulations in the pattern of msp protein expression as do PMCs. The regulated expression of these surface proteins is therefore tightly coupled to the skeletogenic program of differentiation. Skeletogenic SMCs also express surface binding sites for a lectin, wheat germ agglutinin, that normally binds specifically to PMCs (Ettensohn and McClay, 1988), although this lectin may recognize the same collection of msp proteins. The de novo expression of a different PMC-specific molecular marker, a cDNA encoding the spicule matrix protein sm50 (Livingston et al., 1991), has also been demonstrated during SMC conversion (Ettensohn, unpublished observations). Therefore, all available evidence supports the view that the program of SMC gene expression triggered during conversion is similar or identical to the normal skeletogenic program of gene expression exhibited by PMCs. It is not known whether SMCs down-regulate SMC-specific proteins or mRNAs when they activate the skeletogenic program, as there no molecular markers currently available that are completely SMC specific.

Spiculogenesis

Despite the delay in the initiation of skeletogenesis in PMC(–) embryos, the final pattern of the larval skeleton is the same as that of control larvae (Fig. 4). Perhaps even more surprising is the observation that when intermediate numbers of PMCs are present in the blastocoel, converted SMCs cooperate with these cells in the construction of a normal skeleton, even though the endogenous PMCs begin spiculogenesis on their own intrinsic timetable and skeletogenic SMCs are integrated into the skeletal pattern at a much later stage (Ettensohn and McClay, 1988). This regulative behavior has not yet been examined in detail, although it has been observed that when the endogenous PMCs in such an embryo are distributed predominantly in one part of the subequatorial ring pattern, converted SMCs have a tendency to fill in the deficient regions (Ettensohn and McClay, 1988). This flexibility in skeletal patterning is consistent with the fact that even when the number of PMCs is manipulated over a wide range, the cells form a normally proportioned subequatorial ring pattern and an essentially normal skeleton (Ettensohn, 1990a). Such behavior provides an especially striking example of the great flexibility of morphogenetic systems and their ability to reach a relatively constant and precise end state by means of several alternative routes (see Wolpert and Gustafson, 1961).

Signaling competence

One distinctive functional property of PMCs is their ability to suppress the skeletogenic potential of SMCs. This property can be assayed by cell transplantations (Fig. 5), and provides a different functional test of the extent to which converted SMCs adopt the PMC phenotype. To test the signaling competence of skeletogenic SMCs, fluorescently labeled converted cells were injected into PMC(–) recipient embryos and the number of endogenous SMCs that switched fate was determined by immunostaining with a monoclonal antibody that recognizes msp proteins. These studies show that skeletogenic SMCs are highly effective at suppressing the conversion of uncommitted SMCs to the skeletogenic fate. When 20-45 (mean = 32) SMCs are microinjected into recipient embryos, the numbers of SMCs in the recipients that express a skeletogenic fate is reduced by 80-85%, to an average of 10 cells/embryo (n = 17) (Ettensohn and Ruffins, 1992). Although the relative effectiveness of PMCs and converted SMCs in this regard has not yet been carefully analyzed, previous experiments provide some basis for comparison. When 30-40 PMCs are present in the blastocoel, an average of 7 SMCs (mean = 6.8, s.d. = 5.0, n = 4) switch fate (Ettensohn and McClay, 1988). Because the injection of 20-45 skeletogenic SMCs leads to the conversion of an average of 10 cells to a skeletogenic fate, converted SMCs appear to be as effective on a per-cell basis as PMCs at suppressing SMC skeletogenesis. This assay demonstrates that when SMCs switch fate they acquire PMC-specific signaling properties.

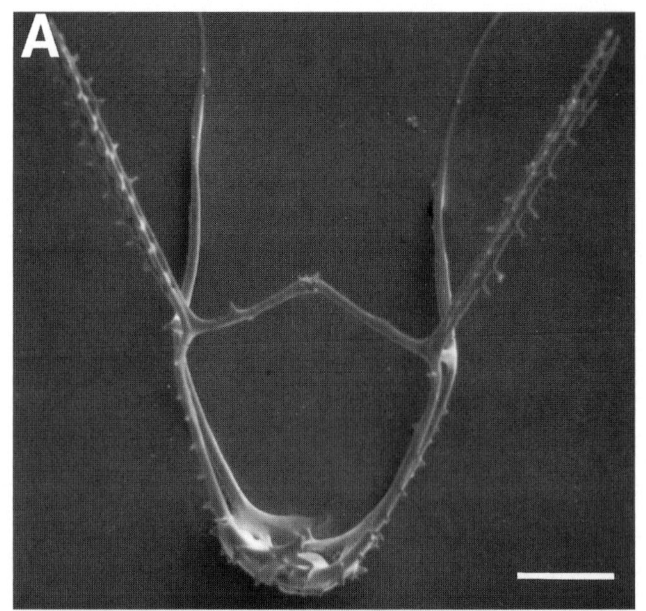

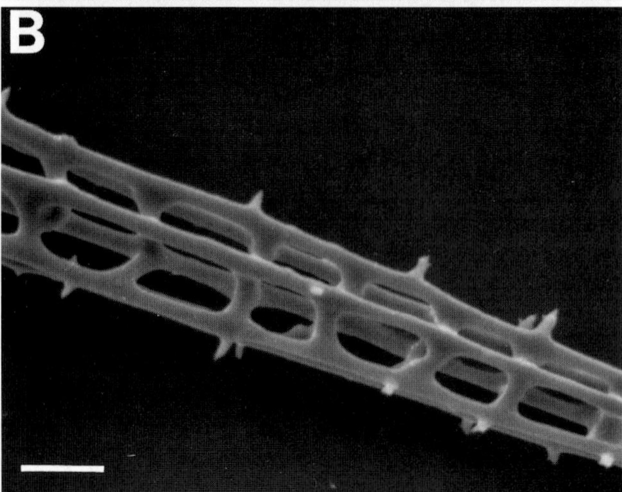

Fig. 4. Scanning electron micrographs of the skeleton formed by converted SMCs in a sand dollar, *Dendraster excentricus*. The final form of the skeleton is the same as that seen in control larvae. Species-specific differences in the skeletal pattern, such as in the fine structure of the spicule rods, are faithfully reproduced by the skeletogenic SMCs. (A) Low magnification view showing the entire skeleton. Scale bar = 50 μm. (B) High magnification view of the anal (postoral) rod formed by the skeletogenic SMCs. The elaborate, fenestrated structure of this skeletal rod is characteristic of *D. excentricus*. Scale bar = 10 μm.

Cellular aspects of the PMC-SMC interaction

The results of cell ablation experiments show that during normal development a signal is transmitted by PMCs that suppresses the skeletogenic potential of SMCs and shunts them into alternative developmental pathways. We have

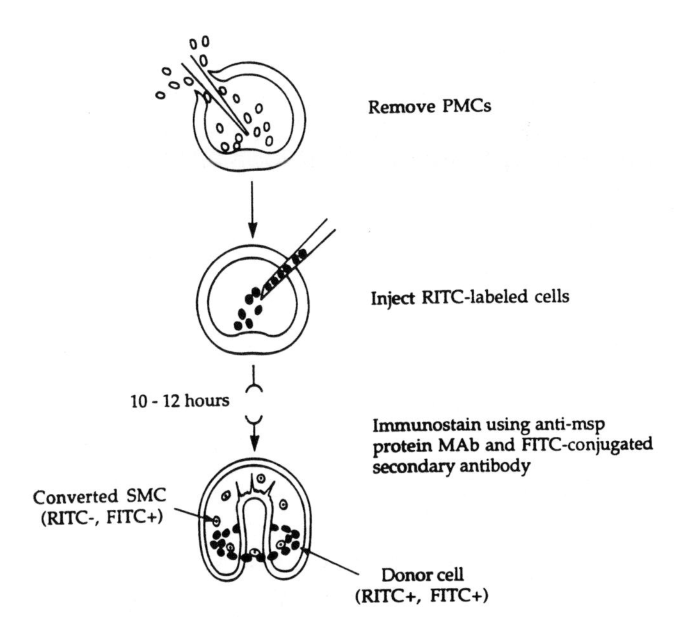

Fig. 5. Assay for mesodermal cell signaling. Rhodamine-labeled PMCs or converted SMCs are microinjected into the blastocoel of a PMC(−) recipient embryo. The number of donor cells and the developmental stage of the recipient can be varied. After SMC conversion, the recipient embryos are processed for immunofluorescence using a Mab specific for the msp proteins and a fluorescein-conjugated secondary antibody. The skeletogenic donor cells fluoresce both green and red, while any converted SMCs derived from the recipient embryo fluoresce green only. By this method the numbers of donor cells and converted SMCs in each embryo can be quantified.

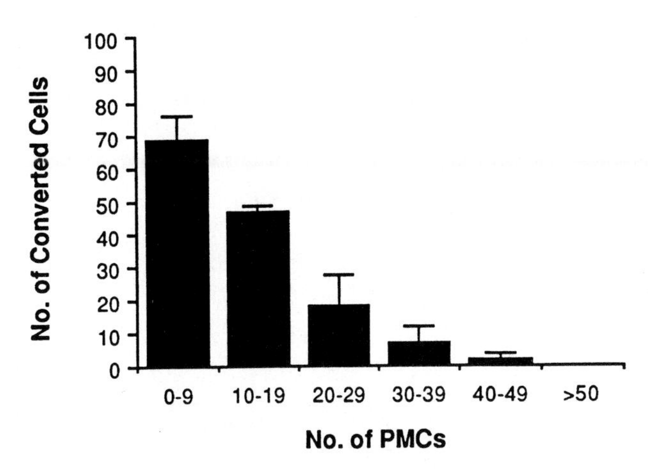

Fig. 7. Quantitative nature of the PMC-SMC interaction. Double-label experiments (Figs. 5 and 6) show that the number of SMCs that switch to a skeletogenic fate is inversely proportional to the number of PMCs in the blastocoel. SMC conversion is completely suppressed when >50 PMCs are present in the blastocoel. For each of the six classes of embryos, 4-8 specimens were scored (35 total embryos). Bars indicate standard deviations (no bar is shown for those embryos with >50 PMCs; of 6 such embryos examined, none had any converted SMCs). Reprinted with permission from Ettensohn and McClay (1988), with permission from Academic Press.

analyzed this interaction at the cellular level, with the belief that such information will provide a context for a molecular analysis of the interaction.

Quantitative aspects of the interaction

One intriguing aspect of the PMC-SMC interaction is its quantitative nature. The experimental strategy shown in Fig. 5 has been used to show that the number of SMCs that express a skeletogenic fate is inversely proportional to the number of PMCs in the blastocoel (Ettensohn and McClay, 1988; Figs. 6, 7). The same approach has been used to show that there is a strict threshhold for conversion; when at least 50 PMCs are present in the blastocoel, SMC fate-switching is completely suppressed. By some mechanism, the embryo 'counts' the number of PMCs in the blastocoel and regulates the number of SMCs that switch fate accordingly.

Timing of the interaction

Because ablation of the PMCs at the early gastrula stage (shortly after PMC ingression) leads to SMC conversion, it may be concluded that the SMCs are still developmentally labile at this stage and that PMCs normally interact with these cells at some time during gastrulation. To investigate the timing of PMC-SMC signaling in more detail, PMCs were eliminated at progressively later developmental stages using a fluorescence photoablation technique (Ettensohn, 1990b; Fig. 1D). These studies show that, even when the PMCs are eliminated at the late gastrula stage, the fate of the SMCs has not been irreversibly specified and these cells switch to a skeletogenic fate (Fig. 8). Photoablation of the PMCs at post-gastrula stages does not result in SMC conversion, suggesting that the SMCs are committed to a non-skeletogenic fate by that time. The fact that SMCs respond to PMC ablation at a late developmental stage, close to the time at which conversion begins as assayed by the expression of msp proteins, might be explained in one of two ways. Either the PMCs do not transmit a signal until the late gastrula stage, or the SMCs are insensitive to the signal until that time (or both). Cell transplantation experiments argue in favor of the latter interpretation. PMCs present in the blastocoel for a short time (3-4 hours) at the start of gastrulation will not suppress SMC skeletogenesis, but will do so effectively if placed in the blastocoel for 3-4 hours at the end of gastrulation (Ettensohn, 1990b). Assuming that the signal produced by the PMCs is the same in both cases, this result indicates that the SMCs are not competent to respond to the signal at the early gastrula stage.

Unidirectional nature of the interaction and the stability of conversion

Once the PMC-derived signal has been received, SMCs are stably committed to expressing a skeletogenic fate. When converted SMCs are microinjected into normal, PMC-containing recipient embryos, they continue to express a skeletogenic phenotype (Ettensohn and Ruffins, 1992). Therefore, although PMCs can suppress the expression of the skeletogenic fate by SMCs, they cannot reverse it once it has taken place. In addition, although PMCs regulate SMC skeletogenesis, there is no indication that they suppress their own skeletogenic differentiation in a similar manner. When large numbers of PMCs (100-150) are introduced into the blastocoel of mesenchyme blastula stage embryos, effec-

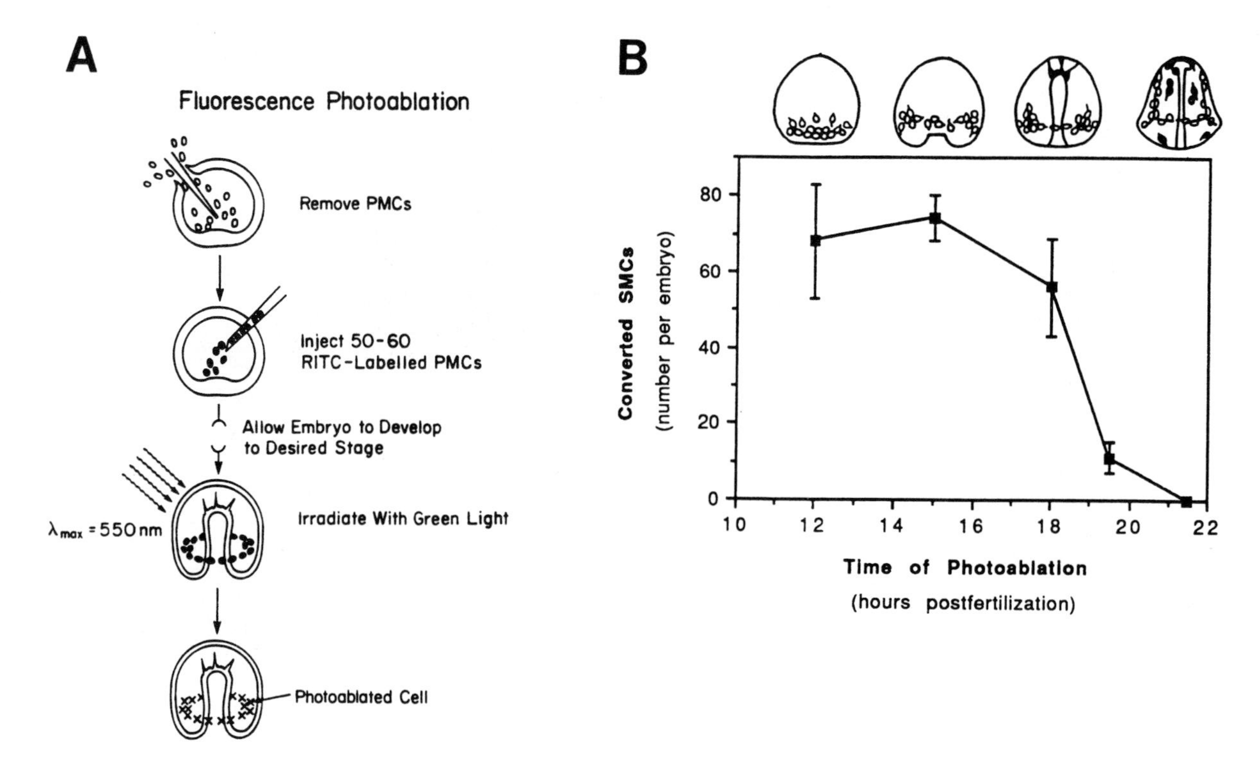

Fig. 8. Timing of the PMC-SMC interaction. (A) The fluorescence photoablation method used to eliminate PMCs after the start of their migration. The entire complement of PMCs is removed at the mesenchyme blastula stage and replaced with an equal number of rhodamine-labeled PMCs. The cells are ablated at the desired stage by prolonged excitation of the fluorochrome using a conventional epifluorescence microscope. (B) Results of fluorescence photoablation experiments. Ablation of the PMCs at or before the late gastrula stage results in SMC conversion. The number of converting cells drops sharply when PMCs are ablated at later developmental stages. Stages of embryos at the time of photoablation are illustrated by the diagrams at top. Bars indicate standard errors (95% confidence limits on the mean). For each developmental stage, 5-11 embryos were scored. Reprinted from Ettensohn (1990b), with permission from the American Assocication for the Advancement of Science.

tively doubling or tripling the normal complement of these cells, all the donor cells take part in skeletogenesis and none are induced to adopt alternative fates (Ettensohn, 1990a). In fact, as discussed above, no experimental conditions have yet been found that can induce cells of the micromere-PMC lineage to adopt a non-skeletogenic fate. The autonomy of PMC fate specification argues strongly that, although signals are transmitted from PMCs to SMCs, the converse is not true.

Lineage, numbers, and fates of the converting cells

Removal of the entire complement of PMCs at the early gastrula stage results in the conversion of 65-75 SMCs to the PMC phenotype. Ablation of skeletogenic SMCs does not result in the recruitment of any additional cells to a skeletogenic fate, demonstrating that by the late gastrula stage the number of SMCs with skeletogenic potential is restricted to 65-75 cells (Ettensohn and Ruffins, 1992).

Fate mapping studies using vital dyes and injected lineage tracers have shown that SMCs are derived from the veg_2 layer of the macromeres (Hörstadius, 1973, Cameron et al., 1991). Two observations, however, led us to test the possibility that descendants of the small micromere daughter cells might also contribute to the population of converting cells. The small micromeres are the siblings of the large micromere daughter cells, the founder cells of the PMC lineage, and are therefore the closest lineal relatives of the PMCs. In addition, these cells inherit the cytoplasm at the extreme vegetal pole of the unfertilized egg, and indirect evidence suggests that skeletogenic determinants are stored in the vegetal cytoplasm (Davidson, 1989). Specific labeling of the small micromere descendants with 5-bromodeoxyuridine after the method of Tanaka and Dan (1990), however, shows that these cells remain associated with the tip of the archenteron during gastrulation and do not contribute to the population of converting SMCs (Ettensohn and Ruffins, 1992). All of the skeletogenic SMCs therefore appear to be derived from the veg_2 tier of blastomeres (Fig. 1).

As noted above, the SMCs give rise to four major cell types (Fig. 9). As one means of gaining information concerning the normal fate(s) of the converting cells, PMC(–) larvae were examined to determine whether specific SMC derivatives were reduced or absent (Ettensohn and Ruffins, 1992). Comparisons of the numbers of muscle cells, coelomic pouch cells, blastocoelar cells, and pigment cells in PMC(–) larvae and controls showed only one significant difference. The number of pigment cells in PMC(–) embryos was reduced from an average of 100 pigment cells in control larvae to 47 cells in PMC(–) larvae, a reduction of more than 50%. Fluorescent cell marking experiments and time-lapse videomicroscopy of gastrulation in

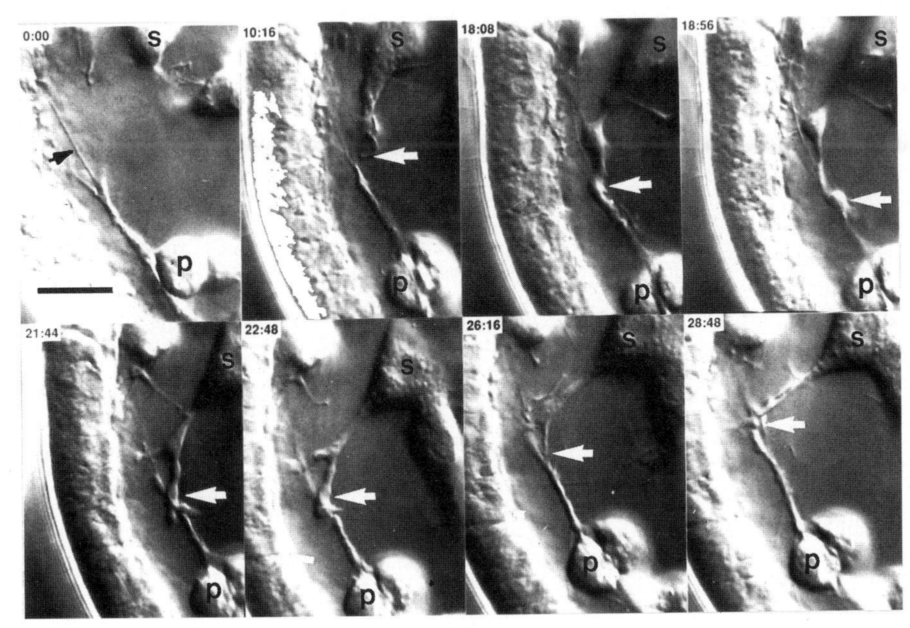

Fig. 10. Filopodial contacts between PMCs and SMCs during gastrulation. Eight frames of a time-lapse video sequence are shown. Numbers at upper left of each frame indicate the time in minutes : seconds. A filopodium of an SMC (s) makes a prolonged contact with a filopodium extending toward the animal pole from a PMC (p) in a ventrolateral cluster. The tip of the SMC filopodium (white arrow) contacts the PMC filopodium and courses down and then back up its length over a period of almost 20 minutes. The PMC filopodium (black arrowhead in panel A) remains essentially immobile during this sequence. The video recording was made using a Zeiss 63X planapochromat lens with differential interference contrast optics, a Hamamatsu Newvicon camera and Argus 10 image processor, and a Panasonic TQ-3038F optical disk recorder. Scale bar = 10 μm.

PMC(–) embryos have shown that, in *Lytechinus variegatus*, a population of prospective pigment cells arises early in gastrulation in which fate specification is independent of PMCs (Ettensohn and McClay, 1988). Larger numbers of SMCs, some of which are also prospective pigment cells, leave the tip of the archenteron later in gastrulation (Ettensohn and Ruffins, 1992). Taken together, this indirect evidence suggests that the converting cells represent a population of late-ingressing SMCs that would otherwise differentiate as pigment cells. According to this view, the PMC-derived signal controls a developmental switch, directing SMCs to adopt a pigment cell phenotype instead of the alternative (default) skeletogenic fate.

Mechanisms of PMC-SMC signaling

PMC migration and patterning can be inhibited in vivo by microinjecting lectins and antibodies that agglutinate these cells into the blastocoel. SMC conversion is nevertheless suppressed, ruling out the possibility that the PMCs act by physically occupying special target sites on the blastocoel wall that promote skeletogenesis (Ettensohn, unpublished observations). Alternative signaling mechanisms include the secretion of soluble factors or extracellular matrix components by the PMCs into the blastocoel, and direct cell-cell contact between the two populations. Repeated microinjection of blastocoel fluid from normal gastrulae into PMC(–) deficient embryos fails to suppress SMC conversion, providing indirect evidence against the former model (Ettensohn, unpublished observations). On the other hand, examples of direct filopodial contact between PMCs and SMCs at the late gastrula stage are relatively easy to document by time-lapse video microscopy, particularly at the tips of the chains of PMCs that extend toward the animal pole from the subequatorial ring (Fig. 10). The duration of such filopodial contacts ranges from a few seconds to as long as 10-15 minutes. It will be important to develop means of preventing normal cell extensions by one or both cell populations to determine whether direct filopodial contact is required to mediate the interaction. A mechanism of signaling based on filopodial extension by SMCs would be consistent with the insensitivity of these cells to the PMC-derived signal at the early gastrula stage, when SMCs exhibit little or no filopodial activity (Gustafson and Wolpert, 1967).

The molecular basis of the signaling mechanism is not unknown. Recent genetic and molecular studies of vulval development in *Caenorhabditis elegans* and photoreceptor cell development in *Drosophila melanogaster* indicate that signal transduction systems based on receptor tyrosine kinases coupled to Ras proteins regulate cell interactions in these systems (Greenwald and Rubin, 1992). Analysis of early inductive cell interactions in the amphibian embryo has provided convincing evidence that members of the fibroblast growth factor (FGF), transforming growth factor beta (TGFb), and/or Wnt gene families mediate these cell-cell signaling mechanisms (Jessell and Melton, 1992). At present, we do not know whether similar molecules play a role in PMC-SMC signaling, although they represent models for potential mechanisms.

Conclusions

The formation of the skeleton by cells of an alternative lineage following PMC ablation is an example of the well-known phenomenon of embryonic regulation. The basis of this regulative system is a unidirectional interaction between a population of skeletogenic cells committed very early in development (PMCs) and a second, more flexible mesodermal lineage (SMCs). The interaction between these two cell populations takes place late in gastrulation, a stage of development that has also been shown to be an important time in the determination of mesodermal cell fates in vertebrate embryos (Ho, this volume). In the sea urchin, athough considerable information has been obtained concerning the cellular aspects of this interaction, its molecular basis remains to be elucidated.

A different question raised by this and other examples of regulative cell interactions is why they should exist at all. The selective pressures that acted during evolution to create development by cellular interactions are unknown (Wolpert, 1990). Whatever its origin, it seems unlikely that the PMC-SMC interaction has persisted because of a selective advantage to the organism. The average number of PMCs per embryo (64 cells) is considerably higher than that required to completely suppress SMC skeletogenesis (50 cells), suggesting that under natural conditions it would be extremely unlikely that an SMC would express a skeletogenic phenotype, a conclusion consistent with fate mapping studies. Nor should the quantitative nature of the interaction be viewed as an adaptive mechanism for closely regulating the number of skeletogenic cells. In fact, the overall pattern of the skeleton is unaffected by the presence of 2-3 times the usual number of PMCs (Ettensohn, 1990a), although presumably there is some minimum number of cells required for skeletogenesis. Instead, the quantitative nature of the PMC-SMC interaction probably reflects the cellular or molecular mechanism of the signaling (e.g., if cell-cell contact is involved, then the probability of contact might be proportional to the number of PMCs in the blastocoel).

Despite the above considerations, this system of cell interactions is widespread among indirect developing echinoids. PMC ablation studies carried out in *L. pictus*, *G. crenularis* and a sand dollar, *Dendraster excentricus*, show that PMC-SMC signaling takes place in these three species as well as in *L. variegatus*, the species that has been used for most studies (Fukushi, 1962; Langelan and Whiteley, 1985; Ettensohn and McClay, 1988; Ettensohn, unpublished observations). Because more 'primitive' indirect developing echinoids lack PMCs and form a skeleton from mesenchyme cells that ingress later in gastrulation (Schroeder, 1981), the evolution of an early-ingressing, skeletogenic cell population may represent an example of developmental heterochrony. Direct developing sea urchins and other classes of echinoderms show great variability in their patterns of skeletogenesis (Korshelt and Heider, 1895; Schroeder, 1981; Wray and McClay, 1988). To begin to understand how the PMC-SMC system evolved, information concerning skeletogenic lineages and mesenchymal interactions in other members of this phylum is needed, especially with respect to those species thought to most closely represent ancestral forms. Further analysis of this system might therefore shed light both on the way evolutionary processes act to modify ontogenetic programs and on the cellular and molecular mechanisms of cell interactions during development.

I am grateful to K. Guss, K. Malinda, E. Ingersoll, and S. Ruffins for valuable comments and suggestions, and to B. Livingston and G. Wessel for providing antibodies and cDNAs. This research was supported by National Institutes of Health Grant HD-24690, a Basil O'Connor Starter Scholar Award from the March of Dimes Foundation, and a National Science Foundation Presidential Young Investigator Award.

References

Burke, R. D. and Alvarez, C. M. (1988). Development of the esophageal muscles in embryos of the sea urchin *Strongylocentrotus purpuratus*. *Cell Tissue Res.* **252**, 411-417.

Cameron, R. A., Fraser, S. E., Britten, R. J. and Davidson, E. H. (1991). Macromere cell fates during sea urchin development. *Development* **113**, 1085-1091.

Carson, D. D., Farach, M. C., Earles, D. S., Decker, G. L. and Lennarz, W. J. (1985). A monoclonal antibody inhibits calcium accumulation and skeleton formation in cultured embryonic cells of the sea urchin. *Cell* **41**, 639-648.

Cox, K. H., Angerer, L. M., Lee, J. J., Davidson, E. H. and Angerer, R. C. (1986). Cell lineage-specific programs of expression of multiple actin genes during sea urchin embryogenesis. *J. Mol. Biol.* **188**, 159-172.

Davidson, E. H. (1989). Lineage-specific gene expression and the regulative capacities of the sea urchin embryo: a proposed mechanism. *Development* **105**, 421-445.

Decker, G. L. and Lennarz, W. J. (1988). Skeletogenesis in the sea urchin embryo. *Development* **103**, 231-247.

Endo, Y. (1966). Development and differentiation. In *Biology for Today*, pp 1-61. Tokyo: Iwanami Shoten.

Ettensohn, C. A. (1990a). The regulation of primary mesenchyme cell patterning. *Dev. Biol.* **140**, 261-271.

Ettensohn, C. A. (1990b). Cell interactions in the sea urchin embryo studied by fluorescence photoablation. *Science* **248**, 1115-1118.

Ettensohn, C. A. (1991a). Primary mesenchyme cell migration in the sea urchin embryo. In *Gastrulation: Movements, Patterns and Molecules*. (ed. R. E. Keller), pp 289-304. New York: Plenum Press.

Ettensohn, C. A. (1991b). Mesenchyme cell interactions in the sea urchin embryo. In *Cell-Cell Interactions in Early Development, 49th Symp. Soc. Dev. Biol.* (ed. J. Gerhart), pp 175-201. New York: Wiley-Liss.

Ettensohn, C. A. and Ingersoll, E. P. (1992). Morphogenesis of the sea urchin embryo. In *Morphogenesis: Analysis of the Development of Biological Structures* (ed. E. F. Rossomando and S. Alexander), pp. 189-263. New York: Marcel Dekker Press.

Ettensohn, C. A. and McClay, D. R. (1986). The regulation of primary mesenchyme cell migration in the sea urchin embryo: transplantations of cells and latex beads. *Dev. Biol.* **117**, 380-391.

Ettensohn, C. A. and McClay, D. R. (1988). Cell lineage conversion in the sea urchin embryo. *Dev. Biol.* **125**, 396-409.

Ettensohn, C. A. and Ruffins, S. (1992). Mesodermal cell interactions in the sea urchin embryo: Properties of skeletogenic secondary mesenchyme cells. *Development* (in press).

Farach-Carson, M. C., Carson, D. D., Collier, J. L., Lennarz, W. J., Park, H. R. and Wright, G. C. (1989). A calcium-binding, asparagine-linked oligosaccharide is involved in skeleton formation in the sea urchin embryo. *J. Cell Biol.* **109**, 1289-1299.

Fukushi, T. (1962). The fates of isolated blastoderm cells of sea urchin blastulae and gastrulae inserted into the blastocoel. *Bull. Marine Biol. Stat. Asamushi* **11**, 21-30.

George, N. C., Killian, C. E. and Wilt, F. H. (1991). Characterization and expression of a gene encoding a 30.6-kDa *Strongylocentrotus purpuratus* spicule matrix protein. *Dev. Biol.* **147**, 334-342.

Gibson, A. W. and Burke, R. D. (1985). The origin of pigment cells in the sea urchin *Strongylocentrotus purpuratus*. *Dev. Biol.* **107**, 414-419.

Greenwald, I. and Rubin, G. M. (1962). Making a difference. The role of cell-cell interactions in establishing separate identities for equivalent cells. *Cell* **68**, 271-281.

Gustafson, T. and Wolpert, L. (1963). Studies on the cellular basis of morphogenesis in the sea urchin embryo. Formation of the coelom, the mouth, and the primary pore-canal. *Exp. Cell Res.* **29**, 561-582.

Gustafson, T. and Wolpert, L. (1967). Cellular movement and contact in sea urchin morphogenesis. *Biol. Rev.* **42**, 441-498.

Harkey, M. A. (1983). Determination and differentiation of micromeres in the sea urchin embryo. In *Time, Space, and Pattern in Embryonic Development* (ed. W. R. Jeffery and R. A. Raff), pp 131-155. New York: Alan R. Liss.

Henry, J. J., Amemiya, S., Wray, G. A. and Raff, R. A. (1989). Early inductive interactions are involved in restricting cell fates of mesomeres in sea urchin embryos. *Dev. Biol.* **136**, 140-153.

Hörstadius, S. (1973). *Experimental Embryology of Echinoderms*. Oxford: Clarendon Press.

Ishimoda-Takagi, T., Chino, I. and Sato, H. (1984). Evidence for the involvement of muscle tropomyosin in the contractile elements of the coelom-esophagus complex of sea urchin embryos. *Dev. Biol.* **105**, 365-376.

Jessell, T. M. and Melton, D. A. (1992). Diffusible factors in vertebrate embryonic induction. *Cell* **68**, 257-270.

Kabakoff, B., Hwang, S. L. and Lennarz, W. J. (1992). Characterization of post-translational modifications common to three primary mesenchyme cell-specific glycoproteins involved in sea urchin embryonic skeleton formation. *Dev. Biol.* **150**, 294-305.

Katoh-Fukui, Y., Noce, T., Ueda, T., Fujiwara, Y., Hashimoto, N., Higashinakagawa, T., Killian, C. E., Livingston, B. T., Wilt, F. H., Benson, S. C., Sucov, H. M. and Davidson, E. H. (1961). The corrected structure of the SM50 spicule matrix protein of *Strongylocentrotus purpuratus*. *Dev. Biol.* **145**, 201-202.

Khaner, O. and Wilt, F. (1991). Interactions of different vegetal cells with mesomeres during early stages of sea urchin development. *Development* **112**, 881-890.

Korschelt, E. and Heider, K. (1895). *Textbook of the Embryology of Invertebrates*. London: Swan Sonnenscheins.

Langelan, R. E. and Whiteley, A. H. (1985). Unequal cleavage and the differentiation of echinoid primary mesenchyme. *Dev. Biol.* **109**, 464-475.

Leaf, D. S., Anstrom, J. A., Chin, J. E., Harkey, M. A., Showman, R. M. and Raff, R. A. (1987). Antibodies to a fusion protein identify a cDNA clone encoding msp 130, a primary mesenchyme-specific cell surface protein of the sea urchin embryo. *Dev. Biol.* **121**, 29-40.

Livingston, B. T., Shaw, R., Bailey, A. and Wilt, F. (1991). Characterization of a cDNA encoding a protein involved in formation of the skeleton during development of the sea urchin *Lytechinus pictus*. *Dev. Biol.* **148**, 473-480.

Livingston, B. T. and Wilt, F. H. (1990). Range and stability of cell fate determination in isolated sea urchin blastomeres. *Development* **108**, 403-410.

Okazaki, K. (1975a). Normal development to metamorphosis. In *The Sea Urchin Embryo: Biochemistry and Morphogenesis.* (ed. G. Czihak), pp 177-232. New York: Springer-Verlag.

Okazaki, K. (1975b). Spicule formation by isolated micromeres of the sea urchin embryo. *Am. Zool.* **15**, 567-581.

Parr, B. A., Parks, A. L. and Raff, R. A. (1990). Promoter structure and protein sequence of msp130, a lipid-anchored sea urchin glycoprotein. *J. Biol. Chem.* **265**, 1408-1413.

Pehrson, J. R. and Cohen, L. H. (1986). The fate of the small micromeres in sea urchin development. *Dev. Biol.* **113**, 522-526.

Schroeder, T. E. (1981). Development of a "primitive" sea urchin (*Eucidaris tribuloides*): irregularities in the hyaline layer, micromeres, and primary mesenchyme. *Biol. Bull. Mar. Biol. Assoc. Woods Hole* **161**, 141-151.

Shimizu-Nishikawa, K., Katow, H. and Matsuda, R. (1990). Micromere differentiation in the sea urchin embryo: immunochemical characterization of primary mesenchyme cell-specific antigen and its biological roles. *Develop., Growth and Differ.* **32**, 629-636.

Solursh, M. (1986). Migration of sea urchin primary mesenchyme cells. In *Developmental Biology: A Comprehensive Synthesis* vol. 2 (ed. L. Browder), pp 391-431. New York: Plenum Press.

Tamboline, C. R. and Burke, R. D. (1992). Secondary mesenchyme of the sea urchin embryo: Ontogeny of blastocoelar cells. *J. Exp. Zool.* **262**, 51-60.

Tanaka, S. and Dan, K. (1990). Study of the lineage and cell cycle of small micromeres in embryos of the sea urchin, *Hemicentrotus pulcherrimus*. *Develop., Growth and Differ.* **32**, 145-156.

Wessel, G. M. and McClay, D. R. (1985). Sequential expression of germ layer specific molecules in the sea urchin embryo. *Dev. Biol.* **111**, 451-463.

Wessel, G. M., Zhang, W. and Klein, W. H. (1990). Myosin heavy chain accumulates in dissimilar cell types of the macromere lineage in the sea urchin embryo. *Dev. Biol.* **140**, 447-454.

Wilt, F. H. and Benson, S. C. (1988). Development of the endoskeletal spicule of the sea urchin embryo. In *Self-Assembling Architecture. 46th Symp. Soc. Dev. Biol.* (ed. J. E. Varner), pp 203-227. New York: Alan R. Liss.

Wolpert, L. (1990). The evolution of development. *Biol. J. Linn. Soc.* **39**, 109-124.

Wolpert, L. and Gustafson, T. (1961). Studies on the cellular basis of morphogenesis of the sea urchin embryo. Development of the skeletal pattern. *Exp. Cell Res.* **25**, 311-325.

Wray, G. A. and McClay, D. R. (1988). The origin of spicule-forming cells in a 'primitive' sea urchin (*Eucidaris tribuloides*), which appears to lack primary mesenchyme cells. *Development* **103**, 305-315.

Development 1992 Supplement, 53-63 (1992)
Printed in Great Britain © The Company of Biologists Limited 1992

A gastrulation center in the ascidian egg

WILLIAM R. JEFFERY

Department of Zoology and Bodega Marine Laboratory, University of California, Davis, PO Box 247, Bodega Bay, CA 94923 USA

Summary

A gastrulation center is described in ascidian eggs. Extensive cytoplasmic rearrangements occur in ascidian eggs between fertilization and first cleavage. During ooplasmic segregation, a specific cytoskeletal domain (the myoplasm) is translocated first to the vegetal pole (VP) and then to the posterior region of the zygote. A few hours later, gastrulation is initiated by invagination of endoderm cells in the VP region of the 110-cell embryo. After the completion of gastrulation, the embryonic axis is formed, which includes induction of the nervous system, morphogenesis of the larval tail and differentiation of tail muscle cells. Microsurgical deletion or ultraviolet (UV) irradiation of the VP region during the first phase of myoplasmic segregation prevents gastrulation, nervous system induction and tail formation, without affecting muscle cell differentiation. Similar manipulations of unfertilized eggs or uncleaved zygotes after the second phase of segregation have no effect on development, suggesting that a gastrulation center is established by transient localization of myoplasm in the VP region. The function of the gastrulation center was investigated by comparing protein synthesis in normal and UV-irradiated embryos. About 5% of 433 labelled polypeptides detected in 2D gels were affected by UV irradiation. The most prominent protein is a 30 kDa cytoskeletal component (p30), whose synthesis is abolished by UV irradiation. p30 synthesis peaks during gastrulation, is affected by the same UV dose and has the same UV-sensitivity period as gastrulation. However, p30 is not a UV-sensitive target because it is absent during ooplasmic segregation, the UV-sensitivity period. Moreover, the UV target has the absorption maximum of a nucleic acid rather than a protein. Cell-free translation studies indicate that p30 is encoded by a maternal mRNA. UV irradiation inhibits the ability of this transcript to direct p30 synthesis, indicating that p30 mRNA is a UV-sensitive target. The gastrulation center may function by sequestration or activation of maternal mRNAs encoding proteins that function during embryogenesis.

Key words: ascidian egg, gastrulation center, protein synthesis, UV irradiation.

Introduction

Classic studies on gastrulation and embryonic axis formation have been carried out in amphibian, teleost fish and bird embryos, which begin gastrulation after the egg has cleaved into thousands of cells (see Trinkaus, 1984 for review). *Drosophila* (Leptin and Grunewald, 1990; Kam et al., 1991) and sea urchins (Hardin, 1987; Ettensohn, 1985), the major invertebrate systems used to study these processes, also begin gastrulation at stages containing a relatively large number of cells. In contrast, gastrulation starts after the fourth or fifth cleavage in some polychaete annelid (Anderson, 1973) and tunicate (Berrill, 1955) embryos. These simple systems provide an opportunity to study gastrulation and axis formation under conditions of low cellular complexity. In addition, the short interval between fertilization and the beginning of gastrulation is favorable for examining the relationship between morphogenesis and earlier developmental events, such as fertilization, ooplasmic segregation and precocious cell determination.

Ascidians begin gastrulation between the sixth and seventh cleavages (110-cell stage) and then undergo a series of morphogenetic movements resulting in the formation of a tadpole larva. The ascidian tadpole larva contains only a few thousand cells and six different cell or tissue types: muscle, mesenchyme, notochord, neural, endoderm and epidermis. Despite this simplicity, ascidian larvae exhibit typical chordate features, including a dorsal nervous system and a tail containing a notochord and flanking bands of striated muscle cells. Because of their determinate cleavage pattern, invariant cell lineage and the existence of localized egg cytoplasmic regions, ascidians are a model system for studying autonomous cell determination (see Venuti and Jeffery, 1989 for review). The cytoplasmic regions of ascidian eggs are thought to contain determinants that cause embryonic cells to adopt a specific fate. For example, muscle determinants may reside in the myoplasm, a cytoplasmic region distributed to the tail muscle cell progenitors during cleavage (Whittaker, 1982). It is also apparent, however, that cell fates can be established by induction in ascidian embryos. The larval brain and its sensory organs are examples of tissues specified by inductive cell interactions (Rose, 1939; Reverberi et al., 1960; Nishida and Satoh, 1989; Nishida, 1991).

In this article, the relationship between ooplasmic segregation, gastrulation and axis formation is examined in ascidian embryos. Evidence is presented that gastrulation is preceded by the establishment of a gastrulation center in fertilized eggs. The gastrulation center may function by sequestering or activating maternal mRNAs encoding proteins that function during morphogenesis.

Experimental section

Ooplasmic segregation

Ascidian eggs are radially symmetric, however, bilateral symmetry is evident shortly after fertilization, coincident with the completion of ooplasmic segregation (Conklin, 1905; Jeffery, 1984a; Sardet et al., 1989). As first described by Conklin (1905), three cytoplasmic regions are distributed in a distinct spatial pattern along the animal-vegetal axis of unfertilized ascidian eggs. In a few ascidian genera, including *Styela* and *Boltenia* (Conklin, 1905; Jeffery, 1984a), these regions contain pigment granules and can be distinguished by color. The animal hemisphere contains the ectoplasm, a clear cytoplasmic region derived from the germinal vesicle during oocyte maturation. The vegetal hemisphere contains the endoplasm, which is filled with yolk granules. Except for a small area around the animal pole (AP), the egg cortex contains the myoplasm (Figs. 1 and 2, Row A), a yellow cytoplasmic region that enters the tail muscle cells during embryogenesis. The description of ooplasmic segregation below will focus on the myoplasm because this region is involved in the establishing the gastrulation center.

The myoplasm is a domain consisting of a peripheral submembrane cytoskeleton and a deeper cortical cytoskeleton (Fig. 2, Row B). The submembrane cytoskeleton stains with actin antibodies and phalloidin, interacts with myosin and can be disrupted by treating extracted eggs with DNase I, suggesting that its integrity is dependent on F-actin (Jeffery and Meier, 1983; Sawada and Osanai, 1985). The cortical cytoskeleton is a three-dimensional network of filaments containing pigment granules, mitochondria, endoplasmic reticulum and maternal mRNA (Jeffery and Meier, 1983; Jeffery, 1984b; Gualtieri and Sardet, 1989), which is attached to the submembrane cytoskeleton and the underlying ectoplasm. NN18, a monoclonal antibody that reacts with vertebrate neurofilaments, stains the cortical cytoskeleton, suggesting that it has structural affinities to intermediate filaments (Swalla et al., 1991). Although both parts of the myoplasmic cytoskeletal domain segregate together after fertilization, the motive force for the first movement is supplied by actin filaments (also see below) in the submembrane cytoskeleton. This was demonstrated by showing that the submembrane domain could segregate without the cortical cytoskeleton after the latter is displaced by low speed centrifugation (Jeffery and Meier, 1984).

The myoplasm segregates in two phases after fertilization (Fig. 2, Row A). During the first phase, a transient cap of yellow cytoplasm is formed in a lobe of cytoplasm near the vegetal pole (VP) of the zygote (Fig. 1). Concomitant with the first phase of segregation, the male pronucleus is

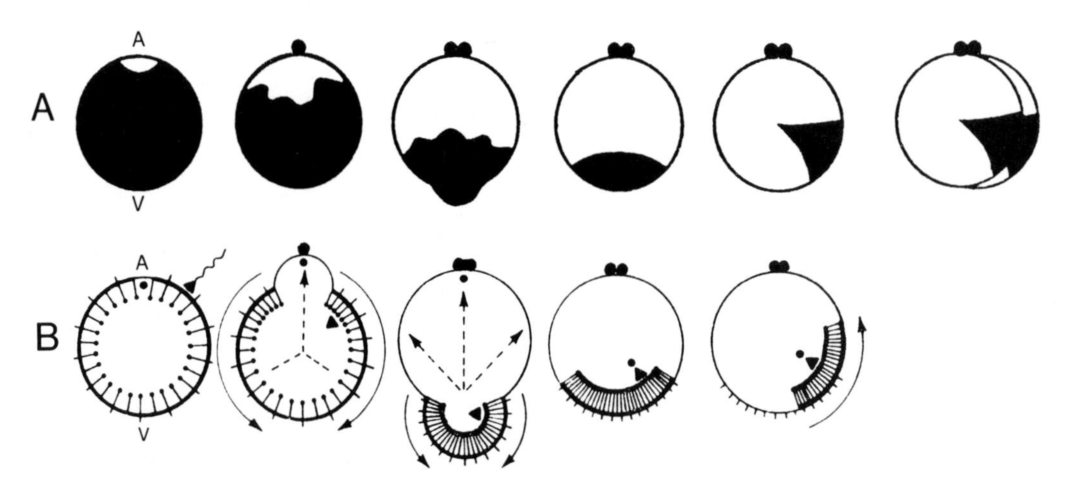

Fig. 2. Diagrams showing changes in the distribution of the myoplasm (Row A) and the myoplasmic cytoskeletal domain (Row B) during ooplasmic segregation in ascidian eggs. Row A. The myoplasm (dark areas) segregates from the egg cortex to the vegetal pole (V) during the first phase of ooplasmic segregation and then is translocated to the posterior of the zygote where it spreads out as the yellow crescent. The yellow crescent is bisected by the first cleavage plane. Row B. The translocation of the myoplasmic cytoskeletal domain in eggs sectioned through the animal-vegetal axis at stages corresponding to A. The filled spheres on the outside of the eggs represent the polar bodies. Small A: animal pole. The thick boundaries represent egg plasma membrane underlain by the sub-membrane cytoskeleton, and the thin boundaries egg plasma membrane without this cytoskeleton. The structures drawn as thin filaments on the outer surface of the plasma membrane represent egg surface components translocated with the underlying myoplasm during the first but not the second phase of segregation. The structures drawn as thin filaments topped with filled spheres on the internal surface of the plasma membrane represent the cortical cytoskeleton and associated organelles that participate in both phases of segregation. The filled spheres inside the egg represent the female pronucleus. The filled triangles represent the sperm head (outside the egg) or the male pronucleus (inside the egg). The arrows with continuous tails outside the egg represent the direction of myoplasmic movement during ooplasmic segregation. The arrows with broken tails on the inside of the egg represent the direction moved by other cytoplasmic components. From Jeffery and Swalla (1990).

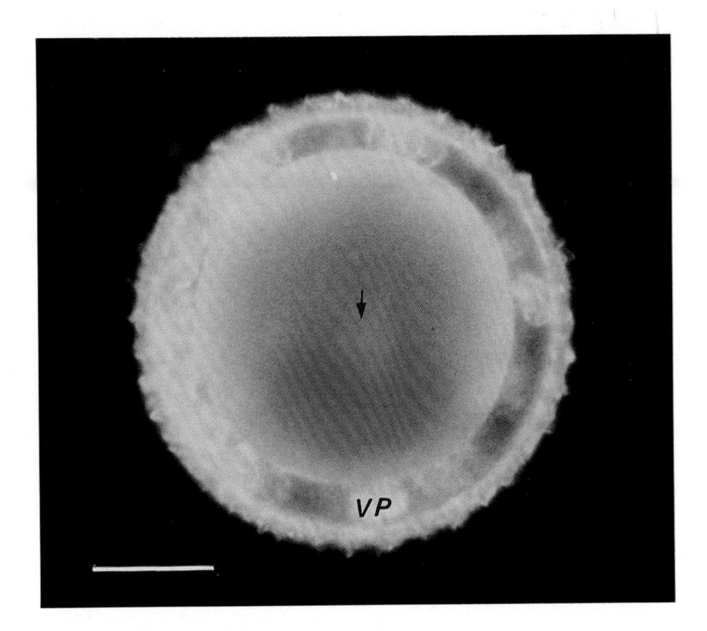

Fig. 1. An *S. clava* egg at completion of the first phase of ooplasmic segregation showing a cap of yellow myoplasm localized at the vegetal pole (VP) and an adjacent clear spot (arrow) representing the site of sperm aster formation. The egg is viewed from a lateral side of the vegetal hemisphere. Scale bar: 50 µm.

translocated into the VP region and a sperm aster appears adjacent to the cap of myoplasm (Fig. 1). The first phase can be blocked by cytochalasin, indicating that it is mediated by actin filaments (Zalokar, 1974; Sawada and Osanai, 1981). During the first phase of segregation, microtubules are rare or disorganized in the myoplasm. After the myoplasm is focussed at the VP, however, it is penetrated by microtubules growing from the sperm aster (Sawada and Schatten, 1988). During the second phase of segregation, most of the myoplasm moves into the sub-equatorial zone where it is extended into a yellow crescent (Fig. 2, Row A). The second phase is blocked by colchicine and likely to be mediated by aster microtubules (Zalokar, 1974; Sawada and Schatten, 1989). As the myoplasm moves into the equatorial zone, it shifts with respect to the cell surface (Fig. 2, Row B). This shift was demonstrated by applying chalk particles to the surface of denuded *Styela plicata* eggs before and during ooplasmic segregation (Bates and Jeffery, 1987). When chalk particles were applied to regions of the cell surface underlain by myoplasm before the first phase of segregation, they were transported into the vegetal hemishpere. In contrast, when chalk particles were applied to the cell surface over the vegetal cap of myoplasm after the first phase was completed, they did not move toward the equator during the second phase. These results suggest that cell surface components undergo the first phase of segregation in concert with the myoplasm, but afterwards remain stationary near the VP.

The axis of bilateral symmetry is visible at the end of the second phase of ooplasmic segregation. The dorsoventral axis approximates the animal-vegetal axis, with the VP region representing the future dorsal side of the embryo. The anteroposterior axis lies perpendicular to the dorsoventral axis; the yellow crescent marks the posterior of the embryo and the site of larval tail formation. The first cleavage furrow bisects the yellow crescent, dividing the embryo into left and right halves through the plane of bilateral symmetry (Fig. 2, Row A). During cleavage, the myoplasm is partitioned to specific blastomeres that generate most of the larval tail muscle cells.

Gastrulation and embryonic axis formation

The morphogenetic events resulting in formation of the tadpole larva include gastrulation, elongation of the anteroposterior axis, neurulation and tail development. Gastrulation involves three steps: invagination of the endoderm, involution of the mesoderm and epiboly of the ectoderm (Conklin, 1905; Mancuso, 1973; Satoh, 1978). The blastula is slightly flattened along the animal-vegetal axis. At the beginning of gastrulation, the vegetal cells contract at their basal margins and expand at their apical margins, which converts them from a columnar to tear-drop shape. The most extreme shape changes occur in the four largest endoderm cells (A7.1 and B7.1 cells; Conklin, 1905), which appear to lead the process of invagination (Figs. 3A, 4A). The invaginating endoderm cells evenually contact ectoderm cells in the AP region on the opposite side of the embryo (Fig. 3B). At the completion of invagination, the blastopore is bordered by presumptive notochord cells on its anterior lip, presumptive mesenchyme on its lateral lips and presumptive muscle cells on its posterior lip (Fig. 3B).

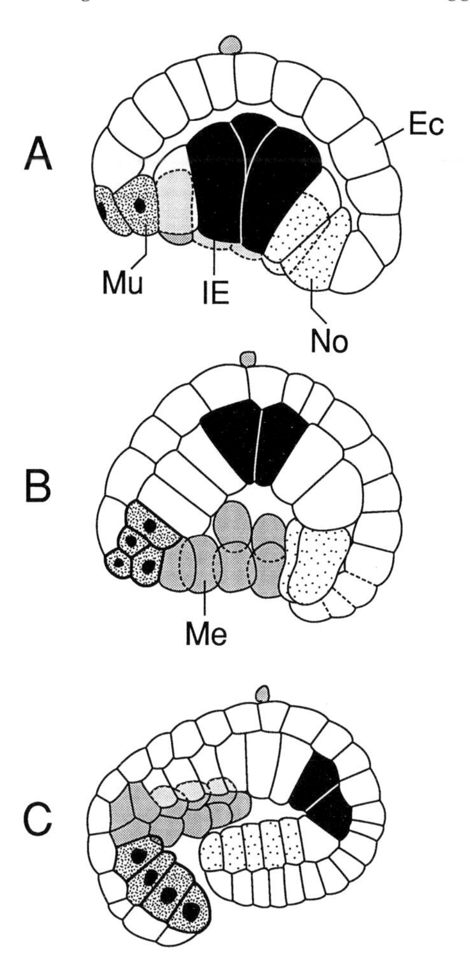

Fig. 3. Diagrammatic representation of cell movements during ascidian gastrulation. The embryos in A-C represent sections through the dorsoventral (animal-vegetal) axis. (A) Shape changes in endoderm cells initiate invagination in the early gastrula. (B) Invagination is completed in the mid-gastrula with contact being established between the endoderm and ectoderm cells in the animal hemisphere, and the mesoderm cells (prospective notochord, mesenchyme and muscle) begin to involute at the lips of the blastopore. (C) The invaginated endoderm cells move anteriorly, the mesoderm is internalized, the ectoderm cells complete epiboly and the embryo begins to elongate along the anteroposterior axis during the late gastrula stage. Gastrulae in A-C are oriented with their anterior poles facing the right and vegetal (dorsal) poles facing down. The small spheres drawn at the top of each embryo are polar bodies. IE: endoderm cells that are first to invaginate. Mu: muscle cells shown with shaded nuclei. Me: mesenchyme cells. No: notochord cells. Ec: ectoderm cells.

The mesoderm involutes over the lips of the blastopore and is replaced by epibolizing ectodermal cells (Fig. 3C).

The mechanism of ascidian gastrulation is poorly understood. However, Whittaker (1973) noted that blastomeres of *Ciona intestinalis* embryos treated with cytochalasin B after the 64-cell stage remain stationary, whereas normal cell distortions and migratory movements can occur in similar embryos treated with colcemid. These observations suggest that microfilaments, rather than microtubules, mediate cell behaviors associated with gastrulation in ascidians. In addition, it is possible that cell division, which contin-

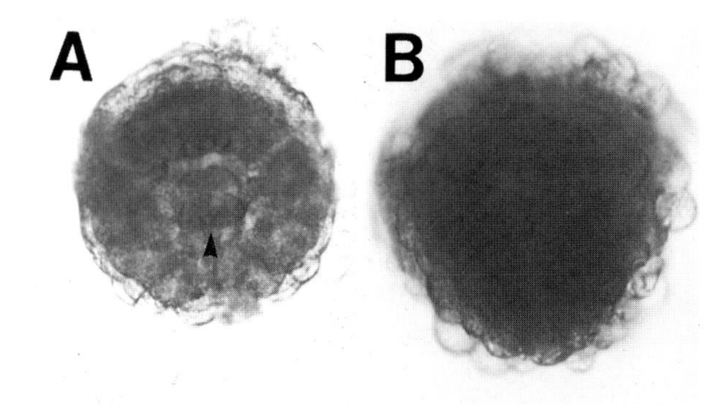

Fig. 4. Photomicrographs of normal (A) and UV-irradiated (B) *S. clava* embryos. (A) A normal embryo viewed from the vegetal side showing the bases of the invaginating A7.1 and B7.1 blastomeres (arrowhead). (B) An embryo that developed from a UV-irradiated zygote at the same stage of development as A showing no invagination. The scale bar is 20 µm; magnification is the same in A and B. From Jeffery (1990a).

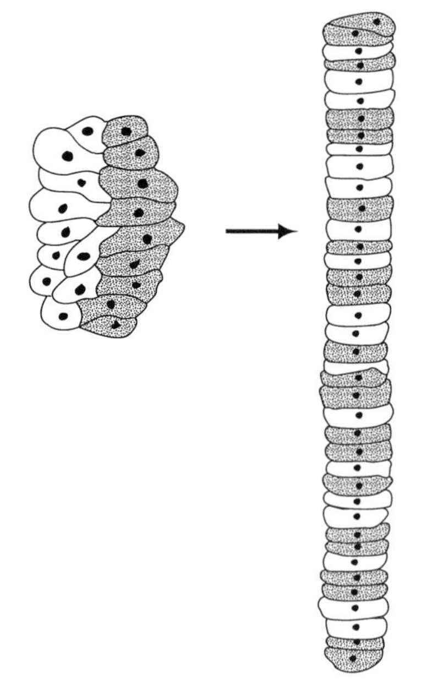

Fig. 5. Diagrammatic representation of cell rearrangements resulting in notochord formation in ascidian embryos. The notochord cells are viewed in frontal sections. Left. A mass of prospective notochord cells at the midline of a neurula. Cells that originate from the right side of the bilaterally-symmetric embryo are shaded, whereas those that originate from the left side are not shaded. Right. The notochord of a tadpole larva after interdigitation of notochord cells originating on either side of the midline and their posterior extension as a column of single cells. Note that cells orginating from either side of the midline are mixed randomly in the notochord (Nishida and Satoh, 1985). In both, orientation is with the posterior pole of the cell mass or notochord facing down.

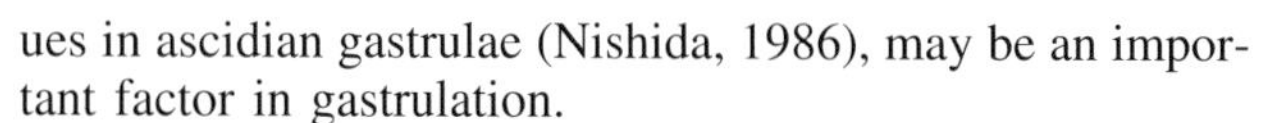

ues in ascidian gastrulae (Nishida, 1986), may be an important factor in gastrulation.

At the end of the gastrula stage, the embryo begins to elongate along its anteroposterior axis. Elongation involves anterior movements of the internalized endoderm cells and posterior movements of mesoderm cells. As the embryo elongates, the neural plate, which has developed in the vegetal hemisphere, invaginates to form the neural tube. The anterior portion of the neural tube later expands to form the larval brain, which ultimately differentiates two pigmented sensory organs. The posterior portion of the neural tube becomes the spinal cord, which runs along the dorsal margin of the tail. A number of investigators have shown that the larval nervous system is induced during the late gastrula stage (Rose, 1939; Reverberi et al., 1960; Nishida and Satoh, 1989). Although the identity of the neural inducer is controversial (Reverberi et al., 1960; Nishida, 1991), inductive activity has been traced to cells arising from the anterior quadrant of the vegetal hemisphere, which underlie the dorsal ectoderm after gastrulation.

The formation of the tail is driven by rearrangements of notochord cells (Conklin, 1905; Cloney, 1964; Miyamoto and Crowther, 1985; Nishida and Satoh, 1985; Swalla and Jeffery, 1990). After gastrulation, a mass of presumptive notochord, several cells wide, is positioned at the posterior midline of the neurula (Fig. 5A). During tail formation, the presumptive notochord cells interdigitate and then extend into a single row of cells that swells to form the notochord (Fig. 5B). During the extension of the larval tail, bands of striated muscle cells differentiate on either side of the notochord.

The gastrulation center

The event that links gastrulation and axis formation with ooplasmic segregation is the establishment of a gastrulation center near the VP of the fertilized zygote. The gastrulation center is defined as a region of the uncleaved egg required for the embryo to gastrulate and develop into a bilaterally symmetric larva.

The first evidence for a specialized role of the vegetal hemisphere in development was provided by microsurgical experiments with *Phallusia mammillata* and *Ascidia malaca* eggs (Ortolani, 1958; Reverberi and Ortolani, 1963). When unfertilized eggs were cut into two fragments through any plane and then fertilized, each sufficiently large fragment was capable of cleavage, gastrulation and development into a normal larva. In contrast, only the vegetal fragments were capable of normal development when the same operation was performed on fertilized eggs. The animal fragments developed into permanent blastula lacking the cell movements associated with gastrulation. These results showed that the vegetal hemisphere is the only part of the fertilized egg with the potential for gastrulation and the formation of a complete larva.

The experiments described above did not discriminate between the developmental potential of different regions of the vegetal hemisphere because *P. mammillata* and *A. malaca* eggs are not endowed with colored cytoplasmic regions that can be used as markers. Additional information about the developmental potential of the vegetal hemisphere was obtained by deleting colored cytoplasmic regions from *Styela plicata* eggs and zygotes (Bates and Jeffery, 1987). The deletion procedure involved: (1) making a hole in the chorion above the region of the egg to be

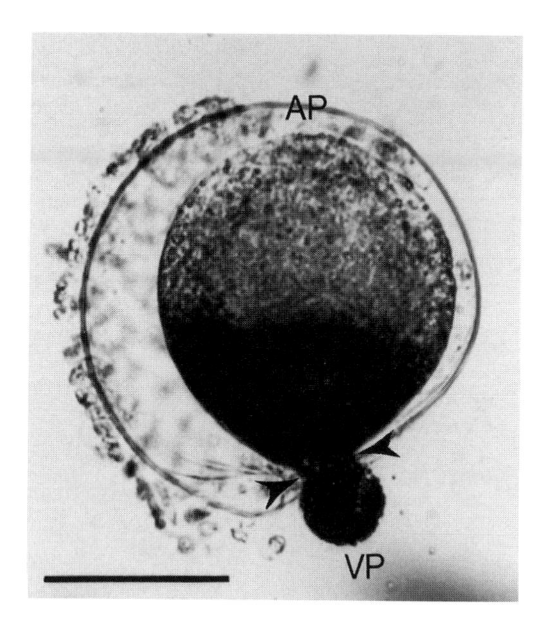

Fig. 6. Deletion of the vegetal pole region of a *S. plicata* zygote shortly after completion of the first phase of ooplasmic segregation. The position of the connection between the extruded anucleate fragment and the egg proper that is severed to produce the deletion is indicated by the arrowheads. The zygote was extracted with Triton X-100 to enhance contrast of the myoplasm (dark vegetal region) for photography. AP: animal pole. VP: vegetal pole. Scale bar: 50 μm. From Bates and Jeffery (1987).

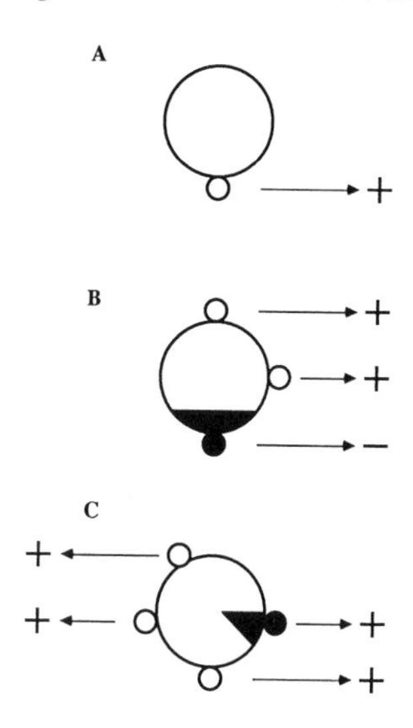

Fig. 7. A summary of deletions carried out in *S. plicata* eggs and zygotes and their effect on gastrulation and normal development. (A) Unfertilized eggs were deleted at the vegetal pole. (B) Zygotes were deleted at the animal pole, in the equatorial region and at the vegetal pole shortly after completion of the first phase of ooplasmic segregation. (C) Zygotes were deleted near the animal pole, in the anterior equatorial region, in the posterior equatorial (yellow crescent) region and at the vegetal pole shortly after completion of the second phase of ooplasmic segregation. The arrows pointing to + or – signs indicate the effect of deletions made at this position on the ability of the egg or zygote to gastrulate and develop into a normal larva: the + sign indicates that gastrulation and normal axis formation occurred, whereas the – sign indicates that gastrulation was blocked. Orientation is with the animal pole on the top and the posterior pole to the right. The shaded areas indicate the position of the myoplasm.

deleted, (2) exerting pressure on the opposite side of the egg to extrude the desired area through the hole and (3) severing the narrow connection between the extrudate and the remainder of the egg. In this way, an anucleate fragment containing plasma membrane, cortex and underlying cytoplasm could be deleted from a specific region of the egg without affecting the viability of the remaining nucleate egg fragment. Fig. 6 shows an example of a deletion in which a small portion of the egg was extruded from a point centered in the myoplasm near the VP.

Fig. 7 summarizes the results of deletions made at different stages of ooplasmic segregation (Bates and Jeffery, 1987). After each deletion, the nucleate fragments were allowed to continue development and assessed for normal cleavage, gastrulation, tail formation, and muscle and sensory cell differentiation. Consistent with the results of earlier experiments, deletion of the VP region of unfertilized eggs did not affect normal development (Fig. 7A). Deletion of as little as 5% of the total cell volume from the VP region after the completion of the first phase of segregation, however, blocked gastrulation, sensory cell induction and embryonic axis formation (Fig. 7B; also see Fig. 8A). Similar to the animal hemisphere fragments described above, the VP deficient embryos developed into permanent blastulae. The permanent blastulae did not form an axis of bilateral symmetry, but some cells expressed acetylcholinesterase (Bates and Jeffery, 1987), a muscle cell marker in ascidians (Whittaker, 1973). The site of the deletion was critical in abolishing gastrulation and axis formation. An effective deletion had to be made within the yellow cap of myoplasm (Fig. 1); deletions at other sites had no effect on development (Figs. 7B and 8). The timing of deletions was also important. Although deletions made at the VP shortly after the first phase of segregation blocked gastrulation, there was no effect on gastrulation when deletions were made at the same site at the yellow crescent stage (Fig. 7C). This result was interpreted to mean that the gastrulation center is dispersed throughout the vegetal hemisphere after sperm aster formation. It is difficult to test this possibility, however, because larger deletions remove the pronuclei, which are localized in vegetal hemisphere during ooplasmic segregation (Conklin, 1905). The results suggest that segregation of myoplasm organizes a gastrulation center in the VP region of the zygote.

The gastrulation center could mediate invagination by entering blastomeres destined to invaginate or blastomeres that ultimately induce other cells to initiate gastrulation. This question was examined by following the fate of cells that inherit the VP region in *S. plicata* embryos (Bates and Jeffery, 1987). As described above, chalk particles placed at the surface of the myoplasm when the first phase of segregation is completed do not move to the posterior pole during yellow crescent formation. By following the distribution of chalk granules placed in this position, the VP region was traced into the first endoderm cells (A7.1 and B7.1 cells) to invaginate at the beginning of gastrulation

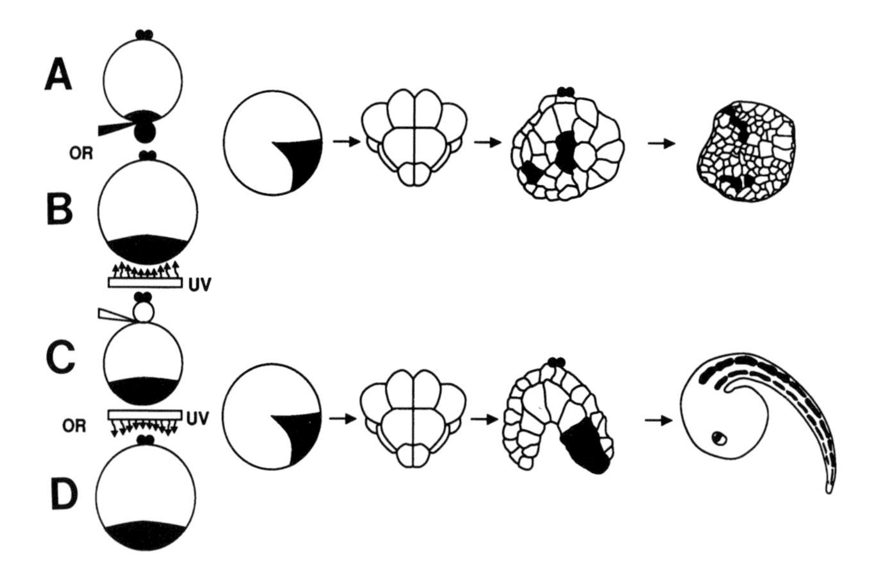

Fig. 8. A comparison of the effects of microsurgical deletion and UV irradiation of the myoplasm on embryonic development. (A) Deletion of the vegetal pole region at completion of the first phase of ooplasmic segregation. (B) UV irradiation of the vegetal pole region at completion of the first phase of ooplasmic segregation. (C) Deletion of the animal pole region at the same time as in A. (D) UV irradiation of the animal pole region at the same time as in B. Stages of normal development (lower row) from left to right: yellow cap, yellow crescent, 16-cell, mid-gastrula and mid-tailbud. Development after vegetal pole deletion or UV irradiation is shown in the upper row. The shaded areas represent myoplasm or myoplasm-containing cells. From Jeffery and Swalla (1990).

(see Fig. 3). Therefore, the gastrulation center is inherited by the cells that initiate gastrulation.

Polarization of the gastrulation center

What factors are responsible for establishing the gastrulation center near the VP? To answer this question, we must identify the cues that polarize the myoplasm during the first phase of ooplasmic segregation.

In his classic paper, Conklin (1905) proposed that the myoplasm is directed to the VP of *Styela partita* eggs because it is the site of sperm entry. In support of this proposal, it was shown that the male pronucleus is localized near the VP after fertilization. An alternate interpretation, however, is that the sperm could been carried to the VP by the advancing myoplasm after entering in another region of the egg. Conklin's observations are consistent with this possibility because his specimens were fixed for cytology relatively late after insemination.

Unfortunately, it is difficult to directly observe the site of fertilization in ascidian eggs because sperm entry is obscured by accessory cells lying between the egg surface and the chorion. Removal of the chorion provides a naked egg, but also prevents fertilization, probably because the follicle is required for sperm activation. Therefore, Bates and Jeffery (1988) tested for a restricted sperm entry site by deleting the VP region of unfertilized *S. plicata* eggs and then determining whether these eggs could be fertilized. The results showed that VP deficient eggs are fertilizable and capable of normal development. In other experiments, Bates and Jeffery (1988) examined the extent of egg surface area capable of sperm entry by inseminating multiple anucleate fragments deleted from unfertilized eggs. It was possible to fertilize three fragments obtained from different regions of the same egg, indicating the sperm entry can occur within a wide area of the egg surface. While these experiments do not exclude a preferred sperm entry site, they suggest that fertilization can occur over most of the egg surface.

The question of a preferred sperm entry site was examined directly by staining fertilized *P. mammillata* eggs with a DNA-specific fluorescent dye (Speksnijder et al., 1989). In this study, the fertilizability problem was solved by activating sperm with chorionated eggs before inseminating the naked eggs. The results showed that fertilization could occur in the animal or vegetal hemisphere, but that the animal hemisphere was the most frequent site of sperm entry (see Fig. 2, Row B). These results do not support Conklin's hypothesis that the myoplasm segregates toward a sperm entry site at the VP.

If sperm entry does not occur at the VP, how is the myoplasm oriented toward this region after fertilization? Current evidence favors the possibility that the direction of myoplasmic segregation is determined by pre-existing polarity along the animal-vegetal axis (Bates and Jeffery, 1988). This idea is supported by experiments in which the direction of myoplasmic segregation was examined in *S. plicata* egg fragments after extrusion of the AP or VP regions of unfertilized eggs. The results showed that myoplasm usually segregates toward the most vegetal region in both animal (Fig. 10) and vegetal fragments. Recent observations suggest that the sperm entry point may bias the direction of myoplasmic segregation in *P. mammillata* eggs (Speksnijder et al., 1990). According to this idea, the myoplasm can be focused up to 50° away from the VP depending on the position of sperm entry in the animal hemisphere. Even if the sperm enters in the vegetal hemisphere, however, the focal point of myoplasmic segregation is always vegetal (Speksnijder et al., 1990), indicating the primary cue for polarization must pre-exist along the animal-vegetal axis.

The identity of the axial polarization cue is currently unknown, although it has been shown that myoplasm will segregate toward an exogenously applied source of Ca^{2+} ionophore (Jeffery, 1982). Thus, a flux of free Ca^{2+}, possibly released from the cortical endoplasmic reticulum (Gaultieri and Sardet, 1989), may orient the gastrulation center toward the VP.

UV inactivation of the gastrulation center

Progress in determining the function of the gastrulation

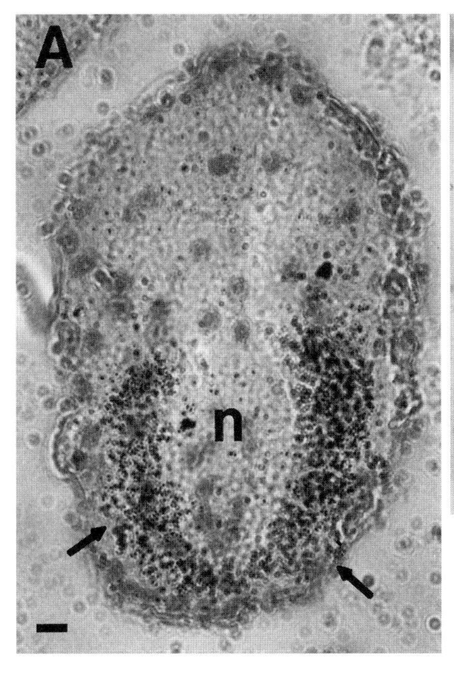

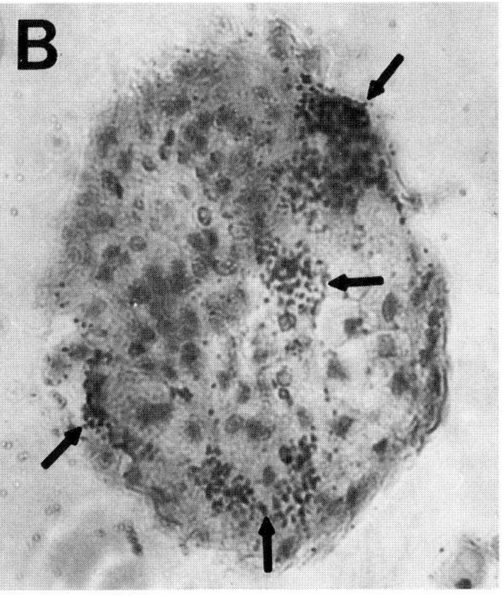

Fig. 9. Muscle cell development assessed by *in situ* hybridization with a muscle actin probe in sections of normal (A) and UV-irradiated (B) *S. clava* embryos. (A) In early tailbud stage embryos, muscle actin mRNA accumulates in muscle progenitor cells (arrows) located on either side of the presumptive notochord (n). (B) In embryos that were UV irradiated shortly after completion of the first phase of ooplasmic segregation, muscle actin mRNA accumulates in patches of muscle progenitor cells (arrows) in an embryo lacking bilateral symmetry. Scale bar: 10 μm; magnification is the same in both frames. From Jeffery (1990a).

center was aided by the development of ultraviolet (UV) irradiation methods to inactivate this region (Jeffery, 1990a). *Styela clava* was the species of choice in these experiments because it produces large numbers of synchronously developing embryos which are useful for molecular analysis. The experiments and results are summarized in Fig. 8. Similarly to cytoplasmic deletion, UV irradiation of the VP region, but not the AP region, prevented gastrulation, brain sensory cell induction and tail formation (Fig. 8B, D; also see Fig. 4B). However, neither cleavage (Fig. 8B) nor muscle cell development, as determined by in situ hybridization of the permanent blastulae with a cloned muscle actin probe (Fig. 9), were affected by UV irradiation of the VP region. These experiments suggest that UV irradiation and cytoplasmic deletion have the similar effects on the function of the gastrulation center.

The UV-sensitivity period for inactivating the gastrulation center was determined by UV irradiating zygotes at various times between fertilization and first cleavage. UV irradiation was effective in blocking the gastrulation and axis formation when administered up to the yellow crescent stage (Fig. 11). These results contrast to those obtained by cytoplasmic deletion, in which gastrulation was inhibited only if the operation was performed at the yellow cap stage or earlier. However, the results are consistent with the possibility that the dispersed gastrulation center is localized

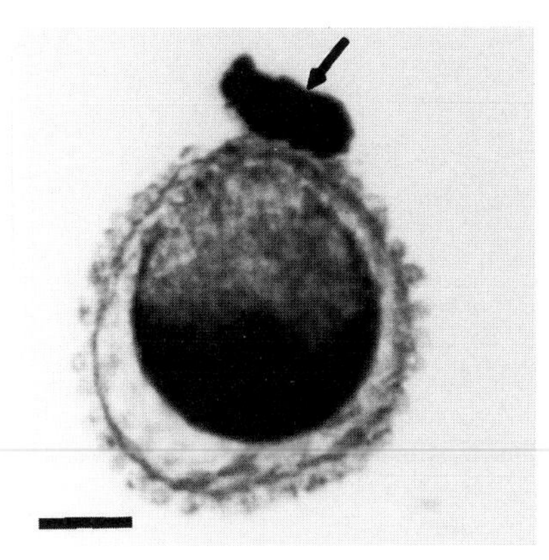

Fig. 10. The direction of myoplasmic segregation is vegetal in animal egg fragments. In this experiment, the vegetal portion of an unfertilized *S. plicata* egg was extruded before marking the animal pole with a chalk particle (arrow) and subsequent insemination. At completion of the first phase of ooplasmic segregation, the zygote was extracted with Triton X-100 to enhance contrast of the myoplasm (dark area), which has segregated to the most vegetal region of the fragment. Scale bar 50 μm. From Bates and Jeffery (1988).

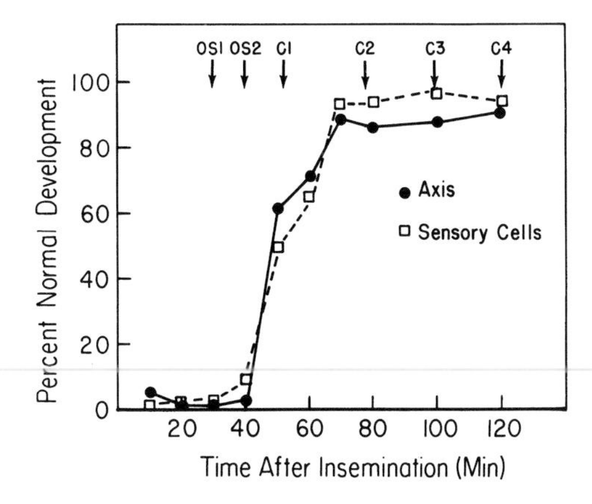

Fig. 11. The UV-sensitivity periods for axis and sensory cell development in *S. clava* embryos. Zygotes were irradiated at various times between insemination and fourth cleavage. Gastrulation was also abolished in these embryos. The timing of various developmental stages is indicated by the arrows. OS 1: completion of the first phase of ooplasmic segregation. OS 2: completion of the second phase of ooplasmic segregation. C1-C4: cleavages 1-4. From Jeffery (1990a).

near enough to the egg surface to be inactivated by UV light.

Molecular analysis of the gastrulation center

Inactivation by UV light permits a molecular analysis of the gastrulation center (Jeffery, 1990b). The first question was whether inactivation of the gastrulation center affects the pattern of embryonic protein synthesis. In these experiments, zygotes were irradiated shortly after completion of the first phase of ooplasmic segregation, allowed to cleave and then incubated with [^{35}S]methionine. Exposure to radioactive amino acids was continued until controls reached the early tailbud stage. Subsequently, proteins were extracted from the UV-irradiated and normal embryos, separated by 2D gel electrophoresis and compared by autoradiography (Fig. 12). Most of the 433 polypeptides detected in the autoradiographs were not affected by UV irradiation. For example, the three major actin isoforms synthesized by *S. clava* eggs were labelled to the same extent in UV-irradiated embryos and controls (Fig. 12). Only about 5% of the polypeptides decreased in labelling or disappeared after UV irradiation. One of the proteins whose synthesis was undetectable after UV irradiation is a 30 kDa molecule (p30). The extreme sensitivity of p30 synthesis to UV light makes it a prime candidate for a protein involved in gastrulation and axis determination.

A series of experiments was performed to determine whether there is a relationship between p30 synthesis and gastrulation (Jeffery, 1990b). The first experiment examined the effect of different UV doses. The results showed that p30 synthesis and gastrulation are abolished by the same UV dose. The next experiment compared the UV-sensitivity periods for p30 synthesis and gastrulation. The results showed that p30 synthesis and gastrulation are sensitive to UV irradiation up to the yellow crescent stage. UV irradiation after the yellow crescent stage showed labelling of p30 was equivalent to that observed in controls. Finally, an experiment was conducted to determine the timing of p30 synthesis during early development. The results showed that p30 synthesis begins at the 16-32 cell stage, peaks

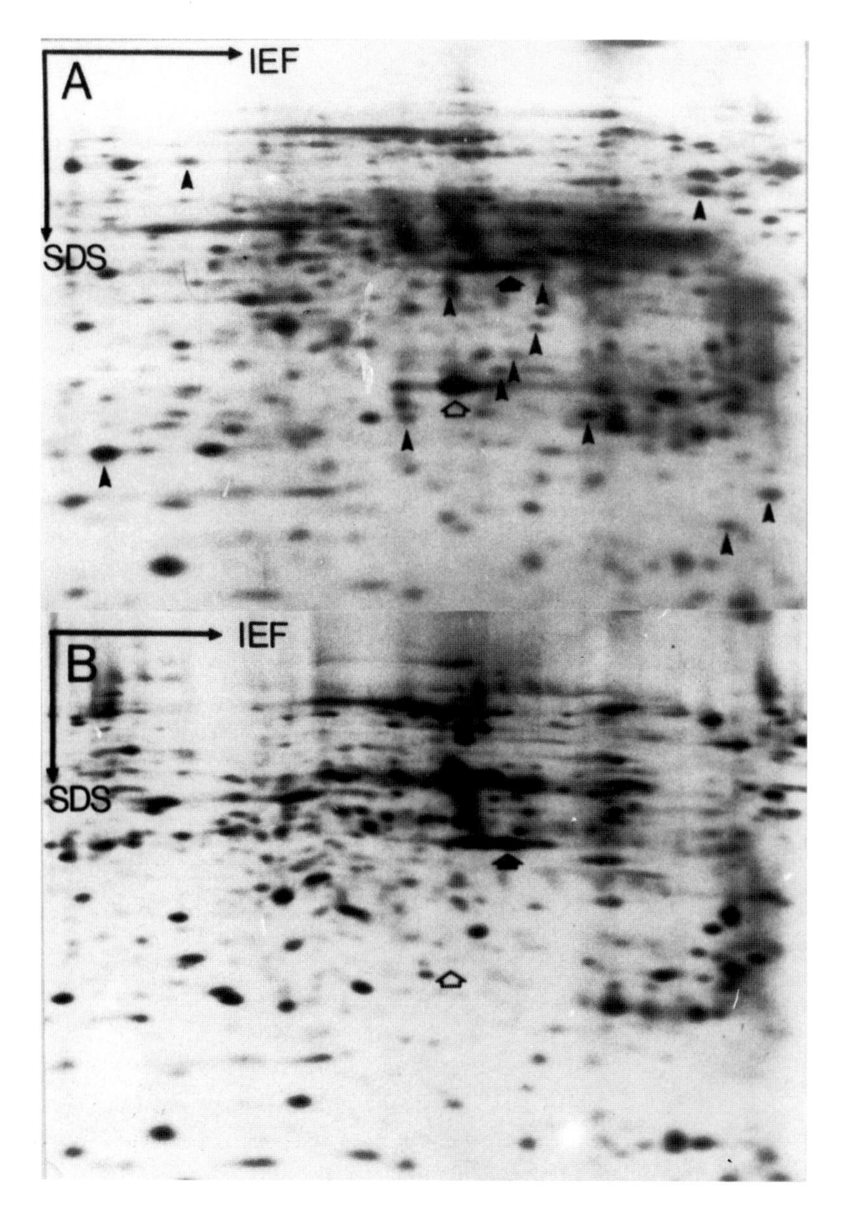

Fig. 12. Autoradiographs of gels containing proteins from normal and UV-irradiated *S. clava* embryos. (A) Proteins from normal embryos labelled with [^{35}S]methionine between the 2-cell and mid-tailbud stages. (B) Proteins extracted from embryos that developed from zygotes UV irradiated shortly after completion of the first phase of ooplasmic segregation and labelled with [^{35}S]methionine from the 2-cell stage until controls reached the mid-tailbud stage. Equal counts were applied to the gels. IEF: direction of isoelectric focussing. SDS: direction of electrophoresis through SDS gel. Solid large arrowhead: position of muscle actin marker. Open large arrowhead: position of p30. From Jeffery (1990b).

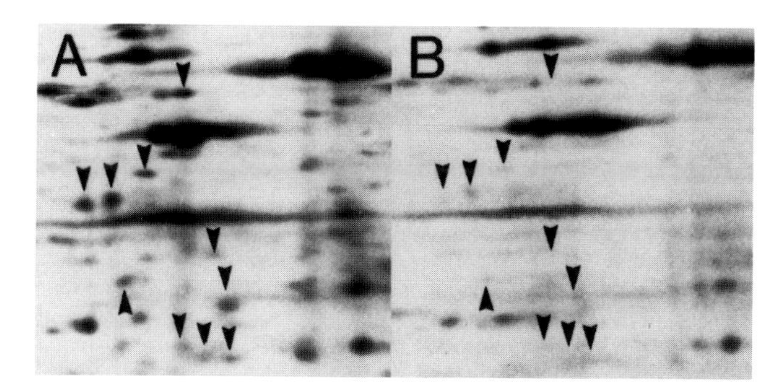

Fig. 13. The effect of UV irradiation on the ability of RNA from uncleaved *S. clava* zygotes to direct protein synthesis in a rabbit reticulocyte lysate. (A) Protein synthesis directed by RNA from normal zygotes. (B) Protein synthesis directed by RNA from UV-irradiated zygotes. p30 and a number of other polypeptides are missing or decreased in B. Upward-facing arrowheads: position of p30. Downward-facing arrowheads: position of other polypeptides whose translation was affected by UV irradiation. Only the p30 region of the gels is shown. Zygotes were UV-irradiated shortly after completion of the first phase of ooplasmic segregation. Equal counts were applied to each gel. From Jeffery (1990b).

during gastrulation and subsides by the early tailbud stage. The timing and UV sensitivity of p30 is consistent with a role in gastrulation.

Additional studies indicate that p30 is a cytoskeletal component (Jeffery, 1991). In these experiments, the distribution of labelled proteins was examined following extraction of embryos with Triton X-100. The results showed that p30 is present in the detergent insoluble fraction, suggesting it is a cytoskeletal component (Jeffery and Meier, 1983). However, p30 could be released from the insoluble to the soluble fraction when detergent extraction was carried out in the presence of DNase I, which specifically depolymerizes actin filaments. Therefore, p30 may be an F-actin binding protein.

Despite its potential role in gastrulation, p30 is not the UV-sensitive target in the gastrulation center: it is does not appear until the end of the UV-sensitive period and the chromophore has an absorbance maximum resembling a nucleic acid rather than a protein (Jeffery, 1991). Maternal RNA is a UV-sensitive target in insect eggs (Kalthoff, 1979). Thus, it was possible that p30 mRNA, rather than p30, is the UV-sensitive target. To determine whether p30 synthesis is directed by maternal mRNA, RNA extracted from unfertilized *S. clava* eggs was translated in a rabbit reticulocyte lysate and the translation products were examined by 2D gel electrophoresis and autoradiography (Jeffery, 1990b). The results showed that egg RNA directed the translation of a protein with the same molecular weight and isoelectric point as p30.

If maternal p30 mRNA is a UV-sensitive target, it should be possible to inactivate it by UV irradiation. To test this possibility, zygotes were irradiated shortly after completion of the first phase of ooplasmic segregation, RNA was extracted from the irradiated eggs and the translation products were compared with those of control eggs. The results showed that p30 and several other proteins were absent from the translation products directed by RNA from UV-irradiated eggs (Fig. 13). Thus, maternal p30 mRNA is a UV target. UV light cannot penetrate deeply into the egg cytoplasm (Youn and Malacinski, 1980), therefore, the suppression of the translation of p30 mRNA and other mRNAs (see Fig. 13) by UV irradiation strongly suggests that they are localized near the surface of the egg, possibly in the gastrulation center.

Concluding remarks

In this article, an invertebrate embryo with a low cell number and a relatively short interval between fertilization and gastrulation is used to investigate the relationship between an egg cytoplasmic region and morphogenesis. The major conclusion is that a gastrulation center is established in ascidian eggs by segregating myoplasm to the VP after fertilization. Later, as the myoplasm moves to the posterior pole, the potential for gastrulation spreads out in the vegetal hemisphere and is inherited by endoderm cells that initiate invagination. In amphibian eggs, a region is established in the vegetal hemisphere after fertilization and is then inherited by endoderm cells that subsequently induce marginal zone cells to begin gastrulation (Gimlich and Gerhart, 1984). In contrast, gastrulation is initiated by invagination of the same cells that inherit the gastrulation center in ascidians.

The experiments described here suggest that endoderm invagination in ascidian embryos is an autonomous process specified by axial determinants residing in the gastrulation center. Autonomous specification of cell fate is known to occur for some of the tail muscle cells in ascidians (see Venuti and Jeffery, 1989 for review). Further experiments are needed to prove that endoderm invagination is also an autonomous process, however. First, it would be necessary to show that endoderm cells undergo the shape changes related to invagination when they are isolated from the embryo. Second, conclusive evidence for axial determinants would require showing that the site of gastrulation could function autonomously after transplantation to another egg. At this time, however, it is possible to conclude that removal or inactivation of the gastrulation center abolishes gastrulation and axis formation, including induction of the nervous system and tail morphogenesis.

Morphogenesis of the ascidian embryo is initiated after VP blastomeres invaginate during gastrulation. The descendants of these and adjacent cells migrate through the embryo and eventually reach a position where they induce animal hemisphere blastomeres to elaborate the brain and sensory organs (Nishida and Satoh, 1989) or undergo morphogentic movements to form the notochord (Conklin, 1905). The experiments described in this article suggest that p30 is associated with the cytoskeleton and that this association is dependent on the integrity of microfilaments. Therefore, p30 may be an actin-binding protein that organizes microfilaments involved in changing the shape and/or motility of the invaginating endoderm cells.

The results support the hypothesis that gastrulation and axis formation may be mediated by maternal mRNA encoding the p30 cytoskeletal protein. During the first phase of ooplasmic segregation, p30 mRNA may be translocated to the VP region with the myoplasm. Then, during the second phase, p30 mRNA probably remains in the VP region after the myoplasm is shifted to the posterior region of the zygote. The localization of p30 mRNA near the VP would place it in the proper location to be inherited and translated in endodermal cells that are first to invaginate and initiate axis formation during gastrulation.

The reason that p30 mRNA and other messages whose translation was abolished are particularly sensitive to UV irradiation may be explained by poor penetration of UV light into the interior of eggs (Youn and Malacinski, 1980). It follows that UV-sensitive mRNAs must be localized near the egg surface. Localized maternal mRNAs have been previously described in the myoplasm of *S. plicata* eggs (Jeffery et al., 1983). Future studies with specific probes for p30 mRNA and protein will be required to determine the spatial distribution of these molecules during development. Experiments are also in progress to identify some of the other UV-sensitive mRNAs that may be localized in the gastrulation center.

This research was supported by NSF (DCB-84116763) and NIH (HD-13970) grants. I thank Dr. B. J. Swalla for critical reading of the manuscript.

References

Anderson, D. T. (1973). *Embryology and Phylogeny in the Annelids and Arthropods.* Oxford: Pergamon Press.

Bates, W. R. and Jeffery, W. R. (1987). Localization of axial determinants in the vegetal pole region of ascidian eggs. *Devl. Biol.* **124**, 65-76.

Bates, W. R. and Jeffery, W. R. (1988). Polarization of ooplasmic segregation and dorsal-ventral axis determination in ascidian embryos. *Devl. Biol.* **130**, 98-107.

Berrill, N. J. (1955). *The Origin of Vertebrates.* Oxford, Clarendon Press.

Cloney, R. A. (1964). Development of the ascidian notochord. *Acta Embryol. Morphol. Exp.* **7**, 111-130.

Conklin, E. G. (1905). The organization and cell lineage of the ascidian egg. *J. Acad. Sci. Natl. Sci. Phila.* **13**, 1-126.

Ettensohn, C. A. (1985). Gastrulation in the sea urchin embryo is accompanied by the rearrangement of invaginating epithelial cells. *Devl. Biol.* **112**, 383-390.

Gimlich, R. L. and Gerhart, J. C. (1984). Early cellular interactions promote embryonic axis formation in *Xenopus laevis. Dev. Biol.* **104,** 117-130.

Gualteiri, R. and Sardet, C. (1989). The endoplasmic reticulum network in the ascidian egg: localization and calcium content. *Biol. Cell* **65**, 301-304.

Hardin, J. D. (1987). Archenteron elongation in the sea urchin embryo is a microtubule-independent process. *Dev. Biol.* **122**, 253-262.

Jeffery, W. R. (1982). Calcium ionophore polarizes ooplasmic segregation in ascidian eggs. *Science* **216**, 545-547.

Jeffery, W. R. (1984a). Pattern formation by ooplasmic segregation in ascidian eggs. *Biol. Bull. mar. biol. lab. Woods Hole* **166**, 277-298.

Jeffery, W. R. (1984b). Spatial distribution of messenger RNA in the cytoskeletal framework of ascidian eggs. *Dev. Biol.* **103**, 482-492.

Jeffery, W. R. (1990a). Ultraviolet irradiation during ooplasmic segregation prevents gastrulation, sensory cell induction and axis formation in the ascidian embryo. *Dev. Biol.* **140**, 388-400.

Jeffery, W. R. (1990b). An ultraviolet-sensitive maternal mRNA encoding a cytoskeletal protein may be involved in axis formation in the ascidian embryo. *Dev. Biol.* **141**, 141-148.

Jeffery, W. R. (1991). Ultraviolet-sensitive determinants of gastrulation and axis development in the ascidian embryo. In *Gastrulation: movements, patterns, and molecules,* (Eds. R. Keller, W. H. Clark, Jr. and F. Griffin), pp. 225-250. New York/London: Plenum Press.

Jeffery, W. R. and Meier, S. (1983). A yellow crescent cytoskeletal domain in ascidian eggs and its role in early development. *Dev. Biol.* **96**, 125-143.

Jeffery, W. R. and Meier, S. (1984). Ooplasmic segregation of the myoplasmic actin network in stratified ascidian eggs. *W. Roux's Arch. Dev. Biol.* **193**, 257-262.

Jeffery, W. R. and Swalla, B. J. (1990). The myoplasm of ascidian eggs: a localized cytoskeletal domain with multiple roles in embryonic development. *Sem. Cell Biol.* **1**, 373-381.

Jeffery, W. R., Tomlinson, C. R. and Brodeur, R. D. (1983). Localization of actin messenger RNA during early ascidian development. *Dev. Biol.* **99**, 408-417.

Kalthoff, K. (1979). Analysis of a morphogenetic determinant in an insect embryo (*Smittia Spec., Chironomidae, Diptera*). In *Determinants of Spatial Organization* (Eds., S. Subtelny and I. R. Konigsberg), pp. 97-126. New York: Academic Press.

Kam, Z., Minden, J. S., Agard, D. A., Sedat, J. W. and Leptin, M. (1991). *Drosophila* gastrulation: analysis of cell shape changes in living embryos by three-dimensional fluorescence microscopy. *Development* **112**, 365-370.

Leptin, M. and Grunewald, B. (1990). Cell shape changes during gastrulation in *Drosophila. Development* **110**, 73-84.

Mancuso, V. (1973). Ultrastructural changes in the *Ciona intestinalis* egg during the stage of gastrula and neurula. *Arch. Biol.* **84**, 181-204.

Miyamoto, D. M. and Crowther, R. J. (1985). Formation of the notochord in living ascidian embryos. *J. Embryol. exp. Morph.* **86**, 1-17.

Nishida, H. (1986). Cell division pattern during gastrulation of the ascidian *Halocynthia roretzi. Dev. Growth Differ.* **28**, 191-201.

Nishida, H. (1991). Induction of brain and sensory pigment cells in the ascidian embryo analyzed by experiments with isolated blastomeres. *Development* **112**, 389-395.

Nishida, H. and Satoh, N. (1985). Cell lineage analysis in ascidian embryos by intracellular injection of a tracer enzyme. II. The 16- and 32-cell stages. *Dev. Biol.* **110**, 440-454.

Nishida, H. and Satoh, N. (1989). Determination and regulation in the pigment cell lineage of the ascidian embryo. *Dev. Biol.* **112**, 355-367.

Ortolani, G. (1958). Cleavage and development of egg fragments in ascidians. *Acta Embryol. Morph. Exp.* **1,** 247-272.

Reverberi, G. and Ortolani, G. (1962). Twin larvae from halves of the same egg in ascidians. *Dev. Biol.* **5**, 84-100.

Reverberi, G., Ortolani, G. and Farinella-Ferruzza, N. (1960). The causal formation of the brain in the ascidian larva. *Acta Embryol. Morphol. Exp.* **3**, 296-336.

Rose, S. M. (1939). Embryonic induction in the ascidia. *Biol. Bull. mar. biol. lab. Woods Hole* **77**, 216-232.

Sardet, C., Speksnijder, J. E., Inoue, S. and Jaffe, L. (1989). Fertilization and ooplasmic movements in the ascidian egg. *Development* **105**, 237-249.

Satoh, N. (1978). Cellular morphology and architecture during early morphogenesis of the ascidian egg: An SEM study. *Biol. Bull. mar. biol. lab. Woods Hole* **155**, 608-614.

Sawada, T. and Osanai, K. (1981). The cortical contraction related to ooplasmic segregation in *Ciona intestinalis* eggs. *W. Roux's Arch. Dev. Biol.* **190**, 201-214.

Sawada, T. and Osanai, K. (1985). Distribution of actin filaments in fertilized eggs of the ascidian *Ciona intestinalis. Dev. Biol.* **111**, 260-265.

Sawada, T. and Schatten, G. (1988). Microtubules in ascidian eggs during meiosis, fertilization and mitosis. *Cell Motil. Cytoskeleton* **9**, 219-230.

Sawada, T. and Schatten, G. (1989). Effects of cytoskeletal inhibitors on ooplasmic segregation and microtubule organization during fertilization and early development of the ascidian *Molgula occidentalis. Dev. Biol.* **132**, 331-342.

Swalla, B. J. and Jeffery, W. R. (1990). Interspecific hybridization between an anural and urodele ascidian: Differential expression of urodele features suggests multiple mechanisms control anural development. *Dev. Biol.* **142**, 319-334.

Swalla, B. J., Badgett, M. and Jeffery, W. R. (1991). Identification of a cytoskeletal protein localized in the myoplasm of ascidian eggs: localization is modified during anural development. *Development* **111**, 425-436.

Speksnijder, J. E., Sardet, C., Inoué, S. and Jaffe, L. F. (1989). Polarity of sperm entry in the ascidian egg. *Dev. Biol.* **133**, 180-184.

Speksnijder, J. E., Sardet, C., and Jaffe, L. F. (1990). The activation wave

of calcium in the ascidian egg and its role in ooplasmic segregation. *J. Cell Biol.* **110**, 1589-1598.
Trinkaus, J. P. (1984). *Cells into Organs. The Forces That Shape the Embryo.* Second Edition. Englewood Cliffs, N. J: Prentice-Hall.
Venuti, J. M. and Jeffery, W. R. (1989). Cell lineage and determination of cell fate in ascidians. *Int. J. Dev. Biol.* **33**, 197-212.
Whittaker, J. R. (1973). Segregation during ascidian embryogenesis of egg cytoplasmic information for tissue-specific enzyme development. *Proc. Nat. Acad. Sci. USA* **70**, 2096-2100.
Whittaker, J. R. (1982). Muscle cell lineage cytoplasm can change the developmental expression in epidermal lineage cells of ascidian embryos. *Dev. Biol.* **93**, 463-470.
Youn, B. W. and Malacinski, G. M. (1980). Action spectrum for ultraviolet irradiation of a cytoplasmic component(s) required for neural induction in the amphibian egg. *J. exp. Zool.* **211**, 369-377.
Zalokar, M. (1974). Effect of colchicine and cytochalasin B on ooplasmic segregation in ascidian eggs. *W. Roux's Arch. Dev. Biol.* **175**, 243-248

Development 1992 Supplement, 65-73 (1992)
Printed in Great Britain © The Company of Biologists Limited 1992

Cell movements and cell fate during zebrafish gastrulation

ROBERT K. HO

Institute of Neuroscience, University of Oregon, Eugene, OR 97403, USA

Summary

The early lineages of the zebrafish are indeterminate and a single cell labeled before the late blastula period will contribute progeny to a variety of tissues. Therefore, early cell lineages in the zebrafish do not establish future cell fates and early blastomeres must necessarily remain pluripotent. Eventually, after a period of random cell mixing, individual cells do become tissue restricted according to their later position within the blastoderm. The elucidation of a fate map for the zebrafish gastrula (Kimmel et al., 1990), has made it possible to study the processes by which cellular identity is conferred and maintained in the zebrafish. In this chapter, I describe single cell transplantation experiments designed to test for the irreversible restriction or 'commitment' of embryonic blastomeres in the zebrafish embryo. These experiments support the hypothesis that cell fate in the vertebrate embryo is determined by cell position. Work on the *spadetail* mutation will also be reviewed; this mutation causes a subset of mesodermal precursors to mismigrate during gastrulation thereby leading to a change in their eventual cell identity.

Key words: *Brachydanio rerio*, commitment, extraembryonic, gastrulation, mesoderm, mutation, zebrafish.

Introduction

Although, the zebrafish has attracted a great deal of recent interest, fish have been around for a very long time, probably about 500 million years, which is considerably longer than any other vertebrate group. As a system for developmental studies, fish embryos have a varied and interesting history deeply rooted in classical embryology (Clapp, 1891; Morgan, 1893; Wilson, 1889). The fish has been a favored organism for observational and experimental studies, in part because egg-laying fish can often produce hundreds of embryos, which develop rapidly and are easy to culture.

Despite this background of study, there is still much about early fish development that remains to be studied or reexamined using modern techniques. For instance, until just a few years ago, an active controversy existed as to whether mesodermal precursors in fish embryos involuted during gastrulation or delaminated from the ectoderm to form the separate germ layers. This particular issue has since been resolved by the use of video time lapse and fluorescent lineage tracer techniques, which showed that embryonic fish cells do, in fact, involute during gastrulation (Wood and Timmermans, 1988; Warga and Kimmel, 1990). However, such confusions may be understandable when one considers that the superclass of *Pisces* is not only the oldest but also, by far, the most diverse vertebrate group. This diversity is partly reflected in the different organizations of various fish embryos ranging from those of the sturgeon, which are opaque and form a 'grey crescent' much like amphibian embryos (Devillers, 1961; Clavert, 1962; Detlaf and Ginsberg, 1954), to those of teleosts, such as the zebrafish, which have evolved unique structures such as the 'yolk syncytial layer' (YSL) (Long, 1983).

However, minor differences aside, gastrulation in fish embryos follows the basic vertebrate pattern. In this chapter, I will outline the cell movements that occur during gastrulation in the zebrafish and how the detailed knowledge of these cell movements has led to the elucidation of a fate map for zebrafish development. I will also describe experiments designed to test for the commitment of cell fate in the zebrafish embryo and how the isolation and characterization of interesting early mutations have given us insights into the conferral of cell identity in the vertebrate embryo.

Gastrulation movements in the zebrafish

The development of the zebrafish for the first embryonic day is shown in Fig. 1. In this section, the cellular movements that occur during gastrulation, namely epiboly, involution, convergence and extension, will be briefly described. One of the earliest movements in the zebrafish is epiboly, which begins at 4h (where h=hours of development at 28.5°C). At this time, the zebrafish embryo has already undergone its 'mid-blastula transition' (Newport and Kirshner, 1982a,b; Kane, 1992) an hour previous and has formed the YSL or yolk syncytial layer of nuclei within the yolk cell underlying the blastoderm (Long, 1980; Kimmel and Law, 1985). Epiboly starts as the blastoderm begins to flatten and expand. This flattening is partly driven by the directed radial intercalations of deeper lying cells into more superficial positions within the blastoderm (Keller et al.,

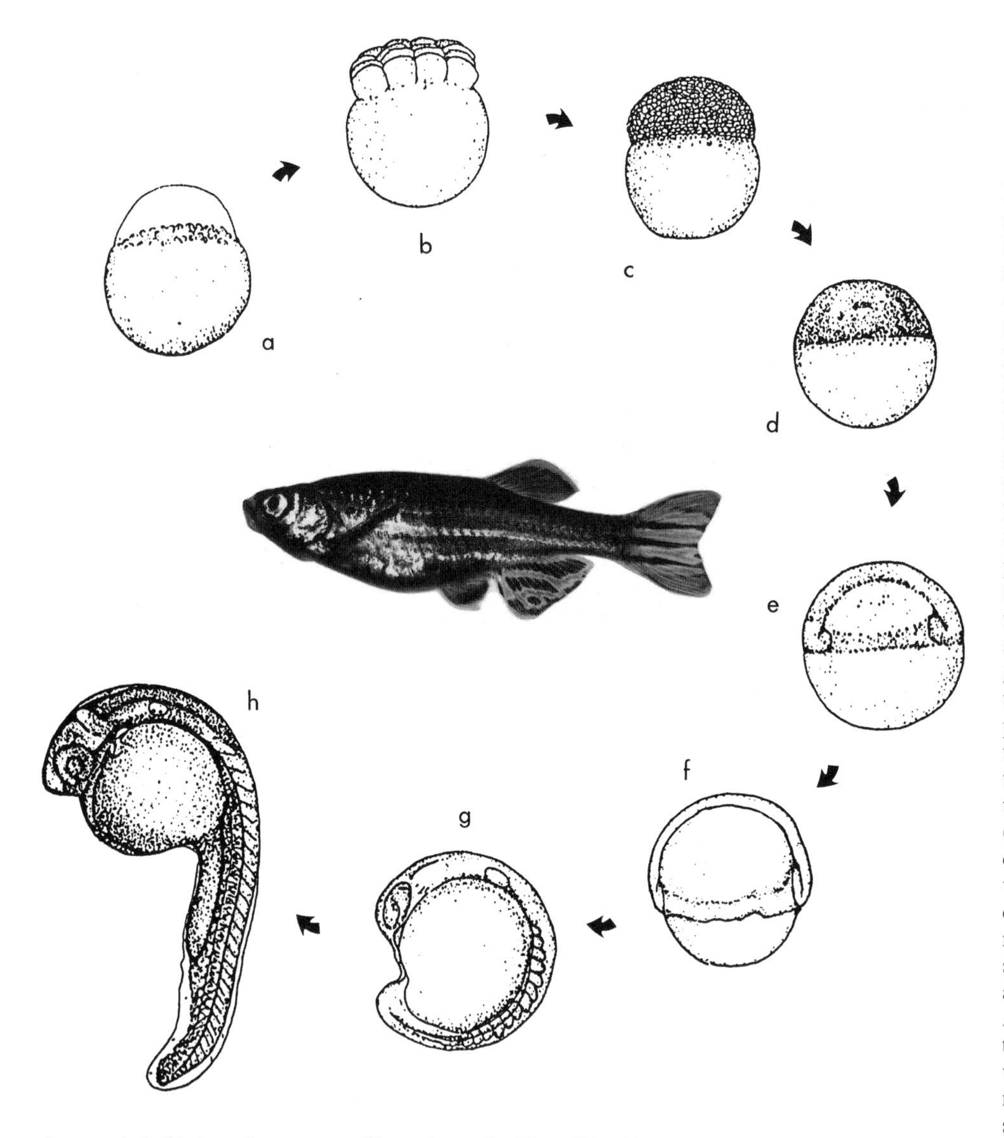

Fig. 1. Life history of the zebrafish. (A) Zygotic stage, (0-0.7h where h=hours of development at 28.5°C) one-celled embryo. In this and all following drawings, the animal pole is towards the top of the page. (B) Cleavage period (1.5h). (C) 1000-cell stage (3h). Beginning of the midblastula transition and the formation of the YSL from marginal blastomeres. (D) Sphere stage (4h). Late blastula period just prior to the onset of epiboly. (E) Shield stage (6h). Cells around the margin have first involuted around the margin to form the germ ring and then converged towards the dorsal side to produce the embryonic shield. In this and all following drawings, the dorsal side of the embryo is to the right. (F) 70%-epiboly stage (7.5h). Epiboly movements continue to draw the blastoderm vegetally around the yolk cell and convergent-extension movements rearrange cells to elongate the shield along the anterior-posterior axis. (G) 14-somite stage (16h). At this time, the optic vesicle and the otic placode are clearly visible. A differentiated notochord is present and the somitic mesoderm organizes into characteristic V-shaped segments of lateral muscle. The tail bud begins to elongate and extend away from the yolk cell. (H) 24h stage. The body plan of the fish is relatively complete and the major organs are clearly visible. Hatching occurs at 48h. In all drawings, the diameter of the yolk cell is approximately 500 μm. The adult zebrafish is approximately 4 cm in length.

1989; Warga and Kimmel, 1990). These cell rearrangements, along with cell shape changes occurring in the yolk cell, cause the blastoderm to thin and spread around the yolk cell in a vegetal direction. The actin-based actions of the YSL also contribute in a major way to this movement as the YSL constricts and 'drags' the margin of the blastoderm down to the vegetal pole (Trinkaus, 1951, 1984). These epiboly movements continue until the end of gastrulation when the blastoderm has entirely engulfed the yolk cell.

As epiboly movements bring the margin of the blastoderm to the equator of the yolk cell, the morphogenetic movements of gastrulation begin. Gastrulation in the zebrafish occurs as the vegetal deep layer cells all around the margin of the blastoderm involute underneath the margin towards the underlying yolk cell surface. After cells involute, they migrate away from the margin region using either the surface of the yolk cell or the overlying cells as substrata. This concerted involution movement forms a bilayered 'germ ring' of an inner 'hypoblast' layer and an overlying 'epiblast' layer of cells.

While epiboly and involution are occurring, both involuting and non-involuting cells of the blastoderm 'converge' to the dorsal side of the embryo to form the embryonic shield. As cells enter the shield, they also intercalate between other cells leading to the lengthening or 'extension' of the embryonic axis in the anteroposterior direction. The convergent extension movements are similar to the processes that have been described in the *Xenopus* embryo, (Keller and Danilchik, 1988; Keller and Tibbets, 1989; Keller et al., 1989) and, as in *Xenopus*, tissues from different areas of the zebrafish embryo undergo convergent extensions to varying degrees, with the dorsal axial tissues undergoing the greatest amount of lengthening.

The zebrafish fate map

Before the onset of gastrulation, the movements of individual cells are somewhat random and unpredictable. This is due, in part, to the radial intercalations of cells that occur during the early stages of epiboly and also to the cryptic orientation of the dorsoventral axis. An early zebrafish blastomere, labeled with lineage tracer dye, will contribute progeny to many diverse tissues of the embryo. Therefore the early lineages of the zebrafish are indeterminate with respect to the future fates of cells (Kimmel and Warga, 1987). For these reasons, it has not been possible to construct an accurate fate map for the zebrafish embryo before gastrulation.

As described in the section above, the movements of cells become more patterned by the onset of gastrulation. Even though the processes of epiboly, involution, convergence and extension are occurring simultaneously, these are directed, non-random morphogenetic movements making the migrations of cells at this stage somewhat more predictable. By cataloguing the positions and later fates of fluorescently labeled cells at gastrulation, Kimmel et al. (1990) elucidated the fate map for the zebrafish shown in Fig. 2. This is the first time in development, just prior to the onset of gastrulation (5.2h, approximately 8000-cell stage) that the fates of individual deep cells can be accurately predicted, (in comparison, a fate map has been reported as early as the 32-cell stage in *Xenopus*, (Dale and Slack, 1987), also the YSL and the enveloping layer cells (EVL) in the zebrafish become lineally restricted around 3.5h and 4h respectively, see below).

To summarize a few points about the zebrafish fate map:

(1) Though not shown in Fig. 2, the YSL and the EVL cells become tissue-restricted lineages before the deep cells. By labeling with lineage tracing molecules (Kimmel and Law, 1985) and observing similarities in cell cycle lengths (Kane, 1992), it was determined that the YSL and EVL tissues became separate tissues and separate mitotic domains during the blastula stage. Both of these lineages are characterized as 'extraembryonic'.

(2) The overall organization of the zebrafish fate map, in terms of tissue-specific fates, topologically resembles fate maps devised in other chordate embryos, such as ascidians and amphibians (Nishida and Satoh, 1983, 1985; Nishida, 1987; Dale and Slack, 1987). For example, ectodermal precursors are located in the more animal pole regions of the blastoderm, whereas mesodermal and endodermal precursors are located in the more vegetal regions of the blastoderm around the margin (Fig. 2). The similarity in the early organization of different chordates is an extremely important point, as it represents the reference point for comparative embryological, molecular and genetic studies amongst vertebrate embryos.

(3) The organization of the zebrafish fate map is best understood in terms of the morphogenetic movements that occur during gastrulation. The position of a cell prior to gastrulation is not only the best predictor of future cell identity, but also indicative of the types of morphogenetic movements that an individual cell is likely to experience during gastrulation. For instance, only deep cells near the margin involute during gastrulation to form the 'hypoblast' germ layer, which gives rise to the mesoderm and endoderm, whereas the non-involuting deep cells form the outer 'epiblast' layer, which gives rise exclusively to ectodermal derivatives.

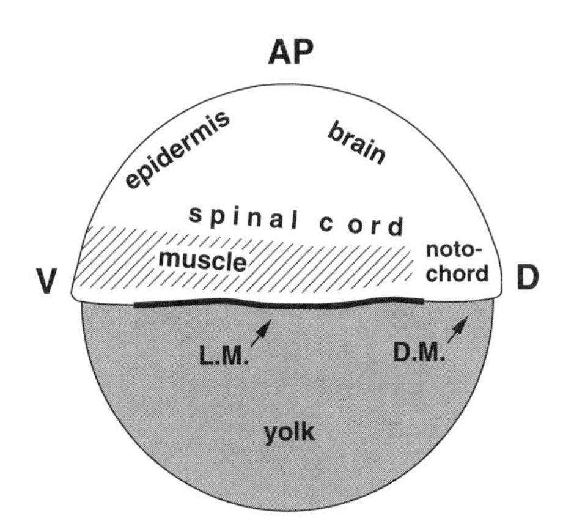

Fig. 2. Fate map of the zebrafish embryo at 50% epiboly, the onset of gastrulation (5.2h). At this stage the blastoderm, in the form of a cup inverted atop the single large yolk cell, contains approximately 8000 cells. The slashed region delineates the mesodermal precursors, which are in the lateral marginal zone, that are affected by the *spadetail* mutation (see text for details). In wild-type embryos, these precursors involute at the lateral margin and converge dorsally to form the paraxial mesoderm, which generates segmental somites. The dorsal marginal zone involutes to form the axial mesoderm, primarily notochord. The non-involuting, animal pole region cells give rise exclusively to ectodermal derivatives. *Abbreviations: D,* dorsal; *V,* ventral; *AP,* animal pole; *LM,* lateral margin; *DM,* dorsal margin.

Cell fate commitment

In many organisms, the earliest precursor cells are thought to be pluripotent, that is, developmentally undefined and capable of expressing a large number of possible phenotypes. This is likely the case for early zebrafish cells, especially considering that before gastrulation the movements of cells appear random and the lineages are indeterminant (Kimmel and Warga, 1987). Eventually, however, individual cells do come to express separate and specific phenotypes. As already described, the elucidation of a fate map for the zebrafish allows one to make an accurate prediction about the future phenotype of an individual cell. However, we have to ask if this lineage restriction is the result of some fundamental change in the developmental program of that particular cell. In other words, simply because a cell during normal development always gives rise to a particular phenotype, does it necessarily follow that the cell is actually restricted to expressing only that phenotype? In fact, it does not. It is possible that, if placed into a different environment, the cell in question could choose to follow a new developmental pathway and express a different fate, where fate is defined as the definitive set of phenotypic characteristics that a cell will eventually come to express.

One of the major problems of developmental biology is knowing when and how an initially pluripotent cell becomes "committed" to expressing a particular fate. Here commitment is defined as the irreversible, autonomous and heritable restriction in the potential of a cell such that it expresses only one fate (or possible set of fates, Stent 1985). Commitment is a fundamental and important concept of developmental biology. It is, however, a concept that is often misused or misunderstood. The state of a cell's commitment to a particular fate can never be inferred through observation alone but requires the experimental manipulation of the embryo under rigorously defined conditions. By far, the most elegant *in vivo* method to study commitment events is the transplantation of a single cell from one area of the embryo into another area (Heasman et al., 1984; Technau and Campos-Ortega 1986a,b). The purpose of these manipulations is to see if the transplanted cell expresses a fate appropriate for its new position, in which case it was still pluripotent and obviously not yet committed, or if it retains the fate of its old position, in which case it was committed to expressing its original fate at the time of the transplantation.

The zebrafish embryo is particularly amenable to this technique as the embryos are easy to manipulate and optically clear. The transplantations of cells between embryos has become a routine procedure in the zebrafish (Eisen, 1990; Ho and Kane, 1990; Hatta et al., 1991). The next two sections describe experiments utilizing single cell transplant techniques to test for commitment in the zebrafish embryo.

Commitment to an extraembryonic fate

The enveloping layer, or EVL, is the second tissue-restricted lineage to become separate in the zebrafish (YSL being the first). The EVL cells have features distinct from the deep cells, such as a flattened epithelial appearance, shared tight junctions (Betchaku and Trinkaus, 1978) and a slower cell cycle (Kane, 1992). Also, whereas the deep cells give rise to future embryonic tissues, the EVL cells develop into an extraembryonic 'periderm', which has been reported to be sloughed off in other fish embryos (Bouvet, 1976). The EVL layer is first formed at the 64-cell stage by the most superficial cells of the blastoderm which surround the internal deep cells. The deep cells always remain interior to the EVL cells and do not contribute progeny to the EVL-derived epithelium. However, EVL cells can contribute progeny to the deep cell layer; in the early blastula, an EVL cell either divides in the plane of the EVL epithelium to form a pair of new EVL cells or it divides perpendicularly to the plane of the epithelium to form one EVL cell and one deep cell.

By the late blastula stage, EVL precursors cease to generate deep cells and become tissue-restricted to generating only EVL cells (Kimmel et al., 1990). At this time, the EVL epithelium is an extraembryonic lineage separate from the deep cell layer tissues of the embryo. Does this lineage restriction represent the irreversible commitment of these cells to an EVL fate? This is an obvious question, as previous to this time the EVL cells were able to generate both EVL cells and deep cells. To test for commitment to the EVL fate, single labeled EVL cells were transplanted into the deep cell layers of an unlabeled host embryo and assayed according to their position and morphology within the embryo after 24h.

Results from these types of experiments are shown in Fig. 3. Fig. 3A shows the result obtained when single late blastula (4h) EVL cells are transplanted into the deep cell layer. Under these conditions the transplanted EVL cells later expressed a typical deep layer fate such as the group of spinal interneurons shown in Fig. 3A. These results showed that when the EVL layer can be first defined as a tissue restricted lineage, individual EVL cells are still pluripotent and therefore not yet committed to expressing only the EVL fate. Interestingly, EVL cells at this time do show a "community effect" (Gurdon, 1988) as a group of four transplanted EVL cells do not transfate when placed into the deep cell layer but retain their EVL phenotype (Fig. 3B). The very different results obtained when a cell is transplanted either singly or in a group, point out the importance of studying the responses of single cells in the absence of all their normal neighbors.

Eventually, EVL cells do become committed to an EVL fate as shown by the single cell transplantations of older EVL cells. EVL cells taken from an embryo at the onset of gastrulation (5.2h) did not transfate to a deep cell phenotype when placed into the deep cell layer. These older transplanted EVL cells either retained an EVL phenotype, indicating that this cell was committed to an EVL fate (Fig. 3C) or formed an isolated epithelial vesicle (Fig. 3D). The formation of an epithelial vesicle, a very abnormal phenotype, is not direct evidence for commitment to an EVL fate but suggests that cells have undergone a restriction in potential and have lost the ability to express a normal deep cell fate.

These experiments show that, when the EVL cells can be first described as lineally restricted, they are uncommitted and able to assume the fate of their closest neighboring cells. However, an hour later in the late blastula, these cells lose their potential to express every type of fate. Just what types of changes occur during this process remains one of the most important and long-lived questions in developmental biology.

Margin cells are uncommitted before gastrulation

As stated previously, the fates of deep cells at the blastula stage cannot be accurately predicted. However, just prior to the onset of gastrulation (5.2h, 50% epiboly), individual deep cells will give rise to progeny that are tissue restricted. For example, cells near the margin have been fate mapped to give rise to mesodermal and endodermal derivatives, and it is only these marginal cells that involute at gastrulation to form the hypoblast germ-layer which comes to lie internal to and underneath the non-involuting, ectodermal epiblast layer (Warga and Kimmel, 1990).

Regions of the embryo now also express specific gene products. One of the earliest expression patterns is exhibited by the zebrafish homologue of the murine T-gene. In the mouse, the T-gene is initially expressed in all mesodermal precursors (Herrmann et al., 1990; Wilkinson et al., 1990) and the pattern of expression in the zebrafish is very similar. By gastrulation, the zebrafish homologue of the T-

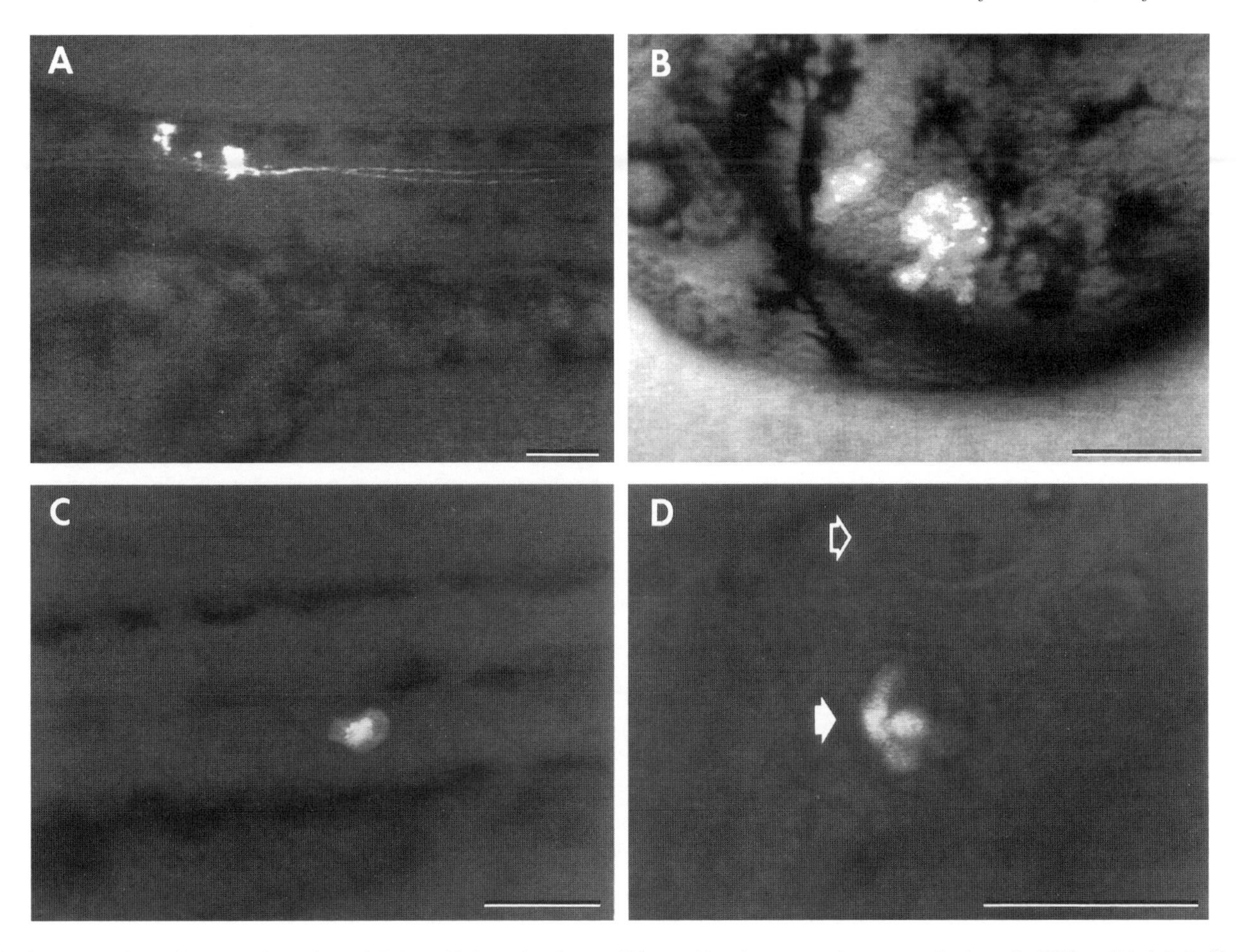

Fig. 3. Results from the transplantation of EVL cells into the deep cell layer. For these experiments, a single early EVL cell is labeled by the intracellular injection of rhodaminated-dextran (ca M_r 40,000). After one or two cell divisions, a single labeled EVL cell is gently drawn up into a tooled microcapillary and expelled directly into the deep cell layer of an unlabeled host embryo, (see Ho and Kane, 1991 for transplantation techniques). (A) A single EVL cell transplanted just after the time of tissue restriction (4h) into the deep cell layer is still pluripotent, uncommitted and able to express a deep cell fate, such as this group of spinal interneurons. (B) In this panel, four labeled EVL cells at the same age as the cell in A had been transplanted into the deep cell layer. These cells exhibited a mass or 'community' effect as they retained an EVL fate. Single EVL cells transplanted from older embryos (after 5h) into the deep cell layers either (C) retained an EVL fate, showing that at this time the transplanted cell was committed to an EVL fate, or (D) expressed an abnormal phenotype, such as this epithelial ball of cells (solid arrow) located near the ear (open arrow). Scale bars = 100µm.

gene is expressed in the marginal cells of the blastoderm (Shulte-Merker, 1992), most of which will involute at gastrulation to form the hypoblast.

So just prior to the onset of gastrulation at 50% epiboly, the margin region of the blastoderm has acquired characteristics that distinguish it from other areas of the embryo. The careful reader should be able to anticipate the next question. Does the acquisition of these traits denote the commitment of the margin region cells to a hypoblast germ-layer fate, or are these cells still developmentally pluripotent? To approach this question, single margin region cells were transplanted into the animal pole region, which gives rise exclusively to ectodermal derivatives. These tranplantations were performed between donor and host embryos at 50% epiboly, which is the first time that the zebrafish fate map can be described. The purpose of these manipulations is to see if the transplanted margin cell later expressed the fate of its new position, namely ectoderm, in which case you would conclude that the margin cell was still pluripotent and therefore not committed to a germ-layer fate. Conversely, if the cell retained its old fate though moved to a new environment, you would conclude that under these experimental conditions the transplanted cell was committed to a mesodermal germ-layer fate.

Under these conditions, most of the cells transplanted at 50% epiboly took on the fate of their new position and formed ectodermal progeny such a epidermis, neurons and retinal tissues (Fig. 4). None of the transplanted cells retained a mesodermal fate though a small percentage of cells (<10%) were characterized as mesenchymal, (i.e. residing in a mesenchymal area of the embryo and undifferentiated at 48h). Therefore, margin cells just prior to gastrulation are not yet committed to a mesodermal germ-layer fate but are pluripotent and still able to express an ectodermal phenotype.

In the preceding section, the EVL cells were shown to become committed to an extraembryonic fate but only after a period in which they were a tissue-restricted, but still

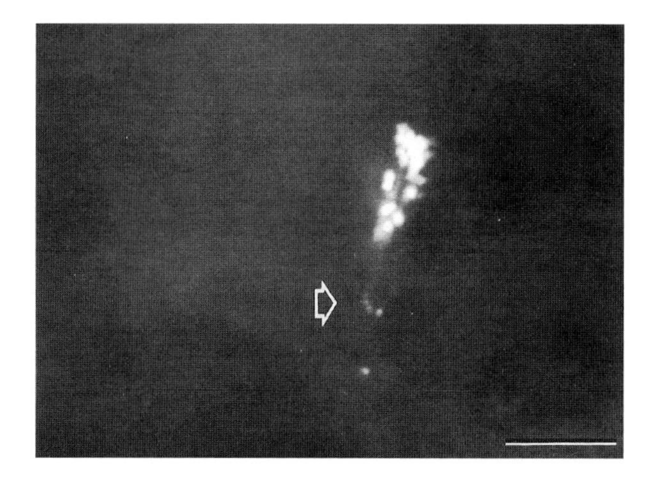

Fig. 4. Transplantation of an early margin cell into the animal pole region. Early margin cells are uncommitted to a mesodermal fate and can still express an ectodermal fate when transplanted into an ectodermal producing region. This transplanted margin cell gave rise to a group of neurons in the brain of a 48h embryo. Small arrow points to the growth cones of the labeled neurons. Scale bar = 50μm.

uncommitted, lineage. A similar phenomenon apparently occurs in the deep layer margin cells. Even though their position within the blastoderm at 50% epiboly is predicative of their future fate, the margin cells were shown to be still uncommitted to a particular phenotype. Thus, by the onset of gastrulation, different cells of the embryo are distinct with respect to their cell potential; the EVL cells are committed at this time, whereas the marginal deep cells are still pluripotent. Thus, the time at which commitment occurs does not appear to be a global feature of the embryo, but different populations may undergo this process at separate times. By analogy to the EVL cells, the deep layer cells may eventually become committed to a specific fate, but this issue remains the focus of future studies.

These experiments showed that transplanting a cell before gastrulation into a different region of the embryo caused that cell to express the phenotype of its new position. Therefore, in this vertebrate embryo, cell position plays a very important role in the determination of future cell fate. Having stressed this point, it seems likely that mechanisms should have evolved to ensure the orderly and even dispersal of cells into the various tissue anlages. As already described, the majority of these cell movements occur during gastrulation, and presumably, any defect that interferes with the normal pattern of gastrulation would have drastic effects upon cell fate and the later patterning of the embryo.

The *spadetail* mutation affects cell movements during gastrulation

One of the advantages of working with the zebrafish is its amenability to genetic analyses (Streisinger, 1981; Walker and Streisinger, 1983) and screens for zygotic lethal mutations that affect the patterning of the embryo are underway. The zygotic lethal *spadetail* mutation is the first identified

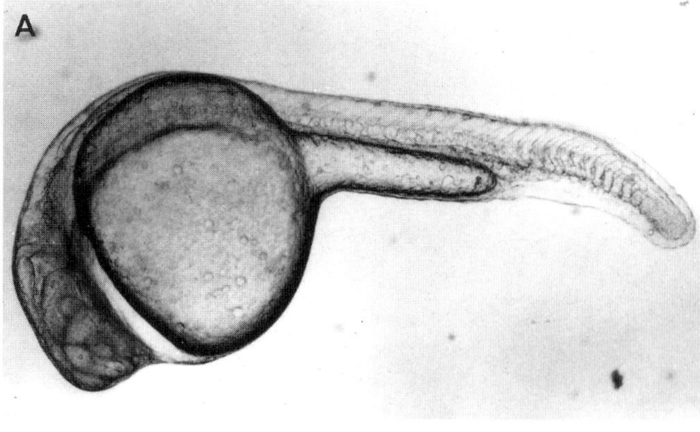

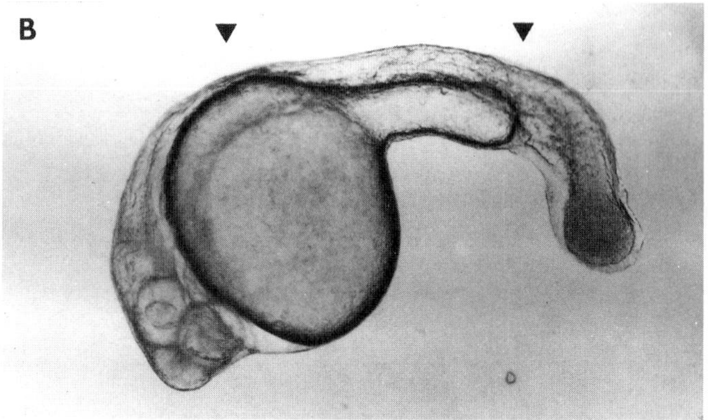

Fig. 5. Comparison of (A) wild-type and (B) *spadetail* zebrafish embryos at 24h. The trunk area of the mutant, delimited by the small arrowheads, is grossly deficient in muscle cells and lacks the segmental chevron-shaped pattern of organized myotomes. Also, note the characteristically bent notochord and excess of cells at the end of the tail. Wild-type embryo is approximately 1.5 mm in length.

zebrafish mutant that affects the movements of cells at gastrulation (Kimmel et al., 1989). At 24h, *spadetail* mutants lack sufficient mesoderm to make somitic segments in the trunk region and also have a very characteristic bulge of excess cells at the end of the tail (Fig. 5).

In wild-type embryos, the precursors of trunk somitic tissue are located within the lateral margin of the blastoderm. Lateral margin precursors normally involute during gastrulation and then converge towards the dorsal side of the embryo. In the *spadetail* mutant, labeled lateral margin cells involute but then fail to converge properly and instead move abnormally towards the vegetal pole which becomes the tail bud. These labeled cells later gave rise to a variety of cell fates including mesenchyme, notochord and muscle cells within the tail; however, the majority of the cells at the end of the tail die after formation of the terminal body segment.

The *spadetail* mutation appears to affect a very specific morphogenetic movement, namely, the convergence of cells into the dorsal axis during gastrulation. To confirm this hypothesis, Don Kane and I used cell transplantation techniques to create genetic mosaics in which we could study the differences between wild-type and mutant cells (Ho and Kane, 1990). We were interested to know which cells of the embryo were being affected by the *spadetail* mutation

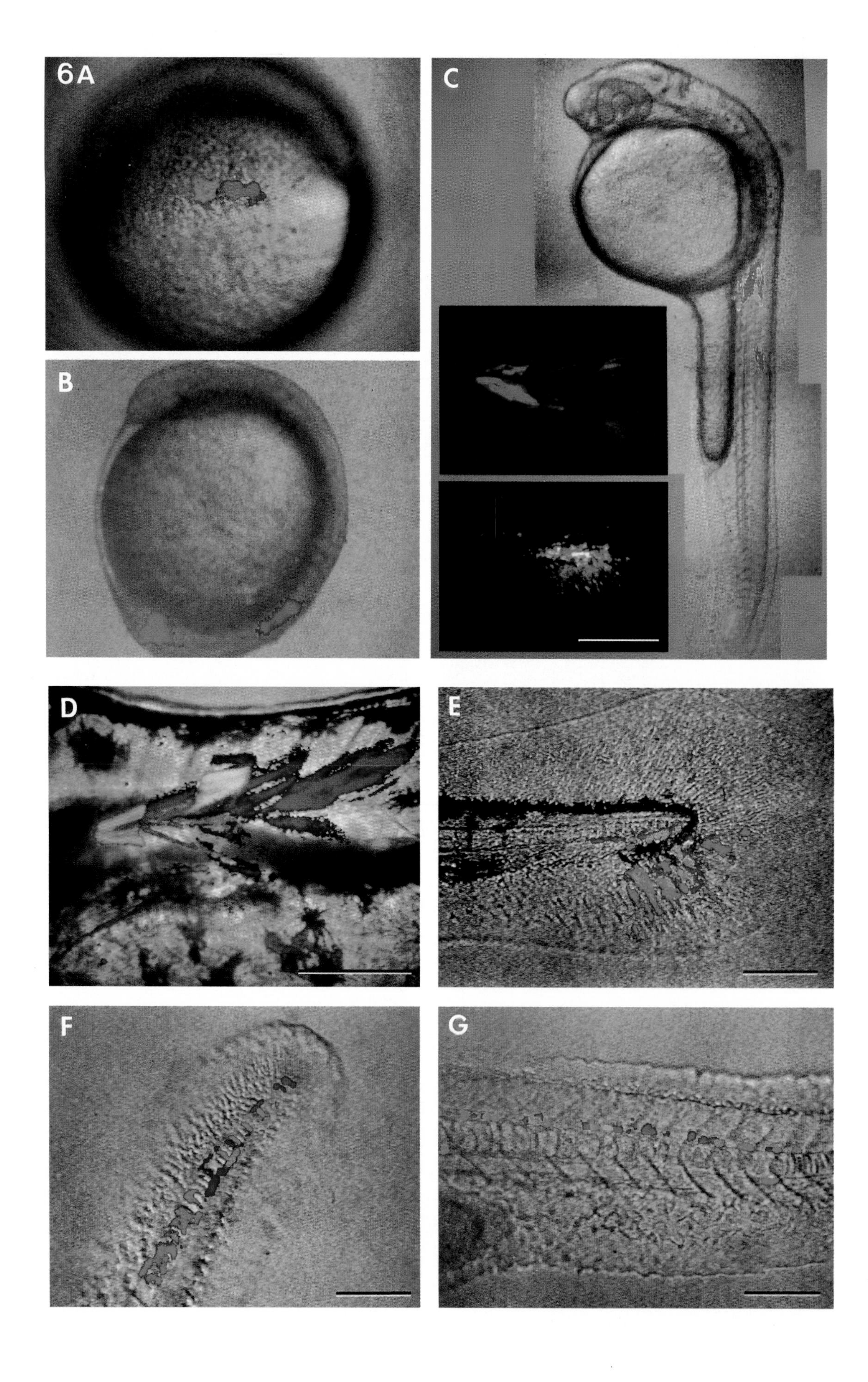
6A
B
C
D
E
F
G

Fig. 6. The *spt-1* mutation specifically and cell-autonomously affects the movements of lateral margin cells. (A) Zebrafish embryo just prior to gastrulation (5h) and immediately following cell transplantation. Two groups of cells from two different donor embryos were transplanted into the lateral marginal zone of a wild-type host embryo (n=11). The green labeled cells came from an embryo mutant for *spadetail* and the red labeled cells came from a wild-type donor. (B) Same embryo after gastrulation (12h). The two groups of cells segregated into different areas of the animal; the red labeled wild-type cells converged to the dorsal axis, whereas the mutant cells moved to the tail bud region. Diameter of the embryo shown in A and B is 500 μm. (C) Same embryo at 30h. The wild-type cells, in the trunk, were anterior to the *spadetail* cells. Wild-type and *spadetail* donor cells showed the same general pattern in both wild-type (n=11) and *spadetail* hosts (n=3). *Spadetail* cells were never observed to be positioned more anteriorly than the wild-type cells (Fisher Sign. Test, P<0.001). Length of the embryo in C is about 1.5 mm. Orientation for A, B and C is animal pole (or anterior) towards the top of the page and dorsal to the right. (C, *upper inset*) Transplanted, red labeled cells of the same embryo at 48h. The transplanted wild-type cells gave rise to striated myotomal muscle cells. (C, *lower inset*) The tail region of the same embryo at 48h showing that the transplanted, green labeled *spadetail* cells gave rise to the fin rays and other associated mesenchymal derivatives. Orientation for the two insets is anterior to the left and dorsal towards the top of the page. Scale bar=100 μm. (D) Co-transplanted wild-type cells did not segregate. When cells from two wild-type embryos were transplanted into the lateral marginal zone of a wild-type host (n=17), the two groups of donor cells migrated together and the intermingled progeny cells differentiated into trunk myotomal muscle by 48h. (E) Co-transplanted mutant cells did not segregate. When both groups of donor cells were of the mutant *spadetail* genotype and transplanted to the lateral marginal zone of a wild-type host (n=3), both groups of cells took up positions within the tail where they formed predominately tail mesenchyme by 48h. (F) The *spadetail* mutation did not affect the movements of precursors at the dorsal margin. Green labeled mutant cells and red labeled wild-type cells did not segregate apart when placed within the dorsal marginal zone (n=3). By 30h both groups of cells gave rise to intermingled notochord cells along the length of the axis, including the tail region as shown here in a wild-type host. (G) The *spadetail* mutation did not affect the movements of ectodermal precursors. Mutant cells and wild-type cells migrated together and formed ectodermal derivatives when placed in the non-marginal zone of a wild-type host (n=25). This figure shows the intermingling of cells in the spinal cord at 24h. Orientation in D, E, F and G is anterior to the left and dorsal towards the top of the page; scale bars = 100 μm. Reprinted by permission of *Nature*.

and, secondly, to describe the nature of the defect. In regards to the latter point, two possibilities presented themselves: the *spadetail* mutation could either be affecting the environmental cues that normally guide cells into their correct pathways or alternatively, the mutation could be directly affecting some function of the cells in a cell-autonomous fashion.

To answer these questions, mixtures of wild-type and mutant cells from differently labeled donor embryos were co-transplanted into the lateral margin region of an unlabeled host embryo before the onset of gastrulation (Fig. 6A). Regardless of the host embryo genotype, the wild-type cells always converged dorsally to form trunk somitic mesoderm whereas the mutant cells failed to migrate with the wild-type cells and instead ended up in the tail region where they gave rise to tail mesenchyme (Fig. 6B,C).

Such extensive separations of neighboring cells do not occur during normal development. In control animals, in which both groups of transplanted cells were of the same genotype (i.e. either wild-type & wild-type or *spadetail* & *spadetail*) both groups of cells migrated together and formed the same derivatives located in the same area of the 24h embryo (Fig. 6D,E). Therefore, our results shown in Fig. 6A-C are due to differences in the behaviors of mutant and wild-type cells. Also, because the mutant phenotype of the *spadetail* cells was not rescued by being transplanted into a wild-type environment, we concluded that the *spadetail* mutation was acting autonomously in the lateral margin precursors.

Furthermore, we were able to delineate the functional boundaries of the *spadetail* gene action. By transplanting cells into various areas of the zebrafish fate map, we determined that the large-scale separation of wild-type and *spadetail* cells only occurred if the transplantation site was the lateral margin region of the blastoderm; *spadetail* cells placed into other areas of the blastoderm always migrated with the co-transplanted wild-type cells and formed the same types of derivatives. Fig. 6F,G shows the results obtained when wild-type and *spadetail* cells were transplanted into the fate map regions that give rise to notochord and spinal cord, respectively. In these host animals, the *spadetail* cells migrated with the co-transplanted wild-type cells and formed the same types of derivatives within the same position.

The functional boundaries of the *spadetail* gene action, as assayed by cell transplantations, (Fig. 2) correlate well with our knowledge of cell fate and cell movements in the embryo at gastrulation. These findings have revealed a surprisingly delicate genetic control of vertebrate gastrulation, as the *spadetail* gene appears to be necessary for the migrations of only those cells located in the lateral margin of the blastoderm, i.e. those cells that both involute and converge during gastrulation to form trunk somitic mesoderm. The *spadetail* gene function does not appear to be necessary in other precursor cells and one important prediction is that these other cells may be using different mechanisms, other than the *spadetail* gene, to migrate and differentiate. The isolation of this early acting, tissue-specific mutation has alerted us to the possible existence of a system of migration-specific gene functions. The study and characterization of the *spadetail* mutation has also opened up a path for a molecular analysis of early cell movements during vertebrate gastrulation, which is a topic about which we still know very little.

Concluding remarks

The morphogenetic movements that occur during zebrafish gastrulation appear similar to the cellular movements described in other vertebrates, especially amphibians. Because fish embryos are optically transparent, one can observe the movements of even the very deepest cells within the intact embryo. The detailed knowledge of these cell movements, coupled with the ability to label early blastomeres with lineage tracing molecules, has led to the elucidation of a detailed fate map for the zebrafish just prior to gastrulation.

The organization of this fate map emphasizes the important role that cell position plays in the determination of cell fate. By 4h, cells around the outside of the blastoderm become lineage restricted to forming EVL cells, whereas the deep layer cells can be fate mapped to germ-layer and tissue-specific fates by 5.2h. However, at the times when the EVL and the deep layer cells can be first described as separate, tissue restricted lineages, they are apparently not yet committed to expressing specific fates but are still pluripotent. Experiments, in which an early single cell is transplanted from one area of the blastoderm into another area, show that changing the position of a cell also changes the identity of that cell. Early EVL cells are able to express a deep cell phenotype if transplanted into the deep cell layers, and early marginal deep cells, fate mapped at the beginning of gastrulation to give rise to mesodermal precursors, can form ectodermal progeny if transplanted into an ectodermal producing region of the blastoderm. Presumably, cells of the embryo eventually become irreversibly restricted to expressing only a single fate, and it was experimentally shown that the EVL cells do become committed to forming only an extraembryonic EVL fate, though somewhat after the time when they can be first fate-mapped as a separate lineage.

The different fate map territories of the blastoderm also represent areas of region-specific morphogenetic movements that will occur during gastrulation. For instance, lateral margin cells, which are shown in Fig. 2 to be muscle precursors, have been shown to first involute and then converge dorsally during gastrulation. This stereotyped pattern of movements places these particular cells into the lateral trunk region of the embryonic axis where they contribute to somitic mesoderm. Cell labeling and cell transplantation experiments have shown that the *spadetail* mutation exclusively and autonomously affects the convergence of these lateral margin precursors, causing them to move abnormally towards the vegetal pole. Consequently, mutant embryos have an excess of precursors in the tail and are very deficient in trunk somitic muscle cells.

Although the nature of the *spadetail* gene function is presently unknown, there are at least two related explanations for why mutant cells move incorrectly. (1) The *spadetail* mutation may have directly changed the identity of the lateral margin cells into tail bud precursors and as a consequence these cells migrated to the vegetal pole, or (2) The *spadetail* mutation may have directly affected some aspect of the lateral margin cells' ability to migrate correctly and, as a consequence, the misdirected cells expressed a fate appropriate for their new position. The first possibility seems somewhat less likely, in light of the single cell transplantation experiments described earlier in this paper. Margin cells were shown to be uncommitted to any particular fate at the beginning of gastrulation and the *spadetail* gene appears to first exert its effects at this same time. Also, control experiments in which *spadetail* cells were transplanted into various non-marginal areas of the blastoderm showed that mutant cells were capable of expressing a variety of cell fates, including muscle cells in non-trunk regions. Thus, it does not seem likely that the *spadetail* mutation is directly causing cells to become committed to a specific fate; however, the action of the *spadetail* gene appears to be a very important step in the process of assigning identities to the lateral margin precursors. As described in the preceding section, the *spadetail* mutation interferes with the migrations of these cells, but a description of the actual cellular defect awaits a molecular characterization of the *spadetail* mutant. The study of the actions of the *spadetail* gene has emphasized the importance of cell position upon cell fate in the vertebrate embryo, which has formed the theme of this paper. Changing the position of early cells, either by physical transplantations or through genetic means, leads to a change in their expressed cell fate.

I would like to thank Drs Marnie Halpern and Charles Kimmel, in whose laboratory these experiments were performed, for comments on this manuscript. Parts of this work were supported by the Helen Hay Whitney Foundation and grant numbers BNS 9009544 and HD224860-06.

References

Betchaku, T. and Trinkaus, J. P., (1978). Contact relations, surface activity, and cortical microfilament of marginal cells of the enveloping later and of the yolk syncytial and yolk cytoplasmic layer of *Fundulus* before and during epiboly. *J. Exp. Zool.* **206**, 381-426.

Bouvet, J. (1976). Enveloping layer and periderm of the trout embryo. *Cell Tiss. Res.* **170**, 367-382

Clapp, C. (1891). Some points in the development of the toad-fish (*Batrachus tau*). *J. Morph.* **5**, 494-501

Clavert, J. (1962). Symmetrization of the egg of vertebrates. *Adv. Morph.* **2**, 27-60

Dale, L. and Slack, J. M. W. (1987). Fate map for the 32-cell stage of *Xenopus laevis*. *Development* **99**, 527-551.

Detlaf, T. A. and Ginsberg, A. (1954). Développement embryonnaire de l'oeuf des acipenserid's et questions concernant leur élevage. *Acad. Sci. U.R.S.S.* **213**

Devillers, C. (1961). Structural and dynamic aspects of the development of the teleostean egg. *Advances in Morphogenesis* **1**,379-429.

Eisen, J. S. (1990). Determination of primary motoneuron identity in developing zebrafish embryos. *Science* **252**,569-572

Gurdon, J. B. (1988). A community effect in animal development. *Nature* **336**,772-774

Hatta, K., Kimmel, C. B., Ho, R. K. and Walker, C. (1991). The *cyclops* mutation blocks specification of the floor plate of the zebrafish central nervous system. *Nature* **350**, 339-341

Heasman, J., Wylie, C. C., Hausen, P. and Smith, J. C. (1984). Fates and states of determination of single vegetal pole blastomeres. *Cell* **37**,185-194

Herrmann, B. G., Labeit, S., Poustka, A., King, T. R. and Lehrach, H. (1990). Cloning of the t-gene required in mesoderm formation in the mouse. *Nature* **343**, 617-622

Ho, R. K. and Kane, D. A. (1990). Cell-autonomous action of zebrafish *spt-1* mutation in specific mesodermal precursors. *Nature* **348**, 728-730.

Kane, D. A. (1992). Zebrafish mid-blastula transition, the onset of zygotic control during development, Ph.D. Thesis, Univ. of Oregon

Keller, R. E. and Danilchik, M. (1988). Regional expression, pattern and timing of convergence and extention during gastrulation of *Xenopus laevis*. *Development* **103**, 193-209.

Keller, R. E. and Tibbets, P. (1989). Mediolateral cell intercalation in the dorsal axial mesoderm of *Xenopus laevis*. *Dev. Biol.* **131**, 539-549

Keller, R. E., Cooper, M. S., Danilchik, M., Tibbetts, P. and Wilson, P. A. (1989). Cell intercalations during notochord development in *Xenopus laevis*. *J. Exp. Zool.* **251**, 134-154

Kimmel, C. B. and Warga, R. (1987). Indeterminate cell lineage of the zebrafish embryo. *Dev. Biol.* **124**, 269-280.

Kimmel, C. B. and Law, R. D. (1985). Cell lineage of zebrafish blastomeres: ii. Formation of the yolk syncytial layer. *Dev. Biol.* **108**, 86-93.

Kimmel, C. B. and Warga, R. M. (1986). Tissue-specific cell lineages originate in the gastrula of the zebrafish. *Science* **231**, 365-368.

Kimmel, C. B., Kane, D. A., Walker, C., Warga, R. W. and Rothman, M. B. (1989). A mutation that changes cell movement and cell fate in the zebrafish embryo. *Nature* **337**, 358-362.

Kimmel, C. B., Warga, R. M. and Schilling, T. F. (1990). Origin and organization of the zebrafish fate map. *Development* **108**, 581-594.

Long, W. L. (1980). Analysis of yolk syncytium behavior in *Salmo* and *Catostomus*. *J. Exp. Zool.* **214**, 323-331.

Long, W. L. (1983). The role of the yolk syncytial layer in determination of the plane of bilateral symmetry in the rainbow trout, *Salmo gairdneri Richardson*. *J. Exp. Zool.* **228**, 91-97.

Long, W. L. (1984). Cell movements in teleost fish development. *Bioscience* **34**, 84-88

Morgan, T. H. (1893). Experimental studies on the teleost eggs. *Anat. Anz.* **8**, 803-814

Newport, J. and Kirschner, M. (1982a). A major developmental transition in early *Xenopus* embryos: I. Characterization and timing of cellular changes at the midblastula stage. *Cell* **30**, 675-686.

Newport, J. and Kirschner, M. (1982b). A major developmental transition in early *Xenopus* embryos: II. Control of the onset of transcription. *Cell* **30**, 687-696.

Nishida, H. and Satoh, N. (1983). Cell lineage analysis in ascidian embryos by intracellular injection of a tracer enzyme. I. Up to the eight-cell stage. *Dev. Biol.* **99**, 382-394

Nishida, H. and Satoh, N. (1985). Cell lineage analysis in ascidian embryos by intracellular injection of a tracer enzyme. II. The 16- and 32 cell-stage. *Dev. Biol.* **110**, 440-454

Nishida, H. (1987). Cell Lineage Analysis in Ascidian Embryos by Intracellular Injection of a Tracer Enzyme. III. Up to the Tissue Restricted Stage. *Dev. Biol.* **121**, 526-541

Schulte-Merker, S., van Eeden, F., Halpern, M., Kimmel, C.B. and Nüsslein-Volhard, C. (1992). The T (*Brachyury*) Gene: Its Role in Vertebrate Embryogenesis. *Abs. Brit. Soc. Dev. Biol.* 75

Stent, G. (1985). The role of cell lineage in development. *Phil. Trans. R. Soc. Lond.* **312**, 3-19

Streisinger, G. W. (1981). Production of clones of homozygous diploid Zebrafish, *Brachydanio rerio*. *Nature* **291**, 293-296.

Technau, G. M. and Campos-Ortega, J. A. (1986a). Lineage analysis of transplanted individual cells in embryos of *Drosophila melanogaster*. II. Comitment and proliferative capabilities of neural and epidermal cell progenitors. *Roux's Arch. Dev. Biol.* **195**,445-454

Technau, G. M. and Campos-Ortega, J. A. (1986b). Lineage analysis of transplanted individual cells in embryos of *Drosophila melanogaster*. III. Commitment and proliferative capabilities of pole cells and midgut progenitors. *Roux's Arch. Dev. Biol.* **195**,489-498

Trinkaus, J. P. (1951). A study of the mechanism of epiboly in the egg of *Fundulus heteroclitus*. *J. Exp. Zool.* **118**,269-320.

Trinkaus, J. P. (1984). Mechanism of *Fundulus* epiboly-a current view. *Am. Zool.* **24**, 673-688.

Walker, C. and Streisinger, G. (1983). Induction of mutations by gamma-rays in pregonial germ cells of zebrafish embryos. *Genetics* **103**, 125-136.

Warga, R. M. and Kimmel, C. B. (1990). Cell movements during epiboly and gastrulation in zebrafish. *Development* **108**, 569-580.

Wilkinson, D. G., Bhatt, S. and Herrmann B. G. (1990). Expression pattern of the mouse t-gene and its role in mesoderm formation. *Nature* **343**, 657-659

Wilson, H. V. (1889). The embryology of the sea bass. *Bull. US Fish Comm.* **9**,209-278

Wood, A. and Timmermans, L. P. M. (1988). Teleost epiboly: a reassessment of deep cell movement in the germ ring. *Development* **102**, 575-585.

Development 1992 Supplement, 75-80 (1992)
Printed in Great Britain

The midblastula transition, the YSL transition and the onset of gastrulation in *Fundulus*

J. P. TRINKAUS

Department of Biology, Yale University, New Haven, Connecticut 06511, USA

Summary

The first signs of cell motility appear in *Fundulus* toward the end of cleavage, after cleavages 11 and 12. When blastomeres cease cleaving, their surfaces undulate and form blebs. At first, these blebbing cells remain in place. Gradually thereafter they begin movement, with blebs and filolamellipodia serving as organs of locomotion. Non-motile cleaving blastomeres have thus differentiated into motile blastula cells. This transformation corresponds to the midblastula transition of amphibian embryos.

Gastrulation in *Fundulus* begins with vegetalward contraction of the external yolk syncytial layer. This causes narrowing of the E-YSL and initiates the epibolic expansion of the blastoderm. Convergent movements of deep cells within the blastoderm begin toward the end of this contraction. The YSL forms as a result of invasion of the yolk cell cytoplasm by nuclei from open marginal blastomeres during cleavage. These YSL nuclei then undergo five metachronous divisions. After this, they divide no more. YSL contraction begins approximately 1.5 hours after cessation of these divisions (21-22°C). This cessation of nuclear divisions is preceded by a gradual decrease in rate. (1) The duration of each succeeding mitosis increases steadily and often some nuclei do not divide at mitosis V. (2) The duration of interphases between succeeding mitoses also increases, but to a much greater degree, and the longest interphase by far is the last one, I-IV, between mitoses IV and V. (3) The mitotic waves responsible for mitosis V move much more slowly than those for the first four mitoses and invariably decelerate. This gradual cessation of YSL nuclear divisions clearly sets the stage for the contraction of YSL cytoplasm and thus the beginning of gastrulation. We call this the YSL transition. It is not to be confused with the midblastula transition, which occurs 3-4 hours earlier. The MBT commences cytodifferentiation; the YSL transition commences morphogenesis.

Key words: *Fundulus,* midblastula transition, yolk syncytial layer, YSL transition, gastrulation.

Introduction

Gastrulation in teleost fishes has two major aspects: movement of so-called deep cells within the blastoderm to congregate and form the embryo and, concomitantly, spectacular epiboly of the blastoderm and its underlying yolk syncytial layer to encompass eventually a large sphere of viscous, fluid yolk. Although these processes take place together, they are conveniently separable for analysis. Deep cells are confined to the space between the monolayered cell surface layer of the blastoderm, the enveloping layer (EVL), and the underlying yolk syncytial layer (YSL), which separates them from the yolk. Their movements are coordinated with the epiboly of the EVL and YSL, but do not contribute to its mechanism (Trinkaus, 1984b). Epiboly proceeds independently of the activities of the deep cells and depends on expansion of the EVL, which in turn depends on expansion of the YSL, with the cooperation of the diminishing yolk cytoplasmic layer (YCL) (Trinkaus, 1984a,b; Betchaku and Trinkaus, 1986).

A fascinating feature of the gastrulation movements of deep cells and the vast epibolic expansion of the EVL and YSL is that they both seem to proceed at about the same rate in different teleost species, willy-nilly, regardless of the size of the egg. In small eggs, like those of the zebrafish or *Serranus* (Wilson, 1889), epiboly is quickly finished and, at closure of the yolk plug, gastrulation of the deep cells is far from completion. In large eggs, like those of the trout (*Salmo*), epiboly is a long process and gastrulation by deep cells is complete and embryo formation is well underway long before closure of the yolk plug. In medium-size eggs, like those of *Fundulus* and *Oryzias*, gastrulation movements of deep cells are essentially complete and embryo formation has just begun at the end of epiboly. If one therefore takes into account specific differences in the size of the yolk sphere, the gastrulation movements of superficially very different teleost eggs are really quite comparable.

In this essay, I shall deal first with the differentiation of motility of the deep cells during and after the cessation of cleavage and then with the events leading to gastrulation: the onset of epiboly of the YSL and EVL and directional movements of the deep cells. The material for all of our work on these subjects has been the embryo of *Fundulus heteroclitus*. For illustrations of the cleavage, blastula and gastrula stages of *Fundulus*, see Armstrong and Child (1965). The YSL and the blastoderm in a late blastula stage are shown in Fig. 1. See also Fig. 2 in Trinkaus (1993).

The differentiation of deep cell motility

The first signs of deep cell motility appear in *Fundulus* when cleavage is nearing its end, during and after cleavages 11 and 12. Although cleavages 11 and 12 differ little if at all from previous cleavages in their duration, they do differ from all preceding cleavages in two other important respects. Cleavage 10 appears to be the last complete cleavage in *Fundulus*, i.e., all blastomeres divide. Cleavage 11 is sometimes incomplete and frequently asynchronous and cleavage 12, the last cleavage, is invariably highly incomplete and highly asynchronous and occasionally does not occur at all. The military precision of the first ten cleavages gradually degenerates to the disarray of the last. Cleavage in *Fundulus* clearly does not cease abruptly, with a bang; it seems rather to grind slowly to a halt.

As cleavage is winding down, the surfaces of many non-cleaving blastomeres begin to bleb. This motile activity, which is totally absent during the first ten cleavages, first appears on the surface of some non-cleaving deep blastomeres during cleavage 11, is frequent just after that, before and during the highly variable 12th cleavage, and seems to involve almost all deep blastomeres soon after. At this point, when speeded up with time-lapse, the blastoderm seems a seething mass of jostling cells. However, when blebbing is observed in detail, it appears gradually. The first blebs are preceded by gentle, slow undulations of the cell surface (Trinkaus, 1973, 1985). Then, some minutes later, a sector of the cell surface explodes to form a hemispheric bleb. As these blebs form, each blebbing cell remains in place. There is no cell movement. Then, gradually, one cell after another begins to move. The prior blebbing activity is a prelude to this movement (Trinkaus, 1973, 1985). With this translocation of the deep cells of the blastula, a new phase of development has begun. Non-motile cleavage blastomeres have differentiated into motile, moving post-cleavage cells. (It is important to note, incidentally, that EVL cells have not been observed to bleb.) The time of the onset of deep cell surface deformation and blebbing is significant for two reasons. (1) It occurs only after the regular, rapid cleavage phase of development has ceased, as is true of many embryos. (2) It often commences within minutes after the cessation of cleavage of a blastomere, whether at cleavage 11 or 12. There is little apparent lag. It seems as if, once a cell's motile system is no longer tied up in rapid cytokinesis, it is quickly available for motile activity, in this case for blebbing (Trinkaus, 1980). This phenomenon corresponds to the so-called midblastula transition that has been intensively studied in amphibian embryos (Signoret and Lefresne, 1971; Johnson, 1976; Gerhart, 1980; Newport and Kirschner, 1982a, b).

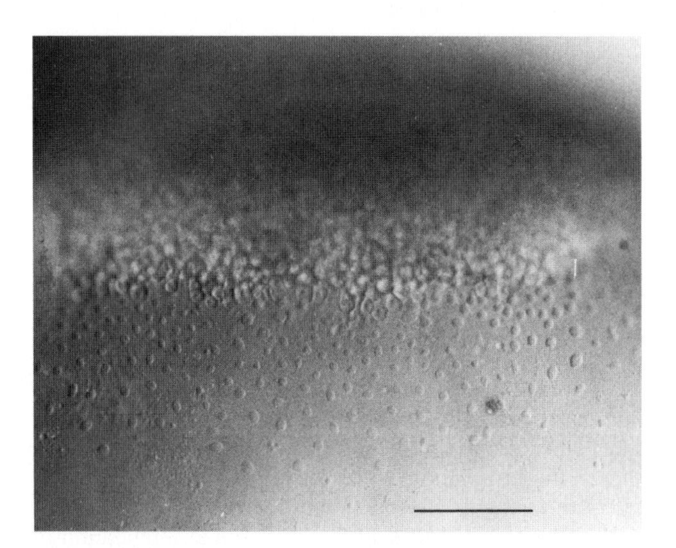

Fig. 1. A definitive yolk syncytial layer (lower part of the micrograph) after the last division of its nuclei. The blastoderm, in a late blastula stage (stage 12), occupies the top half of the micrograph, but only some of its marginal deep cells are in focus. Note the width of the YSL and the approximate spacing of its nuclei. Contraction of the YSL (Fig. 2) begins about an hour and a half after this stage (at 21-22°C). Scale bar equals 200 μm.

Although the MBT of *Fundulus* resembles that of *Xenopus* in so far as the appearance of motility is concerned, as yet we lack the critical biochemical information that research on *Xenopus* has provided (Newport and Kirschner, 1982a, b). The cessation of rapid mitosis no doubt leads to the establishment of the G_1 phase in *Fundulus*, with accompanying commencement or augmentation of transcription, as in *Xenopus*, but study of this in *Fundulus* is for the future.

In the meantime, however, we possess considerable information on the subsequent motile activity of these blebbing cells (Trinkaus, 1973, 1985). About an hour or more after cells begin blebbing, they begin to move, at first occasionally and soon more and more. The blebs become organs of locomotion. Instead of retracting immediately, a bleb persists, shows circus movement and cytoplasm pours into it. The bleb then protrudes more and the rest of the cell follows. We have termed this 'blebbing movement' (see also Trinkaus and Erickson, 1983; Fink and Trinkaus, 1988). Soon after this, some of these blebs spread on the substratum, usually the internal YSL, to form broad, thick lamellipodia and then these lead the cells in movement. Thick filolamellipodia (Trinkaus and Erickson, 1983) soon appear as well. By means of these varied protrusive activities, deep cells eventually move actively within the segmentation cavity. Two aspects of these early movements are noteworthy. (1) The cell movements are apparently random. No directional bias has been observed. (2) Cells begin movement in full sway in a so-called 'blastula' stage, long before the official beginning of gastrulation. The deep cells of *Fundulus* are obviously well prepared to participate in the directional movements of gastrulation well before that stage is reached. The MBT occurs in *Fundulus* 3-4 hours before contraction of the YSL and the beginning of epiboly (see below) and even longer before the gastrulation movements of deep cells. Another interesting feature of the MBT in *Fundulus* is its stability. Once cells differentiate into a motile state and begin blebbing, their adhesive affinity for other cells develops independently of their normal associations in the blastoderm. They do the same when isolated in vitro (Trinkaus, 1963).

Slowing and cessation of nuclear divisions in the YSL set the stage for the onset of gastrulation

During teleost epiboly, the EVL and the YSL spread together to encompass the large, spherical yolk sphere (Wilson, 1889; Stockard, 1915). Once formed, the EVL adheres firmly to the underlying YSL solely by its marginal cells by means of tight junctions, which become more and more extended as tension within the EVL increases with its steady expansion in epiboly (Betchaku and Trinkaus, 1978). In spite of this tension and the extensive junctional complex joining them to each other (Lentz and Trinkaus, 1971) and their high resistance barrier (Bennett and Trinkaus, 1970), individual EVL cells continually rearrange in the plane of the monolayer during epiboly, adjusting to the geometrical problems imposed by the expansion of an originally rather flat monolayered cell sheet over a sphere (Keller and Trinkaus, 1987). Indeed, the meticulous coordination of these rearrangements of the EVL cells, as they actively participate in epiboly, constitutes one of the most precise of all morphogenetic cell movements. But crucial though these beautiful EVL cell rearrangements are for the integrated progress of epiboly, they do not provide the motive force. All evidence points to the yolk syncytial layer. The blastoderm will not undergo epiboly unless it is attached to the YSL. The YSL, in contrast, undergoes complete epiboly in the complete absence of the blastoderm (Trinkaus, 1951). Moreover, it does so faster when relieved of the burden of the clinging blastoderm (Betchaku and Trinkaus, 1978) and will even surge ahead locally, if one small region is surgically freed of the restraint of the EVL (Trinkaus, 1971). Perhaps the most impressive evidence of the epibolic force of the YSL comes at the very beginning of epiboly. Soon after the external YSL (E-YSL) has formed a wide nucleated belt around the perimeter of the blastoderm (Fig. 1 and Trinkaus, 1951), it undergoes a spectacular narrowing (Fig. 2). This is caused by active cytoplasmic contraction (for evidence, see Trinkaus, 1984b). With this contractile narrowing of the E-YSL, the margin of the EVL and the rest of the blastoderm (Fig. 2) and the

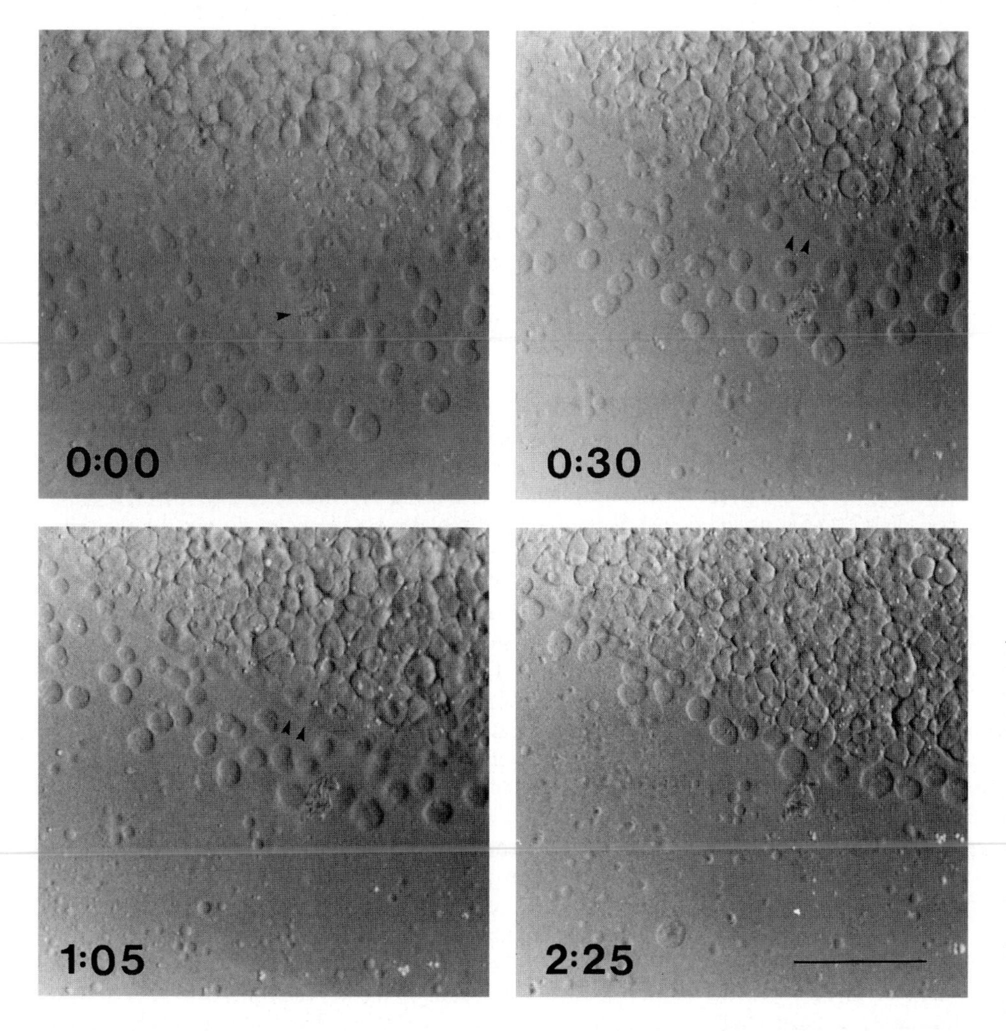

Fig. 2. Photomicrographs illustrating the contraction of the E-YSL, epiboly of the blastoderm, the crowding of the YSL nuclei and the disappearance of the nuclei beneath the blastoderm. 28°C. Time in hours and minutes. Scale bar equals 100 μm. 0:00 The E-YSL has contracted approximately one-third. Some nuclei are now in the I-YSL underneath the blastoderm in the cytoplasm of the I-YSL. Those that remain peripheral to the blastoderm are already crowded and several are in contact with each other. A dust fragment lodged between the egg and the coverslip serves as a stable reference point (arrowhead). The blastoderm is now in stage 13 and epiboly has just begun. 0:30 In this short period, the E-YSL has narrowed a great deal and many nuclei have disappeared beneath the blastoderm. The E-YSL is in its most rapid phase of contraction. Blastoderm epiboly has advanced considerably. The margin of the enveloping layer is evident (arrowheads). Many of the deep cells in the blastoderm are motile. 1:05 Contraction of the E-YSL has continued but its rate now is slower than during the first 30 minutes. Blastoderm epiboly is also moving more slowly. The border of the EVL is visible (arrowheads). The E-YSL nuclei are now densely packed and more have disapeared beneath the blastoderm. The embryo is now in stage 14½ and the germ ring is just evident. Convergence of deep cells has begun. 2:25 The contraction and narrowing of the E-YSL is almost complete. Only an irregular row of YSL nuclei is now visible. The blastoderm has continued to expand, but at a lower rate than previously. The border of the EVL is not visible in this micrograph. A number of motile deep cells engaged in convergent movements of gastrulation (Trinkaus et al., 1992) are now in focus. The embryo is in stage 15.

internal YSL (I-YSL) are pulled vegetalward in a sweeping movement of epiboly (Trinkaus, 1984b, Figs 3 and 4).

This represents the beginning of gastrulation, for not only does EVL and I-YSL epiboly begin but, soon after, the highly motile deep cells undergo involution (Thorogood and Wood, 1987; Wood and Timmermans, 1988; Warga and Kimmel, 1990) and convergence (Sumner, 1904; Oppenheimer, 1936; Pasteels, 1936; Ballard, 1966, 1973; Trinkaus et al., 1992). Deep cells converging in the germ ring are illustrated in Fig. 2 (at 2:25). As the converging deep cells join the embryonic shield, they participate in its anteroposterior extension. The result is a fully formed embryo. Teleost gastrulation thus offers a classic example of convergent extension, a hallmark of vertebrate gastrulation.

Because of the crucial importance of the contraction of the YSL in the onset of *Fundulus* gastrulation, I have recently investigated the complete development of the YSL in detail (Trinkaus, 1990, 1993). Very briefly, the nuclei of the YSL are derived from open, marginal blastomeres of late cleavage stages, as in other teleosts, and then divide five times to form the definitive, wide syncytial layer encircling the blastoderm (Fig. 1). About 1.5 hours after completion of the last (fifth) mitosis, this broad E-YSL begins its contraction, which progresses slowly over a 4-5 hour period and has a number of dramatic results. (1) The E-YSL narrows (Fig. 2). (2) This contractile narrowing causes the initially smooth surface of the E-YSL to buckle, throwing it into complex folds (Betchaku and Trinkaus, 1978). (3) At the margin of the E-YSL, the surface folds become the sites of highly localized programmed endocytosis (Betchaku and Trinkaus, 1986). (4) The cytoplasm of the E-YSL thickens, as it is compressed by its contraction (see Fig. 19 of Betchaku and Trinkaus, 1978). (5) Nuclei of the E-YSL become increasingly crowded as their cytoplasmic environment narrows along with its thickening (Fig. 2). (6) Epiboly of the I-YSL and the blastoderm commences, as their margins are pulled by the contracting E-YSL closer and closer to the E-YSL margin (Fig. 2). (7) YSL nuclei disappear into the I-YSL cytoplasm beneath the advancing blastoderm to form the nucleated I-YSL (Trinkaus, 1993). About two-thirds the way through this YSL contraction, deep cells within the blastoderm abandon their casual, undirected locomotory activities and begin the more disciplined movements of gastrulation.

Unfortunately, we have no information as yet at the molecular level of the activities of the cytoplasmic contractile proteins in the E-YSL during this contractile period. But we do have some useful morphological data. The cortical cytoplasm of the E-YSL, and especially its surface folds, is packed with microfilaments. Since they are the 4-6 nm variety, we presume that they are actin-containing (Betchaku and Trinkaus, 1978).

As indicated, the contraction of the E-YSL commences soon after the cessation of mitotic activity in the YSL. In consequence, details of this cessation deserve close attention. In the first place, the arrest of nuclear divisions after the last mitosis (M-V) is definitive and occurs simultaneously throughout the YSL. Nuclei enter a kind of permanent interphase after M-V. Contraction of the fully nucleated YSL commences about 1.5 hours after M-V and occurs

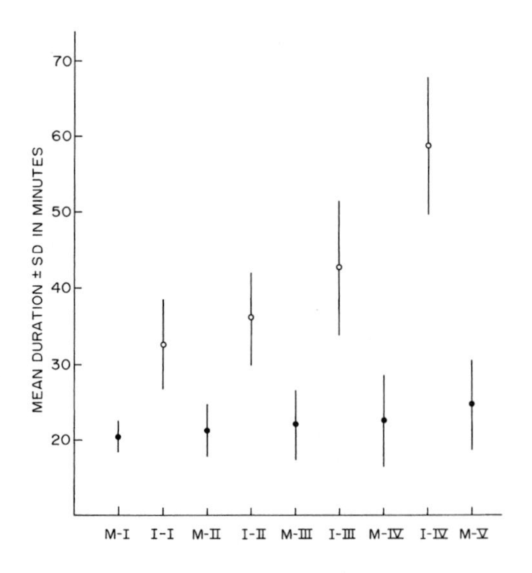

Fig. 3. A graph depicting the duration of the mitoses of the YSL and the interphases between them, as measured in continuous time-lapse video tapes. The duration of the mitoses are represented by solid circles and those of the interphases by open circles. Along the abscissa, each successive mitosis is represented by M-I, M-II, etc. Each successive interphase after each mitosis is represented by I-I (after M-I), I-II (after M-II), etc. M-V is the last mitosis. M-V is often incomplete (not shown on this graph).

essentially simultaneously around the periphery of the blastoderm. There is obviously a close temporal and circumferential relationship between the cessation of YSL nuclear divisions and the contraction of YSL cytoplasm. But the relationship begins prior to this. Although the cessation of mitosis in the YSL is abrupt, the events leading up to it are not. The duration (and variability) of each succeeding YSL mitosis increases gradually from the first one (M-I), the mean of which is 20.3 ± 2.0 minutes, to the last (M-V), the mean of which is 24.7 ± 6.1 minutes (Fig. 3). Moreover, M-V is often incomplete; some nuclei do not divide. Also, and probably more significant in the present context, the duration of the interphases between succeeding mitoses increases, and to a much greater degree (from a mean of 32.4 ± 5.7 minutes for interphase I (I-I) to a mean of 58.6 ± 9.1 minutes for I-IV) (Fig. 3). I-IV occurs between M-IV, the penultimate mitosis, and M-V, the last mitosis. (For quantitative details, see Trinkaus, 1993.) It should be emphasized that I-IV is not only the longest interphase of the series; it is by far the longest.

Variation in another aspect of the YSL mitoses is also relevant. These YSL mitoses are metachronous. Mitotic wave fronts progress through the YSL cytoplasm and stimulate nuclei to enter mitosis. These waves vary greatly in speed and regularity. Some waves move at a uniform rate across the field of observation; others accelerate; still others decelerate. Significantly, the wave fronts for M-V are the only ones that are consistent. They almost always move most slowly and they always decelerate as they move (Trinkaus, 1993). It seems likely that this deficient wave front is part of the cause of the frequent incompleteness of M-V (see Trinkaus, 1993).

There is clearly an attenuation of the mitotic forces of

the YSL with successive mitoses. The causes of this attenuation are of much interest and could well be the vast increase in the nucleocytoplasmic ratio of the YSL, as the number of its nuclei progressively increases (Edgar et al., 1986; see discussion in Trinkaus, 1993). But consideration of this fascinating subject is outside the province of this paper. On the contrary, the results of this attenuation (longer mitoses, longer interphases and fading mitotic waves) are very relevant to the subject at hand - the events leading to contraction of the YSL cytoplasm. In the absence of further analysis, there seem to be two possible reasons for assuming that this slowing of mitotic activity in the YSL forms the basis for its contraction: a diminution of the antagonism between mitotic division and cytoplasmic contractility and the appearance of the G_1 phase of the mitotic cycle, during which transcription and the synthesis of new contractile proteins might occur. These possibilities are not mutually exclusive.

It is well-established that there is an antagonism between mitosis and cell movement (e.g., Trinkaus, 1980; Trinkaus et al., 1992). Since cell movement depends in part on cytoplasmic contractility, there also must be an antagonism between mitosis and the cytoplasmic contractility associated with cell motility. It is not unreasonable, therefore, to propose that when the contractile mitotic machinery of the YSL is released, it should be available for other contractile activities, namely the contraction of the YSL cytoplasm.

It seems certain that the longer interphases between successive mitoses of the YSL, in particular, the last, longest interphase (I-IV) and the permanent interphase after M-V, make possible the establishment and augmentation of G_1 phase. This, of course, would allow the activation and augmentation of transcription and the consequent expression of new gene products, such as the molecular machinery for massive cytoplasmic contraction. Although, unfortunately, there has as yet been no investigation of mRNA during the development of the YSL of *Fundulus*, this matter has been investigated with great care in *Drosophila*, where certain aspects of early syncytial development are remarkably similar to the development of the YSL in teleosts. As in *Fundulus*, the interphases of the last nuclear divisions of *Drosophila* gradually increase in duration, in particular the last one (nuclear cycle 14), whose interphase is much longer than the preceding ones (Foe and Alberts, 1983). This last nuclear division is then followed immediately by an important new cytodifferentiation - cellularization, which in turn is quickly followed by a new morphogenesis - gastrulation. In *Drosophila*, several studies have shown that newly synthesized mRNA is first detectable at nuclear cycle 11 and that transcription increases substantially with each cycle thereafter (for references, see Edgar et al., 1986). This sequence certainly suggests that cellularization is triggered by this mRNA synthesis, especially that which occurs during the long, last nuclear cycle. Since the cellular form changes that lie at the basis of the next event, gastrulation, depend heavily on cytoplasmic contraction (Trinkaus, 1984a), the analogy with *Fundulus* is not completely far-fetched. Obviously, studies like these would be desirable in *Fundulus*.

A puzzling feature of the YSL morphogenetic transition in *Fundulus* is the long lag between the last mitosis of the YSL and the contraction of the YSL cytoplasm - about 1.5 hours at 21-22°C. It could be that the E-YSL cytoplasm is ready to contract much sooner but is restrained to do so by the firmly adherent blastoderm. To test this hypothesis, I removed the blastoderm as quickly as possible after the cessation of the last YSL mitosis. The result was conclusive. The E-YSL quickly contracted and the I-YSL quickly expanded (Trinkaus, 1984b). The result was precocious epiboly of the I-YSL. This experiment indicates that the normal slow delay in the contraction of the E-YSL after cessation of nuclear division is due to inhibitory restraint imposed by the adhering blastoderm (actually the EVL). It would seem, therefore, that the E-YSL actually becomes contractile very soon after cessation of its nuclear divisions, but is normally restrained for an hour or so by the attached blastoderm. In view of this, I suggest that when the E-YSL eventually contracts, after its normal delay, it does so either because of an eventual weakening of the restraint imposed by the blastoderm or because of a sudden increase in its contractile force.

This YSL transition that brings on gastrulation in *Fundulus* naturally reminds one of the famous midblastula transition. However, they are really separate processes. The MBT occurs much earlier in development and involves the cessation of nuclear divisions and cytokinesis in cleaving blastomeres, not the much later cessation of nuclear divisions in a syncytium. In addition, and more importantly, the MBT results in the onset of motility, G_1 and transcription only in individual deep blastomeres, whereas the YSLT results in the onset of global morphogenetic movements of the whole embryonic system. To express it succinctly, the midblastula transition commences cytodifferentiation; the YSL transition commences morphogenesis.

I am indebted to Madeleine Trinkaus for editorial and photographic assistance and to Kurt Johnson, Ray Keller and Charles Kimmel for helpful discussions. This research has been supported by a Merit Award from the NCI of the NIH.

References

Armstrong, P. B. and Child, J. S. (1965). Stages in the normal development of *Fundulus heteroclitus*. *Biol. Bull.* **28**, 143.

Ballard, W. W. (1966). Origin of the hypoblast in *Salmo*. I. Does the blastodisc edge turn inward? *J. Exp. Zool.* **161**, 201-210.

Ballard, W. W. (1973). Morphogenetic movements in *Salmo gairdneri* Richardson. *J. Exp. Zool.* **184**, 381-426.

Bennett, M. V. L. and Trinkaus, J. P. (1970). Electrical coupling between embryonic cells by way of extracellular space and specialized junctions. *J. Cell Biol.* **44**, 592-610.

Betchaku, T. and Trinkaus, J. P. (1978). Contact relations, surface activity, and cortical microfilaments of marginal cells of the enveloping layer and of the yolk syncytial and yolk cytoplasmic layers of *Fundulus* before and during epiboly. *J. Exp. Zool.* **206**, 381-426.

Betchaku, T. and Trinkaus, J. P. (1986). Programmed endocytosis during epiboly of *Fundulus heteroclitus*. *Amer. Zool.* **26**, 193-199.

Edgar, B. A., Kiehle, C. P. and Schubiger, G. (1986). Cell cycle control by the nucleo-cytoplasmic ratio in early Drosophila development. *Cell* **44**, 365-372.

Fink, R. D. and Trinkaus, J. P. (1988). *Fundulus* deep cells: Directional migration in response to epithelial wounding. *Dev. Biol.* **129**, 179-190.

Foe, V. E. and Alberts, B. M. (1983). Studies of nuclear and cytoplasmic behaviour during the five mitotic cycles that precede gastrulation in *Drosophila* embryogenesis. *J. Cell Sci.* **61**, 31-70.

Gerhart, J. G. (1980). Mechanisms regulating pattern formation in the amphibian egg and early embryo. In *Biological Regulation and Development,* Vol. 2, (ed. R.F. Goldberger), pp. 133-316, New York and London: Plenum Press.

Johnson, K. E. (1976). Circus movements and blebbing locomotion in dissociated embryonic cells of an amphibian, *Xenopus laevis. J. Cell Sci.* **22**, 575-583.

Keller, R. E. and Trinkaus, J. P. (1987). Rearrangement of enveloping layer cells without disruption of the epithelial permeability barrier as a factor in *Fundulus* epiboly. *Dev. Biol.* **120**, 12-24.

Lentz, T. L. and Trinkaus, J. P. (1971). Differentiation of the junctional complex of surface cells in the developing *Fundulus* blastoderm. *J. Cell Biol.* **48**, 455-472.

Newport, J. and Kirschner, M. (1982a). A major developmental transition in early Xenopus embryos: I. Characterization and timing of cellular changes at the midblastula stage. *Cell* **30**, 675-686.

Newport, J. and Kirschner, M. (1982b). A major developmental transition in early Xenopus embryos: II. Control of the onset of transcription. *Cell* **30**, 687-696.

Oppenheimer, J. M. (1936). Processes of localization in developing *Fundulus. J. Exp. Zool.* **73**, 405-444.

Pasteels, J. (1936). Etudes sur la gastrulation des vertébrés méroblastiques. I. Téleostéens. *Arch. Biol.* **47**, 205-308.

Signoret, J. et Lefresne, J. (1971). Contribution à l'étude de la segmentation de l'oeuf d'axolotl: I - Définition de la transition blastuléenne. *Ann. Embr. Morph.* **4**, 113-123.

Stockard, C.R. (1915). A study of wandering mesenchymal cells on the living yolk-sac and their developmental products: chromatophores, vascular epithelium and blood cells. *Amer. J. Anat.* **18**, 525-594.

Sumner, F. B. (1904). The study of early fish development. Experimental and morphological. *Wilhelm Roux ArchEntw. Mech. Org.* **17**, 92-149.

Thorogood, P. and Wood, A. (1987). Analysis of *in vivo* cell movement using transparent tissue systems. *J. Cell Sci. Suppl.* **8**, 395-413.

Trinkaus, J. P. (1951). A study of the mechanism of epiboly in the egg of *Fundulus heteroclitus. J. Exp. Zool.* **118**, 269-320.

Trinkaus, J. P. (1963). The cellular basis of *Fundulus* epiboly.
Adhesivity of blastula and gastrula cells in culture. *Dev. Biol.* **7**, 513-532.

Trinkaus, J. P. (1971). Role of the periblast in *Fundulus* epiboly (in Russian), *Ontogenesis* **2**, 401-405.

Trinkaus, J. P. (1973). Surface activity and locomotion of *Fundulus* deep cells during blastula and gastrula stages. *Dev. Biol.* **30**, 68-103.

Trinkaus, J. P. (1980). Formation of protrusions of the cell surface during cell movement. In *Tumor Cell Surfaces and Malignancy* (ed. R.O. Haynes and C.F. Fox). *Progress in Clinical and Biological Research*, **41**, 887-906.

Trinkaus, J. P. (1984a). *Cells into Organs. The Forces that Shape the Embryo.* Second Revised Edition, Prentice-Hall, Inc. Englewood Cliffs, New Jersey, 543 pp.

Trinkaus, J. P. (1984b). Mechanism of *Fundulus* epiboly - a current view. *Amer. Zool.* **24**, 673-688.

Trinkaus, J. P. (1985). Protrusive activity of the cell surface and the initiation of cell movement during morphogenesis. In *Cell Traffic in the Developing and Adult Organism* (ed. G. Haemmerli and P. Sträuli), Basel, Karger. *Exp. Biol. and Med.* **10**, 130-173.

Trinkaus, J. P. (1990). Some contributions of research on early teleost embryogenesis to general problems of development. In *Experimental Embryology in Aquatic Plants and Animals* (ed. Marthy, H.-J.), Plenum Press, New York, pp. 315-327.

Trinkaus, J. P. (1993). The yolk syncytial layer of *Fundulus*. Its origin and history and its significance for early embryogenesis. *J. Exp. Zool.* (in press).

Trinkaus, J. P. and Erickson, C. A. (1983). Protrusive activity, mode and rate of locomotion, and pattern of adhesion of *Fundulus* deep cells during gastrulation. *J. Exp. Zool.* **228**, 41-70.

Trinkaus, J. P., Trinkaus, M. and Fink, R. D. (1992). On the convergent cell movements of gastrulation in *Fundulus*. *J. Exp. Zool.* **261**, 40-61.

Warga, R. M. and Kimmel, C. B. (1990). Cell movements during epiboly and gastrulation in zebrafish. *Development* **108**, 569-580.

Wilson, H. V. (1889). The embryology of the sea bass (*Serranus atrarius*). *Bull. U.S. Fish Commission* **9**, 209-278.

Wood, A. and Timmermans, L. P. M. (1988). Teleost epiboly: a reassessment of deep cell movement in the germ ring. *Development* **102**, 575-585.

Development 1992 Supplement, 81-91 (1992)
Printed in Great Britain © The Company of Biologists Limited 1992

The patterning and functioning of protrusive activity during convergence and extension of the *Xenopus* organiser

RAY KELLER[1], JOHN SHIH[2] and CARMEN DOMINGO[1]

[1]*Department of Molecular and Cell Biology, University of California, Berkeley, Berkeley CA, USA*
[2]*The Beckman Institute, California Institute of Technology, Pasadena CA, USA*

Summary

We discuss the cellular basis and tissue interactions regulating convergence and extension of the vertebrate body axis in early embryogenesis of *Xenopus*. Convergence and extension occur in the dorsal mesoderm (prospective notochord and somite) and in the posterior nervous system (prospective hindbrain and spinal cord) by sequential cell intercalations. Several layers of cells intercalate to form a thinner, longer array (radial intercalation) and then cells intercalate in the mediolateral orientation to form a longer, narrower array (mediolateral intercalation). Fluorescence microscopy of labeled mesodermal cells in explants shows that protrusive activity is rapid and randomly directed until the midgastrula stage, when it slows and is restricted to the medial and lateral ends of the cells. This bipolar protrusive activity results in elongation, alignment and mediolateral intercalation of the cells. Mediolateral intercalation behavior (MIB) is expressed in an anterior-posterior and lateral-medial progression in the mesoderm. MIB is first expressed laterally in both somitic and notochordal mesoderm. From its lateral origins in each tissue, MIB progresses medially. If convergence does not bring the lateral boundaries of the tissues closer to the medial cells in the notochordal and somitic territories, these cells do not express MIB. Expression of tissue-specific markers follows and parallels the expression of MIB. These facts argue that MIB and some aspects of tissue differentiation are induced by signals emanating from the lateral boundaries of the tissue territories and that convergence must bring medial cells and boundaries closer together for these signals to be effective. Grafts of dorsal marginal zone epithelium to the ventral sides of other embryos, to ventral explants and to UV-ventralized embryos show that it has a role in organising convergence and extension, and dorsal tissue differentiation among deep mesodermal cells. Grafts of involuting marginal zone to animal cap tissue of the early gastrula shows that convergence and extension of the hindbrain-spinal cord are induced by planar signals from the involuting marginal zone.

Key words: convergence, extension, morphogenesis, motility, gastrulation, *Xenopus*.

Introduction

We have made significant progress in understanding the cellular basis of the powerful convergence and extension movements that function in gastrulation, neurulation and formation of the vertebrate body axis in amphibians (reviewed in Keller et al., 1991a, b; also see Keller et al., 1992a,b; Keller and Shih, 1992), as well as other vertebrates, including fish (Warga and Kimmel, 1990; Trinkaus et al., 1992) and birds (Schoenwolf and Alvarez, 1989). The cellular behavior underlying these movements in amphibians, their patterning and timing, and the geometry and mechanics of their function are very complex. For details, the reader should consult the papers cited above. We will use this occasion to provide a more accessible overview of the essential findings.

Convergence and extension

The terms 'convergence' and 'extension' were defined in the classical 'Entwicklungsmechanik' period of developmental biology and they describe the narrowing and lengthening of the dorsal tissues of the embryo, including the prospective notochord, somitic mesoderm and posterior neural plate during gastrulation and neurulation (see Keller et al., 1991b for a history) (Fig. 1A). These movements are important because they drive much of gastrulation (Keller, 1986) and constitute a major part of the morphogenetic activity of the Spemann organiser and of the nervous system that the organiser induces (Keller et al., 1991b; Keller et al., 1992a,b). These movements elongate the body axis, thus transforming the spherical egg into the elongate, bilaterally symmetrical vertebrate body plan. These movements are fundamental, general features of morphogenesis throughout the Metazoa, both during early and late stages of development and during regeneration (see Keller, 1987). Lastly, these movements are important because they represent the type of 'mass movement' in which mechanical integration of local cell motility produces forces that change the shape of the cell population, a process about which we know very little, despite its overwhelming contribution to the shaping of the embyro.

Convergence and extension are studied in explants

Although these movements have been described (Vogt, 1929; Keller, 1975, 1976; Jacobson and Gordon, 1976) and analyzed (Keller and Tibbetts, 1989; Keller et al., 1989a, b) in the whole embryo, the cell behaviors and mechanics underlying them can best be seen and analyzed in cultured explants. Convergence and extension of all dorsal tissues are displayed in sandwich explants of the dorsal marginal zone (DMZ) (Keller et al., 1985a, b; Keller and Danilchik, 1988) (Fig. 1B). There are two regions of convergence and extension, one in the involuting marginal zone (IMZ), consisting of prospective notochordal and somitic mesoderm and archenteron roof endoderm, and another in the prospective posterior neural plate, comprising the prospective hindbrain and spinal cord (Fig. 1B). Both show mechanically autonomous, active convergence and extension that are *not* dependent on cell traction on external substrata (Keller et al., 1985a, b). Thus the forces producing these movements must be generated within the explant.

Convergence and extension occur by radial and mediolateral cell intercalation

Convergence and extension in *Xenopus* occur primarily as a result of cell *intercalations* in two directions. These cell intercalations were revealed directly by time-lapse videorecordings of the deep cells in 'open-faced' explants, which expose these cells to observation (Wilson et al., 1989; Wilson and Keller, 1991; Keller et al., 1989a) (Fig. 1C). First, the deep cells intercalate perpendicular to the surface of the embryo such that several layers of deep cells form fewer layers of greater area (*radial intercalation*) (Fig. 2). For reasons not yet understood, the greater area produced by radial intercalation in the marginal zone is expressed solely by *extension* in the animal-vegetal (anterior-posterior) axis (Wilson and Keller, 1991), rather than in all directions, as is the case in epiboly of the animal cap (Keller, 1978, 1980). Following radial intercalation, the deep cells intercalate along the mediolateral axis to form a longer, narrower array (*mediolateral intercalation*), thus producing *convergence* and *extension* (Fig. 2). During this period of deep cell intercalation, the superficial epithelial cells spread, divide and intercalate mediolaterally to form a longer, narrower array, despite the fact that their apices are connected by a junctional complex (see Keller, 1978; Keller and Trinkaus, 1987).

This sequence of radial and mediolateral cell intercalations also produces convergence and extension of the posterior nervous system. By grafting patches of cells labeled with fluorescein dextran amine (FDA) into unlabeled hosts and monitoring their intercalation with unlabelled cells during subsequent stages of development, we showed that convergence and extension of the hindbrain and spinal cord in sandwich explants and in whole embryos also occurs by radial intercalation followed by mediolateral intercalation (Keller et al., 1992a,b). We also used this method to show that convergence and extension of the mesoderm occurs by radial (Keller et al, 1992a) and mediolateral intercalation (Keller et al., 1985a, b; Keller and Tibbetts, 1989) in *whole embryos*. Thus what is observed in the explants is not an artifact of manipulation or culture.

The timing and position of radial and mediolateral intercalation

Radial intercalation and mediolateral intercalation occur in sequence. Radial intercalation begins in the dorsal mesoderm at the early gastrula stage (stage 10) and is largely complete by the early midgastrula (stage 10.5). Then mediolateral intercalation begins, with some overlap, and continues until near the end of neurulation (stage 18). The same sequence occurs in the hindbrain-spinal cord, beginning slightly later than in the mesoderm. In both mesoderm (Wilson et al., 1989; Wilson and Keller, 1991; Shih and

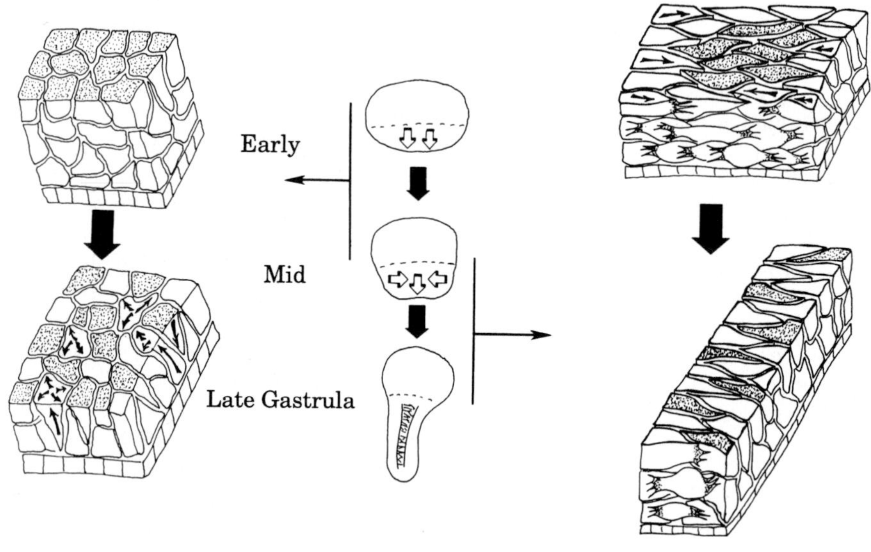

Fig. 2. Diagrams show the cell intercalations underlying convergence and extension of the IMZ in open-faced explants. In the center column, change in the shape of open-faced explants are shown at early, middle and late gastrula stages and the cell behaviors producing these movements are shown on the sides. The inner, deepest surface is shown uppermost and the epithelial layer normally forming the outer surface of the embryo is at the bottom in all diagrams. The open arrows in the center column illustrate the tissue deformation. The small arrows on the cells in the right and left columns indicate the direction of cell movements. Radial intercalation (left) produces thinning and extension in the first half of gastrulation and mediolateral intercalation (right) produces convergence and extension in the second half. Modified from Wilson and Keller, 1991.

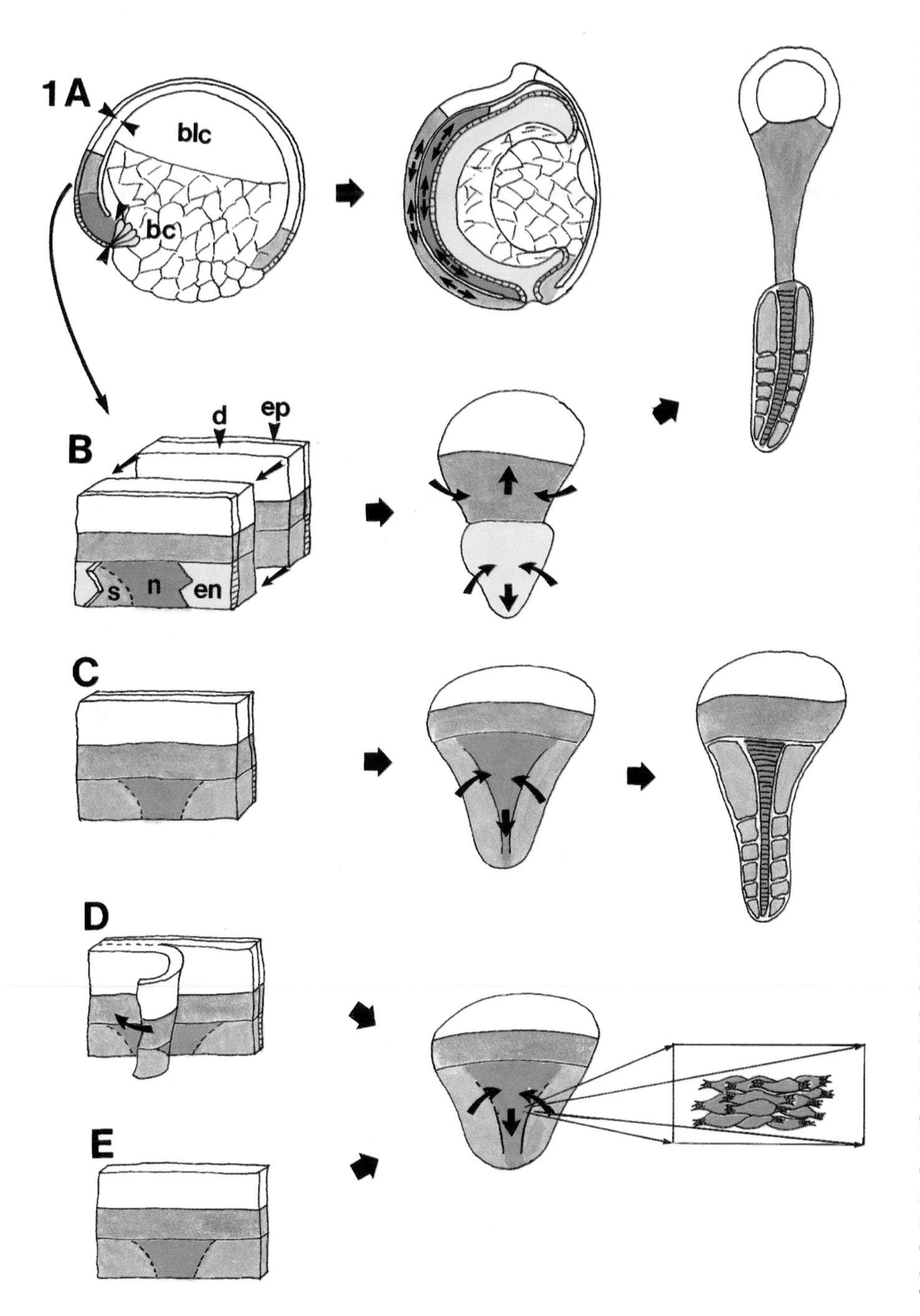

Fig. 1. Schematic diagrams show the convergence and extension of the involuting marginal zone (IMZ), shown in red (mesoderm) and yellow (endoderm), and the posterior neural plate (prospective hindbrain-spinal cord), shown in blue, in whole embryos (A) in sandwich explants (B), in open-faced explants (C), in shaved, open-faced explants (D) and in deep cell explants (E). Arrows in the left column indicate manipulations of the tissue; arrows in the middle column indicate tissue movements. Sagittal views of the early gastrula and neurula show the extraordinary extension of the mesodermal and posterior neural tissues during gastrulation and neurulation (A). Convergence and extension of both tissues occur in sandwich explants (B), made by cutting out the dorsal sector of the early gastrula and sandwiching it with another, identical explant. This explant consists of a superficial epithelium (ep) and a deep, nonepithelial (mesenchymal) region (d). It consists of an animal cap region (white), which normally forms forebrain, a noninvoluting marginal zone (blue), which normally forms hindbrain and spinal cord, and an involuting marginal zone, consisting of a superficial layer of prospective endoderm (en, yellow), and a deep region of prospective notochordal (n, red) and somitic (s, orange) mesoderm. Open-faced explants (C) are designed to expose the deep mesodermal cells to observation and consist of one explant cultured in modified Danilchik's solution under a coverslip conditions under which the IMZ will converge, extend and differentiate into somites and notochord. Shaved open-faced explants are a further modification designed to expose the deep cells next to the overlying endodermal epithelium (D). These are made and cultured like open-faced explants but the innermost layers of deep cells are shaved or peeled off with an eyebrow knife, exposing the deep cells next to the endodermal epithelium to observation and videorecording for analysis of cell behavior at high resolution (box, D). Lastly, deep cell explants are made from stage 10.5 or 11 embryos in which the same tissues are excised and the adjacent sheet of prospective endoderm removed, leaving a layer of deep mesodermal cells. (E). Convergence and extension of the deep mesoderm is largely independent of the overlying epithelium after these stages. The mesoderm is partially involuted at these stages and must be straightened before culture (not shown). Modified from Keller et al., 1991a.

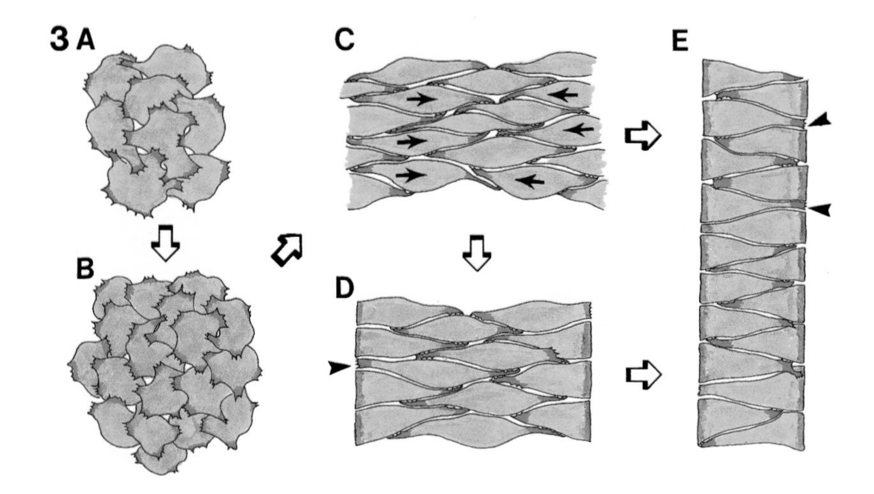

Fig. 3. The cellular behavior and protrusive activity driving mediolateral intercalation of deep mesodermal cells are illustrated. In the early gastrula (stage 10-10.5) the deep mesodermal cells divide once and show rapid protrusive activity (red) oriented in all directions (A,B). At the midgastrula stage (10.5), cell division ceases, protrusive activity slows dramatically and protrusive activity becomes polarized in the form of large, stable protrusions at the medial and lateral ends of the cells (C). These directed, stable protrusions are applied directly to the surfaces of adjacent cells and advance between adjacent cells without contact inhibition. Traction generated in the mediolateral axis by these oriented, invasive protrusions result in mediolateral cell elongation, alignment and intercalation (C). Mediolateral intercalation results in the convergence and extension of the cell population through the rest of gastrulation and neurulation (C-E). At stage 11 and thereafter, the boundary between notochordal and somitic mesoderm forms and inhibits the invasive, protrusive activity where the bipolar cells contact the boundary (blue), making the boundary cells monopolar in protrusive activity. Thus internal cells are captured at the boundary and elongate the boundary while pulling more deep cells to the boundary (D,E) (see Keller et al., 1989a). Active protrusions entering the boundary (pointers, D, E) first bleb, spread on the boundary and then cease all protrusive activity.

Keller, 1992a) and in the neural tissue (Keller et al., 1992a), intercalation begins in the prospective anterior end and proceeds posteriorly. Mediolateral intercalation begins in the *post-involution*, prospective anterior axial mesoderm and proceeds posteriorly, forming a constriction ring at the blastoporal lip that drives involution of the marginal zone (Keller, 1986; Wilson and Keller, 1991). These observations explain how convergence and extension of the IMZ can drive much of gastrulation without participation of other components of the gastrula (Keller et al., 1985a, b). If the continuity of the constriction ring is broken, involution fails and 'ring' embryos form (Schechtman, 1942; Keller, 1984).

Bipolar, mediolaterally directed protrusive activity drives mediolateral cell intercalation, and convergence and extension

We were unable to see how cells generate the forces that move them between one another by videorecording the deep cells at the inner surface of the open-faced explant (Fig. 1C). The deepest mesodermal cells do not participate in *active* mediolateral cell intercalation until later stages (see Keller et al., 1989a; Wilson et al., 1989), because the overlying endodermal epithelium organises mediolateral intercalation from the outside inward (see discussion below and Shih and Keller, 1992a). Thus, to observe the deep cells immediately beneath the epithelium, we developed the 'shaved' explant in which the deepest mesodermal cells in the open-faced explant are shaved off with an eyebrow hair, exposing the ones immediately beneath the epithelium to observation, videorecording and manipulation (Fig. 1D). Alternatively, we made explants of *only* the deep mesoderm of the early midgastrula (stage 10.5), which by this time has been thinned by radial intercalation and is no longer dependent on the overlying epithelium to organise mediolateral intercalation. This explant consists of only deep mesodermal cells and most or all the cells can be seen, thus negating the problem that unseen cells may be generating forces that we attribute to the cells that we observe.

The cell behavior in these two types of explants was recorded with epi-illumination and protrusive activity was recorded with low-light, fluorescence microscopy and image processing (Keller et al., 1989a; Keller et al., 1991a, b; Shih and Keller, 1992b). We will summarize our understanding of the mechanism of deep mesodermal cell intercalation (Fig. 3) and then show examples of evidence supporting these ideas.

Deep mesodermal cells divide once in early gastrulation, while remaining isodiametric and showing rapid, randomly directed protrusive activity (Fig. 3A, B). Then, at the transition of early and midgastrulation (stage 10.5) (Fig. 3C), protrusive activity slows dramatically and becomes directed medially and laterally in the form of large lamelliform and filiform protrusions applied to the surfaces of adjacent cells. The cells elongate, align and intercalate parallel to the mediolateral axis and perpendicular to the anterior-posterior axis (animal-vegetal axis) (Fig. 3C). They then intercalate mediolaterally to form a longer narrower array (Fig. 3C, E).

These events appear directly related to the traction of the large medial and lateral protrusions on adjacent cells. These protrusions advance across the neighboring cells without contact inhibition of movement. When the notochord-somitic mesodermal boundary forms, beginning at stage 11, the bipolar cells in the region of the boundary cease protrusive activity at their boundary-facing margin (Fig. 3D) (Keller et al., 1989a). Protrusions of cells entering the boundary by intercalation (pointers, Fig. 3D,E), first bleb violently, then spread in the plane of the notochord boundary and finally cease all protrusive activity. Thus, they become monopolar and stabilize their position at the boundary while they exert traction on deeper cells (Fig. 3D) (Keller et al., 1989a; Shih and Keller, 1992a). As a result, internal cells are pulled to the boundary and captured there, thus elongating the boundary (Fig. 3D,E). Both the bipolar and monopolar protrusive activity contributes to mediolat-

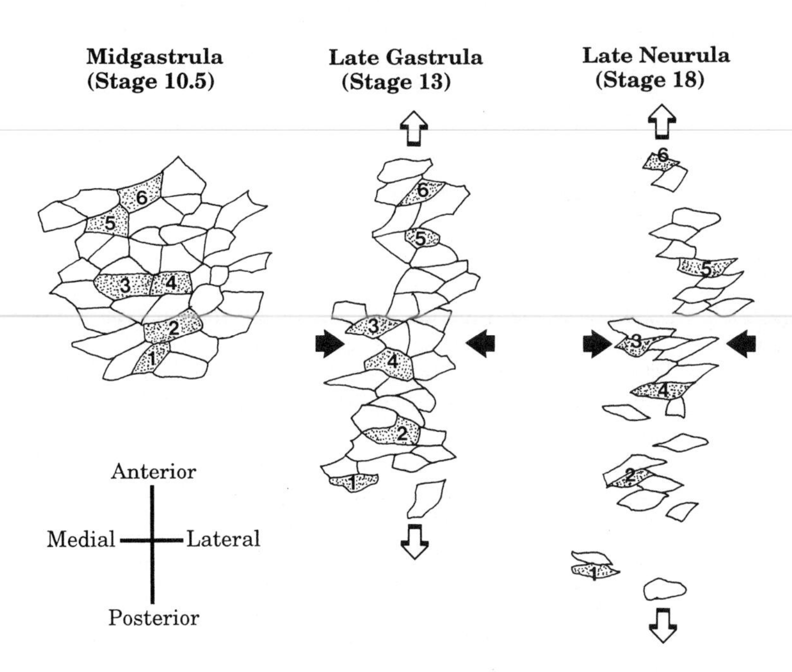

Fig. 4. Cell tracings from a deep cell explant show mediolateral intercalation of deep mesodermal cells during convergence and extension of an explant without the endodermal epithelium. The filled arrows indicate convergence movements and the open arrows indicate extension movements.

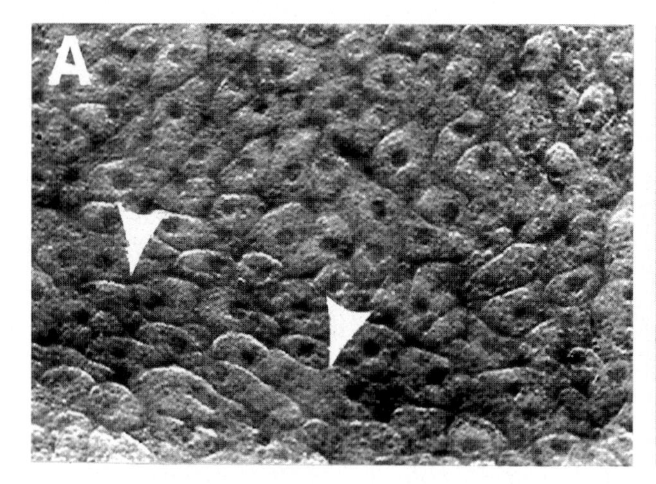

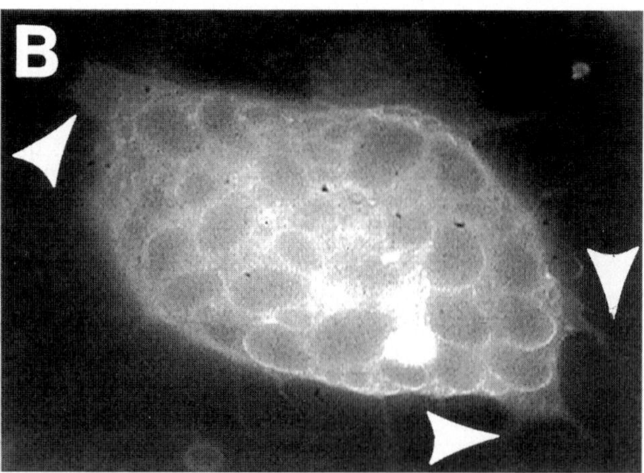

Fig. 5. A print of a videorecording made with epi-illumination shows the mediolateral cell elongation and alignment accompanying mediolateral cell intercalation (below pointers, A). A print of a videorecording made with low-light fluorecence microscopy and image processing of DiI-labeled, intercalating mesodermal cell shows the bipolar morphology, mediolateral elongation (upper left to lower right) and protrusive activity at the medial and lateral ends (pointers, B). Another DiI-labeled cell lies beneath the lower edge of the brightly labeled cell.

eral cell intercalation. The bipolar, medially and laterally directed protrusive activity exerts traction on adjacent cells, elongates them, aligns them and pulls them between one another to form a longer, narrower array prior to notochordal-somitic boundary formation (Fig. 3C-E). This process continues in central regions after the boundary forms, bringing more cells within reach of the boundary, where the monopolar activity and cell capture elongates the boundary.

Tracking of individual cells in time-lapse recordings shows details of cell shape changes and mediolateral intercalations during convergence and extension (Fig. 4). Although cells elongate mediolaterally and shorten anterior-posteriorly, their mediolateral intercalation is sufficient to produce convergence and extension of the tissue (see shaded cells, Fig. 4). The pattern of intercalation is indeterminate, with no precise order of intercalation (Fig. 4). Cells will occasionally change anterior-posterior or mediolateral positions once or even several times in the course of intercalation. Little or no cell division occurs during mediolateral intercalation. Epi-illumination shows the dramatic mediolateral elongation and alignment that accompanies mediolateral intercalation (Fig. 5A). Low-light, fluorescence microscopy and video image processing shows that bipolar protrusive activity, directed medially and laterally, drives the elongation, alignment and intercalation of mesodermal cells (Fig. 5B). Plots of the direction of protrusive activity from recordings of fluorescently labeled cells show protrusive activity is randomly directed in early gastrulation (stage 10-10.25) and is polarized medially and laterally in the second half of gastrulation (stage 10.5-12), when the cells elongate, align and intercalate (Fig. 6). (For details and additional evidence see Keller et al., 1989a, b). Also, computer modelling of rules of cell behavior has proven useful in evaluating the contribution of specific cell behaviors to the cell intercalation, cell shape changes, and convergence and extension of the notochord (Welicky et al., 1991).

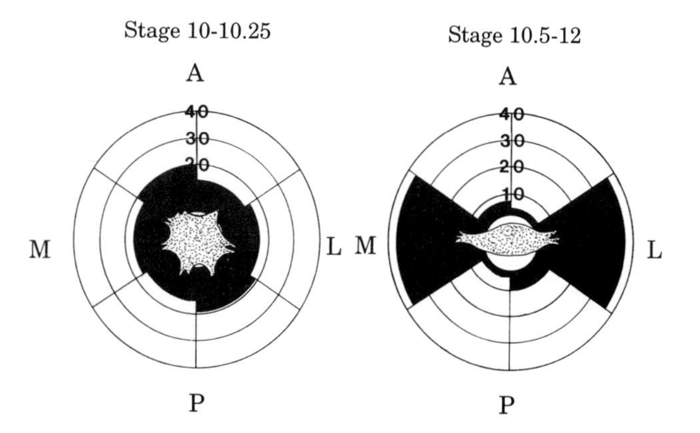

Fig. 6. Plots of percent of protrusive activity in each of 6 sectors around the deep mesodermal cells shows the initial random orientation in the early gastrula (left) and the dramatic restriction of protrusive activity to the lateral and medial ends of the cells after the midgastrula stage when these cells are intercalating mediolaterally (right).

Patterns of mediolateral intercalation behavior (MIB)

The bipolar protrusive activity, cell elongation and cell alignment that precede and accompany mediolateral intercalation are collectively called mediolateral intercalation behavior (MIB) and the expression of MIB can be followed in large explants by following cell elongation and alignment. MIB is expressed progressively across the mesodermal cell population in a pattern that reflects both the function of mediolateral intercalation in gastrulation and the progress of the signalling and response system that organises the IMZ (Shih and Keller, 1992a). The progression of MIB will be described in terms of the anterior-posterior and lateromedial axes of the prospective notochord and somitic mesoderm in large explants of the IMZ (Fig. 7), and MIB

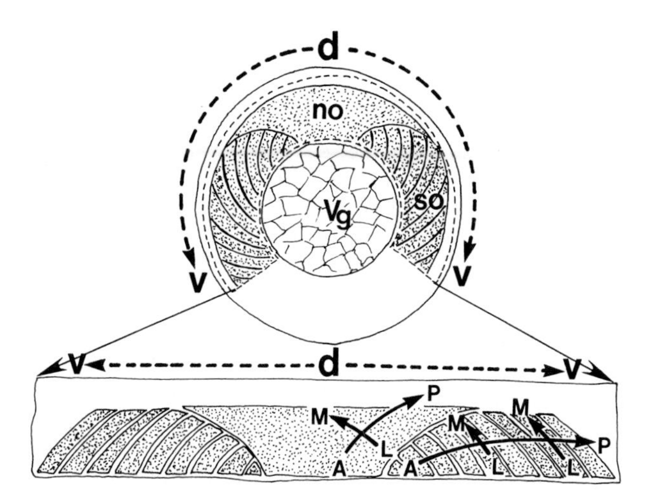

Fig. 7. The top diagram shows the vegetal (Vg) aspect of the early gastrula as it would appear if the dorsal 270 degrees of superficial endoderm were removed, exposing the prospective mesodermal tissues, including the notochord (no) and somitic (so) mesoderm. Patterning of mediolateral intercalation behavior (MIB) was studied in large explants of this mesoderm, shown in the bottom diagram. The patterning of MIB will be discussed in the context of the *prospective* anterior (A) to posterior (P) and lateral (L) to medial (M) axes of the prospective notochordal and somitic mesoderm in these large explants of the marginal zone. The dorsal (d) to ventral (v) axis of the gastrula is indicated by a dashed line. This diagram is based on Keller, 1975, 1976; Wilson et al., 1989 and Keller et al., 1989.

will be represented as elongated and aligned cell outlines (Fig. 8). MIB produces convergence and extension, but in turn convergence and extension modifiy the expression of MIB. Therefore, we will compare MIB expression in explants in which convergence and extension has been blocked by mechanical resistance (Fig. 8A) with explants that converge and extend (Fig. 8B).

MIB is expressed first in an arc at the anterior end of the axial mesoderm, called the vegetal alignment zone (VgAZ). The VgAZ is initiated bilaterally, about 15 degrees from the midline and progresses towards the midline in the early midgastrula (Fig. 8, stage 10.5-10.75). MIB then spreads laterally and ventrally along the vegetal edge of the IMZ (Fig. 8, stage 11), which is an anterior-posterior progression along the lateral edge of the prospective somitic mesoderm (refer to Fig. 7). The boundary of the notochord forms at the midgastrula stage (stage 11), beginning *within* the VgAZ and progressing posteriorly (Fig. 8, stage 11-onward). MIB then spreads posteriorly from its origin in the VgAZ, following the lateral aspect of the notochord, along its boundary (Fig. 8, stage 11-onward). From its origin laterally in the somitic mesoderm, MIB progresses medially, toward the prospective notochord. From its lateral origin in the notochord, it spreads medially towards the center of the notochord territory (Fig. 8, stage 11-onward). Differentiation of the notochord, as indicated by vacuolation and by staining with notochord-specific antibody tor-70 (Domingo, unpublished results), and differentiation of somitic mesoderm, as indicated by segmentation and staining with somite-specific antibody 12-101, follow the pattern of MIB expression, beginning anteriorly, progressing posteriorly along the lateral boundaries and progressing medially from the lateral boundary at any anterior-posterior level.

The expression of MIB and differentiation of cells in the posterior-medial region of both the notochordal and somitic territories is dependent on convergence bringing these cells closer to the lateral boundaries. If the explant encounters too much mechanical resistence, the bipolar protrusive activity results in elongation and alignment of the cells but they do not intercalate and thus convergence and extension do not occur. In this case, those cells farthest from the boundaries in the medial, posterior regions, do not show MIB or differentiate (black dots, Fig. 8A, stage 11-onward). If convergence and extension are allowed to occur, cell elongation and alignment are followed by mediolateral intercalation and convergence and extension, bringing the lateral boundaries of the notochord and somitic mesodermal territories closer to their respective medial cells (Fig. 8B). In this case, the medial cells express MIB and differentiate.

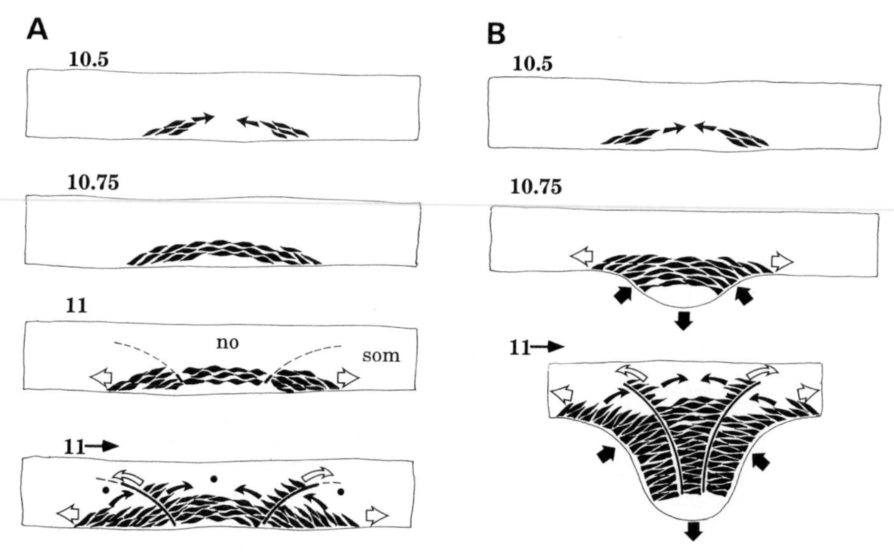

Fig. 8. Patterns of progression of mediolateral intercalation behavior (MIB) are represented by outlines of elongated, aligned cells in explants not allowed to converge and extend (A) and ones that are allowed to converge and extend (B). The numbers above each explant indicate stage of development. The solid arrows inside the explants indicate lateral-medial progression of MIB and the open arrows indicate anterior-posterior progression. The arrows outside the explants show the direction of convergence and extension, when they occur. Prospective notochord (no) and somitic mesoderm (som) and the prospective boundary between them (dashed line) are indicated. Formation of the boundary is indicated by a heavy line. The black dots indicate areas that do not express MIB or differentiate in nonextending, nonconverging explants.

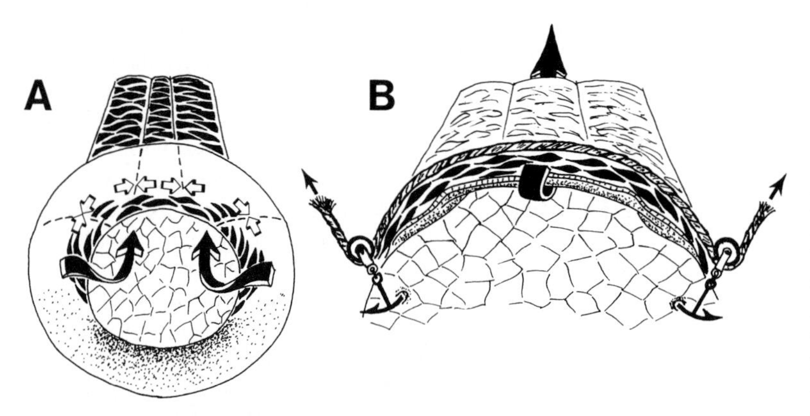

Fig. 9. Posterior progression of mediolateral intercalation in the involuted mesoderm approaches the blastoporal lip (black cell outlines) and forms a constricting arc at the blastoporal lip (open arrows, A). The constricting arc is anchored in the vegetal endoderm at each end and, as it constricts , it functions to roll the pre-involution IMZ over the lip, as indicated by the rope and pulley analogy (B).

Mechanical function of MIB progression in gastrulation

The anterior-posterior progression of MIB has a specific mechanical function in gastrulation. At the beginning of gastrulation, the leading edge of the IMZ involutes as a result of active mesodermal cell migration (Winklbauer, 1990) and bottle cell formation (Hardin and Keller, 1988). Once involution is initiated, MIB expression begins in the anterior region of the IMZ just inside the blastoporal lip and progresses posteriorly to the blastoporal lip (Fig. 9A). At the blastoporal lip, MIB and the resulting intercalation forms a constriction ring, anchored laterally in the vegetal endoderm, that presses on the inner surface of the blastoporal lip (Fig. 9B). As this constriction ring progresses posteriorly in the IMZ, it pulls the remaining IMZ over the lip (Fig. 9B). The function of these arcs during involution is illustrated dramatically by the fact that breaking their continuity in the dorsolateral sector results in the extension of the dorsal sector (notochord) straight out into the medium and the development of somites in situ, without involution, around the margins of the vegetal endoderm (see Keller, 1981, 1984; Schechtman, 1942). Moreover, the constriction forces generated by this posteriorly progressing arc of intercalating cells can produce involution of the remaining IMZ *without* the participation of forces generated by animal cap epiboly, mesodermal cell migration, or neural extension and convergence. These regions can be removed, yet involution of the notochordal and somitic mesoderm goes on unabated (Keller et al., 1985a, b; R. Keller and S. Jansa, unpublished results).

Significance of MIB progression for patterning

Although the anterior-posterior progression of cell behavior and differentiation could be predicted from previous work (see Wilson et al., 1989; Wilson and Keller, 1991), it was unexpected that cell behavior in axial and paraxial mesoderm would progress medially from lateral boundaries. Several facts suggest that the progression reflects the order in which the cells received the signals eliciting these cell behaviors. The simplest explanation of the MIB progression itself is that the cells receive the signals organising MIB in the order of MIB expression. The fact that cells not allowed to approach the boundaries, from which MIB and differentiation begin, do not show expression of MIB or differentiation implies that these boundaries provide the organising signals. Finally, a linear array of early gastrula animal cap cells grafted into the central notochordal territory, express MIB and differentiate from the anterior end posteriorly (C. Domingo, unpublished results). This means that these cells must be receiving organising signals from the notochord field in the order of MIB expression, since these cells do not normally express MIB at all, much less in a progressive pattern. We are currently testing for lateral-medial progression of behavior among strings of animal cap cells grafted into the IMZ in this orientation. Progressive reception of organising signals predicts that commitment of cells to behavioral and histological phenotypes should also be progressive from lateral to medial and anterior to posterior. We are testing this by moving cells from notochordal to somitic territories (and vice versa) and from medial to lateral positions (and vice versa) within each tissue. It is not clear at this point whether the progression of MIB reflects a propagation of a mechanical component of specification or a propagation of a specifying signal per se.

These results suggest a revision of our ideas about the geometry of the signals patterning notochordal and somitic mesoderm. Grafts of dorsal marginal zone or 'organiser' to the ventral marginal zone induce ventral tissues to make dorsal mesodermal and neural tissues (Spemann, 1938; Gimlich and Cooke, 1983). Explants of ventral marginal zone will not make dorsal tissues in isolation (Dale and Slack, 1987), nor will these tissues converge and extend (Keller and Danilchik, 1988). These facts suggest that a dorsalizing signal emanates from the organiser and passes laterally and ventrally in the marginal zone, converting these tissues to dorsal fates (the third signal in the the three-signal hypothesis; see Dale and Slack, 1987). The term 'dorsalizing' is correct in that the dorsalizing signal converts what would have become belly mesodermal derivatives (mesothelium and blood) to somitic tissue. But it is somewhat misleading, since the result of this signal is primarily the addition of prospective *posterior* somites to the segmental plate and thus it acts along the *prospective* anterior-posterior axis rather than along the dorsal-ventral axis of the notochordal and somitic mesoderm (Fig. 7). The actual mediolateral patterning or dorsal-ventral patterning within the prospective notochordal-somitic mesoderm is determined in a *lateral*-to-*medial* order, along arcs emanating from the vegetal edge of the IMZ and arcing toward the midline, parallelling the progress of the MIB expression. The first sign of dorsal behavior is MIB expression at the lateral edge of the somitic territories, which then arcs to the midline. The next patterning event is establishment of the notochordal-somitic boundary transverse to this arc. Finally, cell behavior and differentiation occur from lateral boundaries towards the medial region within both these territories.

The anterior-posterior and lateral-medial progressions could be explained with a predominately anterior-posterior signal radiating from the mid-dorsal region, somewhere in the vegetal endoderm, and passing along the anterior-posterior axes of notochordal and somitic mesoderm (Fig. 7). This signal would intersect with signals that had already set up special competences in the territories destined to be lateral somitic mesoderm and lateral notochordal mesoderm, such that cells in these regions would then organize their own type of behavior and differentiation medially from the boundary. Responses of cells to combinations of growth factors have led to the idea that mesodermal phenotypes may be determined by threshold responses to intersecting concentration gradients of two signals. Response thresholds would thus be expected to give rise to sharp boundaries, such as the notochord-somite boundary discussed (Green, 1992).

It should be noted that the medial edges of only the first 6 or 7 somites meet the notochord. More posterior somites come to lie next to the notochord as the notochord extends in the late gastrula and neurula. The notochord shears posteriorly with respect to the somitic mesoderm and pushs the blastopore posteriorly. As this occurs, the somitic mesoderm lying lateral and ventral to the blastopore swings dorsally alongside the notochord (Wilson et al., 1989; Keller et al., 1989a). These movements, rather than the earlier convergence movements (Fig. 8B), bring the ventral tissues of prospective posterior somitic mesoderm dorsally, next to the notochord (see Figs 5 and 6, Keller et al., 1991b).

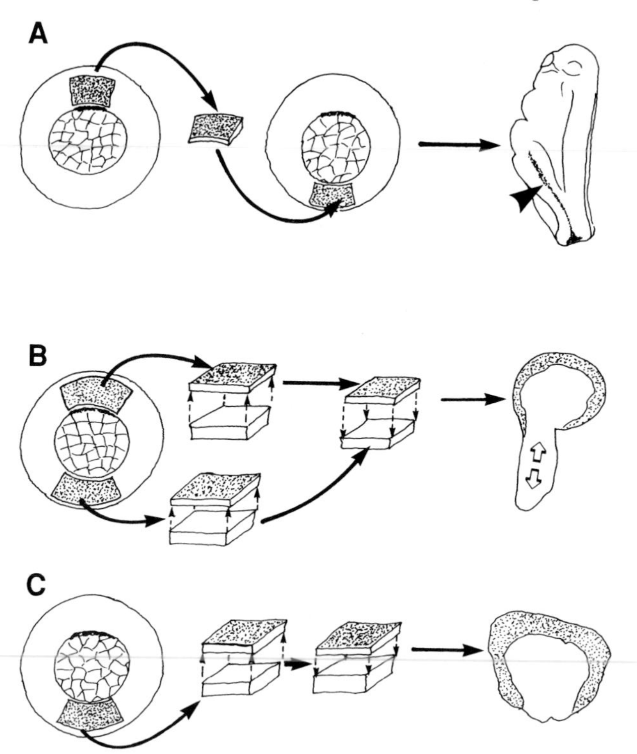

Fig. 10. Grafts of dorsal epithelium of the early gastrula to the ventral side of another early gastrula result in formation of a secondary axis (pointers) on the ventral side of the host embryo (A). Recombinant explants of dorsal epithelium and ventral deep cells converge, extend (open arrows) and differentiate dorsal tissues (B) whereas if the same tissues of the ventral marginal zone are separated and recombined, no convergence and extension or dorsal tissues develop (C). Modified from Keller et al., 1991b.

Epithelial mesenchymal interactions controlling mesodermal cell intercalation

The endodermal epithelium of the marginal zone is essential for the deep mesodermal cells to develop the ability to undergo mediolateral intercalation independently. If the endodermal epithelium is removed from explants in the early gastrula stage, convergence and extension do not occur (Wilson, 1990). After the early gastrula stage (stage 10.5), the deep cells are capable of independent convergence and extension by mediolateral cell intercalation without the epithelium (Fig. 4). The epithelial layer is not necessary for production of mechanical forces since deep cells alone can converge and extend after the early gastrula stage. The epithelial layer appears to provide the deep cells with essential support or instruction on how to do mediolateral cell intercalation. Grafts of the dorsal epithelial layer of the early gastrula to the ventral marginal zone of normal animals result in the formation of a second set of posterior axial structures, including notochord, somitic mesoderm and neural tissue (Fig. 10A) (Shih and Keller, 1992c). Also, embryos ventralized with UV irradiation can be rescued to

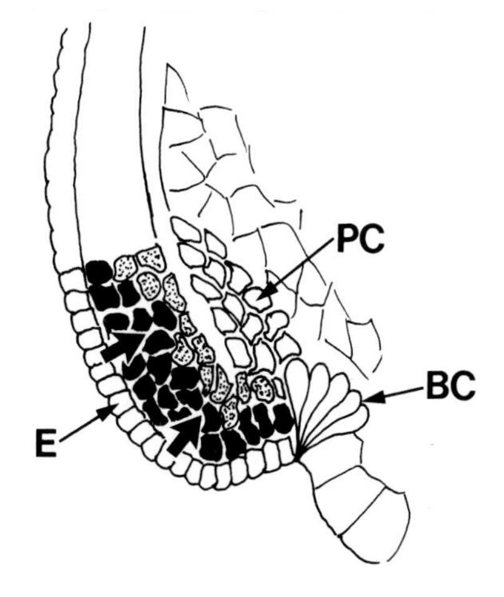

Fig. 11. Our concept of how the dorsal epithelial cells organise mediolateral intercalation is summarized diagrammatically. The dorsal epithelium interacts with the outermost mesodermal cells in the early gastrula (arrows) making it possible for them to do mediolateral cell intercalation at some later time (dark cells). Eventually the remaining, deeper mesodermal cells acquire the same ability (gray cells), either by interacting with the epithelium as they move closer to it by radial intercalation, or by interaction with other deep cells already capable of mediolateral intercalation. The prechordal cells, leading the migration of the mesoderm, are deprived of epithelial influence, having lost contact with their corresponding epithelium, the bottle cells (BC), during the formation of the bottle cells (see Hardin and Keller, 1988) and thus they never express mediolateral cell intercalation behavior.

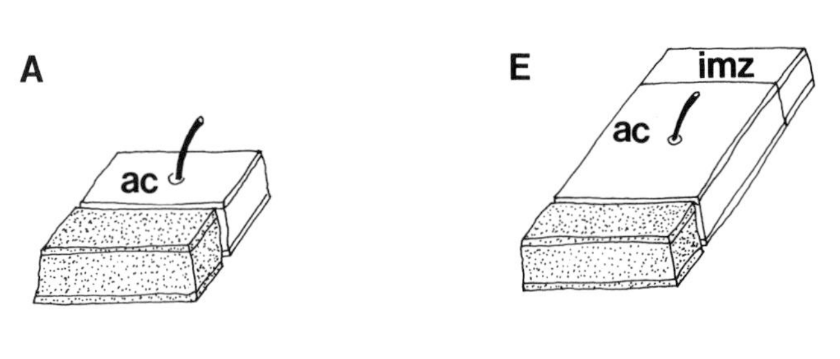

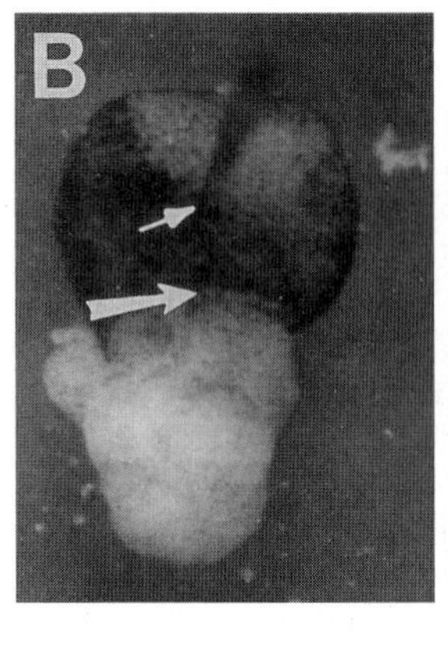

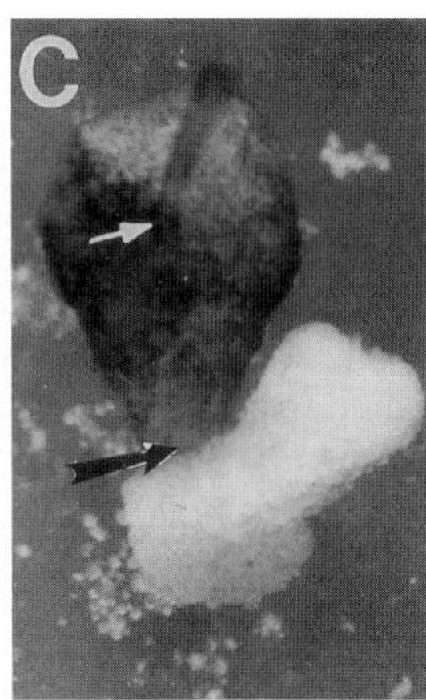

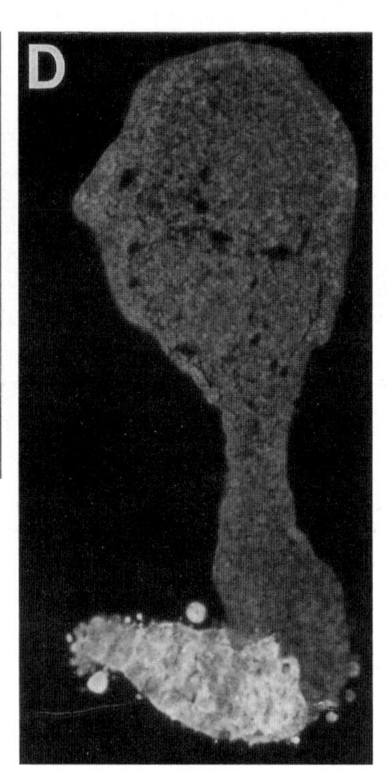

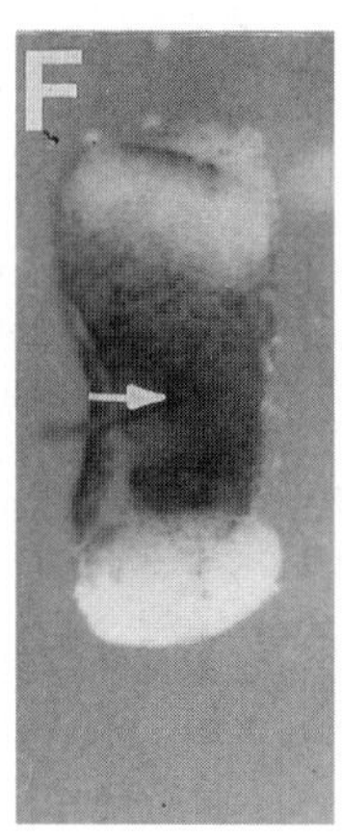

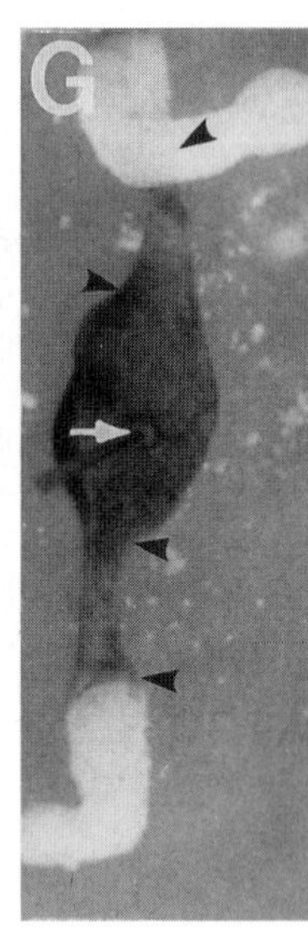

Fig. 12. Planar induction of the convergence and extension of the hindbrain-spinal cord is illustrated. The animal ends of explants of IMZ labeled with fluorescein dextran amine (shaded, A) were grafted to the edges of explants of the animal cap (ac), skewered to the substratum with an eyebrow hair (A). Convergence and extension of the animal cap tissue between the inducing IMZ and the eyebrow hair followed (B,C), whereas this tissue never converges and extends when explanted alone (not shown). Sections show that no labeled inducing tissue had invaded the animal cap region, proving that no vertical inductive interactions between mesoderm (IMZ) and the ectodermal animal cap had occurred (D). When a second IMZ (shaded) was added to the edge of the animal cap, opposite the native IMZ (E), a second region of convergence and extension appeared opposite the usual one, but with opposite polarity (F,G). All grafts were taken from the early gastrula stage. Modified from Keller et al., 1992b.

form posterior axial structures with grafts of dorsal epithelium. In both cases, unlabelled host tissues participated in convergence and extension and differentiated into dorsal axial structures as indicated by tissue markers. Grafts of dorsal epithelium to ventral explants results in convergence, extension and differentation of dorsal tissues in the explant (Fig. 10B), whereas control ventral explants do none of these things (Fig. 10C).

We envision the dorsal epithelium interacting with the underlying deep cells early in gastrulation (prior to stage 10.5) and either instructing or supporting the commitment of the shallowest deep cells to MIB expression (black shading, Fig. 11) from the outside inward (Fig. 11, arrows). Later, deeper mesodermal cells become committed to MIB expresssion (stippled cells). The prechordal mesoderm (unshaded cells, PC; Fig. 11), which lost contact with the epithelium during bottle cell formation (Hardin and Keller, 1988), are never committed to MIB expression, perhaps because of this loss of contact with the epithelium at the beginning of gastrulation.

Planar induction of neural convergence and extension

The convergence and extension of the posterior neural plate was not expected in explants (Keller et al., 1985a, b), since no mesoderm underlies this tissue prior to explantation. Thus neural-type convergence and extension could not have occurred by the classical vertical route of neural induction but must involve signals passing within the plane of the dorsal tissue of the embryo (see Keller and Danilchik, 1988) (*planar induction* see Phillips, 1991, for a review). However, planar induction is difficult to demonstrate for several reasons. Time-lapse tracing of the prospective areas of the neural anlagen show that the entire hindbrain-spinal cord of the neural plate is very broad and very, very short at the onset of gastrulation, and elongates tremendously by radial and mediolateral intercalation during late gastrulation and neurulation (Keller et al., 1992a). Marking experiments show that it is quickly underlain by the involuting mesoderm (Keller et al., 1992a). Moreover, tests of planar induc-

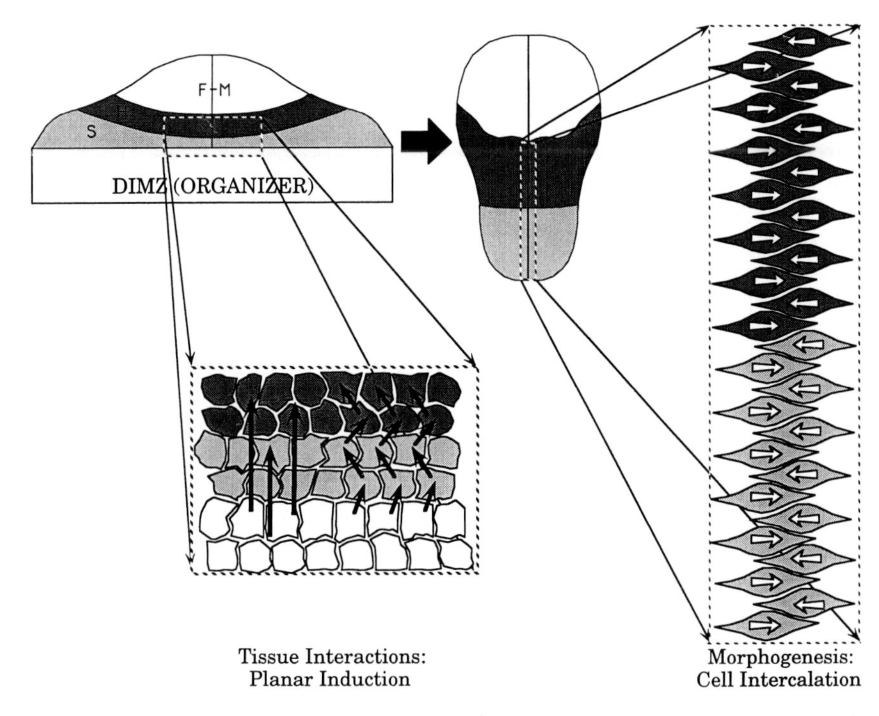

Fig. 13. Schematic diagrams show the significance of planar induction of cell intercalation in neural plate morphogenesis. The background of the figure (top) shows the changes in shape of the the prospective neural plate from the early gastrula through stage 15, including the forebrain and midbrain (F-M), hindbrain (H) and spinal cord (S) from stage 10+ to 15, as if they were flattened out in a plane instead of being wrapped around one side of sphere. Note the convergence and extension of the nervous system, particularly the hindbrain and spinal cord. The foreground (bottom) represents the underlying cellular events in induction and execution of the convergence and extension movements in the hindbrain and spinal cord (Keller et al., 1992a). Planar signals produced by the dorsal involuting marginal zone (DIMZ) act on a few rows of ectodermal cells animal to them in the first part of gastrulation (left). Planar signals could come from the mesoderm directly (long arrows), or relayed through the ectoderm by homeogenetic propagation of the inductive signal (multiple arrows). The morphogenetic response of the ectoderm to planar induction is to undergo radial and mediolateral cell intercalation, to produce a longer, thinner and narrower array of cells (right). Figure and legend from Keller et al., 1992b.

tion depend on removing all leading edge mesoderm from the inner surface of the explant, lest these invasive cells migrate animally in the core of the prospective neural tissues. There they are strung out along the anterior-posterior axis by convergence and extension and thus provide vertical interactions unbeknownst to the investigator (see Fig. 10, Keller, 1991). We solved these problems and demonstrated planar induction of neural convergence and extension by grafting labeled dorsal IMZ (organiser) to unlabeled animal cap tissue (Fig. 12A). The unlabelled animal cap tissue converges, extends and differentiates as neural tissue (Fig. 12B,C). The labeled IMZ (organiser) tissue retains its planar apposition throughout the experiment, thus precluding vertical interactions (Keller et al., 1992b) (Fig. 12D). Adding another IMZ to the opposite side of the animal cap results in two regions of polarized, neural convergence and extension, one induced by the original IMZ and another induced by the added IMZ (Fig. 12D-G) (Keller et al., 1992b). Cell labeling experiments show that neural convergence and extension take place by the same sequence of radial and mediolateral cell intercalation seen in the mesodermal tissue (Keller et al., 1992a). These two studies (Keller et al. 1992a, b) show how short-range planar signals can induce a large part of the nervous system. Planar signals delivered to the posterior edge of the prospective nervous system in the late blastula and early gastrula act over a short distance to induce cell intercalation. Cell intercalation then greatly elongates and narrows the prospective hindbrain-spinal cord (Fig. 13). The fact that planar signals do not have to act over more than a few cell diameters to induce neural convergence and extension does not preclude action of these or related signals over longer distances, perhaps through the entire brain. Planar signals also induce expression of anterior-posterior markers in their normal order (Doniach et al., this volume). The induced convergence and extension also bears its normal anterior-posterior polarity (Keller et al., 1992b).

Misconceptions

It is commonly thought that convergence and extension imply that ventral tissue moves dorsally. This does happen but not as part of the initial convergence. Between the early midgastrula (stage 10.5) and the late gastrula (stage 11.5), time-lapse recordings show that convergence brings the dorsal tissue closer to the ventral tissue and extension moves the dorsal blastoporal lip across the yolk plug, the combination of these movements resulting in closure of the blastoporal lips on the ventral aspect of the vegetal endoderm (see Keller and Danilchik, 1988). Convergence of dorsal tissue towards the ventral side is also evident in giant explants of the entire dorsal to ventral extent of the marginal zone (Keller and Danilchik, 1988). After stage 11.5, the blastoporal region is pushed ventrally by the extending notochord (see Figure 1, Keller and Danilchik, 1988) and in explants the mesoderm of the ventral marginal zone that forms the posterior somites is seen streaming around both sides of the blastoporal region and thus 'converging' on a dorsal position next to the notochord (see Wilson et al., 1989; Keller et al., 1989a; Figures 5, 6, Keller et al., 1991b). Thus this late 'convergence', which does actually sweep ventral material dorsally, has a different cellular and mechanical basis than the earlier convergence, which brings all points on the arc closer together, thus pulling the dorsal midline closer to the vegetal endoderm at the lateral ends of the arcs (see Fig. 9, above).

It is likewise a misconception that mesodermal cells 'migrate' dorsally because more cells find themselves in the dorsal midline than were there originally. The bipolar protrusive activity described above pulls the cells between one another along the *full length* of the arcs, sweeping from their anchorage in the vegetal endoderm on one side to their corrresponding anchorage on the other side. Cells have never been observed to migrate 'dorsally'. In fact, migration in either direction results when convergence and extension are failing rather than succeeding. When convergence and extension are blocked by mechanical resistence, the cells 'walk' out on one another's surfaces to greater than normal lengths (aspect ratios of 6 or 7) and then they break loose at one end or the other and 'migrate'. This results in mediolateral exchange of places but no effective intercalation (Shih and Keller, 1992b). These behaviors in *Xenopus* contrast with those in the teleost fish, where convergence movements do involve directed migration (Trinkaus et al., 1992).

Conclusions

Our recent work has focused on the cellular and mechanical basis of the powerful convergence and extension movements of the notochordal and somitic mesoderm and the hindbrain-spinal cord tissue of *Xenopus*, and the tissue interactions that regulate these behaviors. The general conclusion of the work described above, and other work as well (see Adams et al., 1990; Koehl et al., 1990; Hardin, 1990), is that relatively simple routines of cell motility and behavior, and rules of contact interaction, acting in the complex geometry and mechanical environment of biological (embryonic) tissues, produce unexpected and extraordinary mechanisms of translating forces developed by individual cells into mass tissue deformations in early embryogensis. These are unexpected and extraordinary only in that they have been largely ignored in modern developmental and cell biology. It is now clear that a large part of the illusive 'mechanism' of morphogenesis lies in the cytomechanics of cell populations, and it is in this arena that the significance and function of individual cells and specific molecules will be defined and given meaning.

We thank Dr Connie Lane, Dr Amy Sater, and Dr Jeremy Green for their insightful comments and suggestions.

References

Adams, D., Keller, R. and Koehl, M. A. R. (1990). The mechanics of notochord elongation, straightening, and stiffening in the embryo of Xenopus laevis. *Development* **110**, 115-130.

Dale, L. and Slack, J. M. W. (1987). Regional specification within the mesoderm of early embryos of *Xenopus laevis. Development* **100**, 279-295.

Doniach, T., Phillip, C. and Gerhart, J. (1992). Planar induction of position-specific homeobox genes in the neurectoderm of Xenopus laevis. *Science,* in press.

Doniach, T., Zoltewicz, S. and Gerhart, J. (1992) Induction of anteroposterior neural pattern in Xenopus by planar signals. *Development Supplement* (this volume).

Gimlich, R. and Cooke, J. (1983). Cell lineage and induction of second nervous systems in amphibian development. *Nature* **30**, 471-473.

Green, J. B. (1992). The role of thresholds and mesoderm inducing factors in axis patterning in Xenopus. In *Formation and Differentiation of the Early Embryonic Mesoderm* (eds. Lash, J. and Bellairs, R.) New York: Plenum Press, in press.

Hardin, J. (1990). Context-sensitive cell behaviors during gastrulation. In *Control of Morphogenesis by Specific Cell Behaviors.* (eds. Keller, R. and Fristrom, D.) *Seminars in Develop. Biol.* **1**, 335-345.

Hardin, J. and Keller, R. (1988). The behavior and function of bottle cells during gastrulation of *Xenopus laevis. Development* **103**, 211-230.

Jacobson, A. and Gordon, R. (1976). Changes in the shape of the developing vertebrate nervous system analyzed experimentally, mathematically, and by computer simulation. *J. Exp Zool.* **197**, 191-246.

Keller, R. E. (1975). Vital dye mapping of the gastrula and neurula of Xenopus laevis. I. Prospective areas and morphogenetic movements of the superficial layer. *Dev. Biol.* **42**, 222-241.

Keller, R. E. (1976). Vital dye mapping of the gastrula and neurula of Xenopus laevis. II. Prospective areas and morphogenetic movements of the deep layer. *Dev. Biol.* **51**, 118-137.

Keller, R. E. (1978). Time-lapse cinemicrographic analysis of superficial cell behavior during and prior to gastrulation in Xenopus laevis. *J. Morph.* **157**, 223-248.

Keller, R. E. (1981). An experimental analysis of the role of bottle cells and the deep marginal zone in gastrulation of Xenopus laevis. *J. Exp. Zool.* **216**, 81-101.

Keller, R. E. (1984). The cellular basis of gastrulation in Xenopus laevis: active post-involution convergence and extension by mediolateral interdigitation. *Am. Zool.* **24**, 589-603.

Keller, R. E. (1986). The cellular basis of amphibian gastrulation. In *Developmental Biology: A Comprehensive Synthesis. Vol. 2. The Cellular Basis of Morphogenesis.* (ed. L. Browder) pp. 241-327. New York: Plenum Press.

Keller, R. E. (1987). Cell rearrangement in morphogenesis. *Zool. Sci.* **4**, 763-779.

Keller, R. (1991). Early embryonic development of Xenopus laevis. In Xenopus laevis: *Practical Uses in Cell and Molecular Biology* (ed. B. Kay and B. Peng) *Methods in Cell Biology* **36**. pp.61-113. San Diego: Academic Press.

Keller, R. E., Danilchik, M., Gimlich, R. and Shih, J. (1985a). Convergent extension by cell intercalation during gastrulation of Xenopus laevis. In *Molecular Determinants of Animal Form* (ed. G. M. Edelman), pp. 111-141. San Diego: Academic Press.

Keller, R. E., Danilchik, M., Gimlich, R. and Shih, J. (1985b). The function of convergent extension during gastrulation of *Xenopus laevis. J. Embryol. Exp. Morph.* **89 Supplement**,185-209.

Keller, R. E. and Trinkaus, J. P. (1987). Rearrangement of enveloping layer cells without disruption of the epithelial permeability barrier as a factor in Fundulus epiboly. *Dev. Biol.* **120**, 12-24.

Keller, R. E. and Danilchik, M. (1988). Regional expression, pattern and timing of convergence and extension during gastrulation of *Xenopus laevis. Development* **103**, 193-210.

Keller, R. E. and Tibbetts, P. (1989). Mediolateral cell intercalation in the dorsal axial mesoderm of Xenopus laevis. *Dev. Biol.* **131**, 539-549

Keller, R., Cooper, M. S., Danilchik, M., Tibbetts, P. and Wilson, P. A. (1989a). Cell intercalation during notochord development in Xenopus laevis. *J. Exp. Zool.* **251**, 134-154.

Keller, R. E., Shih, J. and Wilson, P. A. (1989b). Morphological polarity of intercalating deep mesodermal cells in the organizer of Xenopus laevis gastrulae. In *Proceedings of the 47th Annual Meeting of the Electron Microscopy Society of America.* San Francisco: San Francisco Press. p. 840.

Keller, R., Shih, J. and Wilson, P. A. (1991a). Cell motility, control and function of convergence and extension during gastrulation of Xenopus. In *Gastrulation: Movements, Patterns, and Molecules.* (ed. R. Keller, W. Clark and F. Griffin) pp. 101-119. New York: Plenum Press.

Keller, R., Shih, J., Wilson, P. A. and Sater, A. K. (1991b). Patterns of cell motility, cell interactions, and mechanism during convergent extension in Xenopus. In *Cell-Cell Interactions in Early Development. Society for Developmental Biology, 49th Symposium* (ed. G. C. Gerhart) pp. 31-62. New York: Wiley-Liss.

Keller, R. and Shih, J. (1992). Mediolateral cell intercalation of mesodermal cells in the Xenopus laevis gastrula. In *Formation and Differentiation of the Early Embryonic Mesoderm* (ed. J. Lash and R. Bellairs) New York: Plenum Press, in press.

Keller, R., Shih, J. and Sater, A. (1992a). The cellular basis of the convergence and extension of the Xenopus neural plate. *Develop. Dynamics* **193**, 199-217.

Keller, R. E., Shih, J., Sater, A. K. and Moreno, C. (1992b). Planar induction of convergence and extension of the neural plate by the organizer of Xenopus. *Develop. Dynamics* **193**, 218-234.

Koehl, M. A. R., Adams, D. and Keller, R. (1990). Mechanical development of the notochord in Xenopus early tail-bud embryos. *Biomechanics of Active Movement and Deformation of Cells. NATO ASI Series. Vol. H 42* (ed. N. Akkas) pp. 471-485. Berlin: Springer-Verlag.

Phillips, C. (1991). Effects of the dorsal blastoporal lip and the involuted dorsal mesoderm on neural induction in Xenopus laevis. In *Cell-Cell Interactions in Early Development. Society for Developmental Biology, 49th Symposium* (ed. G. C. Gerhart) pp. 93-107. New York: Wiley-Liss.

Schechtman, A. M. (1942). The mechanics of amphibian gastrulation. I. Gastrulation-producing interactions between various regions of an anuran egg (Hyla regilia). *Univ. Calif. Publ. Zool.* **51**, 1-39.

Schoenwolf, G. C. and Alvarez, I. C. (1989). Roles of neuroepithelial cell rearrangement and division in shaping the avian neural plate. *Development* **106**, 427-439.

Shih, J. and Keller, R. (1992a). Patterns of cell motility in the organiser and dorsal mesoderm of *Xenopus. Development* **116**, in press.

Shih, J. and Keller, R. (1992b). Cell motility driving mediolateral intercalation in explants of *Xenopus. Development* **116**, in press.

Shih, J. and Keller, R. (1992c). The epithelium of the dorsal marginal zone of *Xenopus* has organiser properties. *Development* **116**, in press.

Spemann, H. (1938). *Embryonic Development and Induction.* New Haven: Yale University Press.

Trinkaus, J. P., Trinkaus, M. and Fink, R. (1992). On the convergent cell movements of gastrulation in Fundulus. *J. Exp. Zool.* **261**, 40-61.

Vogt, W. (1929). Gestaltungsanalyse am Amphibienkeim mit örtlicher Vitalfärbung. II. Teil. Gastrulation und Mesodermbildung bei Urodelen und Anuren. *Wilhelm Roux Arch. EntwMech. Org.* **120**, 384-706.

Warga, R. and Kimmel, C. (1990). Cell movements during epiboly and gastrulation in zebrafish. *Development* **108**, 569-580.

Weliky, M., Minsuk, S., Oster, G. and Keller, R. (1992). The mechanical basis of cell rearrangement. II. Models for cell behavior driving notochord morphogenesis in *Xenopus laevis. Development* **113**, 1231-1244.

Wilson, P. (1990). The development of axial mesoderm in Xenopus laevis. PhD. Dissertation, University of California, Berkeley.

Wilson, P. A., Oster, G. and Keller, R. (1989). Cell rearrangement and segmentation in *Xenopus*: direct observation of cultured explants. *Development* **105**, 155-166.

Wilson, P. A. and Keller, R. E. (1991). Cell rearrangement during gastrulation of *Xenopus*: direct observation of cultured explants. *Development* **112**, 289-305.

Winklbauer, R. (1990). Mesodermal cell migration during Xenopus gastrulation. I. Interaction of mesodermal cells with fibronectin. *Dev. Biol.* **142**, 155-168.

Development 1992 Supplement, 93-97 (1992)

Fate mapping the neural plate and the intraembryonic mesoblast in the upper layer of the chicken blastoderm with xenografting and time-lapse videography

HILDE BORTIER[1,*] and L. C. A. VAKAET[2]

[1]*Laboratory of Embryology, University of Gent, Godshuizenlaan 4, B-9000 Belgium*
[2]*Laboratory for Experimental Cancerology, University Hospital, Gent, Belgium*

*Author for correspondence

Summary

The disposition of the Anlage fields of the neural plate and the intraembryonic mesoblast in the upper layer of the chicken blastoderm was studied at the primitive streak stage prior to the regression of Hensen's node (stages 5V to 6V, L. Vakaet (1970) *Arch. Biol.* 81, 387-426). Chicken blastoderms were cultured by New's technique on a mixture of thin egg white and agar. The anterior half of the deep layer was reflected with a tungsten needle. A circular fragment of the upper layer was punched out with a pulled out Pasteur pipette and discarded. It was replaced with an isotopic and isopolar piece of quail upper layer that was punched out with the same pipette. The deep layer was replaced and the chimeras were reincubated for 24 hours. The xenografts were followed with time-lapse videography. After fixation, the quail cells were located using Le Douarin's quail nucleolar marker technique. Integrating the observations with time-lapse videography and the results of Feulgen stained sections, we have drawn a new fate map of the disposition of the Anlage fields in the upper layer of the chicken blastoderm at stages prior to the regression of Hensen's node (stages 5V to 6V). The disposition of the neural plate and of the notochord, somites, nephrotome and lateral plates was therefore determined before the Anlage fields are morphologically discernible. The pathway of the fields in the upper layer towards their disposition was documented with time-lapse videography in chimeric chicken blastoderms that developed normally.

Key words: fate map, time-lapse videography, xenografts, chicken blastoderm.

Introduction

The neural plate of avian embryos becomes visible at stage 7V (Vakaet, 1970) corresponding to stage 4 of Hamburger and Hamilton (1951). Pasteels (1937) and Malan (1953) used vital staining to construct fate maps in prestreak and primitive streak stages (stages 0V to 6V). Due to diffusion and fading of the marks, this technique has limited value for determining the final destination of the marked fields. The maps of Spratt (1952) are based on carbon marking at primitive streak and headfold stages (stages 7V to 9V). Carbon marks on the upper side of the upper layer are not reliable because the carbon particles can remain unattached or move randomly. At stages prior to the regression of Hensen's node (which starts at stage 7V corresponding to stage 4 of Hamburger and Hamilton), the disposition of the prospective neural plate and the intraembryonic mesoblast has not been exactly determined as appears from the differences between the maps of Rawles (1936), Rudnick (1944), Waddington (1952), Rosenquist (1966) and Vakaet (1984). We have studied the Anlage fields of both the neural plate and the intraembryonic mesoblast in the upper layer in stages prior to the regression of Hensen's node (5V to 6V). At stage 5V, upper layer cells start to ingress through the primitive streak. At the beginning of stage 6V, the primitive groove is fully developed. The ingression of the endoblast is completed by the end of stage 6V (Vakaet, 1962). We used time-lapse videography to follow the chick-quail xenografts. After 24 hours of culture, the grafts and their position within the embryo were determined by using the nucleolar marking technique (Le Douarin, 1973). By combining xenografting and registration of the morphogentic movements of the grafts with videography, we have drawn a new fate map of the neural plate and the intraembryonic mesoblast before they are morphologically discernable.

Materials and methods

Chicken eggs (White Rock, from the Rijksstation voor Pluimveeteelt, Merelbeke, B-9820) and quail eggs (from laboratory stock) were incubated at 38°C for 15 hours to obtain stage 5V to 6V blastoderms. They were cultured by New's (1955) technique, except that the substratum used was a mixture of 25 ml thin egg white and a gel made of 150 mg Bacto-agar Difco (Detroit, Michigan) in 25 ml Ringer's solution, instead of pure thin egg white. This semi-solid medium allowed microsurgery and further culturing on the same substratum.

For microsurgery, we used a Pasteur pipette with a tip diameter of 0.20 mm to 0.25 mm. The experiments consisted in punching out and discarding a circular fragment of the upper layer. It was replaced with an isotopic and isopolar piece of quail upper layer that was punched out with the same pipette (Bortier and Vakaet, 1992). After the intervention a Polaroid photograph was taken using an M8 stereomicroscope (Wild, Heerbrugg, CH-9435) at magnification 25×. The culture vessels were covered with a glass lid, sealed with melted paraffin and further incubated at 38 ± 0.5°C.

For time-lapse videography (Bortier and Vakaet, 1987) we used an M8 stereomicroscope (Wild, Heerbrugg, CH-9435), on top of which a WV-1850 camera (Panasonic, Osaka, Japan) was mounted in a plexiglass incubator at 38 ± 1°C (Fig.1). The camera was linked to a U-matic video-recorder VO-5850P (Sony, Tokyo, Japan) through an animation control unit EOS AC580 (EOS Electronics A.V., Barry, South Glamorgan). A time date generator WJ-810 (Panasonic, Osaka, Japan) displayed the chronological information on a WV-5340 monitor (Panasonic, Osaka, Japan). One video-image was recorded every 30 seconds, yielding an acceleration of 750× at normal projection speed (25 images/second). To reduce the temperature gradient due to the light that is transmitted from below, an open transparent well with a planparallel bottom was filled with distilled water and placed under the culture vessel. The cooling of the bottom of the vessel thus provided was usually sufficient to prevent damping of the glass lid. If necessary, a thin layer of glycerin was spread on the inside of the glass lid.

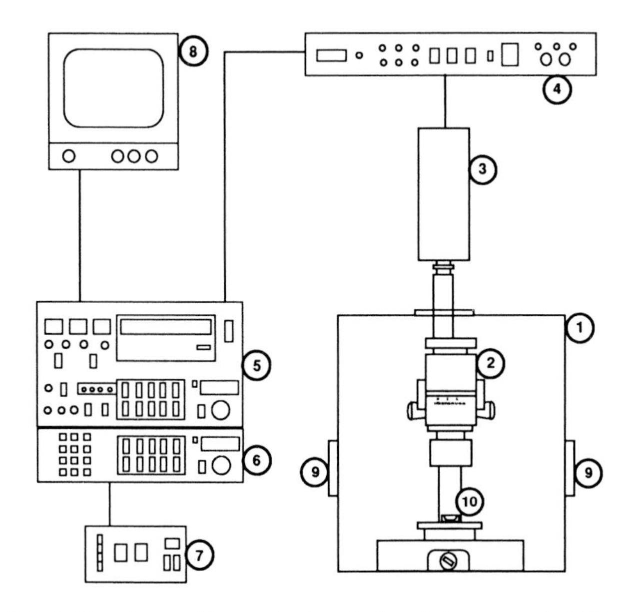

Fig. 1. Time-lapse videography installation. 1. Incubator; 2. Wild M8 stereomicroscope; 3. WV-1850 camera (Panasonic, Osaka, Japan); 4. Time date generator WJ-810 (Panasonic, Osaka, Japan); 5. U-matic video-recorder VO-5850P (Sony, Tokyo, Japan); 6. Animation control unit EOS AC580 (Electronics, Barry, South Glamorgan); 7. Interval timer (Vel, Leuven, B-3030); 8. WV-5340 monitor (Panasonic, Osaka, Japan); 9. Entrances; 10. Culture vessel.

All chimeras were reincubated for 24 hours. The culturing was interrupted by fixing the blastoderms in a mixture of absolute alcohol, formaldehyde 4%, acetic acid (75:20:5, v:v:v). After paraffin embedding and sectioning, they were stained after Feulgen and Rossenbeck (1924).

Results

The results are the integration of time-lapse videography and histology from 72 chimeras that developed normally by in ovo standards. The fate map is drawn for a chicken blastoderm at stage 6V, prior to the regression of Hensen's node.

Videography

Before healing, the quail grafts unroll at their edges and thereby enlarge. After healing they are distinguishable in the host as clearly delineated patches. This is due to the fact that the quail cells stay together and that they contain less yolk than the chicken cells.

Grafts anterior to a transverse line through Hensen's node can be followed up to stage 9V (headfold stage). Only grafts in a small area anterior to this line seem to move into the neural plate. To do so, they move as a field towards the midline. Concomitantly the more anterior grafts extend in an anterior direction, the more posterior grafts in a posterior direction. The neural plate therefore extends as a whole in an anterior as well as in a posterior direction (Fig. 2B). The plate becomes visible due to its convergence. After elevation of the neural lips, the whole neural plate is no longer in focus and the grafts can not be followed further. Grafts outside the neural plate form sharply delineated fields.

All grafts posterior to Hensen's node migrate towards the primitive streak in a rostrolateral to mediocaudal direction. They form patches that elongate in the direction of their migration. As they arrive at the primitive streak, the grafts extend in length. After ingression, mesoblast cells can no longer be followed. Sometimes a graft seen as a delineated patch, partly disappears and partly stays in the upper layer.

Histology

Histology of the grafts was performed at later stages than could be followed with videography. This enabled us to recognize the nature of the tissues formed from the grafts.

We found that the transverse line through Hensen's node in stage 6V constitutes a border between the neural plate and the intraembryonic mesoblast. Grafts anterior to this line only formed neural tissue. Grafts caudal to this line only yielded intraembryonic mesoblast. Grafts that partly became neural tissue and partly mesoblast had been inserted on the border between both fields, as could be seen with videography.

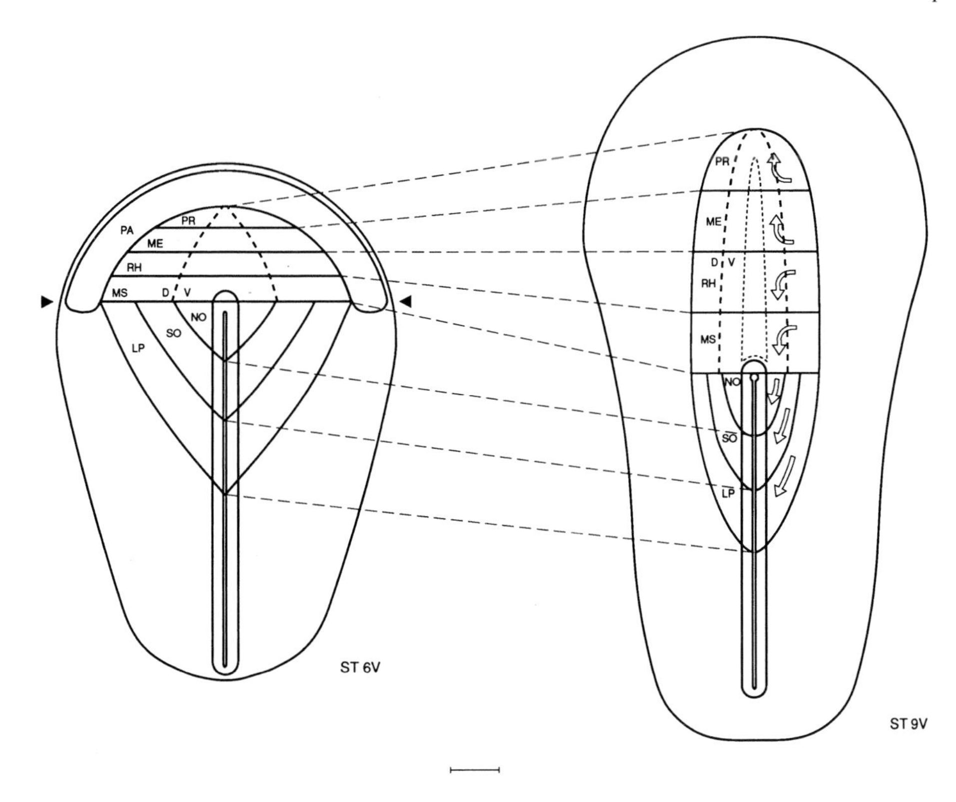

Fig. 2. Neural plate and intraembryonic mesoblast in stages 6V and 9V on dorsal views. (A) Disposition of the Anlage fields in stage 6V. A transverse line (between the arrowheads) through Hensen's node forms the border between neural tissue and intraembryonic mesoblast. PA, proamnion; PR, prosencephalon; ME, mesencephalon; RH, rhombencephalon; MS, spinal cord. A dotted line marks the border between dorsal and ventral neural structures; D, dorsal; V, ventral; NO, notochord; SO, somites; LP, lateral plate. (B) Disposition of the Anlage fields in Stage 9V. In the neural plate, the arrows indicate the convergence-extension movements within these fields. In the mesoblast area, the arrows show the convergence-regression movements of the Anlage fields. Scale bar : 200 μm.

Anterior to the transverse line through Hensen's node lies the neurectoblast area. Its peripheral border is situated at the inner arc of the endophyllic crescent, which marks the proamnion posteriorly.

(1) Grafts inside the inner arc of the proamnion were found in the nervous system. The most anterior grafts are found in prosencephalon, more posterior grafts successively in mesencephalon and rhombencephalon, while the most posterior grafts that formed neural tissue were found in the spinal cord (Fig. 2A). As observed with videography, each graft in the neural plate forms a field that extends lengthwise. From stage 9V on, the posterior extension of the grafts was longer in the rhombencephalon and especially in the spinal cord. Grafts just rostral to Hensen's node eventually extend throughout the whole spinal cord (Rosenquist, 1966; Schoenwolf et al., 1989). Lateral grafts in the neural plate mark the dorsal parts of the neural tube, medial grafts mark ventral parts. The limit between dorsal and ventral lateral plates is indicated by a dotted line on Fig. 2.

(2) Grafts on the inner arc of the endophyllic crescent could be traced back in the rim of the neural plate, in neural crest and in head epiblast.

(3) Grafts outside the inner arc of the endophyllic crescent were found in epiblast-derived structures of the head (epiblast of the pharyngeal membrane, adenohypophysis, primordium of the lens placode, auditory placode) and extraembryonic epiblast of head and trunk (amnion).

The part of the upper layer caudal to the transverse line through Hensen's node at stage 6V contains presumptive intraembryonic mesoblast. In this area, the prospective notochord, somites and lateral plates occupy three fields (Fig. 2A). In a mediorostral triangle, we found future notochord. Bilaterally flanking the notochord, we found a strip of future somitic tissue. A third strip flanked the somite area on both sides. It yielded nephrotome and lateral plate. The ingression of these fields at stage 5V has been described (Bortier and Vakaet, 1992).

Discussion

Part of the discrepancies between earlier presumptive fate maps is due to the vagueness of the stage of the blastoderms. Incubation hours can not be relied on for staging. For this reason, Vakaet (1970) defined ten stages between the unincubated and the headfold chicken blastoderm. He added three stages between stages 3 and 4 of Hamburger and Hamilton (1951). In stage 4V, the streak is rod-like and elongating; in stage 5V, it is elongating and grooving and, in stage 6V, it ends its elongation and is fully grooved. In stage 7V (stage 4 of Hamburger and Hamilton), regression of the anterior half of the streak takes place.

Another reason for the discrepancies between earlier presumptive fate maps is the use of different markers. Vital dyes (Pasteels, 1937 and Malan, 1953) fade away. Carbon marks (Spratt, 1952) are not reliable. In tritiated thymidine-labeled grafts (Rosenquist, 1966), only 50-70% of the graft nuclei were marked. Xenografting after Le Douarin (1973) is reliable cell per cell. Moreover, the grafts can be followed with time-lapse videography during their movements in the upper layer. Therefore, the movements need not be reconstructed afterwards, as they are observed.

The neural plate (Fig. 2A) does not extend caudally beyond a transverse line through Hensen's node. At stage 6V, it is remarkably small and occupies only a segment of a disk. It is limited anteriorly by the inner arc of the proam-

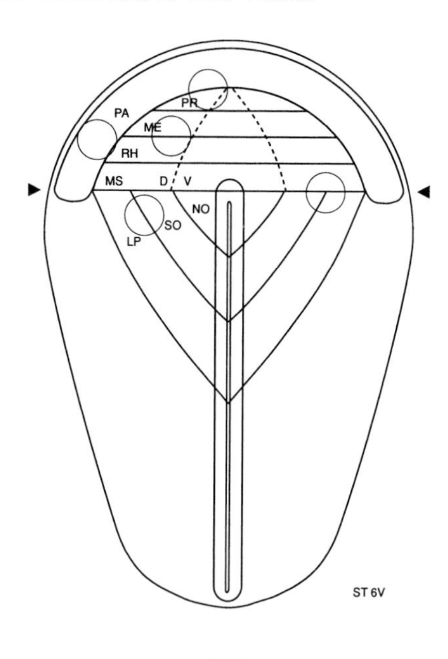

Fig. 3. Some examples of grafts to illustrate that one graft can mark several structures. Scale bar : 100 µm. Legends as in Fig. 2.

nion. Its posterior border is a transverse line with the intraembryonic mesoblast. This is in contradiction with the fate maps by Rawles (1936), Rudnick (1944), Spratt (1952), Rosenquist (1966) and Vakaet (1984). All these maps show a prospective neural plate with caudally directed wing-like extensions.

As demonstrated with videography, these fields converge. This reduces the width of the neural plate (compare Fig. 2A and 2B). Concomitantly the fields elongate in an anterior and posterior direction from the time Hensen's node starts to regress (convergence-extension as coined by Pasteels, 1937). In Fig. 2A, we have ascribed tentatively the same anteroposterior widths to the prosencephalon, mesencephalon, rhombencephalon and to the spinal cord. Moreover, we assumed in Fig. 2B that the elongation of these fields is similar up to stage 9V, the last presomitic stage. After stage 9V, the length over which grafts extend posteriorly is longer in the rhombencephalon and even more in the spinal cord.

In the region behind Hensen's node, the mesoblast cells ingress through the regressing anterior half of the primitive streak (Fig. 2B). Some grafts partly ingress and partly stay in the upper layer as described with videography. This illustrates the advantage of making grafts with a diameter of about 200 µm. Due to their size, they usually mark more than one Anlage field (Fig. 3). This marking of at least two Anlage fields shows that some fields have a border in common. Small grafts that only mark one Anlage field do not show its topographical disposition.

Our fate map complements the results of Couly and Le Douarin (1985, 1987, 1988) who studied with xenografts the neural plate and its periphery in presomitic and early somitic stages. At these stages the neural tube is already closing. Our xenografts were made in stage 5-6V blastoderms in which the topographical situation of the Anlage fields was not yet predictable. The resulting fate map shows the disposition of the Anlage fields prior to their final morphogenetic movements in the upper layer. The most extensive of these movements are: (1) the convergence-extension of the neural plate that is situated at stage 6V in a small segment of the area pellucida anterior to Hensen's node, (2) the backward movement of the border between the neural plate and the intraembryonic mesoblast that corresponds to the regression of Hensen's node and (3) the convergence-regression of the intraembryonic mesoblast fields and the shrinking of their surface by ingression of cells through the primitive streak.

This work was supported by a grant of the National Fund for Scientific Research (NFSR.) (3.9001.87), Brussels/Belgium. H. Bortier is a Senior Research Assistant of the NFSR., Brussels/Belgium. The authors thank G. Van Limbergen, P. Martens and R. Mortier for excellent technical assistance.

References

Bortier, H. and Vakaet, L. (1987). Videomicrography in the study of morphogenetic movements in the early chick blastoderm. *Cell Diff.* **20** Suppl., 114S.

Bortier, H. and Vakaet, L. (1992). Mesoblast anlage fields in the upper layer of the chicken blastoderm at stage 5V. In *Formation and Differentiation of Early Embryonic Mesoderm*, (eds. J. W. Lash and R. Bellairs), New York : Plenum Publishing Corporation (in press)

Couly, G. F. and Le Douarin, N. M. (1985). Mapping of the early neural primordium in quail-chick chimeras. I. Developmental relaionships between placodes, facial ectoderm, and prosencephalon. *Dev. Biol.* **110**, 422-439.

Couly, G. F. and Le Douarin, N. M. (1987). Mapping of the early neural primordium in quail-chick himeras. II. The prosencephalic neural plate and neural folds : Implications for the genesis of cephalic human congenital abnormalities. *Dev. Biol.* **120**, 198-214.

Couly, G. F. and Le Douarin, N. M. (1988). The fate map of the cephalic neural primordium at the presomitic to the 3-somite stage in the avian embryo. *Development* **103 Suppl.**, 101-113.

Feulgen, R. and Rossenbeck, H. (1924). Mikroskopisch-chemischer Nachweis einer Nucleinsäure vom typus der Thymonucleinsäure und die darauf beruhende elektive Färbung von Zellkernen in mikroskopischen Präparaten. *Hoppe Seylers Z. Physiol. Chem.* **135**, 203-252.

Hamburger, V. and Hamilton, H. L. (1951). A series of normal stages in the development of the chick embryo. *J. Morphol.* **88**, 49-92.

Le Douarin, N. M. (1973). A Feulgen positive nucleolus. *Exp. Cell Res.* **77**, 459-468.

Malan, M. E. (1953). The elongation of the primitive streak and the localization of the presumptive chorda-mesoderm on the early chick blasoderm studied by means of coloured marks with nile blue sulphate. *Arch. Biol.* **64**, 149-182.

New, D. A. T. (1955). A new technique for the cultivation of the chick embryo in vitro. *J. Embryol. Exp. Morphol.* **3**, 326-331.

Pasteels, J. (1937). Etudes sur la gastrulation des vertébrés méroblastiques III. Oiseaux IV. Conclusions générales. *Arch. Biol.* **48**, 381-488.

Rosenquist, G. C. (1966). A radioautographic study of labeled grafts in the chick blastoderm. Development from primitive-streak stages to stage 12. *Contrib. Embryol. Carnegie Inst.* **38**, 71-110.

Rawles, M. E. (1936). A study in the localization of organ-forming areas in the chick blastoderm of the head-process stage. *J. Exp. Zool.* **72**, 271-315.

Rudnick, D. (1944). Early history and mechanics of the chick blastoderm. *Quart. Rev. Biol.* **19**, 187-212.

Schoenwolf, G. C., Bortier, H. and Vakaet, L. (1989). Fate mapping the avian neural plate with quail/chick chimeras : Origin of prospective median wedge cells. *J. Exp. Zool.* **249**, 272-278.

Spratt, N. T. (1952). Localization of the prospective neural plate in the early chick blastoderm. *J. Exp. Zool.* **120**, 109-130.

Vakaet, L. (1962). Some new data concerning the formation of the definitive endoblast in the chick embryo. *J. Embryol. Exp. Morph.* **10**, 38-57.

Vakaet, L. (1970). Cinephotomicrographic investigations of gastrulation in the chick blastoderm. *Arch. Biol.* **81**, 387-426.

Vakaet, L. (1984). Early development of birds. In *Chimeras in Developmental Biology* (eds. N. Le Douarin and A. Mc. Laren), pp. 71-88. London : Academic Press.

Waddington, C. H. (1952). *The Epigenetics of Birds,* pp. 30-31, London : Cambridge Univ. Press.

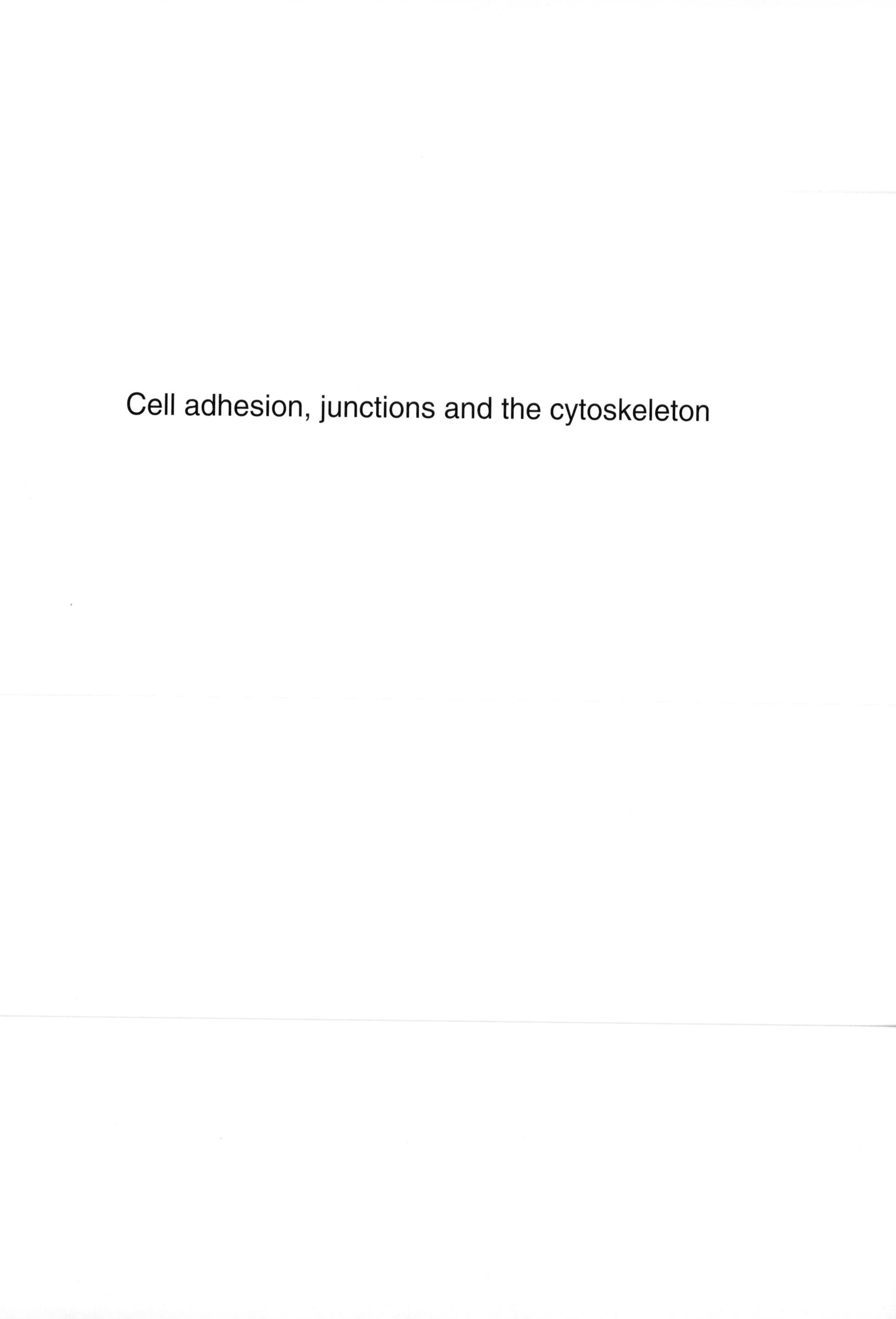

Cell adhesion, junctions and the cytoskeleton

Development 1992 Supplement, 99-104 (1992)
Printed in Great Britain © The Company of Biologists Limited 1992

NCAM and its polysialic acid moiety: a mechanism for pull/push regulation of cell interactions during development?

U. RUTISHAUSER

Departments of Genetics and Neuroscience, Case Western Reserve University, 10900 Euclid Ave., Cleveland OH 44106, USA

Summary

Many cell adhesion molecules have a distinct pattern of expression and well-defined role in cell-cell recognition. In contrast, NCAM is broadly expressed and perturbations of its function affect many diverse aspects of embryonic development. Evidence has been obtained suggesting that the molecule and its polysialic acid moiety serve not only to contribute to specific interactions, but also to regulate overall cell-cell apposition. In this latter mode, the molecule can have both a positive and a negative effect on a wide variety of contact-dependent cellular events.

Key words: neural cell adhesion molecule, regulation of cell interactions, adhesive preferences, axon outgrowth.

Introduction

Being the first cell adhesion molecule to be identified in vertebrate embryos, NCAM tended to be considered as a candidate receptor in nearly every cell-cell interaction that occurs during development. The molecule itself encouraged this practice by its expression on diverse cell types in many tissues, and observations that antibodies that block NCAM-associated adhesion interfere with a wide variety of cellular events. Subsequently, other adhesion molecules were discovered that were more specifically associated with particular events. Eventually it became clear that the role of NCAM in these same events needed re-evaluation, and a rationale provided for the extraordinarily broad influence of this molecule. This chapter reviews the biological role of NCAM from this perspective, focussing on its ability to act both as a mediator of specific interactive preferences, and as a general regulator of cell-cell contact.

NCAM expression in development

The broad distribution of NCAM in vertebrate embryos includes the primitive neural ectoderm of the neural plate, transient expression in morphogenetically active structures (such as notochord, somites, placodes, epidermis, mesenchyme, mesonephros), a uniform and persistent presence on neurons, and a spatially and temporally regulated expression by glial and muscle cells (Thiery et al., 1982; Jacobson and Rutishauser, 1986; Keane et al., 1988; Grumet et al., 1982; Silver and Rutishauser, 1984; Maier et al., 1986). Therefore, a substantial portion of embryonic tissues are composed of cells that produce, or whose precursors have produced, detectable amounts of NCAM. On this basis alone, it is clear that NCAM by itself is not well suited for identification of individual cells or specification of precise cell-cell interactions.

NCAM is a complex molecule with multiple functions

There is a tendency to label a molecule's function in the framework of its initial discovery, which for NCAM was as a homophilic cell adhesion molecule. It is now clear from its large size and diverse structural domains and variants (see Rutishauser, 1991, for review), that this simplistic view will require expansion. For example, the splicing of mRNA encoding both intracellular and extracellular regions of the polypeptide, an affinity for heparan sulfate, the remarkable postranslational addition of polysialic acid, and evidence that the molecule can influence second messenger systems, all need to be considered in a comprehensive view of NCAM function. As yet such an integration of NCAM functions has not been made at the molecular level, and its observed biological effects undoubtedly reflect a mixture of different mechanisms.

For these reasons, the ensuing paragraphs will use the relatively vague terms 'interaction' or 'recognition' to refer to the combined functions of NCAM, with the specific term 'adhesion' reserved for the physical association of membranes via receptors. Thus in schematic figures depicting cell-cell interactions, the illustration of membranes becoming more closely apposed can also be viewed in the larger context of interaction.

NCAM as a recognizer: combination and competition with other CAMs

In some biological situations, NCAM does in fact behave like a simple homophilic recognition molecule. That is, a phenomenon involving a specific interaction between two NCAM-positive cells appears to reflect directly the adhesion properties of this molecule. In most of these cases, the term 'specific' is operationally defined by the ability to inhibit the phenomenon with antibodies against NCAM. Leaving aside the more artificial cell aggregation assays, these situations have been best documented in the developing nervous system, including fasciculation of retinal ganglion cell neurites in the optic nerve (Thanos et al., 1984), migration of growth cones on cellular substrata both in vitro (Rutishauser et al., 1983; Bixby et al., 1988; Covault et al., 1987; Neugebauer et al., 1988) and in vivo (Silver and Rutishauser, 1984), innervation of muscles by motorneurons both in development (Landmesser et al., 1988) and regeneration (Rieger et al., 1988), and paralysis-induced sprouting (Booth et al., 1990).

While NCAM function in these cases appears relatively straightforward, it is important to emphasize that the molecule's influence is only appreciated in terms of the simultaneous action of other CAMs. For neuronal processes, these include the axon-associated CAMs (L1, F11, neurofascin, etc.) as well as the more broadly expressed cadherin and integrin families. For example, in studies of axon-axon bundling or growth cone migration in vitro, the action of individual CAMs appears to be additive or synergistic with others, relative contributions varying for different cell types and developmental ages (Bixby et al., 1987; Neugebauer et al., 1988; Rathjen et al., 1987; Tomaselli et al., 1988; Bixby et al., 1988; Drazba and Lemmon, 1990). When axon bundling and growth combine in the context of in vivo innervation, they tend to oppose each other as illustrated in Fig. 1. In the case of motor innervation of the chick hindlimb, the competition, which is most easily described in terms of adhesion but may be more complex, appears to be between L1's promotion of fasciculation and NCAM's support of nerve growth on muscle (Landmesser et al., 1988). While less information is available, it is reasonable to suspect that such principles also apply in the proposed role of NCAM in optic nerve bundling (Thanos et al., 1984) and guidance along epithelial endfeet (Silver and Rutishauser, 1984).

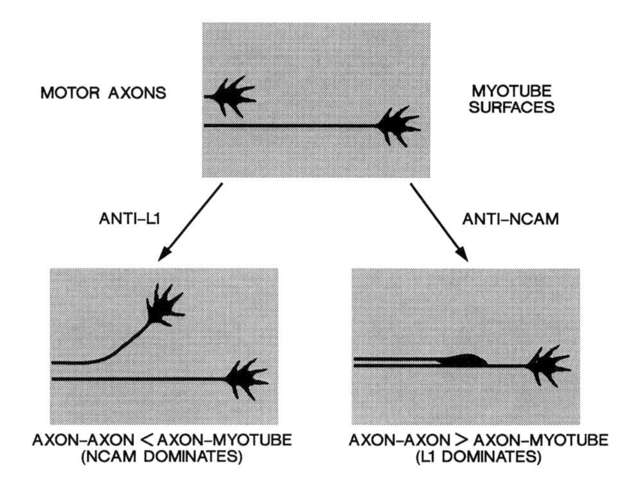

Fig. 1. Competition between CAMs in the outgrowth and bundling of axons. The top part of the drawing depicts one axon that has extended over the surface of a myotube, while a second has just begun to grow. As the second axon grows, a dominance of L1-mediated axon-axon interactions over NCAM-mediated axon-muscle interactions (lower right) tends to produce elongation along the first axon. Conversely, if axon-myotube interactions are stronger (lower left), the two axons will diverge. These different patterns can be manipulated experimentally by the addition of antibodies that specifically block L1 or NCAM adhesion.

Regulation of non-NCAM interactions by NCAM

Other studies suggest that NCAM's influence goes beyond a tug-of-war between CAMs. For example, antibodies against NCAM can inhibit cell-cell interactions whose intrinsic mechanism does not directly involve NCAM itself: gap junctional communication in the neural plate (Keane et al., 1988), contact-dependent regulation of neurotransmitter enzymes (Acheson and Rutishauser, 1988), and possibly cadherin function (Rutishauser et al., 1988; Knudsen et al., 1990). These findings are accompanied by an increasing awareness that adhesion can trigger other cellular events, even avoidance reactions between neurites (Kapfhammer and Raper, 1987; Kapfhammer et al., 1986) and de-adhesion (Dustin and Springer, 1989). The central theme is that adhesion can be a regulatory element that promotes recognition or interaction, but is not the whole process in of itself.

This permissive influence is most easily described in the case of junctional communication (Keane et al., 1988). NCAM is an early positive marker for neural induction (Jacobson and Rutishauser, 1986), and its expression appears to precede junctional communication. In addition, there is a precise correlation between the ability of two cells to exchange a fluorescent dye and the presence of NCAM on their surfaces. Finally, whereas the block of junctional communication by the *src* gene product does not prevent NCAM expression, the addition of anti-NCAM Fab to histotypic cultures of neuroepithelial cells delays the establishment of extensive communication among cells that express NCAM. Since there is no reason to believe that NCAM itself contributes to the structure of the channels, it is more likely that, in neuroepithelium, NCAM-mediated adhesion is required to allow sufficient interaction between cells for the assembly of stable junctions (Fig. 2).

If such a role represents the simple augmentation of cell-cell apposition, then NCAM's function should be readily replaced by other CAMs. In fact, transfected E-cadherin (L-CAM) (Musil et al., 1990) also has been shown to be capable of promoting formation of junctions. Moreover, gap junctions are capable of assembly in tissues that do not express NCAM.

Such a two-step mechanism for cell interaction has some useful attributes. It provides for additional specificity through independent regulation of the CAM and the information-generating entities. That is, a cell can decide not only what array of specific communications are desirable,

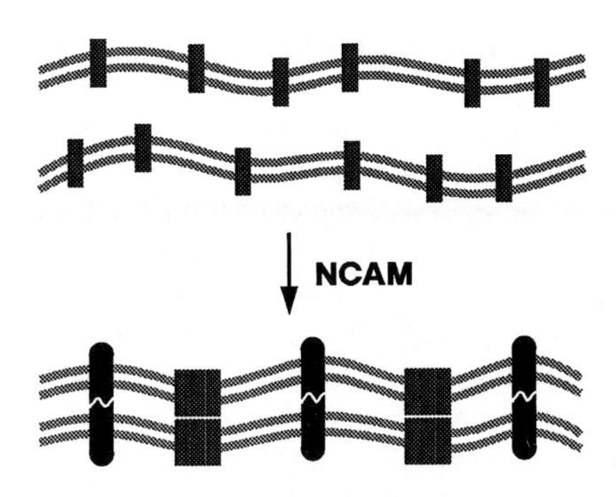

Fig. 2. Proposed mechanism for positive regulation of junctional communication between cells by NCAM-mediated adhesion. Two membranes containing junctional subunits (rectangles) are not sufficiently apposed (top) to allow the subunits to assemble into a stable junction. When NCAM is expressed, the resulting homophilic adhesion promotes junctional assembly (bottom).

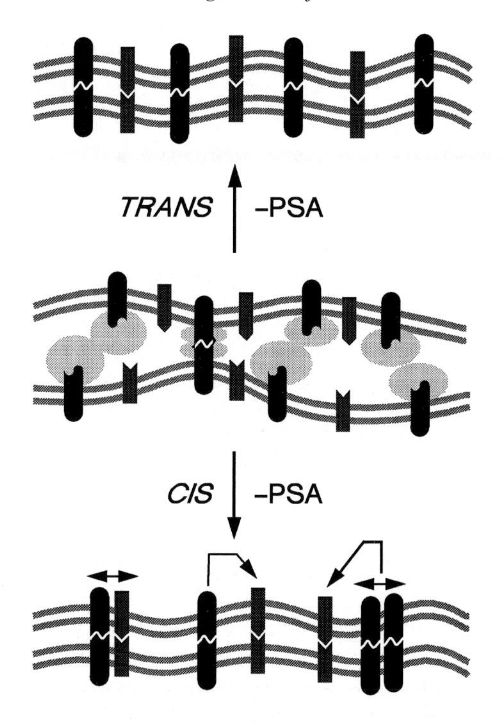

Fig. 3. Different modes by which removal of PSA (shaded elipses) could affect cell-cell interactions. Top: loss of PSA removes a physical impediment to membrane apposition and thereby enhances the *trans* encounter between receptors on apposing cells. Bottom: loss of PSA augments *cis* interactions among receptors which in turn promote cell-cell interaction. Such *cis* effects could include heterophilic or homophilic receptor clustering (double arrows) with NCAM, and could generate intracellular signals (single bent arrows) as well.

but also when and where this ensemble is functionally engaged. There is also the potential for biosynthetic economy, since signals can often be generated with far fewer molecules than are required for macroscopic adhesion between cells.

NCAM polysialic acid: turning pull into push

Theoretical models and indirect evidence for multi-step recognition have been around for decades. The demonstration that NCAM can serve in such a capacity is therefore a welcome but not surprising finding. What is more novel and unanticipated is the fact that the molecule, through an unusual form of glycosylation, can provide for negative as well as positive regulation.

Perhaps the most remarkable structural variant of NCAM is its post-translational modification with polysialic acid (PSA) (see Rutishauser, 1989, for review). For nearly 10 years, it has been known that the presence of this very large linear homopolymer on NCAM decreases the molecule's ability to promote adhesion (Cunningham et al., 1983; Hoffman and Edelman, 1983; Rutishauser et al., 1985). In the context of the positive regulatory function for NCAM described above, PSA could therefore serve as a negative regulator of other cell interactions through direct effects on NCAM.

However, this negative regulation also can occur in the absence or independent of known NCAM-binding functions. For example, the presence of PSA on the surface of a cell can inhibit its attachment to a laminin substratum, a process that is blocked by antibodies against laminin but not by antibodies against NCAM (Acheson et al., 1991). Even agglutination of membrane vesicles by plant lectins is reduced by this large polymer (Rutishauser et al., 1988). Another surprising finding is that removal of PSA from cells can enhance the function of other CAMs even more so than NCAM itself. This is particularly true for the L1 adhesion molecule (Acheson et al., 1991), the consequences of which have important biological consequences as described below.

In considering mechanisms by which PSA may act in molecules other than NCAM itself, two distinct modes must be considered: those in which PSA would impede *trans* interactions between receptors on apposing cells, and those that would involve a change in *cis* interactions between receptors on the same cell (Fig. 3). The *trans* mode could include not only the specific inhibition of NCAM-NCAM homophilic interaction (Hoffman and Edelman, 1983), but also, if overall membrane apposition is affected, the interaction of other receptors as well (Rutishauser et al., 1988). The *cis* mode is more complex, in that it could reflect changes in an adhesion-promoting interaction of NCAM with other receptors on the same cell (Kadmon et al., 1990), an indirect augmentation of interactions as a result of intracellular signalling, or clustering of NCAMs within the plane of the membrane (Singer, 1992).

At present, there is too little information to choose definitively among these possibilities. The importance of *cis* interactions among membrane components has been established in a variety of systems, and the presence of the bulky and charged PSA moiety could easily disrupt binding of proteins to NCAM within the plane of the cell membrane. Some studies argue against mechanisms involving intracellular signalling or specific *cis* interactions between NCAM and other receptors. For example, PSA can alter the adhesion of purified membranes vesicles (Rutishauser et al., 1985), and of reconstituted bilayers containing only lipid

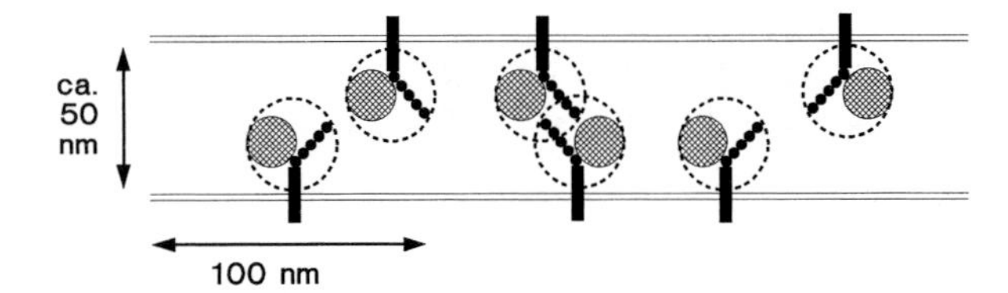

Fig. 4. The density and excluded volume of polysialylated NCAM are sufficient to affect overall membrane-membrane contact. This figure is a scale representation of the interaction of two cell membranes, each bearing polysialylated NCAM at the density measured on live cells (Yang et al., 1992). The size of the NCAM polypeptide was estimated from electron micrographs and the increase in radius of hydration of the polysialylated molecule. The dashed circle around each NCAM represents the ability of the molecule to rotate rapidly in the plane of the membrane.

and NCAM (Hoffman and Edelman, 1983). These findings, however, do not exclude the possibility that *cis* interactions among NCAMs might be a requisite for both NCAM-mediated adhesion between membrane vesicles, and a consequent signalling to other receptors in live cells (Fig. 3, bottom right). Evidence for cooperativity among NCAMs has in fact been observed with both membrane vesicles and cells transfected with NCAM cDNA (Hoffman and Edelman, 1983; Doherty et al., 1990b).

The *trans* mechanism, in which PSA impedes cell-cell contact, has two critical predictions: that enough space is occupied by PSA to affect overall membrane-membrane apposition, and that intercellular space is actually changed upon removal of PSA. Recently, evidence to this effect has been obtained (Yang et al., 1992). The abundance of NCAM in the membrane (about 1 molecule per 100×100 nm patch of bilayer) and the large hydrated volume of PSA (over ten times that of an equivalent mass of protein) is compatible with overall steric effects (Fig. 4). Furthermore, the distance between apposed cell membranes upon specific enzymatic removal of PSA decreases by about 25%. Such a decrease in space is equivalent to about two immunoglobulin domains and would be expected to have a substantial effect on interaction between such domains on apposing cells.

The potential value of regulation at the level of overall membrane-membrane contact is that changes in the closeness, extent, or duration of apposition can be a basis for selection between different interactions. Some receptors may be too small or too large to work effectively at a particular membrane-membrane distance; others may be both big and flexible enough to minimize the effects of this variable. Similarly, the surface area and stability of contacts can be factors depending on the number of molecular bonds required, or the speed of effective interaction. Although the model is easier to illustrate schematically as a complete disengagement in the presence of PSA (Fig. 3), much of its attractiveness lies in the ability to affect interactions partially as well as selectively. In this manner, a cell could choose advantageous interactions (or avoid deleterious ones), and the ability of PSA to promote axonal elongation on cellular substrata may well reflect this capacity (Landmesser et al., 1990; Doherty et al., 1990a).

PSA as an in vivo regulator

Whatever the exact mechanism underlying PSA's effect on cell interactions, there is substantial evidence that this type of regulation is an important developmental parameter in living embryos. PSA is highly regulated during development, with patterns quite distinct to those of NCAM itself (except of course that PSA expression requires NCAM expression). The original documentation of naturally occurring variation in NCAM sialylation focussed on the state of the molecule isolated from the brains of embryonic and adult tissues (Chuong and Edelman, 1984). This pattern of temporal regulation was found in other tissues as well, leading to the description of heavily sialylated NCAM as the embryonic form and the less sialylated molecule as the adult form. Additional studies, however, have revealed that low PSA NCAM also predominates very early in development (Sunshine et al., 1987), that the level of PSA is regulated independently in many tissues (Rougon et al., 1982; Chuong and Edelman, 1984; Boisseau et al., 1991) and can vary even within the same cell (Schlosshauer et al., 1984), and that high PSA NCAM persists in some adult tissues such as the olfactory system (Chuong and Edelman, 1984) and hypothalamus (Theodosis et al., 1991). It should also be mentioned that although PSA is most thoroughly studied in neural tissues, it is also abundant and regulated in many non-neural structures, including skeletal muscle, heart, kidney, and mesenchymal tissue.

To date, almost all functional characterization of PSA has been with neurons. As discussed above and illustrated in Fig. 1, L1 and NCAM appear to compete with each other to produce patterns of motorneuron innervation of muscle (Landmesser et al., 1988). Nevertheless, in the naturally occurring variations in pattern (age-related changes, fast versus slow fibers), neither of these molecules appears to be regulated in its expression. Instead, it is the PSA content of NCAM on axons that is developmentally correlated with in vivo patterns, and removal of this carbohydrate with a PSA-specific endoneuraminidase produces a predictable change in these patterns (Landmesser et al., 1990). Moreover, these predictions center on modulation of L1's function, not NCAM's, which as described above is consistent with an overall effect of PSA on membrane apposition.

This analysis of PSA regulation in vivo has also revealed an unexpected influence of the carbohydrate on synaptic activity-dependent sprouting and branching of motor neurons (Landmesser et al., 1990). Broadly stated, it appears that PSA can serve as a molecular link between synaptic activity and changes in the pattern of axonal innervation. The experimental system used was again innervation of the chick limb, and combined the analysis of L1, NCAM and PSA with blockade of synaptic activity by the neurotoxin curare. As in normal development, neither NCAM nor L1 expression is correlated with the increased sprouting and branching produced by curare, whereas PSA is dramatically increased on axonal surfaces of curare-treated animals. A causal relationship between activity, PSA and innervation pattern (Fig. 5) is supported by the observation in vivo that the morphological effects of curare on innervation are reversed by co-administration of an endoneuraminidase that specifically destroys PSA.

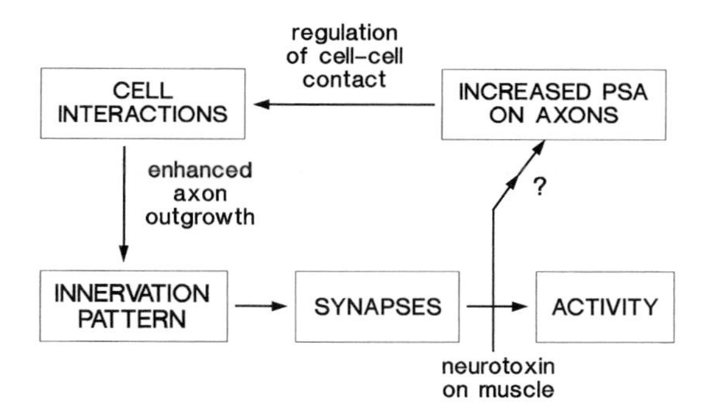

Fig. 5. Schematic representing the functional relationship between the molecular, cellular and physiological parameters that contribute to synaptic activity-dependent remodelling of muscle innervation. In vivo studies (Landmesser et al., 1990) have revealed that enhanced PSA expression after blockade of synapses leads directly to alteration of cell interactions involved in this remodelling.

Conclusion

It is now clear that NCAM did not evolve to serve a specific recognition function, such as axon guidance or synaptogenesis. Instead the molecule appears to part of the machinery that mediates and/or regulates a fundamental cellular property, the ability to form close contacts with other cells. As with other basic cell behaviors, such as mitosis, migration and differentiation, this parameter is incorporated into a myriad of physiological contexts both within and outside the nervous system. NCAM function must therefore be closely integrated into cellular biochemistry through which it is coordinated with other major cell processes. Such coordination can be achieved through a number of mechanisms involving second messengers, postranslational modification and cytoskeletal elements as well as gene transcription.

References

Acheson, A. and Rutishauser, U. (1988). NCAM regulates cell contact-mediated changes in choline acetyltransferase activity of embryonic chick sympathetic neurons. *J. Cell Biol.* **106**, 479-486.

Acheson, A., Sunshine, J. L. and Rutishauser, U. (1991). NCAM polysialic acid can regulate both cell-cell and cell-substrate interactions. *J. Cell Biol.* **114**, 143-153.

Bixby, J. L., Pratt, R. S., Lilien, J. and Reichardt, L. F. (1987). Neurite outgrowth on muscle cell surfaces involves extracellular matrix receptors as well as Ca^{2+}-dependent and -independent cell adhesion molecules. *Proc. Natl. Acad. Sci. USA* **84**, 2555-2559.

Bixby, J. L., Lilien, J. and Reichardt, L. F. (1988). Identification of the major proteins that promote neuronal process outgrowth on Schwann cells in vitro. *J. Cell Biol.* **107**, 353-361.

Boisseau, S., Nedelec, J., Poirier, V., Rougon, G. and Simonneau, M. (1991). Analysis of high PSA N-CAM expression during mammalian spinal cord and peripheral nervous system development. *Development* **112**, 69-82.

Booth, C. M., Kemplay, S. K. and Brown, M. C. (1990). An antibody to neural cell adhesion molecule impairs motor nerve terminal sprouting in a mouse muscle locally paralyzed with botulinum toxin. *Neurosci.* **35**, 85-91.

Choung, C.-M. and Edelman, G. M. (1984). Alterations in neural cell adhesion molecules during development of different regions of the nervous system. *J. Neurosci.* **4**, 2354-2368.

Covault, J., Cunningham, B. and Sanes, J. (1987). Neurite outgrowth on cryostat sections of innervated and denervated skeletal muscle. *J. Cell Biol.* **105**, 2479-2488.

Cunningham, B. A., Hoffman, S., Rutishauser, U., Hemperly, J. J. and Edelman, G. M. (1983). Molecular topography of the neural cell adhesion molecule N-CAM: surface orientation and location of sialic acid-rich and binding regions. *Proc. Natl. Acad. Sci. USA* **80**, 3116-3120.

Doherty, P., Cohen, J. and Walsh, F. S. (1990a). Neurite outgrowth in response to transfected N-CAM changes during development and is modulated by polysialic acid. *Neuron* **5**, 209-219.

Doherty, P., Fruns, M., Seaton, P., Dickson, G., Barton, C. H., Sears, T. A., and Walsh, F. S. (1990b). A threshold effect of the major isoforms of NCAM on neurite outgrowth. *Nature* **343**, 464-466.

Drazba, J. and Lemmon, V. (1990). The role of cell adhesion molecules in neurite outgrowth on Muller cells. *Dev. Biol.* **138**, 82-93.

Dustin, M. L. and Springer, T. A. (1989). T-cell receptor cross-linking transiently stimulates adhesiveness through LFA-1. *Nature* **341**, 619-624.

Grumet, M., Rutishauser, U. and Edelman, G. M. (1982). N-CAM mediates adhesion between embryonic nerve and muscle cell *in vitro*. *Nature* **295**, 693-695.

Hoffman, S. and Edelman, G. M. (1983). Kinetics of homophilic binding by embryonic and adult forms of the neural cell adhesion molecule. *Proc. Natl. Acad. Sci. USA* **80**, 5762-5766.

Jacobson, M. and Rutishauser, U. (1986). Induction of neural cell adhesion molecule (NCAM) in Xenopus embryos. *Dev. Biol.* **116**, 524-531.

Kadmon, G., Kowitz, A., Altevogt, P. and Schachner, M. (1990). The neural cell adhesion molecule N-CAM enhances L1-dependent cell-cell interactions. *J. Cell Biol.* **110**, 193-208.

Kapfhammer, J. P. and Raper, J. A. (1987). Collapse of growth cone structure on contact with specific neurites in culture. *J. Neurosci.* **7**, 201-212.

Kapfhammer, J. P., Grunewald, B. E. and Raper, J. A. (1986). The selective inhibition of growth cone extension by specific neurites in culture. *J. Neurosci.* **6**, 2527-2534.

Keane, R. W., Mehta, P. P., Rose, B., Honig, L. S., Loewenstein, W. R. and Rutishauser, U. (1988). Neural differentiation, NCAM-mediated adhesion and gap junctional communication in neuroectoderm. *J. Cell Biol.* **106**, 1307-1308.

Knudsen, K. A., McElwee, S. A. and Myers, L. (1990). A role for the neural cell adhesion molecule, NCAM, in myoblast interaction during myogenesis. *Dev. Biol.* **138**, 159-168.

Landmesser, L., Dahm, L., Schultz, K. and Rutishauser, U. (1988). Distinct roles for adhesion molecules during innervation of embryonic chick muscle. *Dev. Biol.* **103**, 645-670.

Landmesser, L., Dahm, L., Tang, J. and Rutishauser, U. (1990). Polysialic acid as a regulator of intramuscular nerve branching during embryonic development. *Neuron* **4**, 655-667.

Maier, C. E., Watanabe, M., Singer, M., McQuarrie, I. G., Sunshine, J. and Rutishauser, U. (1986). Expression and function on neural cell adhesion molecule during limb regeneration. *Proc. Natl. Acad. Sci. USA* **83**, 8395-8399.

Musil, L. S., Cunningham, B. A., Edelman, G. M. and Goodenough, D. A. (1990). Differential phosphorylation of the gap junction protein connexin43 in junctional communication-competent and -deficient cell lines. *J. Cell Biol.* **111**, 2077-2088.

Neugebauer, K. M., Tomaselli, K. J., Lilien, J. and Reichardt, L. F. (1988). N-cadherin, NCAM, and integrins promote retinal neurite outgrowth on astrocytes in vitro. *J. Cell Biol.* **107**, 1177-1187.

Rathjen, F. G., Wolff, J. M., Frank, R., Bonhoeffer, F. and Rutishauser, U. (1987). Membrane glycoproteins involved in neurite fasciculation. *J. Cell Biol.* **104**, 343-353.

Rieger, F., Nicolet, M., Pincon-Raymond, M., Murawsky, M., Levi, G., and Edelman, G. M. (1988). Distribution and role in regeneration of N-CAM in the basal laminae of muscle and Schwann cells. *J. Cell Biol.* **107**, 707-719.

Rougon, F., Deagostini-Bazin, H., Hirn, M. and Goridis, C. (1982). Tissue and developmental stage-specific forms of a neural cell surface antigen linked to differences of glycosylation of a common polypeptide. *EMBO J.* **1**, 1239-1244.

Rutishauser, U. and Edelman, G. M. (1980). Effects of fasciculation on the outgrowth of neurites from spinal ganglia in culture. *J. Cell Biol.* **87**, 370-378.

Rutishauser, U., Grumet, M. and Edelman G. M. (1983). N-CAM mediates initial interactions between spinal cord neurons and muscle cells in culture. *J. Cell Biol.* **97**, 145-152.

Rutishauser, U., Watanabe, M., Silver, J., Troy, F. A. and Vimr, E. R. (1985). Specific alteration of NCAM-mediated cell adhesion by an endoneuraminidase. *J. Cell Biol.* **101**, 1842-1849.

Rutishauser, U., Acheson, A., Hall, A. K., Mann, D. M. and Sunshine, J. (1988). The neural cell adhesion molecule (NCAM) as a regulator of cell-cell interactions. *Science* **240**, 53-57.

Rutishauser, U. (1989). Polysialic acid as a regulator of cell interactions. In *Neurobiology of Glycoconjugates* (eds. R. U. Margolis and R. K. Margolis). New York: Plenum Publishing, pp 367-382.

Rutishauser, U. (1991). Neural cell adhesion molecule and polysialic acid. In *Receptors for Extracellular Matrix,* (eds. J.A. McDonald and R.P. Mecham), San Diego, CA: Academic Press, Inc., pp 131-156.

Schlosshauer, B., Schwartz, U., and Rutishauser, U. (1984). Topological distribution of different forms of NCAM in the developing chick visual system. *Nature* **310**, 141-143.

Silver, J. and Rutishauser, U. (1984). Guidance of optic axons in vivo by a preformed adhesive pathway on neuroepithelial endfeet. *Dev. Biol.* **106**, 485-499.

Singer, S. J. (1992). Intercellular communication and cell-cell adhesion. *Science* **255**, 1671-1677

Sunshine, J., Balak, K., Rutishauser, U. and Jacobson, M. (1987). Changes in NCAM structure during vertebrate neural development. *Proc. Natl. Acad. Sci. USA* **84**, 5986-5990.

Thanos, S., Bonhoeffer, F., and Rutishauser, U. (1984). Fiber-fiber interaction and tectal cues influence the development of the chicken retinotectal projection. *Proc. Natl. Acad. Sci. USA* **18**, 1906-1910.

Theodosis, D. T., Rougon, G. and Poulain, D. A. (1991). Retention of embryonic features by an adult neuronal system capable of plasticity: Polysialylated neural cell adhesion molecule in the hypothalamo-neuohypophysial system. *Proc. Natl. Acad. Sci. USA* **88**, 5494-5498.

Thiery, J.-P., Duband, J.-L., Rutishauser, U. and Edelman G. M. (1982). Cell adhesion molecules in early chicken embryogenesis. *Proc. Natl. Acad. Sci. USA* **79**, 6737-6741.

Tomaselli, K. J., Neugebauer, K. M., Bixby, J. L., Lilien, J. and Reichardt, L. F. (1988). N-cadherin and integrins: Two receptor systems that mediate neuronal process outgrowth on astrocyte surfaces. *Neuron* **1**, 33-43.

Yang, P., Yin, X. and Rutishauser, U. (1992). Intercellular space is affected by the polysialic acid content of NCAM. *J. Cell Biol.* **116**, 1487-1496.

Development 1992 Supplement, 105-112 (1992)
Printed in Great Britain © The Company of Biologists Limited 1992

Epithelial differentiation and intercellular junction formation in the mouse early embryo

TOM P. FLEMING*, QAMAR JAVED and MARK HAY

Department of Biology, Biomedical Sciences Building, University of Southampton, Bassett Crescent East, Southampton SO9 3TU, UK

*Author for correspondence

Summary

Trophectoderm differentiation during blastocyst formation provides a model for investigating how an epithelium develops in vivo. This paper briefly reviews our current understanding of the stages of differentiation and possible control mechanisms. The maturation of structural intercellular junctions is considered in more detail. Tight junction formation, essential for blastocoele cavitation and vectorial transport activity, begins at compaction (8-cell stage) and appears complete before fluid accumulation begins a day later (approx 32-cell stage). During this period, initial focal junction sites gradually extend laterally to become zonular and acquire the peripheral tight junction proteins ZO-1 and cingulin. Our studies indicate that junction components assemble in a temporal sequence with ZO-1 assembly preceding that of cingulin, suggesting that the junction forms progressively and in the 'membrane to cytoplasm' direction. The protein expression characteristics of ZO-1 and cingulin support this model. In contrast to ZO-1, cingulin expression is also detectable during oogenesis where the protein is localised in the cytocortex and in adjacent cumulus cells. However, maternal cingulin is metabolically unstable and does not appear to contribute to later tight junction formation in trophectoderm. Cell-cell interactions are important regulators of the level of synthesis and state of assembly of tight junction proteins, and also control the tissue-specificity of expression. In contrast to the progressive nature of tight junction formation, nascent desmosomes (formed from cavitation) appear mature in terms of their substructure and composition. The rapidity of desmosome assembly appears to be controlled by the time of expression of their transmembrane glycoprotein constitutents; this occurs later than the expression of more cytoplasmic desmosome components and intermediate filaments which would therefore be available for assembly to occur to completion.

Key words: mouse preimplantation embryos, cell polarity, trophectoderm, inner cell mass, tight junctions, ZO-1, cingulin, desmosomes.

Introduction

Mammalian development begins with three consecutive and binary cell diversification events that are controlled by cell-cell interactions and give rise to the embryonic and extraembryonic lineages of the conceptus (Fig. 1). The first diversification takes place during cleavage where one group of cells differentiate into the outer trophectoderm epithelium of the blastocyst while the other group of cells differentiate into the inner cell mass (ICM). Secondly, as trophectoderm generates the blastocoele by vectorial fluid transport, this epithelium segregates into polar (proliferative) and mural (non-proliferative) subpopulations. Thirdly, the ICM segregates into primary endoderm (epithelium; at blastocoele interface) and primary ectoderm (ICM core), the latter forming the foetal lineages after implantation. These diversification processes generate firstly a radial polarity in the embryo and subsequently the embryonic-abembryonic axis (discussed in Johnson et al., 1986a).

In this paper, we are concerned with the first segregation step, which underpins all subsequent morphogenesis. There has been a number of recent reviews on the mechanisms and characteristics of trophectoderm differentiation and tissue diversification in the mouse embryo, some of which focus on specific issues (Johnson and Maro, 1986; Johnson et al., 1986a; Wiley, 1987; Fleming and Johnson, 1988; Maro et al., 1988; Pratt, 1989; Kimber, 1990; Wiley et al., 1990; Fleming, 1992). Here, we shall briefly review some of the main features and proposed models for early development before considering in more detail our more recent work on the lineage-specific maturation of the trophectoderm intercellular junctional complex. Junction maturation is clearly essential for nascent epithelial cells to integrate as a functional tissue and, we argue, is therefore crucial for the sequence of differentiative decisions outlined above.

Epithelial biogenesis: a brief overview

The earliest stage at which trophectoderm differentiation is detectable is compaction in the 8-cell embryo. Each blastomere becomes adhesive and polarises along an apicobasal axis (apical on outer embryo surface). Polarisation embraces an extensive reorganisation of cytocortical and cytoplasmic domains of the cell. For example, in the cytoplasm, the distribution of actin filaments, microtubules and their organising centres, clathrin and endosomal organelles are all affected, generally by accumulating in the apical area. In the cortex, polarity results in the formation of an apical pole of microvilli, a basolateral distribution of uvomorulin (E-cadherin) responsible for cell-cell adhesion, and basolateral gap and nascent tight junctions. Various actin-associated cytocortical proteins also polarise and there is limited evidence that certain membrane proteins become distributed asymmetrically (reviewed in Johnson and Maro, 1986; Johnson et al., 1986a; Fleming and Johnson, 1988; Wiley et al., 1990; Fleming, 1992).

Biosynthetic inhibitors have been used to investigate the temporal regulation of compaction. Collectively, these studies indicate that compaction takes place at the 8-cell stage largely in response to post-translational changes in components synthesised as early as the 2-cell stage (about 24 hours earlier), rather than by contemporary transcriptional or translational activity (Kidder and McLachlin, 1985; Levy et al., 1986). Cell contact signalling at the time of compaction appears to regulate the orientation and spatial features of polarisation in blastomeres that are already in a permissive state for this cue (Ziomek and Johnson, 1980; Johnson and Ziomek, 1981a). The cell contact signal is most likely to be mediated by uvomorulin homotypic binding, although, in the absence of this signal, polarisation can still take place but later than normal, presumably as a result of a non-specific cue (Peyrieras et al., 1983; Johnson et al., 1986b; Blaschuk et al., 1990). Evidence from different experimental approaches indicate that, upon receipt of the uvomorulin inductive signal, aspects of cytocortical polarisation are initiated as a primary response and lead secondarily to polarisation within the cytoplasm (reviewed in Johnson and Maro, 1986; Johnson et al., 1986a; Fleming and Johnson, 1988; Fleming, 1992).

Two mechanisms have been proposed for the establishment of a stable primary axis of polarity within the cytocortex resulting from uvomorulin adhesion. One involves the generation of a transcellular ion current driven by membrane channels and pumps (influx apically, efflux basolaterally) that would cause charged molecules in the membrane and cytoplasm, and ultimately organelles, to be mobilised, leading to a polarised state. Circumstantial evidence in support of this model has been reported (Nuccitelli and Wiley, 1985; Wiley and Obasaju, 1988, 1989; Wiley et al., 1990, 1991). A second mechanism proposed to explain axis determination in polarising blastomeres involves the phosphatidylinositol (PI) second messenger system. There is evidence that protein kinase C activation is required both for initiating uvomorulin adhesivity at compaction and for establishing cytocortical polarity in blastomeres (Bloom, 1989; Winkel et al., 1990). Moreover, specific protein phosphorylation events coincide with compaction (Bloom and McConnell, 1990; Bloom, 1991). It is possible that both a transcellular ion current and protein kinase C activation may integrate to achieve polarisation (discussed in Fleming, 1992).

In succeeding 16- and 32-cell cycles, daughter cells inheriting the apical domain of polarised 8-cell blastomeres (see Fig. 1) continue to differentiate into a functional epithelium, achieved by the late 32-cell stage when cavitation occurs. This progression embraces various cellular features, including maturation of cell-cell adhesion systems and junction formation (see later), basement membrane formation, cytoplasmic reorganisation, and the establishment of membrane polarity. These topics have been reviewed in detail elsewhere (Johnson et al., 1986a; Fleming and Johnson, 1988; Fenderson et al., 1990; Kimber, 1990; Wiley et al., 1990; Fleming, 1992). In consequence, the nascent trophectoderm at the late 32-cell stage acquires the capacity for vectorial fluid and selective molecular transport. Blas-

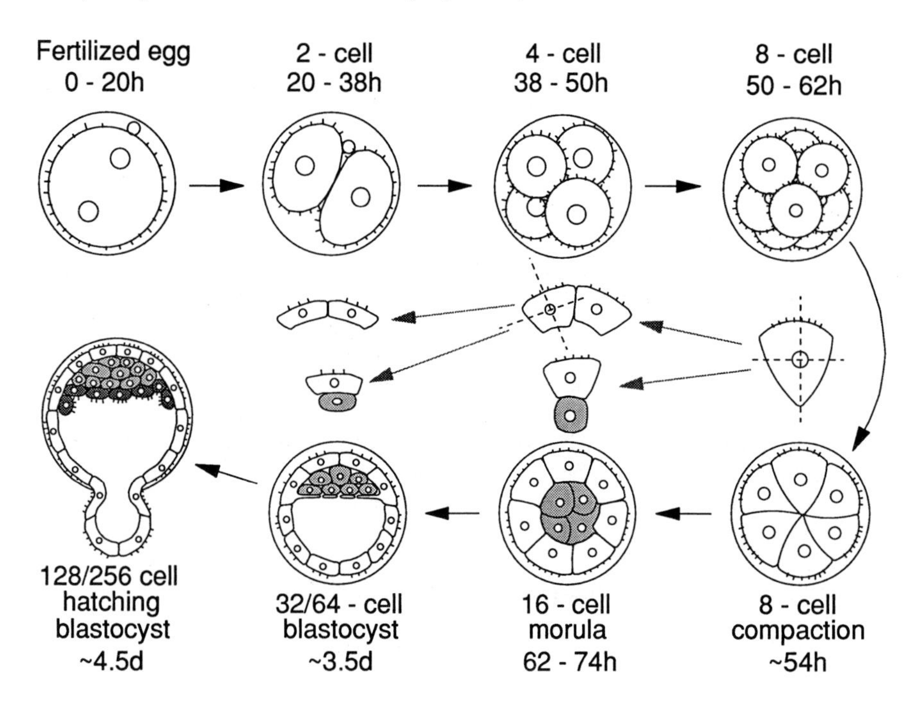

Fig. 1. A schematic representation of mouse preimplantation development. Embryo stage and time elapsed since fertilization are indicated. In the centre, the conservative and differentiative division planes (dotted lines) of polar 8- and 16-cell blastomeres are shown, which lead to distinct trophectoderm and ICM cell lineages. Shaded cells belong to the ICM lineage, which, in the hatching blastocyst, differentiates primary endoderm at the blastocoele surface (darker shading). Reprinted from Fleming (1992) with permission of publishers, Chapman and Hall.

tocoele formation is largely a result of Na^+ influx at the apical membrane via several transporters, including the Na^+/H^+ exchanger and Na^+-channel (Manejwala et al., 1989), and active Na^+ efflux basally, mediated by functional Na^+,K^+-ATPase (reviewed in Wiley, 1987), thereby generating passive water flow in the apicobasal direction. Regulation of this important early function of the epithelium is indeed complex. Regulatory processes include biosynthetic control of Na^+,K^+-ATPase expression (Gardiner et al., 1990; Kidder and Watson, 1990; Watson et al., 1990b) and subsequent basal polarisation (Watson et al., 1990a), the formation of a permeability seal to prevent blastocoele leakage (Magnuson et al., 1978; Van Winkle and Campione, 1991), and a physiological control mediated by cAMP (Manejwala et al., 1986; Manejwala and Schultz, 1989).

Intercellular junction formation

The junctional complex of epithelial cells is a fundamental characteristic of this phenotype, essential for polarised cellular function and tissue formation (see reviews by Staehelin, 1974; Garrod and Collins, 1992). The complex consists of an apicolateral tight junction (zonula occludens), an intermediate zonula adherens, and lateral membrane gap and desmosomal junctions, each possessing distinct structural and molecular properties. In the mouse embryo, the formation of these specialised junctions is temporally regulated and, with the exception of gap junctions, tissue-specific. Gap junctions form from the 8-cell stage (Lo and Gilula, 1979; Goodall and Johnson, 1984), following expression of different junctional components either from maternal genes or during early cleavage (McLachlin et al., 1983; McLachlin and Kidder, 1986; Barron et al., 1989; Nishi et al., 1991). Inhibition of gap junctional communication at the time of compaction by different means fails to prevent cell polarisation (Goodall, 1986; Levy et al., 1986) but does inhibit the adhesion and integration of such cells into the morula and blastocyst (Lee et al., 1987; Buehr et al., 1987; Bevilacqua et al., 1989). The zonula adherens is not a prominent junction type in the embryo and its maturation has received little attention. However, the adherens junction components, vinculin and α-actinin, begin to localise apicolaterally from the compaction stage (Lehtonen and Reima, 1986; Lehtonen et al., 1988; Reima, 1990). Tight junction and desmosome formation have been studied in greater detail and are discussed below.

Tight junctions

The tight junction is a belt-like structure around the cell apex where the membranes of adjacent epithelial cells are closely apposed, and possibly partially fused. The freeze-fracture image of the tight junction, composed of an anastomosing network of intramembraneous cylindrical fibrils (P-face) and complementary grooves (E-face) has been well documented (eg, Staehelin, 1974). These junctions restrict uncontrolled paracellular transport between mucosal and serosal compartments by occlusion of the intercellular space, thereby contributing in large part to the transepithelial electrical resistance. They are also thought to act as a barrier to the mixing of apical and basolateral integral membrane proteins and exoplasmic leaflet lipids, thereby helping to preserve membrane polarity. For recent reviews of tight junction structure and function, see Stevenson et al. (1988) and Cereijido (1991).

Tight junctions are multimolecular complexes that are as yet poorly understood in terms of their composition (reviewed in Anderson and Stevenson, 1991). The integral membrane component providing the freeze-fracture image is undefined. However, a total of four peripheral membrane proteins have been proposed as tight junction components, two of which have been characterised in some detail. ZO-1, originally identified in mouse liver membrane fractions, was the first protein reported as a ubiquitous and specific component of the tight junction in a variety of epithelia (Stevenson et al., 1986). ZO-1 is a high relative molecular mass ($215\text{-}225\times10^3$) protein that, from extraction and immunogold studies, is avidly associated with the cytoplasmic face of the junction, positioned very close to the membrane domain (Anderson et al., 1988; Stevenson et al., 1989). The biophysical properties of ZO-1 suggest that it is an asymmetric monomer and is phosphorylated on serine residues (Anderson et al., 1988). Cingulin (140×10^3 M_r) is a second ubiquitous component of tight junctions from various sources, originally identified in avian intestinal epithelium (Citi et al., 1988, 1989). A purified 108×10^3 M_r polypeptide region of cingulin occurs as a heat-stable elongated homodimer organised in a coiled-coil configuration similar to the myosin rod. Cingulin, like ZO-1, is a phosphoprotein, and the assembly of both molecules at the tight junction may be regulated by kinase activity (Denisenko and Citi, 1991; Nigam et al., 1991). Cingulin is localised more cytoplasmically than ZO-1 at the tight junction of various epithelia following double labelling immunogold analysis (Stevenson et al., 1989). Other likely tight junction-associated proteins are a 192×10^3 M_r polypeptide in rodent liver recognised by monoclonal antibody BG9.1 (Chapman and Eddy, 1989) and 160×10^3 M_r protein that co-immunoprecipitates with ZO-1 from MDCK cells (Gumbiner et al., 1991). In addition, actin filaments have been visualised in ultrastructural preparations at the cytoplasmic face of tight junctions (Madara, 1987).

Temporal expression and assembly of tight junction proteins

The development of the tight junction in mouse embryos has been examined structurally and immunologically. Freeze-fracture and thin-section studies have shown that at compaction in the 8-cell embryo, focal apicolateral contact sites are formed exhibiting an intramembraneous particle organisation and associated cytoplasmic density typical of tight junctions. During the morula stage, outer blastomeres gradually acquire a belt-like tight junction organisation as focal sites extend laterally, becoming facial and then zonular in appearance before cavitation begins (Ducibella and Anderson, 1975; Ducibella et al., 1975; Magnuson et al., 1977; Pratt, 1985). Immunolocalisation of ZO-1 follows a similar pattern to these structural events. ZO-1 sites first appear at compaction as a series of punctate foci distributed along the apicolateral contact region; in morulae, these sites become a series of discontinuous lines before appearing zonular (Fleming et al., 1989). In the blastocyst, trophectoderm cells are bordered by a prominant belt-like ZO-1 distribution (Fig. 2). Immunoblotting and cellular experiments involving biosynthetic inhibitors suggest that ZO-1 is first synthesised from the late 4-cell/early 8-cell stage, although it appears that the gene is transcribed earlier in the third cell cycle (Fleming et al., 1989). See Fig. 3 for a summary of tight junction protein expression patterns during early development.

The expression and membrane assembly of cingulin in the early embryo shows a pattern both complex and quite distinct from that of ZO-1 (Javed et al., 1992; Fleming et al., 1992). The complexity of cingulin expression is due mainly to the fact that the protein is synthesised not only by the embryonic genome but also by the maternal genome. One consequence of this is that cingulin localisation is not restricted to developing tight junctions, a factor that must be borne in mind when considering biosynthetic control of junction assembly. Maternally expressed cingulin is detectable as a 140×10^3 M_r protein in unfertilised eggs following either immunoblotting or immunoprecipitation. Synthesis of maternal cingulin is unaffected by fertilisation but runs down after first cleavage, corresponding with the time of global maternal transcript degradation (Schultz, 1986), and the initiation of embryonic gene activity (Flach et al., 1982). Maternal cingulin is not associated with tight junctions but is localised in the egg cytocortex as a uniform submembraneous layer in the microvillous domain of the egg, being depleted or absent in the smooth membrane domain overlying the metaphase II spindle (Fig. 3). We envisage that maternal cingulin may function during oogenesis as a cytocortical component in the interaction between cumulus cell processes and the oocyte surface where uvomorulin (Vestweber et al., 1987), and gap and desmosome junctions (Anderson and Albertini, 1976) have been identified previously. Indeed, cingulin is detectable at both the cumulus and oocyte sides of this contact site in ovarian preovulatory follicles (Fleming et al., 1992).

Cingulin synthesis from the embryonic genome occurs at very low levels during early cleavage (approx. tenfold less than in unfertilised eggs) but is enhanced significantly at compaction in the 8-cell embryo (Javed et al., 1992). This enhancement therefore occurs later than the period identified for ZO-1 synthesis (see above). The time at which cingulin is detectable at the developing tight junction is also later than that of ZO-1, in most embryos (or synchronised cell clusters) being during the 16-cell stage (Fleming et al., 1992). This delay suggests that junction maturation at the molecular level is a sequential process, at least partly controlled by the varied time of synthesis of different junction components.

Although such a model is attractive, we must also consider whether maternal cingulin contributes to tight junction formation in the early embryo. This possibility is validated by the facts that proteins in the egg or embryo tend to have a long half-life (eg, Merz et al., 1981; Barron et al., 1989) and that cytocortical cingulin (characteristic of the egg) is also detectable during cleavage up to the morula stage (Fleming et al., 1992). Significantly, cytocortical cingulin is associated preferentially with the outer, more microvillous, membranes of the embryo (e.g., apical poles of 8- and 16-cell blastomeres, see Fig. 3), which, in undisturbed embryos, would be relatively older membrane inherited from the egg (see Pratt, 1989). However, two factors argue against maternal cingulin contributing to junction development in the embryo. First, pulse-chase metabolic labelling and immunoprecipitation data suggest that newly synthesised maternal cingulin is turned over quite rapidly ($t_{\frac{1}{2}}$~4 hours) compared with that of embryonic cingulin at the blastocyst stage ($t_{\frac{1}{2}}$~10 hours), and is therefore unlikely to persist during the 24-36 hour period between the decline of maternal expression (2-cell) and the time of cingulin assembly at junctions. Second, cytocortical cingulin during early cleavage does not appear to be a stable component of the membrane but rather characterises those sites where microvillous growth takes place. Thus, if 4-cell or early 8-cell embryos are disaggregated and then reaggregated such that original cell orientations are randomised, by the time of compaction (6-12 hours later), cytocortical cingulin is localised not at its original site but at the new outer surface where microvillous poles develop. Moreover, prior to compaction, the low level of embryonic cingulin turns over with similar rapid kinetics ($t_{\frac{1}{2}}$~4 hours) to maternal cingulin.

Collectively, these and other experiments indicate that the cytocortex represents a labile site of assembly that is available for cingulin expressed either from maternal or embryonic genomes. However, it is conceivable that this site may still act as a pool for tight junctional assembly during the morula period. This is unlikely since tight junctional assembly of cingulin is sensitive to cycloheximide treatment (Fleming et al., 1992), indicating incorporation exclusively of newly synthesised protein. The function of cytocortical cingulin during cleavage is therefore elusive but may be explained by the retention of binding sites associated with microvillous membrane that were required for cumulus cell interactions earlier in development. Cytocortical cingulin persists up until the late morula/early blastocyst stage when endocytic activity at the apical membrane results in the internalisation and gradual disappearance (and presumably degradation) of this site (Fig. 3; Fleming et al., 1992).

Regulation of tight junction protein membrane assembly

Although biosynthetic events appear to influence the sequence of molecular assembly at the tight junction, cell-cell contacts clearly provide spatial control of assembly. Thus, if uvomorulin-mediated cell adhesion at compaction is inhibited by specific antibody incubation or extracellular calcium depletion, ZO-1 assembly at the membrane is delayed and, significantly, is randomly distributed rather than apicolateral with respect to cell contact position (Fleming et al., 1989). A permissive role for uvomorulin adhesion in ZO-1 assembly has also been identified in MDCK cells (Siliciano and Goodenough, 1988) and Caco-2 cells (Anderson et al., 1989), and is likely to reflect the need for close and stable cell contact for tight junction formation to occur (Gumbiner et al., 1988). Our recent finding that cingulin synthesis is enhanced significantly at compaction (Javed et al., 1992), and that ZO-1 protein level in Caco-2 cells increases following provision of cell adhesion (Anderson et al., 1989), both suggest that cell adhesion may promote the expression of tight junctional components as well

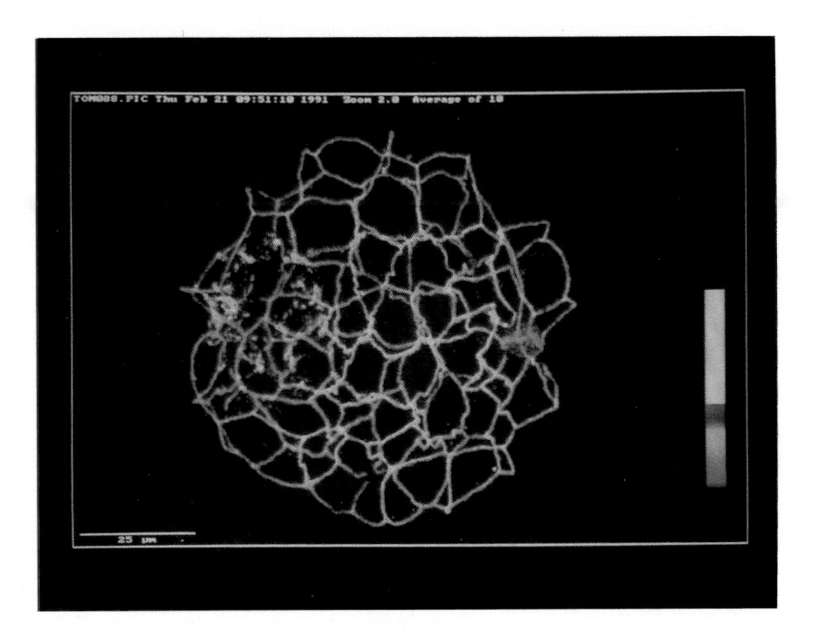

Fig. 2. ZO-1 immunolocalisation in a complete blastocyst viewed by confocal microscopy. Intensity of ZO-1 immunofluorescence is indicated by the colour code (white, highest; blue, lowest). Note the zonular network of ZO-1 that is associated with the trophectoderm (from S. Spong and T. Fleming, in preparation).

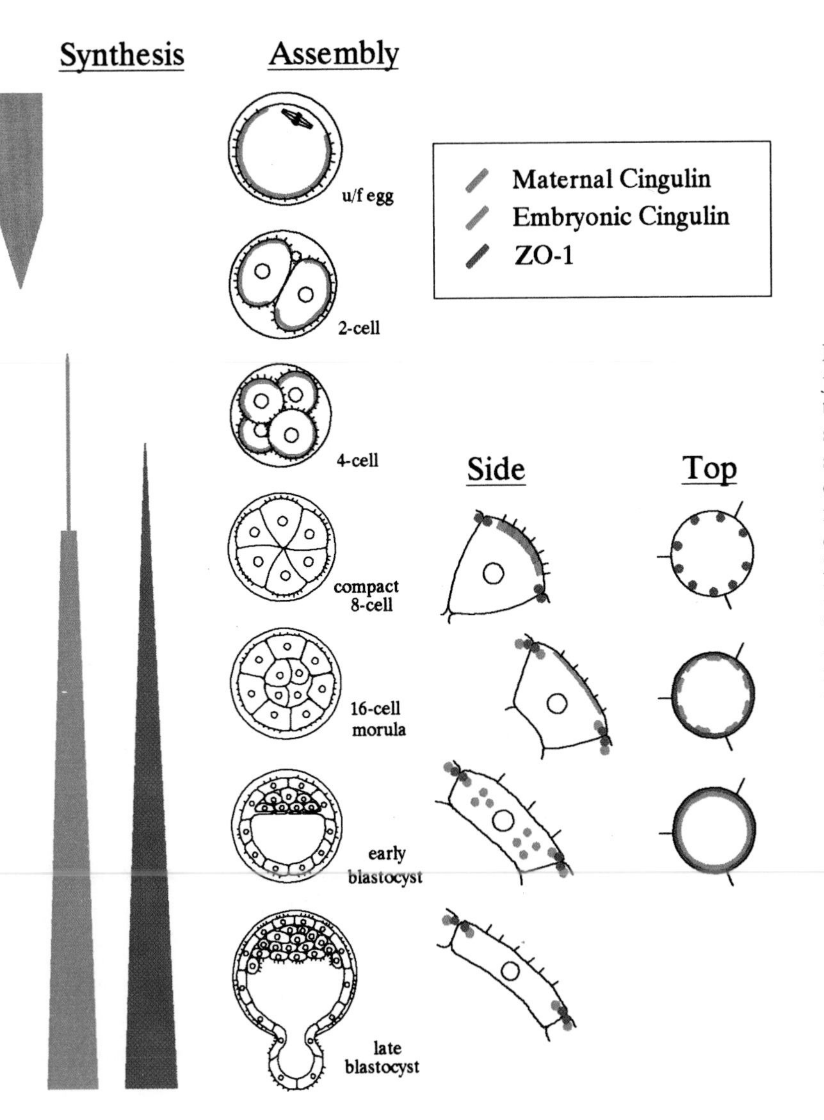

Fig. 3. Summary model of tight junction protein expression in the embryo. Relative levels of synthesis at each stage are shown on the left, measured directly (immunoprecipitation) for maternal and embryonic cingulin, but only estimated for ZO-1 from immunoblotting and localisation studies involving cycloheximide treatments. Assembly sites are shown on the right, with side and top views of the trophectoderm lineage at later stages enlarged. Cytocortical sites have been excluded from top views for clarity. The heterogeneous pattern in the apical cytocortex at compaction is intended to represent the run down of maternal cingulin and its replacement by embryonic cingulin. Cytoplasmic foci of embryonic cingulin in blastocysts represent endocytosed cytocortical sites. See text for details.

as their membrane assembly. In addition to cell adhesion, we have identified three other requirements for membrane assembly of ZO-1 in the embryo (Fleming et al., 1989; Fleming and Hay, 1991). First, assembly is dependent upon intact microfilaments but not microtubules. Second, the adjacent cell must be equally competent to form tight junctions (demonstrated in heterogeneous 4-cell + 8-cell couplets), suggesting that apicolateral assembly is dependent upon molecular interactions that traverse the intercellular boundary. Third, a contact-free membrane surface, but not necessarily apical in terms of molecular character, must be preserved to maintain ZO-1 membrane assembly. This last requirement appears to contribute to tissue-specific control of ZO-1 expression in the embryo and is discussed below.

In addition to the regulation of tight junction formation per se, it is also important to understand how this process integrates with other features of epithelial maturation occurring in the trophectoderm lineage. Since experimental evidence has suggested that cell polarisation at compaction is established initially in the cytocortex and leads secondarily to polarisation within the cytoplasm (see earlier), it is therefore of interest to determine whether the specification of the tight junctional domain in the cytocortex may be a primary event in cell polarisation. The close temporal correlation between the onset of tight junction formation and microvillous polarisation at compaction (Ducibella and Anderson, 1975; Fleming et al., 1989) is consistent with an important role for the tight junction in global cell polarisation. However, it has been possible in two experimental situations to test this proposition. First, in the 4-cell + 8-cell heterogeneous couplets referred to above, tight junction formation is perturbed (assayed by ZO-1 assembly) yet the establishment of uvomorulin adhesion and microvillous polarisation in the 8-cell blastomere is not affected. Second, cycloheximide treatment of synchronised 8-cell couplets can inhibit expression and assembly of ZO-1, again without disturbing the onset of adhesion and microvillous polarisation (Fleming et al., 1989). Thus, tight junction formation appears not to be an essential causal event in the generation of a polarised phenotype in the embryo, a conclusion consistent with experimental evidence derived from other epithelial systems (Vega-Salas et al., 1987; McNeill et al., 1990).

Another regulative feature of tight junction development in the embryo is that during blastocyst formation these junctions are tissue-specific, being confined to the trophectoderm lineage. In our analysis of cingulin synthesis during early development, we have indicated that net synthesis rates are also tissue-specific, being up to fifteen-fold higher in the trophectoderm than in the ICM of expanding blastocysts (Javed et al., 1992). To investigate the cellular mechanisms underlying tissue-specificity of tight junction formation and related synthetic activity, we have monitored ZO-1 immunolocalisation during and after cell divisions leading to the distinct lineages of the blastocyst (Fleming and Hay, 1991). These lineages are established by differentiative divisions of polarised 8- and 16-cell blastomeres such that daughter cells inherit either the apical (polar; trophectoderm lineage) or the basal (nonpolar; ICM lineage) region of their parent cell (Johnson and Ziomek, 1981b; Pedersen et al., 1986; Fleming, 1987; see Fig. 1). During differentiative divisions, ZO-1 is inherited by both polar and nonpolar daughter cells and localises to the periphery of the contact site. Once nonpolar cells are fully enclosed by polar cells (as occurs in the embryo interior), their ZO-1 sites are destabilised, first by dispersing into punctate foci distributed randomly on the cell membrane and then disappearing altogether. Thus, loss of cell contact asymmetry initiates ZO-1 down-regulation, a process that is fully reversible by re-exposing nonpolar cells to conditions of asymmetric contacts (Fleming and Hay, 1991). By culturing ICMs isolated from early blastocysts in the presence of biosynthetic inhibitors, it has been possible to investigate the level at which ZO-1 down-regulation is controlled. Data from these experiments indicate that the ICM lineage retains transcripts for ZO-1 that can be utilised to re-express the protein rapidly upon restoration of cell contact asymmetry. Down-regulation at the translational rather than the transcriptional level may be advantageous for the developing ICM in terms of processing efficiency if such transcripts become available for primary endoderm differentiation at the blastocoele interface (Fleming and Hay, 1991).

Desmosomes

The disc-shaped desmosome junction is characterised by membrane-associated cytoplasmic plaques positioned in register in apposed cells and to which intermediate filaments attach, and an intercellular adhesive domain possessing a dense midline. These junctions are thought to contribute to epithelial tissue formation and integration. Desmosomes are composed of numerous interacting components that can be broadly classified into desmosomal proteins (dp) and transmembrane glycoproteins (dg). The principal constituents have been characterised and include dp1 and dp2 (250 and 215×10^3 M_r, desmoplakins), dp3 (83×10^3 M_r, plakoglobin), dg1 ($175\text{-}150\times10^3$ M_r, desmoglein I) and dg2 and dg3 (115 and 107×10^3 M_r, desmocollins) (reviewed in Garrod et al., 1990; Schwarz et al., 1990). Dp and dg constituents occupy distinct morphological domains of the desmosome complex. In bovine epidermis, dg2 and dg3 are located mainly in the intercellular space, dg1 spans the intercellular space and the plaque, dp3 is in the plaque, and dp1 and dp2 are at the cytoplasmic side of the plaque where intermediate filaments attach (Miller et al., 1987). Although desmosome assembly has been studied in considerable detail in cultured epithelial cells by manipulating extracellular calcium levels (reviewed in Garrod et al., 1990; Garrod and Collins, 1992), little is known of how assembly might be regulated in vivo, as part of a morphogenesis programme.

In the mouse embryo, desmosomes first appear in the trophectoderm once cavitation is initiated (32- to 64-cell stage) (Ducibella et al., 1975; Magnuson et al., 1977; Fleming et al., 1991). Ultrastructurally, nascent desmosomes appear mature, containing an intercellular midline organisation, cytoplasmic plaques and associated cytokeratin filaments. The latter are first formed earlier during cleavage, predominantly in the 16-cell embryo (Chisholm and Houliston, 1987). The mechanisms controlling desmosome biogenesis have been studied by monitoring the expression of the principal desmosomal proteins and glycoproteins using specific antibodies (Fleming et al., 1991). Dp1+2, dg1, and dg2+3 are all first detectable by immunofluorescent labelling at the 32-cell stage, specifically in the trophectoderm after cavitation has begun, at punctate sites along apposed lateral membranes. Dp3 is also first detectable at the 32-cell stage, but initially is not tissue-specific, is linear in appearance at membrane contact sites, and is evident before cavitation. This distinction is transitory and may reflect the wider distribution of dp3 at non-desmosomal locations.

These results imply that nascent desmosomes in the embryo contain all the principal molecular components identified for these junctions, consistent with their mature appearance ultrastructurally. However, no indication is evident for the timing of desmosome assembly at this stage of development. Immunoprecipitation of desmosome constituents from staged embryos has shown that, in contrast to membrane assembly events, their time of initial synthesis is quite variable. Thus, dp3 synthesis is detectable at least from the 8-cell stage, dp1+2 from the 16-cell stage, and dg1 and dg2+3 from cavitation onwards (Fleming et al., 1991). This temporal sequence implicates a key role for desmosomal glycoprotein expression (either transcription or translation) in junction assembly. The intriguing correlation between cavitation and desmosome formation awaits further investigation. It will be of interest to establish the causal sequence if any in this relationship.

Conclusions

The mouse embryo continues to be an exciting model in which to study mammalian cell differentiation. Trophectoderm biogenesis takes place rather slowly with long cell cycles and can do so in the face of considerable experimental abuse, an ideal combination for the mechanistic approach. It is also a 'real' tissue and, despite the limitation of material for biochemical studies, offers some opportunities not available in parallel epithelial cell culture systems. For example, the focus in this review has been on cell junctions, and it is clear that consideration of maturational events along a time axis can be instructive for identifying how assembly might be controlled. At the molecular level, it appears that the tight junction is formed sequentially, with the assembly of membrane and membrane-associated constituents (intramembraneous particles, ZO-1) preceding that of more cytoplasmic constituents (cingulin). Moreover, the assembly of ZO-1 at compaction coincides with an upsurge in the synthesis of cingulin. Are these events causally linked? In contrast, the desmosome appears to form at the membrane in a single step, and may be regulated by the sequential expression of intrinsic components, with the last to be synthesised in this case being the membrane constituents (dgs) which trigger assembly. Our future goals will include the inhibition of specific proteins from participating in assembly events and monitoring the consequences for junction formation and early morphogenesis.

We are grateful for the financial support provided by The Wellcome Trust for research in our laboratory.

References

Anderson, E. and Albertini, D. F. (1976). Gap junctions between the oocyte and companion follicle cells in the mammalian ovary. *J. Cell Biol.* **71**, 680-686.

Anderson, J. M. and Stevenson, B. R. (1991). The molecular structure of the tight junction. In *Tight Junctions* (M. Cereijido, ed.), pp. 77-90, Florida: CRC Press.

Anderson, J. M., Stevenson, B. R., Jesaitis, L. A., Goodenough, D. A. and Mooseker, M. S. (1988). Characterization of ZO-1, a protein component of the tight junction from mouse liver and Madin-Darby canine kidney cells. *J. Cell Biol.* **106**, 1141-1149.

Anderson, J. M., Van Itallie, C. M., Peterson, M. D., Stevenson, B. R., Carew, E. A. and Mooseker, M. S. (1989). ZO-1 mRNA and protein expression during tight junction assembly in Caco-2 cells. *J. Cell Biol.* **109**, 1047-1056.

Barron, D. J., Valdimarsson, G., Paul, D. L. and Kidder, G. M. (1989). Connexin32, a gap junction protein, is a persistent oogenetic product through preimplantation development of the mouse. *Devel. Genet.* **10**, 318- 323.

Bevilacqua, A., Loch-Caruso, R. and Erickson, R. P. (1989). Abnormal development and dye coupling produced by antisense RNA to gap junction protein in mouse preimplantation embryos. *Proc. Natl. Acad. Sci. USA* **86**, 5444-5448.

Blaschuk, O. W., Sullivan, R., David, S. and Pouliot, Y. (1990). Identification of a cadherin cell adhesion recognition sequence. *Dev. Biol.* **139**, 227-229.

Bloom, T. L. (1989). The effects of phorbol ester on mouse blastomeres: a role for protein kinase C in compaction? *Development* **106**, 159-171.

Bloom, T. L. (1991). Experimental manipulation of compaction of the mouse embryo alters patterns of protein phosphorylation. *Molec. Reprod. Devel.* **28**, 230-244.

Bloom, T. L. and McConnell, J. (1990). Changes in protein phosphorylation associated with compaction of the mouse preimplantation embryo. *Molec. Reprod. Devel.* **26**, 199-210.

Buehr, M., Lee, S., McLaren, A. and Warner, A. (1987). Reduced gap junctional communication is associated with the lethal condition characteristic of DDK mouse eggs fertilized by foreign sperm. *Development* **101**, 449-459.

Cereijido, M. (ed.) (1991). *Tight Junctions.* Boca Raton, Florida: CRC Press.

Chapman, L. M. and Eddy, E. M. (1989). A protein associated with the mouse and rat hepatocyte junctional complex. *Cell Tisue Res.* **257**, 333-341.

Chisholm, J. C. and Houliston, E. (1987). Cytokeratin filament assembly in the preimplantation mouse embryo. *Development* **101**, 565-582.

Citi, S., Sabanay, H., Jakes, R., Geiger, B. and Kendrick-Jones, J. (1988). Cingulin, a new peripheral component of tight junctions. *Nature* **333**, 272-276.

Citi, S., Sabanay, H., Kendrick-Jones, J. and Geiger, B. (1989). Cingulin: characterization and localization. *J. Cell Sci.* **93**, 107-122.

Denisenko, N. and Citi, S. (1991). Cingulin phosphorylation and localization in confluent MDCK cells treated with phorbol esters and protein kinase inhibitors. *J. Cell Biol.* **115**, 480a.

Ducibella, T., Albertini, D. F., Anderson, E. and Biggers, J. D. (1975). The preimplantation mammalian embryo: characterization of intercellular junctions and their appearance during development. *Dev. Biol.* **45**, 231-250.

Ducibella, T. and Anderson, E. (1975). Cell shape and membrane changes in the 8-cell mouse embryo: prerequisites for morphogenesis of the blastocyst. *Dev. Biol.* **47**, 45-58.

Fenderson, B. A., Eddy, E. M. and Hakomori, S. (1990). Glycoconjugate expression during embryogenesis and its biological significance. *BioEssays* **12**, 173-179.

Flach, G., Johnson, M. H., Braude, P. R., Taylor, R. and Bolton, V. N. (1982). The transition from maternal to embryonic control in the 2-cell mouse embryo. *EMBO J.* **1**, 681-686.

Fleming, T. P. (1987). A quantitative analysis of cell allocation to trophectoderm and inner cell mass in the mouse blastocyst. *Dev. Biol.* **119**, 520-531.

Fleming, T. P. (1992). Trophectoderm biogenesis in the preimplantation mouse embryo. In *Epithelial Organization and Development* (T.P. Fleming, ed.), pp. 111-136, London: Chapman and Hall.

Fleming, T. P., Garrod, D. R. and Elsmore, A. J. (1991). Desmosome biogenesis in the mouse preimplantation embryo. *Development* **112**, 527-539.

Fleming, T. P. and Hay, M. J. (1991). Tissue-specific control of expression of the tight junction polypeptide ZO-1 in the mouse early embryo. *Development* **113**, 295-304.

Fleming, T. P., Hay, M., Javed, Q. and Citi, S. (1992). Localization of tight junction protein cingulin is temporally and spatially regulated during early mouse development. *Development* submitted for publication.

Fleming, T. P. and Johnson, M. H. (1988). From egg to epithelium. *Ann. Rev. Cell Biol.* **4**, 459-485.

Fleming, T. P., McConnell, J., Johnson, M. H. and Stevenson, B. R. (1989). Development of tight junctions *de novo* in the mouse early embryo: control of assembly of the tight junction-specific protein, ZO-1. *J. Cell Biol.* **108**, 1407-1418.

Gardiner, C. S., Williams, J. S. and Menino, A. R. (1990). Sodium/potassium adenosine triphosphatase α- and β-subunit and α-subunit mRNA levels during mouse embryo development in vitro. *Biol. Reprod.* **43**, 788-794.

Garrod, D. R., Parrish, E. P., Mattey, D. L., Marston, J. E., Measures, H. R. and Vilela, M. J. (1990). Desmosomes. In *Morphoregulatory Molecules* (G. M. Edelman, B. A. Cunningham and J. P. Thiery, eds.), pp. 315-339, Neurosciences Institute Publications Series, Chichester: John Wiley and Sons.

Garrod, D. R. and Collins, J. E. (1992). Intercellular junctions and cell adhesion in epithelial cells. In *Epithelial Organization and Development* (T. P. Fleming, ed.), pp. 1-52, London: Chapman and Hall.

Goodall, H. (1986). Manipulation of gap junctional communication during compaction of the mouse early embryo. *J. Embryol. exp. Morph.* **91**, 283-296.

Goodall, H. and Johnson, M. H. (1984). The nature of intercellular coupling within the preimplantation mouse embryo. *J. Embryol. exp. Morph.* **79**, 53-76.

Gumbiner, B., Lowenkopf, T. and Apatira, D. (1991). Identification of a 160kDa polypeptide that binds to the tight junction protein ZO-1. *Proc. Natl. Acad. Sci., USA* **88**, 3460-3464.

Gumbiner, B., Stevenson, B. and Grimaldi, A. (1988). The role of the cell adhesion molecule uvomorulin in the formation and maintenance of the epithelial junctional complex. *J. Cell Biol.* **107**, 1575-1587.

Javed, Q., Fleming, T. P., Hay, M. and Citi, S. (1992). Tight junction protein cingulin is expressed by maternal and embryonic genomes during early mouse development. *Development* submitted for publication.

Johnson, M. H., Chisholm, J. C., Fleming, T. P. and Houliston, E. (1986a). A role for cytoplasmic determinants in the development of the mouse early embryo? *J. Embryol. exp. Morph.* **97**(Supplement), 97-121.

Johnson, M. H. and Maro, B. (1986). Time and space in the mouse early embryo: a cell biological approach to cell diversification. In *Experimental Approaches to Mammalian Embryonic Development* (J. Rossant and R. A. Pedersen, eds.), pp. 35-66, New York: Cambridge University Press.

Johnson, M. H., Maro, B. and Takeichi, M. (1986b). The role of cell adhesion in the synchronisation and orientation of polarisation in 8-cell mouse blastomeres. *J. Embryol. exp. Morph.* **93**, 239-255.

Johnson, M. H. and Ziomek, C. A. (1981a). Induction of polarity in mouse 8-cell blastomeres: specificity, geometry and stability. *J. Cell Biol.* **91**, 303-308.

Johnson, M. H. and Ziomek, C. A. (1981b). The foundation of two distinct cell lineages within the mouse morula. *Cell* **24**, 71-80.

Kidder, G. M. and McLachlin, J. R. (1985). Timing of transcription and protein synthesis underlying morphogenesis in preimplantation mouse embryos. *Dev. Biol.* **112**, 265-275.

Kidder, G. M. and Watson, A. J. (1990). Gene expression required for blastocoel formation in the mouse. In *Early Embryo Development and Paracrine Relationships (UCLA Symposia on Molecular and Cellular Biology, New Series* Vol. 117) (S. Heyner and L. M. Wiley, eds.), pp. 97-107, New York: Alan R. Liss.

Kimber, S. J. (1990). Glycoconjugates and cell surface interactions in pre- and peri-implantation mammalian embryonic development. *Int. Rev. Cytol.* **120**, 53-167.

Lehtonen, E., Ordonez, G. and Reima, I. (1988). Cytoskeleton in preimplantation mouse development. *Cell Differentiation* **24**, 165-178.

Lehtonen, E. and Reima, I. (1986). Changes in the distribution of vinculin during preimplantation mouse development. *Differentiation* **32**, 125-134.

Lee, S., Gilula, N. B. and Warner, A. E. (1987). Gap junctional communication and compaction during preimplantation stages of mouse development. *Cell* **51**, 851-860.

Levy, J. B., Johnson, M. H., Goodall, H. and Maro, B. (1986). Control of the timing of compaction: a major developmental transition in mouse early development. *J. Embryol. exp. Morph.* **95**, 213-237.

Lo, C. W. and Gilula, N. B. (1979). Gap junctional communication in the preimplantation mouse embryo. *Cell* **18**, 399-409.

Madara, J. L. (1987). Intestinal absorptive cell tight junctions are linked to cytoskeleton. *Am. J. Physiol.* **253**, C171-C175.

Magnuson, T., Demsey, A. and Stackpole, C. W. (1977). Characterization of intercellular junctions in the preimplantation mouse embryo by freeze-fracture and thin-section electron microscopy. *Dev. Biol.* **61**, 252-261.

Magnuson, T., Jacobson, J. B. and Stackpole, C. W. (1978). Relationship between intercellular permeability and junction organization in the preimplantation mouse embryo. *Dev. Biol.* **67**, 214-224.

Manejwala, F. M., Cragoe, E. J. and Schultz, R. M. (1989). Blastocoel expansion in the preimplantation mouse embryo: role of extracellular sodium and chloride and possible apical routes of their entry. *Dev. Biol.* **133**, 210-220.

Manejwala, F. M., Kali, E. and Schultz, R. M. (1986). Development of activatable adenylate cyclase in the preimplantation mouse embryo and a role for cyclic AMP in blastocoel formation. *Cell* **46**, 95-103.

Manejwala, F. M. and Schultz, R. M. (1989). Blastocoel expansion in the preimplantation mouse embryo: stimulation of sodium uptake by cAMP and possible involvement of cAMP-dependent protein kinase. *Dev. Biol.* **136**, 560-563.

Maro, B., Houliston, E. and de Pennart, H. (1988). Microtubule dynamics and cell diversification in the mouse preimplantation embryo. *Protoplasma* **145**, 160-166.

McLachlin, J. R., Caveney, S. and Kidder, G. M. (1983). Control of gap junction formation in early mouse embryos. *Dev. Biol.* **98**, 155-164.

McLachlin, J. R. and Kidder, G. M. (1986). Intercellular junctional coupling in preimplantation mouse embryos: effect of blocking transcription or translation. *Dev. Biol.* **117**, 146-155.

McNeill, H., Ozawa, M., Kemler, R. and Nelson, W. J. (1990). Novel function of the cell adhesion molecule uvomorulin as an inducer of cell surface polarity. *Cell* **62**, 309-316.

Merz, E. A., Brinster, R. L., Brunner, S. and Chen, H. Y. (1981). Protein degradation during preimplantation development of the mouse. *J. Reprod. Fertil.* **61**, 415-418.

Miller, K., Mattey, D., Measures, H., Hopkins, C. and Garrod, D. (1987). Localisation of the protein and glycoprotein components of bovine nasal epithelial desmosomes by immunoelectron microscopy. *EMBO J.* **6**, 885-889.

Nigam, S. K., Denisenko, N., Rodriguez-Boulan, E. and Citi, S. (1991). The role of phosphorylation in development of tight junctions in cultured renal epithelial (MDCK) cells. *Biochem. Biophys. Res. Comm.* **181**, 548-553.

Nishi, M., Kumar, N. M. and Gilula, N. B. (1991). Developmental regulation of gap junction gene expression during mouse embryonic development. *Dev. Biol.* **146**, 117-130.

Nuccitelli, R. and Wiley, L. (1985). Polarity of isolated blastomeres from mouse morulae: detection of transcellular ion currents. *Dev. Biol.* **109**, 452-463.

Pedersen, R. A., Wu, K. and Balakier, H. (1986). Origin of the inner cell mass in mouse embryos: cell lineage analysis by microinjection. *Dev. Biol.* **117**, 581-595.

Peyrieras, N., Hyafil, F., Louvard, D., Ploegh, H. L. and Jacob, F. (1983). Uvomorulin: a non-integral membrane protein of early mouse embryo. *Proc. Natl. Acad. Sci. USA* **80**, 6274-6277.

Pratt, H. P. M. (1985). Membrane organization in the preimplantation mouse embryo. *J. Embryol. exp. Morph.* **90**, 101-121.

Pratt, H. P. M. (1989). Marking time and making space: chronology and topography in the early mouse embryo. *Int. Rev. Cytol.* **117**, 99-130.

Reima, I. (1990). Maintenance of compaction and adherent-type junctions in mouse morula-stage embryos. *Cell Differen. Devel.* **29**, 143-153.

Schultz, R. (1986). Utilization of genetic information in the preimplantation mouse embryo. In *Experimental Approaches to Mammalian Embryonic Development* (J. Rossant and R. A. Pedersen, eds.), pp. 239-266, New York: Cambridge University Press.

Schwarz, M. A., Owaribe, K., Kartenbeck, J. and Franke, W. W. (1990). Desmosomes and hemidesmosomes: constitutive molecular components. *Ann. Rev. Cell Biol.* **6**, 461-491.

Siliciano, J. D. and Goodenough, D. A. (1988). Localization of the tight junction protein, ZO-1, is modulated by extracellular calcium and cell-

cell contact in Madin-Darby canine kidney epithelial cells. *J. Cell Biol.* **107**, 2389-2399.

Staehelin, L. A. (1974). Structure and function of intercellular junctions. *Int. Rev. Cytol.* **39**, 191-283.

Stevenson, B. R., Anderson, J. M. and Bullivant, S. (1988). The epithelial tight junction: structure, function and preliminary biochemical characterization. *Molec. Cell. Biochem.* **83**, 129-145.

Stevenson, B. R., Heintzelman, M. B., Anderson, J. M., Citi, S. and Mooseker, M. S. (1989). ZO-1 and cingulin: tight junction proteins with distinct identities and localizations. *Am. J. Physiol.* **257**, C621-C628.

Stevenson, B. R., Siliciano, J. D., Mooseker, M. S. and Goodenough, D. A. (1986). Identification of ZO-1: a high molecular weight polypeptide associated with the tight junction (zonula occludens) in a variety of epithelia. *J. Cell Biol.* **103**, 755-766.

Van Winkle, L. J. and Campione, A. L. (1991). Ouabain-sensitive Rb^+ uptake in mouse eggs and preimplantation conceptuses. *Dev. Biol.* **146**, 158-166.

Vega-Salas, D. E., Salas, P. J., Gundersen, D. and Rodriguez-Boulan, E. (1987). Formation of the apical pole of epithelial (Madin-Darby canine kidney) cells: polarity of an apical protein is independent of tight junctions while segregation of a basolateral marker requires cell-cell interactions. *J. Cell. Biol.* **104**, 905-916.

Vestweber, D., Gossler, A., Boller, K. and Kemler, R. (1987). Expression and distribution of cell adhesion molecule uvomorulin in mouse preimplantation embryos. *Dev. Biol.* **124**, 451-456.

Watson, A. J., Damsky, C. H. and Kidder, G. M. (1990a). Differentiation of an epithelium: factors affecting the polarized distribution of Na^+,K^+-ATPase in mouse trophectoderm. *Dev. Biol.* **141**, 104-114.

Watson, A. J., Pape, C., Emanuel, J. R., Levenson, R. and Kidder, G. M. (1990b). Expression of Na,K-ATPase α and β subunit genes during preimplantation development of the mouse. *Devel. Genet.* **11**, 41-48.

Wiley, L. M. (1987). Trophectoderm: the first epithelium to develop in the mammalian embryo. *Scanning Microsc.* **2**, 417-426.

Wiley, L. M., Kidder, G. M. and Watson, A. J. (1990). Cell polarity and development of the first epithelium. *BioEssays* **12**, 67-73.

Wiley, L. M., Lever, J. E., Pape, C. and Kidder, G. M. (1991). Antibodies to a renal Na^+/glucose cotransport system localize to the apical plasma membrane domain of polar mouse embryo blastomeres. *Dev. Biol.* **143**, 149-161.

Wiley, L. M. and Obasaju, M. F. (1988). Induction of cytoplasmic polarity in heterokaryons of mouse 4-cell-stage blastomeres fused with 8-cell- and 16-cell-stage blastomeres. *Dev. Biol.* **130**, 276-284.

Wiley, L. M. and Obasaju, M. F. (1989). Effects of phlorizin and ouabain on the polarity of mouse 4-cell/16-cell stage blastomere heterokaryons. *Dev. Biol.* **133**, 375-384.

Winkel, G. K., Ferguson, J. E., Takeichi, M. and Nuccitelli, R. (1990). Activation of protein kinase C triggers premature compaction in the four-cell stage mouse embryo. *Dev. Biol.* **138**, 1-15.

Ziomek, C. A. and Johnson, M. H. (1980). Cell surface interactions induce polarisation of mouse 8-cell blastomeres at compaction. *Cell* **21**, 935-942.

Development 1992 Supplement, 113-118 (1992)

The relationship of gap junctions and compaction in the preimplantation mouse embryo

DAVID L. BECKER[1], CATHERINE LECLERC-DAVID[2] and ANNE WARNER[1]

[1]*Department of Anatomy and Developmental Biology, University College London, Gower Street, London, WC1E 6BT, UK*
[2]*Centre de Biologie du Developpement, URA, CNRS 675, 118, Rue de Narbonne, 31062, Toulouse-Cedex, FRANCE*

Summary

In the mouse embryo, gap junctions first appear at the 8-cell stage as compaction is about to take place. Compaction of the embryo is important for the differentiation of the first two cell types; the inner cell mass and the trophectoderm. Our studies examine the contribution of gap junctional communication at this stage of development. We have characterised the normal sequence of appearance of gap junction protein and its distribution. The extent of communication as shown by the passage of dye between cells has been recorded in both normal embryos and embryos treated with drugs that influence gap junctional communication. Comparisons have been made with embryos that express a lethal gap junction defect and attempts were made to rescue such embryos by increasing their gap junction communication.

Key words: mouse embryo, gap junction, connexin 43, compaction.

Introduction

In placental mammals, gastrulation of the embryo is necessarily preceded by events that ensure the formation not only of the embryo itself, but also the structures involved in implantation and subsequent generation of the supportive tissues; the extraembryonic membranes and the placenta. During this early phase of development, cells generated by cleavage of the fertilized egg are separated into embryonic and extraembryonic lineages. This process constitutes the first differentiative step in the development of the mammalian embryo.

Cells produced by the first three cleavages initially contact each other only loosely and are not linked by any specialised intercellular contacts. However, during the 8-cell stage, cellular inter-relations change dramatically. The cells compact down on each other, minimising the intercellular space. They become polarized as tight junctions form between them at their outer edges and they begin to communicate with each other through the intercellular structure, the gap junction. Both tight and gap junctions are important for polarization and compaction, which ensure the subsequent establishment of the separate identities of outer cells, which will form the trophectoderm, and inner cells, which will give rise to the embryo proper. Since gastrulation is restricted to cells of the embryo, the processes that control polarization and compaction during the preimplantation stages take on crucial importance. This article examines evidence implicating gap junctional communication during these stages. We include also a brief description of recent work suggesting that gap junctions are regulated in response to permeability changes.

How tight junctions and gap junctions contribute to the morphogenetic movements of gastrulation has not yet been investigated for any species. One can predict from the nature of the invagination through the primitive streak that all close cellular interactions must be modified. The tight junctions that maintain the integrity of the epithelium must break down. Transient gap junction formation between cells destined to form the mesoderm layer of the embryo is likely during invagination, but how this might contribute to cellular rearrangements achieved by gastrulation remains to be investigated.

The time course of appearance of gap junctions in the mouse embryo

Prior to the 8-cell stage, cells may communicate via cytoplasmic bridges or mid bodies remaining after cleavage. Such communication is characterised by the ability to pass large molecules, such as Horse Radish Peroxidase, between cells. In contrast, intercellular communication through gap junctions will only allow the transfer of small molecules such as the dye Lucifer Yellow. Gap junctional communication begins for the first time at the 8-cell stage, as the embryo begins to compact (Lo and Gilula, 1979; Goodall and Johnson, 1982; 1984). From this time onwards gap junctional communication is found between all cells of the preimplantation embryo, except during division, when gap junctions (and other junctions) dissociate and communication ceases until division is over (Goodall and Maro, 1986). mRNAs for gap junction protein connexin 43 are first expressed at the 4-cell stage and gap junction plaques appear during the 8-cell stage (Dulcibella et al., 1975; Barron et al., 1989; Nishi et al., 1991). Gap junctions are constructed from a highly conserved family of proteins, termed connexins (Beyer et al., 1990), with the basic structure and assembly being the same for all members. A hexameric arrangement of the connexins creates a central channel, through which communication can take place when two hexamers in opposing membranes are linked. The evidence so far suggests that the first gap junctions in the early mouse embryo are constructed from connexin 43.

Gap junctions first appear in the compacting 8-cell-stage embryo as punctate spots between the cells at points of contact. Fig. 1A shows a confocal optical section through an early 8-cell embryo stained with an antipeptide antibody to an amino acid sequence unique to connexin 43 (Harfst et al., 1990; Gourdie et al., 1990). This illustrates the punctate intercellular staining, characteristic of gap junctions. Control competition experiments on heart sections indicate that gap junction labelling with this antibody is eliminated by the addition of the corresponding free peptide. Antibodies to peptides unique to other connexins (e.g. connexin 32: Evans and Rahman, 1989; Rahman and Evans, 1991; Evans, W.H., Green, C.R. and Warner, A.E., in preparation) do not stain at this early stage, indicating that connexin 43 is the predominant gap junction protein. Fig. 1B, C show projections of two compacting 8-cell embryos, reconstructed from a series of confocal optical sections. One embryo (B) has not yet begun to compact and no gap junctions are present. The second embryo (C) has begun to compact and punctate junctions, located in the central region of the embryo, are clearly visible. At this stage, functional communication, mapped with Lucifer Yellow transfer, may occur between as few as two cells or as many as eight, depending on the accretion rate of gap junction protein into plaques in individual samples. As compaction takes place, adjacent cell membranes flatten against each other and the junction plaques increase in both size and number. This is accompanied by a shift of gap junction structures from a predominantly central location to a peripheral position within the junctional complex, where tight junctions form (see Fleming, this volume). Fig. 1D shows a surface view of a compacted embryo illustrating how peripheral spots gradually zip up to form a belt surrounding each cell as compaction is completed.

After division to 16 cells, some progeny lie entirely within the embryo and from this time on inner and outer cells have different destinations: inner cells become the inner cell mass which will form the embryo proper, whereas outer cells form trophectoderm which contributes to the extraembryonic membranes. Trophectodermal cells are linked to each other by gap junctions in stripes along the outermost limiting membrane (Fig. 1D), in close association with the tight junctions (Magnuson et al., 1977; Fleming, this volume). Junctions between inner cells and between inner and outer cells remain in the form of punctate spots.

Cavitation of the embryo, marking the appearance of the nascent blastocoele, depends on the integrity of the tight junction, which allows fluid generated by the action of the sodium pump to accumulate in the intercellular spaces. The third intercellular adhesive structure, the desmosome, appears at cavitation (see Fleming, this volume). The differential distribution of gap junctions between trophectoderm and inner cells is retained throughout blastocoele expansion and is most striking in the expanded blastocyst. Fig. 1E and F shows a projection of confocal optical sections through the surface of an expanded blastocyst where the thin trophectoderm cells are clearly linked by belt gap junctions (E). In contrast, a projection of confocal sections taken through the centre of the embryo, reveals numerous punctate junctions within the inner cell mass (F).

Gap junctions and the maintenance of compaction

The importance of tight junctions for the establishment and maintenance of the compacted state is self evident; the separate identity of inner and outer cells is thereby established and a tight adhesive link between cells is clearly an essential prerequisite for the retention of blastocoel fluid. However, a role for gap junctional communication is not so obvious. Nevertheless, two lines of evidence suggest that gap junctions also may be important for maintaining compaction.

The first comes from experiments in which antibodies against gap junction protein were injected into one cell either at the 8-cell stage, after communication had been established or into one cell at the 2-cell stage, well before gap junctions appear (Lee et al., 1987). The antibody prevented the transfer from cell to cell of small molecules and ions, while preimmune serum was without effect. At the 8-cell stage, gap junction antibody injection led to the decompaction and extrusion of the non-communicating cell or cells while the rest of the embryo maintained compaction and continued to develop. At the 2-cell stage, cell division continued without interruption. However, the progeny of the antibody-injected cell failed to take part in compaction while the progeny of the uninjected cell compacted with a normal time course (see Table 1).

The second line of evidence arises from a study of DDK defective embryos. The DDK syndrome is generated when DDK females mate with males from another strain. DDK/DDK progeny are normal whereas 90-95% of DDK/cross embryos express the defect and are destined to die before the expanded blastocyst stage (Wakasugi, 1973; Buehr et al., 1987). The defect is characterized by the gradual decompaction of cells, beginning at the 16-cell stage and continuing through to the expanded blastocyst. Measurements of the efficiency of gap junctional communication were made by injecting Lucifer Yellow into one cell of fully compacted 8-cell embryos and determining the time taken for the last cell to become visibly fluorescent. DDK/DDK embryos showed transfer times that were no different from other strains (median 92 seconds) while DDK/C3H embryos were significantly slower (139 seconds).

One way of testing for a link between gap junctional communication and compaction is to manipulate gap junc-

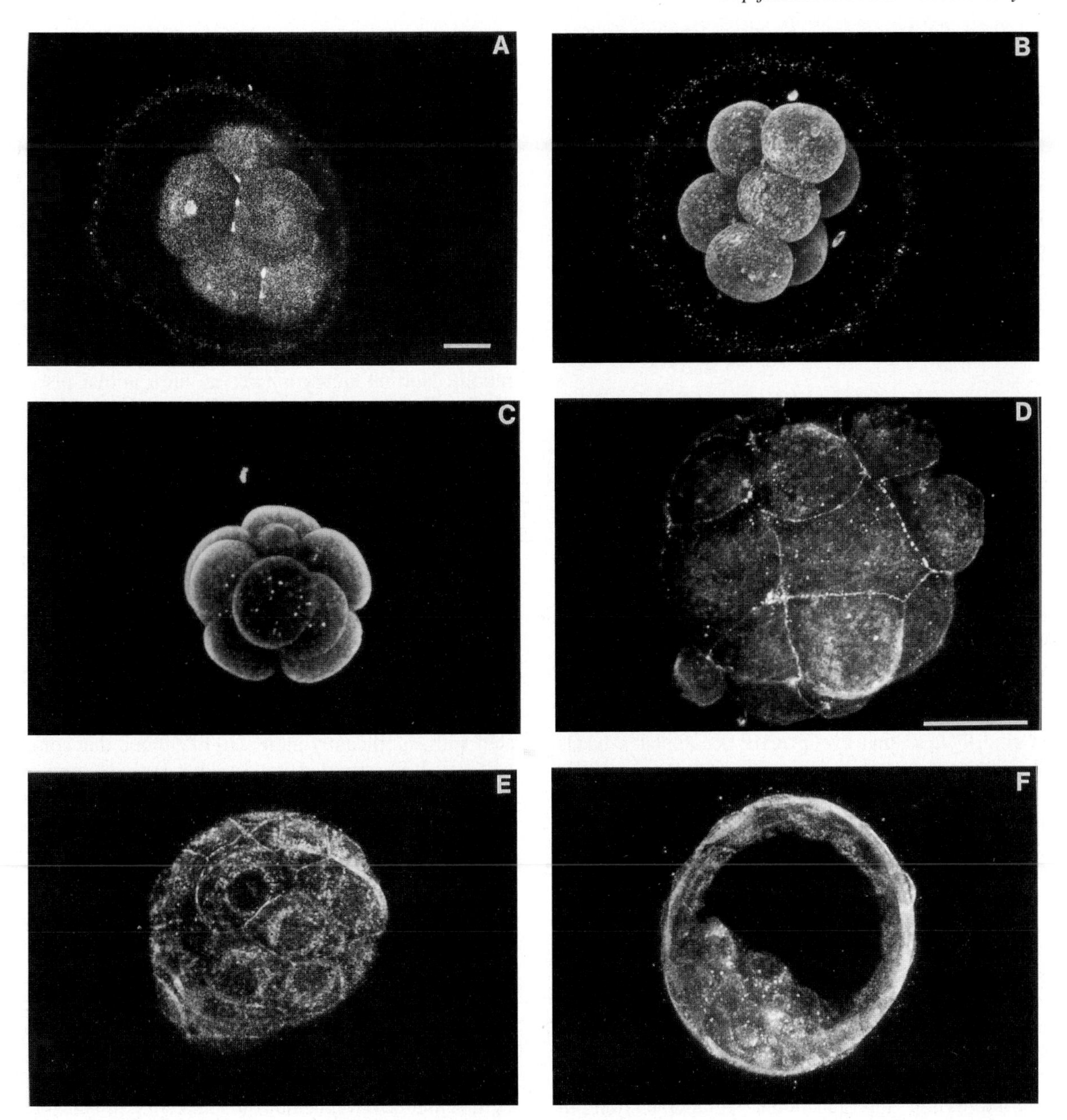

Fig. 1. A-F shows images of mouse embryos immuno labelled for the gap junction protein connexin 43. These were constructed by the use of a laser scanning confocal microscope. (A) Single optical section through the centre of a compacting 8-cell-stage embryo. The punctate intercellular labelling of the gap junction protein is clearly visible. (B) Complete reconstruction of a series of optical sections (34) taken through an embryo that has just reached the 8-cell stage. Note the very round cells and the absence of connexin 43 labelling. (C) Complete reconstruction of a compacting 8-cell-stage embryo showing the punctate labelling of all gap junctions. Note the restriction of junctions to the centre of the embryo. (D) Projection of an almost completely compacted embryo. Note the punctate gap junctions lining up along the outermost region of intercellular contact. As compaction progresses the junctions become smaller and closer together creating linear labelling zipping up the cells. (E) A projection of the top eight optical sections of an expanded blastocyst which shows the linear labelling of the gap junctions between the trophectoderm cells. (F) A projection of 10 optical sections through the centre of the same embryo, to show the inner cell mass where the intercellular labelling is in the form of punctate spots. Scale bars are 25 μm. Bar in A also applies to B,C,E and F.

tion permeability and observe the developmental consequences. Gap junction permeation is highly sensitive to intracellular pH (Turin and Warner, 1977, 1980; Bennett et al., 1978). The effect is both rapid and completely reversible and a relatively small change in pH_i will alter substantially junctional conductance because the relationship between conductance and pH_i is very steep. Incubation in a solution containing mM concentrations of either a weak acid or weak base provides a convenient way to lower or raise intracellular pH and decrease (acid) or increase (base) gap junction conductance.

The weak base methylamine produced a rapid speeding

Table 1. *Gap junction permeability and the maintenance of compaction*

Strain	Treatment	Junctions	Compaction maintained	n/N
B10CF1	None	Normal	91%	39/43
B10CF1	Pre Immune IgGs	Normal	100%	11/11
B10CF1	Gap junction Ab	Blocked	23%	4/17
B10CF1	Butyrate	Reduced	40%	12/30
DDK/DDK	None	Normal	91%	99/109
DDK/DDK	Butyrate	Reduced	63%	25/40
DDK/C3H	None	Reduced	10%	17/188
DDK/C3H	Methylamine	Increased	31%	21/67
DDK/C3H	Cyclic AMP	Increased	21%	14.67

of junctional transfer in both control (DDK/DDK) and defective (DDK/C3H) embryos (Buehr et al., 1987). A relatively short (6 hours) treatment of DDK/C3H zygotes from the 8-cell stage, which is before decompaction begins, improved significantly the survival to the blastocyst stage (see Table 1), but did not affect controls.

An alternative way of improving gap junction conductance is to increase intracellular levels of cyclic AMP (David et al., in preparation). This was achieved either by exposure to a membrane-permeable derivative of cyclic AMP (1 mM) or by treatment with a reagent such as Forskolin (100 μM), which activates adenylate cyclase (Hax et al., 1974) so that cyclic AMP accumulates inside the cell. Both treatments produce a significant increase in gap junction permeation in control and DDK/C3H (defective) embryos, although DDK/C3H embryos respond relatively slowly (> 30 minutes), perhaps because their internal pH is intrinsically low (see below). When DDK/C3H embryos are treated with a membrane-permeable derivative of cyclic AMP for 6 hours beginning at the 8-cell stage, survival to the expanded blastocyst stage is increased (Table 1), although the effect is not as marked as for methylamine. Since cyclic AMP increases junction permeation without a change in intracellular pH, this result supports the view that the efficiency of junctional communication is the dominant factor.

Since improving gap junctional communication in embryos destined to decompact can restore normal development, we asked whether reducing junctional permeability in normal embryos can reproduce the DDK phenotype (David et al., in preparation). The weak acid butyrate lowers intracellular pH by 0.2 to 0.5 of a pH unit in control embryos where pH_i normally lies close to neutrality. When B10CF1 embryos were treated with 10 mM butyrate, the fall in intracellular pH reduced junctional transfer. The degree to which junctional communication slowed was variable, probably because of differing rates of entry of butyrate and intracellular buffering. After 60 minutes exposure to butyrate, 50% (n=12) took more than 25 minutes to transfer dye and were considered to demonstrate complete block. With longer exposure times, the proportion showing complete block increased and after 2 hours all the embryos tested failed to transfer Lucifer yellow. The developmental consequences were tested on embryos at the 8-cell stage and treatment with butyrate lasted 4 to 6 hours, exactly the same interval as for methylamine and cyclic AMP. The embryos were returned to normal medium and scored for maintenance of compaction 24 and 48 hours later. Measurements of intracellular pH showed that even after long incubation periods pH_i was restored rapidly to neutrality. Butyrate-treated embryos reproduced the DDK syndrome; a proportion (Table 1) failed to maintain compaction and did not reach the expanded blastocyst stage.

The finding that lowering internal pH in normal embryos, thereby reducing the permeability of gap junctions, induces the DDK defective phenotype raises the interesting possibility that DDK/C3H embryos might be characterized by naturally low intracellular pH. We used the fluorescent dye BCECF to measure intracellular pH (David et al., in preparation). Normal embryos have an intracellular pH close to pH 7.0. However, pH_i in DDK/C3H embryos is significantly lower (mean pH 6.7). Thus, poor gap junctional communication in DDK/C3H embryos probably derives from low pH rather than any alteration in the properties of the gap junctions. This finding has interesting consequences for understanding the basis of the DDK defect since it suggests that the primary defect may come from defective pH regulating mechanisms.

Table 1 summarizes the various manipulations that influence junctional permeability together with the consequences for the maintenance of compaction. There is a clear correlation between junctional communication and compaction. Reduced gap junctional communication is associated with significantly increased likelihood that compaction will fail. It is important to note that both decompaction of normal embryos and improved compaction in DDK/C3H embryos were achieved by relatively short treatments at the 8-cell stage, almost 24 hours before the developmental consequences were apparent.

The regulation of gap junctions

The consequences of manipulating gap junction permeability are expressed as subsequent alterations in embryonic development, manifested long after junctional permeability has returned to normal. This raises the interesting possibility that the developmental outcome may reflect regulation of gap junctions in response to the imposed permeability change. This possibility is reinforced by the observation that junction permeability must be held at an altered level for a minimum of 4 to 6 hours to see long-term effects. Thus Buehr et al. (1987) noted that rescue of DDK/C3H embryos to the expanded blastocyst stage required at least 6 hours treatment with methylamine and that embryos could only respond if treatment was completed before the morphological consequences of the defect became manifest. We have recently begun to analyse the long-term consequences of such treatments (Becker, D. L., Green, C. R. and Warner, A. E. unpublished data).

Since intracellular pH is naturally low and junction permeability reduced in DDK/C3H embryos, we have begun by examining the distribution of gap junctions in DDK defective progeny. Fig. 2 shows an example of a DDK/C3H embryo at the 16-cell stage stained with an antibody to connexin 43. It is clear that there are many punctate gap junctions, indeed a greater number than normally found in con-

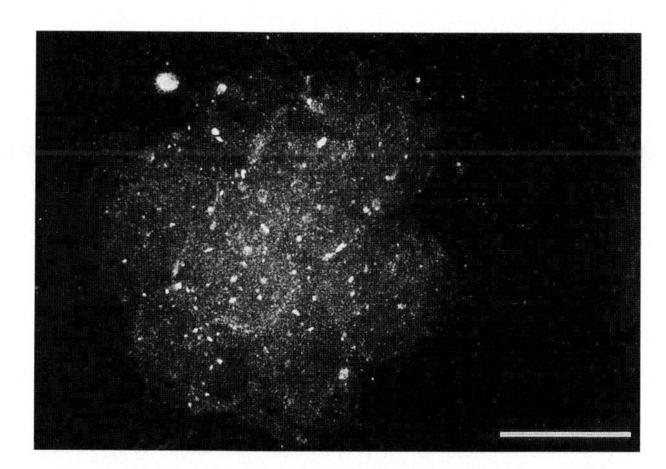

Fig. 2. Projection of a series of optical sections reconstructing a poorly compacted DDK/C3H morula which has been immunolabelled against the gap junction protein connexin 43. The gap junction labelling is heavy and in the form of large punctate spots. Note the absence of linear bands of label between the outermost cells of the presumptive trophectoderm. Scale bar 25 μm.

trols of the same age. Furthermore, the junctions exist entirely as punctate spots apparently randomly distributed along the intercellular opposing membranes. The translocation to form belt gap junctions within the junctional complex between outermost cells has not occurred, implying that cell polarization is lost as the defect is expressed. A small proportion of DDK/C3H embryos showed a normal distribution of gap junctions; presumably these embryos fall into the class of DDK/C3H embryos that are destined to survive and form normal blastocysts (approximately 10%).

We mimicked the DDK syndrome by exposing normal embryos to butyrate for 18 hours and examined gap junction distribution as soon as treatment ended. Most of these embryos fail to compact and the distribution of gap junctions matched that observed in the majority of DDK/C3H embryos. Outer cells failed to develop belt gap junctions between them and all cells were linked by numerous, large, punctate gap junctions. As with DDK/C3H embryos a very small proportion reached the expanded blastocyst stage; these embryos proceed through the preimplantation stages at a slower rate than controls.

The consequences of increasing gap junctional communication have as yet only been studied in normal embryos treated with Forskolin (100 μM), which raises intracellular cyclic AMP and speeds junctional communication. Although the number of embryos so far examined is small, the pattern of gap junction location matches that of normal controls. However, double blind analysis suggests that such embryos develop more rapidly than untreated controls and appear to express fewer gap junctions than normal.

The results of these preliminary experiments are summarized in Table 2 and suggest that gap junctions up-regulate when junctional permeability is low and down-regulate when junctional permeability is high. The mechanism that triggers these alterations in gap junctions remains to be explored.

Table 2. *Gap junction regulation in response to permeability change*

Strain	Treatment	Junctions permeability	Junctions number
B10CF1	None	normal	normal
B10CF1	Forskolin	high	reduced
DDK/C3H	None	low	increased

Future directions

These experiments raise many questions. It will be necessary to follow gap junction protein synthesis to define the minimum interval required to trigger change. One interesting issue that inevitably emerges is the inter-relation between the various intercellular junctions that link cells of the preimplantation embryo. Compaction is normally accompanied by the generation of tight junctions and an embryo that fails to maintain the compacted state must lack functionally effective tight junctions. It is already known that antibody-induced dissociation of tight junctions can induce decompaction without influencing gap junctional communication (Goodall, 1986). One clear message that emerges from our experiments is that the converse is unlikely to be true. With the increasing availability of antibodies and probes for following the consequences of these manipulations for tight junctions and desmosomes as well as gap junctions, it should now be possible to follow all three junctional contacts and see whether they are regulated in a coordinated manner. The extension of such studies into the postimplantation stages that include gastrulation may remain difficult in the mouse because of the constraints imposed by the culture of embryos outside the uterine environment, which is necessary for experimental investigation of these problems. However, work on other vertebrates should prove more feasible and provide information that can illuminate our understanding of the way in which close cellular interactions contribute to gastrulation in the mouse.

We are grateful to our colleagues for allowing us to describe the results of unpublished experiments. This work was made possible by grants from the Wellcome Trust to A. E. W. and to C. R. Green and A. E. W. A. E. W. thanks the Royal Society for their support.

References

Barron, D. J., Valimarsson, G., Paul, D. L. and Kidder, G. M. (1989). Connexin 32, a gap junction protein is a persistent oogenetic product through preimplantation development of the mouse. *Dev. Genetics* **10**, 318-323.

Bennett, M. V. L., Brown, J. E., Harris, A. L. and Spray, D. C. (1978). Electrotonic junctions between Fundulus blastomeres: reversible block by low intracellular pH. *Biol. Bull.* **155**, 428-429.

Beyer, E. C., Paul, D. L. and Goodenough, D. A. (1990). connexin family of gap junctions. *J.Membrane Biol.* **116**, 187-194.

Buehr, M. , Lee, S., McLaren, A. and Warner, A. (1987). Reduced gap junctional communication is associated with the lethal condition characteristic of DDK mouse eggs fertilized by foreign sperm. *Development* **101**, 409-459

Ducibella, T., Albertini, D. F., Anderson, E. and Biggers, J. D. (1975). Intercellular junctions and their appearance during development. *Dev. Biol.* **45**, 231-250.

Evans, W. H., and Rahman, S. (1989). Gap junctions and intercellular communication: topology of the major junctional protein of rat liver. Membrane Structure and Dynamics **17**, 983-985.

Goodall, H. (1986). Manipulation of gap junctional communication during compaction of the early mouse embryo. *J. Embryol. Exp. Morph.* **91**, 283-296.

Goodall, H. and Johnson, M. H. (1982). Use of carboxy fluorescein dicacetate to study formation of permeable channels between mouse blastomeres. *Nature* **295**, 524-526

Goodall, H. and Johnson, M. H. (1984). The nature of intercellular coupling within the preimplantation mouse embryo. *J. Embryol. Exp. Morph.* **79**, 53-76.

Goodall, H. and Maro, B. (1986). Major loss of technical coupling during mitosis in early mouse embryo. *J. Comparative Biol.* **102**, 568-575.

Gourdie, R. G., Harfst, E., Severs, N. J. and Green, C. R. (1990). Cardiac gap junctions in rat ventricle: localization using site directed antibodies and laser scanning confocal microscopy. *Cardioscience* **1**, 75-82.

Harfst, E., Severs, N. J. and Green, C. R. (1990). Cardiac myocyte gap junctions: evidence for a major connexin protein with an apparent relative molecular mass of 70 000. *J. Cell Sci.* **96**, 591-604.

Hax, W. M. A., Van Venrooij, G. R. P. M. and Vossenberg, J. B. J. (1974). Cell communication: a cyclic-AMP mediated phenomenon. *J. Membrane Biol.* **19**, 253-266.

Lee, S., Gilula, N. B. and Warner, A. (1987). Gap junctional communication and compaction during preimplantation stages of mouse development. *Cell* **51**, 851-860.

Lo, C. W. and Gilula, N. B. (1979). Gap junctional communication in the preimplantation mouse embryo. *Cell* **18**, 399-409.

Magnuson, T., Demsey, A. and Stackpole, C. W. (1977). Characterization of intercellular junctions in the preimplantation mouse embryo by freeze-fracture and thin-section electron microscopy. *Dev. Biol.* **61**, 252-261.

Nishi, M., Kumar, N. M. and Gilula, N. B. (1991). Developmental regulation of gap junction gene expression during mouse embryonic development. *Dev. Biol.* **146**, 117-130.

Rahman, S. and Evans, W. H. (1991). Topography of connexin32 in rat liver gap junctions. *J. Cell Biol.* **100**, 567-578.

Turin, L. and Warner, A. (1977). Carbon dioxide reversibly abolishes ionic communication between cells of early amphibian embryo. *Nature* **270**, 56-57.

Turin, L. and Warner, A. (1980). Intracellular pH in early Xenopus embryos: its effect on current flow between blastomeres. *J. Physiol.* **300**, 489-504.

Wakasugi, N. (1973). Studies on fertility of DDK mice: reciprocal crosses between DDK and C57BL/6J strains and experimental transplantation of the ovary. *J. Reprod. Fert.* **33**, 283-291.

Development 1992 Supplement, 119-125 (1992)

The role of intermediate filaments in early *Xenopus* development studied by antisense depletion of maternal mRNA

JANET HEASMAN, NICHOLAS TORPEY* and CHRIS WYLIE

Wellcome/CRC Institute, University of Cambridge, Tennis Court Road, CAMBRIDGE CB2 1QR, UK

*Current Address: St. George's Hospital Medical School, Cranmer Terrace, LONDON SW17 ORE, UK

Summary

The effects of depleting a maternal cytokeratin mRNA on the developing embryo are described. Cytokeratins are members of the intermediate filament family of cytoskeletal proteins, and are expressed in a cortical network of the superficial cytoplasm of the oocyte. After fertilisation, a new cortical network is built up, which comes to occupy only the most superficial cells of the blastula. The maternal cytokeratin mRNA is abundantly translated, both during oogenesis, and during oocyte maturation and after fertilisation. Depletion of the mRNA results in depletion of the cortical filaments at the blastula stage and leads to gastrulation abnormalities. We discuss the various possible control experiments required for antisense oligo depletion studies and the implications of these results for cytokeratin function.

Key words: maternal mRNA, antisense oligos, cytokeratin, *Xenopus*.

Introduction

During oogenesis, the *Xenopus* oocyte grows to a size of 1.2 mm and accumulates a large store of nutrients, mRNAs and proteins. It has long been recognized that this store is used during the early stages of development, particularly as zygotic transcripts are not expressed until after the mid-blastula transition (MBT), which occurs approximately at the 4000-cell stage. However, the precise roles of individual mRNAs and proteins are far from clear. The pre-MBT phase is a critical time in development, during which axes are established, adhesive differences appear and the earliest cell and tissue types are determined. Defining the functions of individual maternal proteins and mRNAs would be of considerable importance in understanding the mechanism of early development.

Since it is not at the moment possible to generate germ-line mutations in vertebrates, nor to carry out genetic manipulations in *Xenopus*, a number of strategies have been used to interfere with the function of individual proteins or mRNAs. One approach has been to inject antibodies specific for particular proteins into fertilized eggs or 2-cell-stage embryos (Warner et al., 1984; Wright et al., 1989, Klymkowsky 1992). This method may have a number of problems, including the accessibility of the protein, the likely mosaicism of effect due to the uneven distribution of the injected antibody and possibly the toxic nature of high local concentrations of immunoglobulins.

An alternative to the use of antibodies is to swamp the wild-type protein with excess of a mutated form (Herskowitz, 1987). This dominant negative strategy has been used successfully to study the function of fibroblast growth factor receptor in mesoderm induction (Amaya et al., 1991). Another strategy is to target and block the function of maternal mRNAs rather than proteins. This method will not work if the store of maternal protein is in itself sufficient for the needs of the embryo until the MBT. However, many gene products are stored both as mRNAs and proteins, suggesting that the protein levels have to be topped up by new translation as development proceeds. Attempts to deplete mRNAs in fertilized eggs by injecting antisense RNA were unsuccessful for endogenous transcripts, for reasons which remain obscure (Melton, 1985; Bass and Weintraub, 1987; Kimelman and Kirschner, 1989).

A more successful way of causing the degradation of specific mRNAs is by injecting antisense oligodeoxynucleotides (oligos) into oocytes or embryos. This approach was developed in oocytes by Dash et al. (1987), and tested in both oocytes and eggs by Shuttleworth et al. (1988). The optimum length of oligo both in terms of depletion activity and of depletion specificity was found to be a 14-18 mer. The method has now been quite widely used in oocytes to study the function of a variety of exogenous membrane proteins (Prives and Foukal, 1992). Antisense oligos have also proved successful tools in understanding the role of the proto-oncogene c-*mos* in meiotic maturation in oocytes (Sagata et al., 1988; Freeman et al., 1990), and of cyclins in mitosis in cell-free egg extracts (Minshull and Hunt, 1987).

In contrast, there have been relatively few functional studies using oligos injected into fertilized eggs (Dagle et al., 1990). This may be because egg injections present a number of problems compared to oocytes. Oligos are unstable once injected and degrade more quickly in fertilized eggs compared to oocytes (Shuttleworth and Colman, 1988; Dagle et al., 1990). Furthermore, oligos and their degradation products are known to carry non-specific toxicity (Shuttleworth et al., 1988; Smith et al., 1990; Woolf et al., 1990) and fertilized eggs and embryos are more sensitive to this than oocytes (Woolf et al., 1990). One further problem of injections into eggs is that of mosaicism of effect due to the uneven distribution of the oligo as cleavage proceeds. Some of these problems may be avoided by injecting oligos into ovarian oocytes, allowing the target mRNA to be depleted and the oligo degraded, and then fertilizing them.

This review outlines the progress that we have made towards developing this method of analysis of maternal mRNA function. A similar approach (Kloc et al., 1989) showed that an asymmetrically distributed maternal mRNA xlgv7 was not required for normal development in *Xenopus*. We initially targetted mRNAs coding for intermediate filament proteins. The basic cytokeratin XCK1(8) is abundant in the oocyte and embryo, both as a protein (Franz et al., 1983; Gall et al., 1983; Godsave et al., 1984) and as an mRNA (Franz and Franke, 1986; Torpey et al., 1992a). Also the immunolocalization of XCK1(8) has been well described (Torpey et al., 1992b) and we have an antibody that recognizes this protein specifically (Torpey et al., 1992b). This keratin is synthesized abundantly both in the oocyte, and during maturation and early development (Torpey et al., 1992a), suggesting that the maternal store of protein may require supplementation by translation of the stored maternal mRNA. Finally the role of the intermediate filament proteins in early development is not clear, making this an important protein to target.

Injection of oligos into oocytes

Oocytes are manually defolliculated from freshly dissected ovary taken from an anaesthetized female. While collagenase treatment removes follicle cells more quickly than the manual method, such treatment also renders the oocytes unfertilizable.

For all the mRNAs that we have targetted, we needed to synthesize a panel of six or more oligos complementary to different parts of the RNA sequence, to find one that would deplete the mRNA. Fig. 1 shows an example of the panel prepared for cytokeratin XCK1(8) mRNA chosen randomly from the sequence. We have so far found no particular pattern of sequence similarity or position that correlates with the ability to cause RNA depletion. This variability of oligo effectiveness has been noted by others (Baker et al., 1990), and presumably reflects the secondary structure of the target RNA making only some parts of the sequence available for oligo hybridization.

Oligos are normally injected into the equatorial region of oocytes although we have not found a difference in effectiveness when injections are into the vegetal pole. Other authors have commented on variability of depletion of mRNAs according to the site of injection (Shuttleworth et al., 1988). Fig. 2A and B are RNAase protection assays

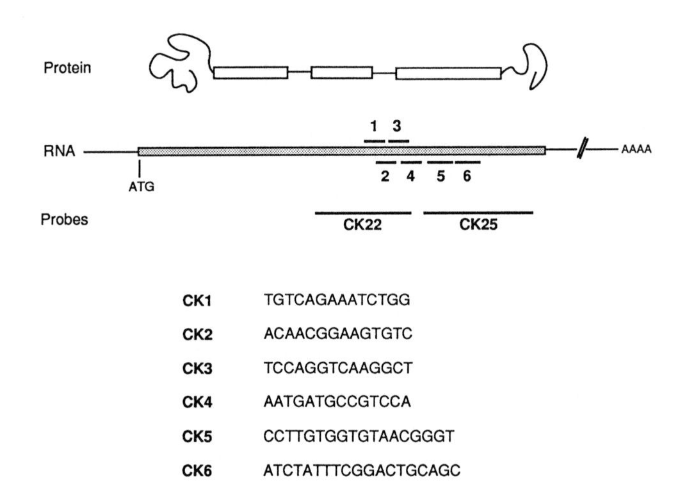

Fig. 1. The relative positions and sequences of the 6 oligos used to to try to deplete cytokeratin XCK1(8) mRNA. Oligos 1-4 are all within the linker1b/2 domain and oligos 5 and 6 are in coil 2. The relative positions of the two RNAase protection probes used in Fig. 2 are also shown.

showing an experiment to compare the ability of six oligos complementary to the XCK1(8) mRNA. Only one of the six oligos (CK5) was effective in depleting the mRNA at doses of 10 ng. This amount of oligo is uncomfortably close to the 15-20 ng dose, which we and others (Woolf et al., 1990) have shown to cause non-specific toxicity in embryos. So to improve the efficiency and reduce the toxicity of oligo treatment, we made a modified form of CK5 (Integrated DNA Technology). This modification was suggested and tested by Dagle et al. (1990). It involves converting the first and last five phosphodiester bonds to methoxyethyl phosphoramidate linkages, which are not substrates for RNAase H. This leaves the central part of the oligo still available for cleavage, but the molecule is more resistant to exonuclease activity (Dagle et al., 1990). They showed that modified oligos are still detectable 40 minutes after injection into oocytes and 30 minutes after injection into fertilized eggs, in contrast to a 10 minute maximum for unmodified oligos. In our hands, such modified forms of oligos are 10-20 times more efficient than their unmodified counterparts, and can thus be injected in amounts ranging from 0.2 to 4 ng to cause substantial (greater than 95%) depletion of target mRNAs (Fig. 2). Once the mRNA in full-grown oocytes is depleted, it is not resynthesized over a period of 60 hours (Fig. 3).

Fertilization of ovarian oocytes

Two methods have been described to fertilize ovarian oocytes (Heasman et al., 1992) neither of which is easy but both of which have given some success in fertilizing oligo-injected oocytes (Kloc et al., 1989; Torpey et al., 1992a). We routinely use the oocyte transfer technique described previously (Holwill et al., 1987). Briefly, this involves maturing the cultured and injected oocytes with progesterone (1 μM) for 8 hours and then vital staining control

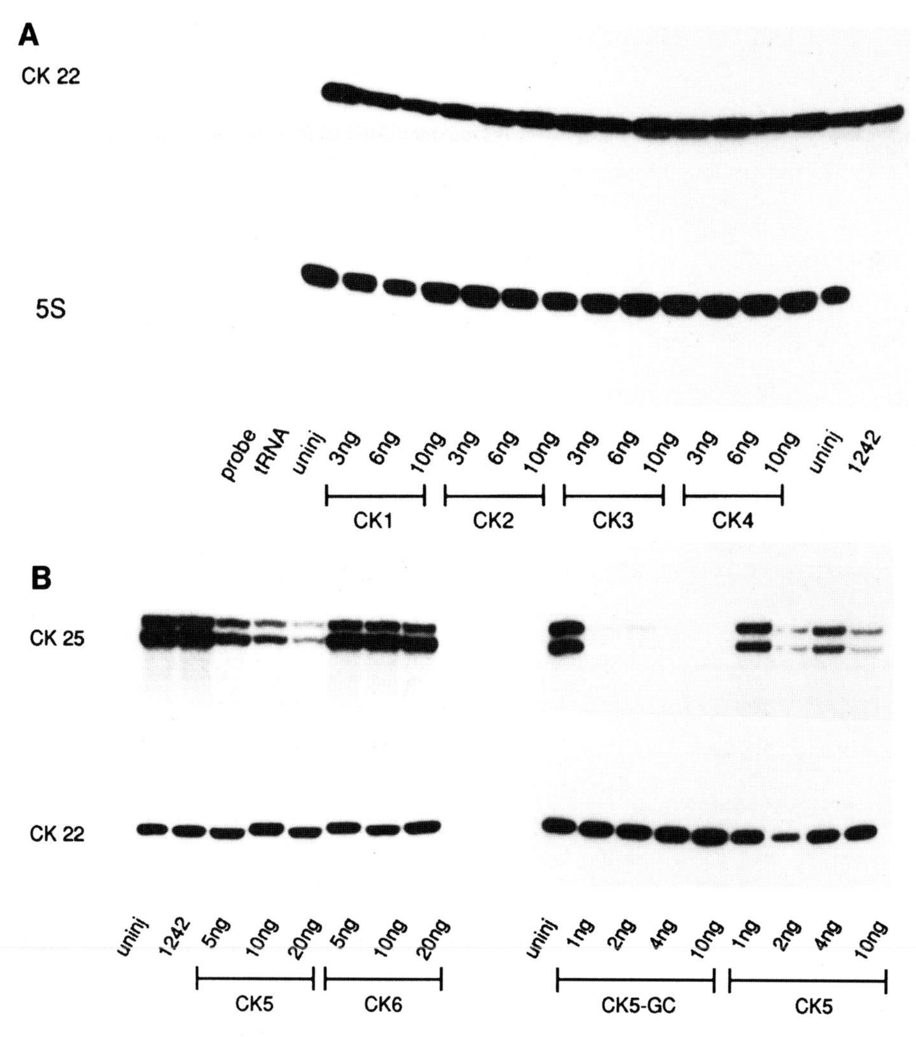

Fig. 2. The effects of oligos on cytokeratin XCK1(8) mRNA. (A) RNAase protection assay using the CK22 probe on RNA extracted from oocytes that had been injected with oligos 1-4 of Fig. 1. The oligos are ineffective at depleting the XCK1(8) mRNA. A small sample of the RNA extracted from each batch was removed and analysed with a 5s RNA probe to control for RNA recovery. (B) RNAase protection assay using the CK25 probe (which spans the target site of the oligos) on oocytes that were injected either with unmodified oligos 5 (CK5) and 6, or with the phosphoramidate-modified form of CK5 (CK5-GC). Oligo 6 does not deplete the mRNA for XCK1(8), but oligo 5 is effective at doses of 10-20 ng. The modified form of CK5 is effective in doses of 1 ng. The CK 22 probe does not span the target sites of these oligos but a position towards the 5′ end of the mRNA. Note that there is very little degradation of the 5′ fragment of the cleaved mRNA.

and experimental groups with vital dyes. The coloured oocytes are then transferred into an anaesthetized host which has been stimulated with human chorionic gonadotrophin to lay eggs. The host female recovers in anaesthetic-free water and lays the coloured eggs 3-5 hours later. The eggs are then fertilized using a sperm suspension.

This method of fertilization is very variable in its success rate, the main variability being in the status of the oocytes themselves. It is essential to use full-grown stage-6 oocytes which have not become over-ripe, but this is generally difficult to assess. However, a high-yield experiment will result in groups of 30-40 fertilized eggs for each experimental batch.

The merits of injection of oligos into oocytes rather than into fertilized eggs

Given the difficulty of fertilizing ovarian oocytes, we wondered if phosphoramidate-modified oligos might be sufficiently efficient in denaturing mRNAs to allow effective depletion without toxicity after injection of small quantities into fertilized eggs. We were encouraged by the result that 10 ng of anti-cyclin oligo had a specific and non-toxic effect in 2-cell-stage embryos (Dagle et al., 1990). In a series of experiments, we monitored the effect of different doses of a modified CK5 injected into 2- and 4-cell-stage embryos both on mRNA levels (Fig. 4) and on embryonic development (Fig. 5). We found that the modified oligos were very efficient. For example, 0.8 ng of modified CK5 oligo injected into 2 cells of the 4-cell-stage embryo reduces the levels of cytokeratin mRNA to 10-20% of control levels, while 0.4 ng is sufficient to produce the same effect when injected into fertilized eggs (Fig. 4). Unfortunately this oligo, together with other unrelated modified oligos (specific for MyoD and vimentin mRNA) injected at the same concentrations, cause similar abnormalities (Fig. 5). On this basis, we judge that the phenotype (irregular cell size due to abnormal cleavages, patchy pigment and lack of adhesion leading to pregastrula arrest) is due to non-specific effects. A similar conclusion was drawn in a recent study using phosphorothioate-modified oligos (Woolf et al., 1990). Eggs developing from modified oligo-injected oocytes in the same dose range (1-5 ng) do not show this phenotype. Since oocytes seemed so much less liable to non-specific effects, we concentrated on these in our mRNA depletion studies.

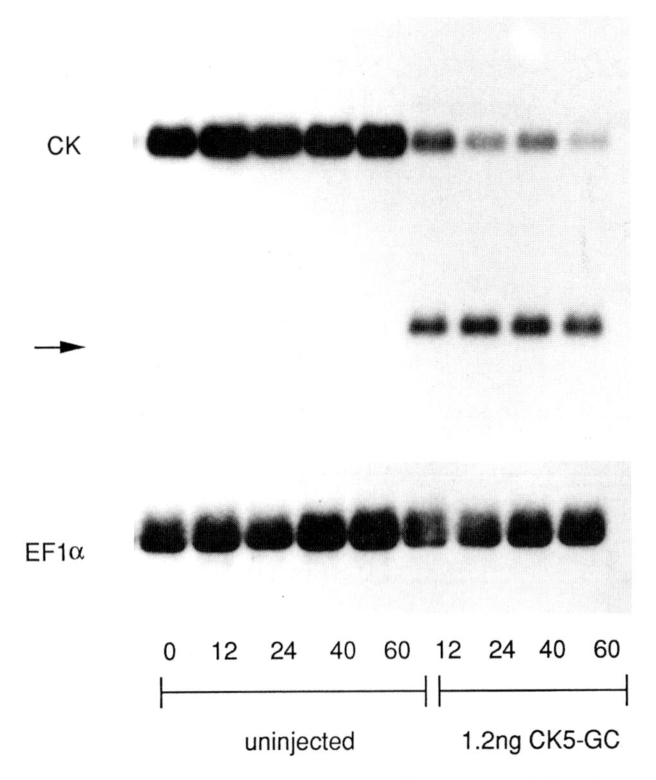

Fig. 3. A northern blot of RNA extracted from oocytes injected with 1.2 ng CK5-GC and frozen at intervals. The blot was probed with a random primed probe for XCK1(8). A group of control uninjected oocytes were analysed in parallel. Arrow indicates the stable 5′ fragment of the cleaved XCK1(8) mRNA.

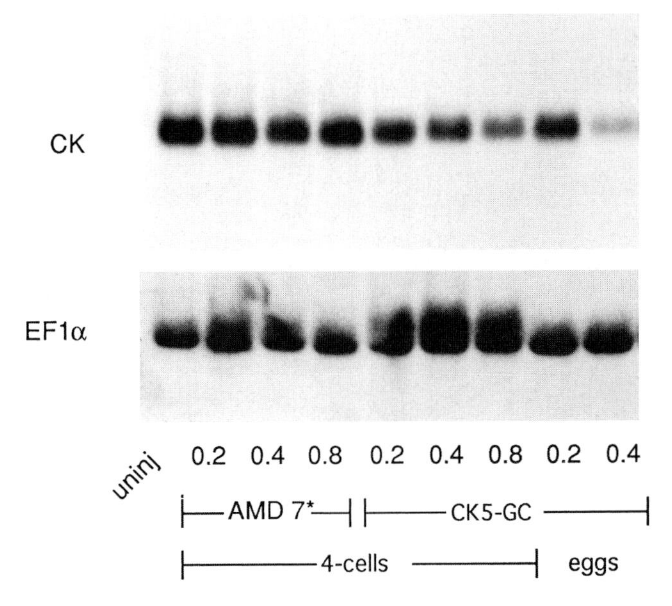

Fig. 4. A northern blot of RNA extracted from 4-cell-stage embryos or fertilized eggs injected either with various doses of modified CK5 oligo (CK5-GC) or with a modified oligo which depletes the mRNA for an unrelated protein MyoD (AMD7*). The blot was probed with a random primed probe for XCK1(8).

The effects of depleting mRNA in ovarian oocytes

One very good indication of the effectiveness of an oligo knock-out of a specific mRNA is to show that once the RNA has been degraded there is an effect on the total amount of target protein and on the levels of new protein synthesis compared to other proteins. We have shown this in the case of XCK1(8) mRNA in a number of ways.

Firstly, we cultured modified CK5-injected oocytes for increasing lengths of time, and the effect of the depletion of XCK1(8) mRNA on cytokeratin protein was monitored in two ways, immunocytochemically, using an XCK1(8) specific antibody, and by 2-dimensional fluorography (Torpey et al., 1992a). Keratin filaments are normally most abundant in the cortices of oocytes where they maintain a cortical lattice for at least 3 days in culture. In contrast, oocytes that have been injected with CK5 oligo and cultured for 60 hours show no visible network. Interestingly, these cytokeratin filaments are obligate heteropolymers of acidic (XLK3a/19) and basic (XCK 1(8)) keratin subunits. We wondered if the oligo was also depleting the acidic cytokeratin mRNA. However, on reprobing a northern blot of CK5 injected-oocyte RNA with an acidic keratin-specific probe, we found that the XLK3a/19 m RNA level was unaffected. When we compared the amounts of newly synthesized XCK1(8) and other proteins by 2-dimensional fluorography of uninjected and CK5-injected oocytes, we found that a protein in the correct position to be the acidic partner of XCK1(8) also showed reduced synthesis after CK5 oligo injection. Therefore, it seems likely that the loss of the basic cytokeratin protein destabilizes the acidic protein, as has been described in other systems for a number of cases of keratin heteropolymers (Domenjoud et al., 1988; Knapp and Franke, 1989; Lu and Lane, 1991).

A second strategy, which is proving successful in showing that depleting mRNA levels has a specific effect on the protein product, is to metabolically label oligo-injected and control oocytes with [^{35}S]methionine for 24 hours and then immunoprecipitate the protein with an appropriate specific antibody.

The phenotypes of maternal mRNA-depleted embryos: how can we know they are specific?

The denoument of these experiments is to fertilize a batch of embryos that has had a target mRNA-depleted as oocytes, and to look for developmental abnormalities. In the light of the known potential toxicity of oligos and the inherent variability between batches of oocytes, it is essential to know that any effect is specifically the result of the absence of the target mRNA and its protein product. This can be achieved by a combination of controls.

(1) Non-specific toxicity

There is clear evidence from this and other work that oligos both in their modified and unmodified forms are toxic above certain concentrations. The toxic phenotype varies from complete arrest at fertilization, to a slowing and abnormal pattern of cleavage, combined with changes in pigment distribution and loss of adhesion, culminating in arrest, generally before gastrulation. Interestingly oligos very seldom

Fig. 5. Injection of modified oligos into embryos causes non-specific abnormalities. Embryos were injected with modified oligos at the 4-cell-stage (2 of 4 cells injected, 0.4 ng each blastomere), and photographed at the late blastula stage. Oligos used were: (A) AMD7*, a modified oligo that depletes MyoD mRNA, (B) Oct, an irrelevent modified oligo, (C) 1242, a modified oligo that depletes vimentin mRNA and (D) CK5-GC, which depletes CK5 mRNA. Patches of cells showing slowed or arrested cleavage and abnormal pigmentation are visible in all cases, although they are most obvious in 1242 injected embryos. Uninjected or water-injected embryos do not show this effect.

prevent maturation or activation. The basis of this toxicity is unknown, but may be due to the high levels of deoxynucleotides released by degradation of the oligo, or by secondary structures that interfere with normal metabolic reactions. When used in very high amounts (greater than 20 ng) oligos also suppress general levels of protein synthesis. This can be checked for by ^{35}S labelling and avoided by using small amounts of modified oligos. We have also ruled out the possibility that a contaminant of the oligo synthesis is the cause of toxicity in the following way: when sense and antisense oligos are synthesized in the same way and injected together into oocytes there is no increase in abnormal phenotype. Indeed for CK5 oligo directed against the cytokeratin mRNA the phenotype is rescued, suggesting the specificty of the knock-out. The potential problem of oligo toxicity is avoided in our experiments by using oocytes rather than fertilized eggs for injections, and by keeping the oligo dose as low as possible by using phosphoramidate-modified oligos. Using this regime, none of the oligos show non-specific toxicity at doses of 0.2-4 ng.

(2) Non-specific RNA knock-out

It seems quite feasible that one oligo complementary to a target mRNA may also by chance hybridize to a totally different mRNA. This can best be controlled for by having more than one oligo of different sequences complementary to different parts of the target mRNA. If they all cause the same phenotype then the chance of them all also recognizing the same non-target species is vanishingly small. Similarly, irrelevant modified oligos particularly ones that target other mRNAs are also used to show the sequence specificity of the phenotype. Also reprobing blots with probes for related mRNAs to the target mRNA can be reassuring about the specificity of a knock-out. Finally, a good test of a specific effect is to show that depletion of the oligo results in a reduction of the target protein levels in the developing embryos. This can be achieved by immunocytochemistry (Torpey et al., 1992a).

Oocyte variability

The age-old cure for oocyte variability is to repeat the experiment on many batches with large numbers. Every oligo experiment is accompanied by a northern blot to show the mRNA is depleted in each case, and great care is taken to ensure that no oocytes miss injection.The final test of the specificity of a phenotype is to rescue it by injecting the full-length mRNA back into the oocyte. This in itself is not without difficulty, particularly when modified oligos have been used, which remain sufficiently active to degrade injected mRNA some time after their (the oligos') introduction.

Phenotypes and conclusions

One reassuring outcome of these experiments is that oligos targetted on different maternal mRNAs cause different phenotypes in development. Preliminary results suggest that vimentin mRNA is required in the early cleavage stages, while EP cadherin mRNA is needed to top up the adhesive protein levels by the mid-blastula stage (our unpublished observations). In contrast, cytokeratin mRNA-depleted oocytes, which have a large store of protein do not show a visible phenotype until the late blastula stage (Torpey et al., 1992a), although immunocytochemical studies show that keratin filaments are much shorter and disorganized by the mid-blastula stage. At the beginning of stage 9 (late blastula) in the control batches, the cortical cells flatten and compact over the embryonic surface in a movement that may mark the beginning of epiboly, and resembles compaction in mouse embryos. In the XCK1(8) depleted embryos, all the surface cells remained raised up, giving a 'cobblestone' appearance. The cell division rate and pigmentation pattern remained unaffected. A second feature of the phenotype is that experimental embryos lose the ability to heal a wounded surface, a feat that is usually completed within an hour for a small cut in the surface of a control oocyte. Finally the XCK1(8) depleted embryos also have a defect at gastrulation. If the vitelline membrane is left in place, the cells that normally invaginate, evaginate as a sheet advancing over the animal cap. If the vitelline membrane is removed, the embryos form classic exogastrulae. This effect is not seen when irrelevant oligos are injected and is almost completely rescued by co-injection of sense and antisense CK5.

These results point to a role for the cytokeratin filaments in producing and maintaining an integrated cortical shell around the embryo. Without them the epithelial surface is destabilized and cannot support normal gastrulation movements. A rather similar result has come from work on the mouse and human cytokeratin 14. Several pieces of work have shown that lack of this keratin, either in transgenic mice (Vassar et al., 1991) or in human hereditary conditions (Coulombe et al., 1991; Lane et al., 1992), results in loss of mechanical strength in the skin.

In conclusion, there are now several means of depleting or functionally perturbing the mRNAs and/or proteins inherited from the oocyte by the early embryo. There are many obvious candidates for targetting, including the cytoskeleton, cell surface and the signalling molecules responsible for early cell specification.

We are extremely grateful for finacial support from the Wellcome Trust. Thanks to Kim Goldstone and Lucinda Vickers for technical assistance.

References

Amaya, E., Musci, T. J. and Kirschner, M. W. (1991). Expression of a dominant negative mutant of the FGF receptor disrupts mesoderm formation in *Xenopus* embryos. *Cell* **66**, 257-270.

Baker, C., Holland, D., Edge, M. and Coleman, A. (1990). Effects of oligo sequence and chemistry on the efficiency of oligodeoxyribonucleotide-mediated mRNA cleavage. *Nucl. Acids Res.* **18**, 3537-3543.

Bass, B. and Weintraub, H. (1987). A developmentally regulated activity that unwinds DNA duplexes. *Cell* **48**, 607-613.

Coulombe, P. A., Hutton, M. E., Letai, A., Hebert, A., Paller, A. S. and Fuchs, E. P. (1991). Point mutations in human keratin 14 genes of epidermolysis bullosa simplex patients: genetic and functional analyses. *Cell* **66**, 1301-1311.

Dagle, J. M., Walder, J. A. and Weeks, D. L. (1990). Targetted degradation of mRNA in *Xenopus* oocytes and embryos directed by modified oligonucleotides: studies of An2 and cyclin in embryogenesis. *Nucl. Acids Res.* **18**, 4751-4757.

Dash, P., Lotan, I., Knapp, M., Kandel, E. R. and Goelet, P.(1987). Selective elimination of mRNAs *in vivo*: Complementary oligodeoxynucleotides promote RNA degradation by an RNase H-like activity. *Proc. Natl. Acad. Sci. USA* **84**, 7896-7900.

Domenjoud, L., Jorcano, J. L., Breuer, B. and Alonso, A. (1988). Synthesis and fate of keratins 8 and 18 in non-epithelial cells transfected with cDNA. *Exp. Cell Res.* **179**, 352-361.

Franz, J. K. and Franke, W. W. (1986). Cloning of cDNA and amino acid sequences of a cytokeratin expressed in oocytes of *Xenopus laevis*. *Proc. Natl. Acad. Sci. USA* **83**, 6475-6479.

Franz, J. K., Gall, L.,Williams, M. A., Picheral, B. and Franke, W. W. (1983). Intermediate-size filaments in a germ cell: expression of cytokeratins in oocytes and eggs of the frog *Xenopus*. *Proc. Natl. Acad. Sci. USA* **80**, 6254-6258.

Freeman, R. S., Kanki, J. P., Ballantyne, S. M., Pickham, K. M. and Donoghue, D. J. (1990). Effects of the v-mos oncogene on Xenopus development: meiotic induction in oocytes and mitotic arrest in cleaving embryos. *J. Cell Biol.* **111**, 533-541.

Gall, L., Picheral, B. and Gounon, P. (1983). Cytochemical evidence for the presence of intermediate filaments and microfilaments in the egg of *Xenopus laevis*. *Biol. Cell* **47**, 331-342.

Godsave, S. F., Wylie, C. C., Lane, E. B. and Anderton, B. H. (1984). Intermediate filaments in the *Xenopus* oocyte: the appearance and distribution of cytokeratin-containing filaments. *J. Embryol. Exp. Morph.* **83**, 157-167.

Heasman, J., Holwill, S. and Wylie, C. C. (1992). Fertilisation of cultured *Xenopus* oocytes and use in studies of maternally inherited molecules. *Methods in Cell Biol.* **36**, 214-231.

Herskowitz, I. (1987). Functional inactivation of genes by dominant negative mutations. *Nature* **329**, 219-222.

Holwill, S., Heasman, J., Crawley, C. R. and Wylie, C. C. (1987). Axis and germ line deficiencies caused by UV irradiation of *Xenopus* oocytes cultured *in vitro*. *Development* **100**, 735-743.

Kimelman, D. and Kirschner, M. W. (1989). An antisense mRNA directs the covalent modification of the transcript encoding fibroblast growth factor in *Xenopus* embryos. *Cell* **59**, 687-696.

Kloc, M., Miller, M., Carrasco, A. E., Eastman, E. and Etkin, L. (1989). The maternal store of the xlgv7 mRNA in full-grown oocytes is not required for normal development in *Xenopus*. *Development* **107**, 899-907.

Knapp, A. C. and Franke, W. W. (1989). Spontaneous losses of control of cytokeratin gene expression in transformed, non-epithelial human cells occurring at different levels of regulation. *Cell* **59**, 67-79.

Lane, E. B., Rugg, E. I., Navsaria, H., Leigh, I. M., Heagerty, A. H. M., Ishida-Yamamoto, A. and Eady, R. J. (1992). A mutation in the conserved helix termination peptide of keratin 5 in hereditary skin blistering. *Nature* **356**, 244-246.

Lu, X. and Lane, E. B. (1991). Retrovirus-mediated transgenic keratin expression in cultured fibroblasts: specific domain functions in keratin stabilisation and filament formation. *Cell* **62**, 681-696.

Melton, D. A. (1985). Injected antisense RNAs specifically block messenger translation *in vivo*. *Proc. Natl. Acad. Sci. USA* **82**, 144-148.

Minshull, J. and Hunt, T. (1987). The use of single-stranded DNA and RNAse H to promote quantitative 'hybrid arrest of translation' of mRNA/DNA hybrids in reticulocyte cell-free translations. *Nuc. Acids Res.* **16**, 6433-6451.

Prives, C. and Foukal, D. (1992). Use of oligodeoxynucleotides for antisense experiments in *Xenopus laevis* oocytes. *Methods in Cell Biology* **36**, 185-210.

Sagata, N., Oskarsson, M., Copeland, T., Brumbaugh, J. and Vande Woude, G. F. (1988). Function of *c-mos* proto-oncogene in meiotic maturation in *Xenopus* oocytes. *Nature* **335**, 519-525.

Shuttleworth, J. and Colman, A. (1988). Antisense oligonucleotide-directed cleavage of mRNA in *Xenopus* oocytes and eggs. *EMBO J.* **7**, 427-434.

Shuttleworth, J., Matthews, G., Dale, L., Baker, C. and Colman, A. (1988). Antisense oligodeoxynucleotide-directed cleavage of maternal mRNA in *Xenopus* oocytes and embryos. *Gene* **72**, 267-275.

Smith, R. C., Bement, W. M., Dersch, M. A., Dworkin-Rastl, E., Dworkin, M. B. and Capco, D. G. (1990). Nonspecific effects of oligodeoxynucleotide injection in *Xenopus* oocytes: a reevaluation of previous D7 mRNA ablation studies. *Development* **110**, 769-779.

Torpey, N., Wylie, C. C. and Heasman, J. (1992a). Function of a maternal cytokeratin in *Xenopus* development. *Nature* **357**, 413-415.

Torpey, N. P., Heasman, J. and Wylie, C. C. (1992b). Distinct distribution of vimentin and cytokeratin in *Xenopus* oocytes and early embryos. *J. Cell Sci.* **101**, 151-160.

Vassar, R., Coulombe, P. A., Degenstein, L., Albers, K. and Fuchs, E. (1991). Mutant keratin expression in transgenic mice causes marked abnormalities resembling a human genetic skin disease. *Cell* **64**, 365-380.

Warner, A. E., Guthrie, S. C. and Gilula, N. B., (1984). Antibodies to gap junctional protein selectively disrupt junctional communication in the early amphibian embryo. *Nature* **311**, 127-131.

Woolf, T., Jennings, C., Rebagliati, M. and Melton, D. A., (1990). The stability, toxicity and effectiveness of unmodified and phosphorothioate antisense oligodeoxynucleotides in *Xenopus* oocytes and embryos. *Nucl. Acids Res.* **18**, 1763-1769.

Wright, C. V., Cho, K. W. Y., Hardwicke, J., Collins, R. H. and DeRobertis, E. M.,(1989). Interference with the functions of a homeobox gene in *Xenopus* embryos produces malformations of the anterior spinal cord. *Cell* **59**, 81-93.

Mesoderm induction and origins of the embryonic axis

Development 1992 Supplement, 127-136 (1992)
Printed in Great Britain

Mesoderm-inducing factors and the control of gastrulation

J. C. SMITH and J. E. HOWARD

Laboratory of Developmental Biology, National Institute for Medical Research, The Ridgeway, Mill Hill, London NW7 1AA, UK

Summary

One of the reasons that we know so little about the control of vertebrate gastrulation is that there are very few systems available in which the process can be studied in vitro. In this paper, we suggest that one suitable system might be provided by the use of mesoderm-inducing factors. In amphibian embryos such as *Xenopus laevis*, gastrulation is driven by cells of the mesoderm, and the mesoderm itself arises through an inductive interaction in which cells of the vegetal hemisphere of the embryo emit a signal which acts on overlying equatorial cells. Several factors have recently been discovered that modify the pattern of mesodermal differentiation or induce mesoderm from presumptive ectoderm. Some of these mesoderm-inducing factors will also elicit gastrulation movements, which provides a powerful model system for the study of gastrulation, because a population of cells that would not normally undertake the process can be induced to do so. In this paper, we use mesoderm-inducing factors to attempt to answer four questions. How do cells know when to gastrulate? How do cells know what kind of gastrulation movement to undertake? What is the cellular basis of gastrulation? What is the molecular basis of gastrulation?

Key words: *Xenopus*, gastrulation, mesoderm induction, activin, BMP-4, FGF, Wnt, integrins.

Introduction

While descriptions of cell behaviour during vertebrate gastrulation have become exquisitely detailed and accurate (see chapters by Bortier, Ho, Keller, Lawson and Stern), an understanding of the process at the levels of cell and molecular biology is lacking. In this article, we ask to what extent this problem might be approached by the use of mesoderm-inducing factors. In amphibian embryos such as *Xenopus laevis*, gastrulation is driven by cells of the mesoderm, and the mesoderm itself arises through an inductive interaction in which cells of the vegetal hemisphere of the embryo emit a signal that acts on overlying equatorial cells (Fig. 1; see review by Smith, 1989). In the last five years, several factors that induce mesoderm from presumptive ectoderm, or which modify the pattern of mesodermal differentiation, have been discovered. These include members of the TGFb superfamily such as activins A and B and bone morphogenetic protein 4 (BMP-4), members of the fibroblast growth factor family such as bFGF or eFGF, and members of the Wnt family such as Xwnt-8 (see reviews by Smith, 1989, 1992; Whitman and Melton, 1989; Dawid and Sargent, 1990; and papers by Smith and Harland, 1991; Sokol et al., 1991; Chakrabarti et al., 1992; Köster et al., 1991; Dale et al., 1992; Jones et al., 1992).

The types of mesoderm induced by each factor, and their effects in combination with each other, differ. Thus activin induces a range of mesodermal cell types, from organizer to tail, according to its concentration (Green et al., 1990; Green and Smith, 1990); BMP-4 induces posterior/ventral mesoderm (Köster et al., 1991; Dale et al., 1992; Jones et al., 1992) and is capable of 'ventralizing' the response to activin (Dale et al., 1992; Jones et al., 1992); FGF also induces posterior/ventral mesoderm but has little ventralizing activity when applied in combination with activin (Kimelman and Kirschner, 1987; Slack et al., 1987; Green et al., 1990; Cooke, 1989). Members of the Wnt family do not induce mesoderm from presumptive ectoderm, but animal caps derived from embryos that have received injections of Xwnt-8 RNA respond to bFGF by forming dorsal, rather than ventral, mesoderm (Christian et al., 1992). Injection of Wnt RNA is also capable of inducing complete secondary axes in *Xenopus* embryos and of rescuing embryos that have been made completely ventral by ultra-violet light irradiation of their vegetal hemispheres before first cleavage (Smith and Harland, 1991; Sokol et al., 1991).

Previously, we have shown that mesoderm-inducing factors cause ectodermal tissue to undergo gastrulation movements (Symes and Smith 1987; Cooke and Smith, 1989; Smith et al., 1990). In this paper, we take advantage of this observation to attempt to answer four questions. How do cells know when to gastrulate? How do cells know what kind of gastrulation movement to undertake? What is the cellular basis of gastrulation? What is the molecular basis of gastrulation? We begin with a brief description of normal gastrulation, based on the work of Keller (1991; and this volume), which highlights the processes that we seek to understand.

Xenopus gastrulation

Gastrulation in *Xenopus* converts the radially symmetrical

blastula into a three-layered structure with anteroposterior and dorsoventral axes. The first external sign of gastrulation is visible about 10 hours after fertilization, when the blastoporal pigment line forms on the dorsal side of the embryo. The pigment line arises because bottle cells in this region undergo apical constriction and, as this happens, they displace the deep mesoderm associated with them inwards and upwards, thus initiating involution. Although bottle cell formation and involution begin on the dorsal side of the embryo, the movements then spread laterally and ventrally, reaching the ventral side of the embryo about 2 hours later.

Once involution has begun, gastrulating cells exhibit three distinct types of behaviour. The first mesoderm to involute undergoes *cell migration*, using the fibrillar matrix on the inner surface of the blastocoel roof as a substratum. On the dorsal side of the embryo, these cells eventually contribute to the head mesoderm (see Winklbauer and Selchow, 1992). Cell migration, however, contributes little to the total force required for gastrulation; if the blastocoel roof, the substratum for migration, is removed, gastrulation proceeds relatively normally (Keller et al., 1985). The main driving force for gastrulation derives from later-involuting cells which, on the dorsal side of the embryo, go on to form notochord and somite. These cells undergo 'convergent extension', during which the marginal zone constricts to form a smaller ring (*convergence*) and lengthens along the anteroposterior axis (*extension*). The greatest degree of convergent extension occurs in this dorsalmost section of the marginal zone; lateral and ventral regions converge to similar extents, but the degree of extension decreases in progressively more ventral regions, such that cells in the ventral midline can be said only to converge (Keller and Danilchik, 1988).

Convergence and extension are driven by cell intercalation (see Keller, 1991). Convergence occurs predominantly through radial intercalation, while extension is the result of both radial and mediolateral intercalation. These processes can be studied in isolated explants of the marginal zone region, where, in contrast to the ventral marginal zone, the dorsal marginal zone undergoes dramatic elongation. A detailed description of cell intercalation during convergent extension is given in the chapter by Keller.

In what follows, we discuss to what extent the use of mesoderm-inducing factors might bring about an understanding of these different aspects of gastrulation.

The timing of gastrulation

The timing of gastrulation in normal *Xenopus* development is remarkably precise; members of a batch of embryos fertilized at the same time will form a dorsal blastopore lip, after about 10 hours of development, within 20 minutes of each other (see review by Cooke and Smith, 1990). The nature of the timer responsible for this synchrony is unknown. One obvious suggestion, that it is timed by reference to the mid-blastula transition (MBT), is not correct, because if the timing of the MBT is changed, in axolotl, newt, sturgeon or *Drosophila* embryos, by changing the nuclear to cytoplasmic ratio, gastrulation still starts at the normal time (see Cooke and Smith, 1990; Yasuda and Schubiger, 1992). Other potential timers have also been ruled out; it is not, for example, simply a question of counting numbers of cell cycles, because gastrulation will proceed on time if cell division is inhibited at the late blastula stage (Cooke, 1973). Furthermore, the 'pseudogastrulation' movements displayed by some unfertilized amphibian eggs begin at the same time, and take the same time, as the normal movements in their fertilized siblings. The 'timer' that controls these movements must be cytoplasmic and independent of transcription, because they will occur in enucleated oocytes (see review by Satoh, 1982), and this is consistent with results in *Caenorhabditis elegans*, which show that gastrulation movements start at the normal time in the presence of α-amanitin (results of L. Edgar, cited by Yasuda and Schubiger, 1992). Unfortunately, in spite of all this circumstantial, albeit interesting, evidence, there is no idea as to the molecular basis of the gastrulation clock.

Is it important that gastrulation is so accurately timed a process? In normal development, gastrulation begins at dorsal lip of the blastopore and movements spread around to the ventral side. Surprisingly, this aspect of the timing of gastrulation does not seem to be essential. If the dorsoventral timing of gastrulation is reversed, by applying a temperature gradient across the embryo, development is normal and resulting embryos are indistinguishable from controls (Black, 1989). Under conditions where the metabolic rate is uniform throughout the embryo, however, there is evidence that the proper timing of gastrulation *is* important in setting up the correct spatial pattern of gene expression and differentiation.

Ultraviolet irradiation of the vegetal hemisphere of the fertilized *Xenopus* egg results in embryos that lack anterior and dorsal structures (see chapter by Gerhart, this volume, and Cooke and Smith, 1987). Such embryos begin gastrulation movements later than controls, with extreme 'ventralized' cases beginning gastrulation at the time of appearance of the ventral blastopore lip in synchronous normal embryos. By contrast, embryos treated with LiCl at the 32-cell stage develop as extreme anterior-dorsal body patterns. Gastrulation movements in these embryos begin on time, but the blastopore lip appears synchronously around the entire marginal zone, rather than spreading from the dorsal to the ventral side. It is difficult to answer the question of whether these abnormal temporal patterns of gastrulation are an early manifestation of the ventralized and dorsalized phenotypes of the embryos or whether they are the cause of them. One experiment, however, is consistent with the suggestion that the total time that cells spend gastrulating influences their anteroposterior pattern of differentiation. Treatment of gastrulating embryos with polysulphonated compounds such as trypan blue or sodium suramin arrest convergent extension movements, and the resulting tadpoles have body axes that are truncated at different anteroposterior levels depending on the time of application of the compound; the earlier the treatment, the more posterior the body axis that develops, whereas if treatment is delayed until the end of gastrulation, the embryos that develop are normal (Gerhart et al., 1989). These results are consistent with the suggestion that the shorter the time that cells spend gastrulating, whether because they start late or because their

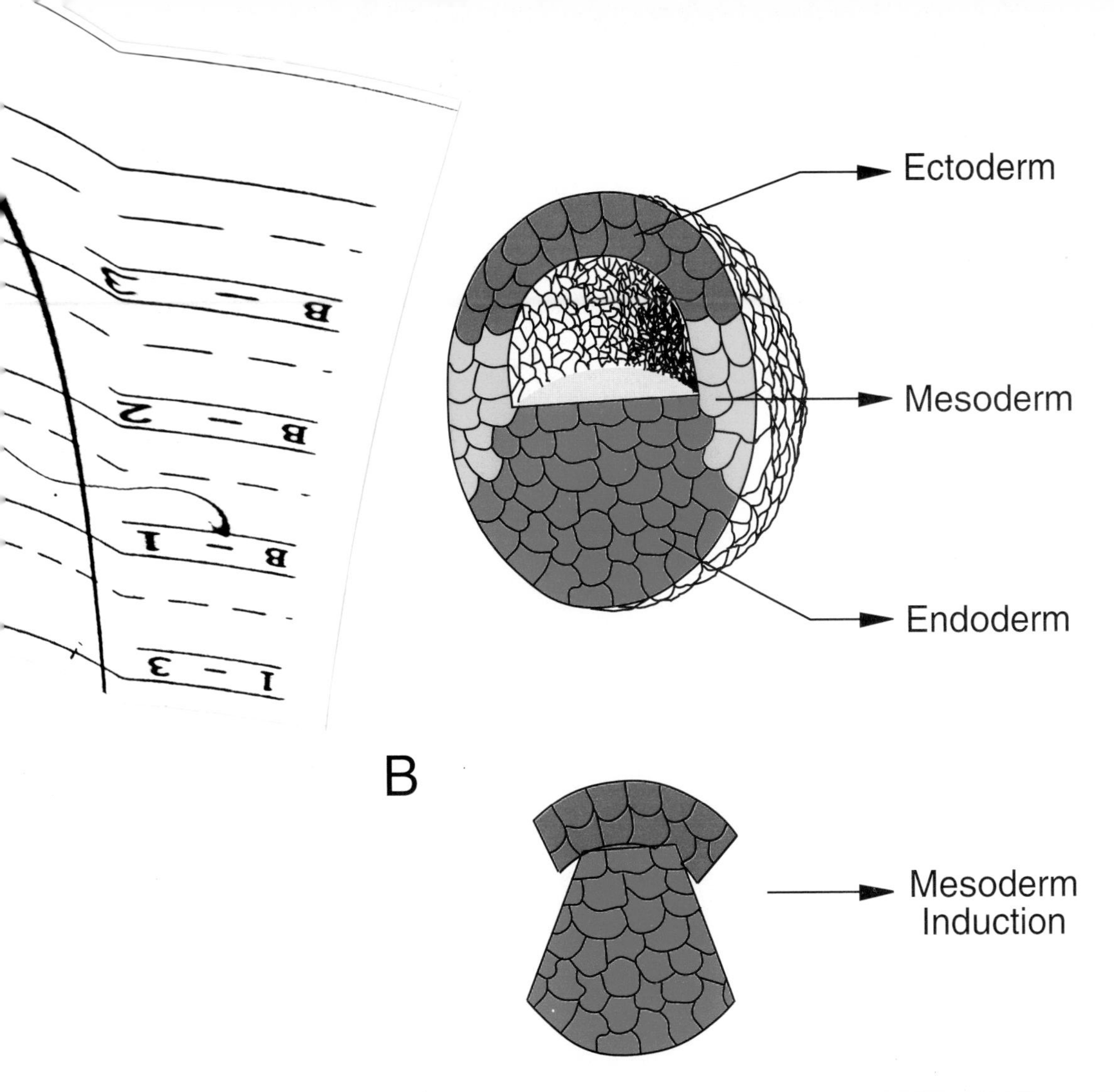

Fig. 1. Mesoderm induction occurs when a signal from cells of the vegetal hemisphere act on overlying equatorial cells. (A) In normal development ectoderm derives from cells of the animal cap (red), mesoderm from the equatorial region (yellow), and endoderm from the vegetal pole (blue). (B) Mesoderm induction can be demonstrated by apposing presumptive ectoderm with presumptive endoderm; some of the ectodermal cells are induced to form mesodermal cell types such as muscle and notochord.

movements
ventral the ce

Mesoderm-induc
What can mesode
timing of gastrulatio
with activin A causes
elongation, very similar
marginal zone regions (S
et al., 1985 for the behav
Additional evidence that the
lation movements comes from
of elongation of the induced an
dorsal marginal zone regions are s
cantly, that they begin at the same t
of the gastrulation-like movements in
is independent of the stage at which the
to activin (Symes and Smith, 1987), and
another suggestion for the timing of gastru
occurs a certain time after mesoderm induction.

It is harder to judge the time of onset of gastrulation-like movements in response to FGF, because, consistent with the suggestion that FGF induces ventral-type mesoderm (Green et al., 1990), the gastrulation movements seen in response to this factor are less dramatic. This makes it more difficult to decide when they begin. An alternative approach to this problem was introduced by Cooke and Smith (1989), who microinjected mesoderm-inducing factors into the blastocoels of host embryos. This causes the cells of the roof of the blastocoel to undergo a transformation in behaviour that resembles the transformation seen in the mesoderm of the marginal zone. By fixing and dissecting embryos at different times after injection, it is relatively simple to judge the time of onset of gastrulation-like movements. When activin was injected into the blastocoel, irrespective of the time of injection (over a 3.5 hour range), or the concentration of factor (over a 250-fold range), this transformation began at the same time, shortly after the onset of gastrulation at the dorsal side of the marginal zone. This result confirms that of Symes and Smith (1987), but also indicates that concentration of inducing factor is also irrelevant to the timing of gastrulation. When FGF is injected into the blastocoel, the time of the transition in cell behaviour is also independent of the time of injection and of the concentration of factor. However, this time is approximately 1.5 hours later than that seen in response to activin, and corresponds more to the time of appearance of the ventral lip of the blastopore (Cooke and Smith, 1989). It is possible, therefore, that at least one aspect of the control of the timing of gastrulation is influenced by the nature of the factor(s) that induce the mesoderm, the tissue that drives gastrulation.

What do these experiments with mesoderm-inducing factors tell us about the timing of gastrulation? Firstly, they shed no light on the nature of the cytoplasmic clock, save to indicate that the clock is present and running even in cells of the animal pole of the embryo, which would not normally need to refer to it. They do, however, indicate that gastrulation is not timed by reference to the time of mesoderm induction nor to the concentration of mesoderm-inducing factor. This latter point is of interest because dif-

pression of cell adhesion molecules (see
low).

Specification of cellular behaviour during gastrulation

At least three distinct types of cellular behaviour drive gastrulation: cell migration, convergence and extension. Different types of behaviour are followed by cells in different regions of the embryo at different times. How do cells 'know' which type of behaviour they should undertake? We have studied the responses made by animal pole tissue to different mesoderm-inducing factors, alone and in combination, to discover how, in principle, cell behaviour might be specified. We compare the gastrulation-like responses made by the cells with the types of tissue induced by each factor or factors.

Convergent extension

Our experiments have not yet allowed us to study convergence and extension separately, so here we refer only to convergent extension. The cells that undertake the most vigorous convergent extension are those of the dorsal marginal zone, which go on to form head mesoderm and notochord, and which express genes such as *goosecoid* (Cho et al., 1991). Notochord, and *goosecoid* expression, are induced from animal pole tissue by activin (Green et., 1990; Cho et al., 1991) and, similarly, activin induces dramatic convergent-extension-like movements from this tissue (Symes and Smith, 1987). As discussed above, these movements also resemble those of convergent extension in that they begin at about the same time that convergent extension begins in the normal embryo. By contrast, FGF does not induce dramatic elongation from animal pole regions, although it does elicit a more subtle change in shape (Slack et al., 1987), which might be interpreted as being due to convergence in the absence of extension. This is, of course, consistent with the proposed role of FGF as an inducer of ventral/posterior mesoderm (Green et al., 1990; Amaya et al., 1991).

One way in which the extent, if not the nature, of these movements might be modified in different regions of the embryo is through different concentrations of inducing factor; lower concentrations of activin, for example, cause less elongation than higher ones (see above). Another possibility is that additional factors modify the putative activin and FGF signals that operate in the embryo. Recently, two such additional signals have been discovered. mRNA for the *Xenopus* homologue of bone morphogenetic protein 4 (XBMP-4) is expressed at low levels maternally, and the gene is then strongly activated throughout the embryo at the mid-blastula transition (Köster et al., 1991; Dale et al., 1992). Injection of mRNA for XBMP-4 into the fertilized egg of *Xenopus* causes the resulting embryo to become severely 'ventralized', a phenotype that may be due to the abilities of XBMP-4 to induce strong expression of *Xhox3* (see Ruiz i Altaba and Melton, 1989), and to override the effect of activin, causing animal caps incubated even in high concentrations of activin to differentiate as ventral cell types (Dale et al., 1992; Jones et al., 1992). This ventralization of the response to activin is also observed at earlier stages, during gastrulation, where the activin-induced elongation of animal pole regions is strongly inhibited by XBMP-4 (Fig. 2). This effect may be important in regulating the extent and duration of convergent extension, but without detailed knowledge of the spatial and temporal distribution of XBMP-4 protein, it is not possible to speculate further.

Another factor that can influence convergent extension is Xwnt-8, a member of the Wnt family of growth factors. Injection of mRNA encoding int-1, Xwnt-8 or wingless protein into the ventral side of *Xenopus* embryos results in duplication of the embryonic axis, while injection into the dorsal side has little effect (McMahon and Moon, 1989; Smith and Harland, 1991; Sokol et al., 1991; Christian et al., 1991; Chakrabarti et al., 1992; reviewed by Smith, 1992). In this respect, the effect of Xwnt-8 is the opposite of that of XBMP-4, in that it 'dorsalizes' ventral mesoderm; consistent with this, if animal caps derived from embryos that have received injections of Xwnt-8 RNA are treated with FGF, they undergo dramatic elongation, as if they had been treated with activin (Christian et al., 1991). Again, the significance of this observation in the absence of knowledge about the spatial distribution of Xwnt-8 protein is not clear, but it does define another route by which gastrulation might be regulated. Finally, as the transcriptional responses to different inducing factors are elucidated, it may become possible to manipulate gastrulation movements by

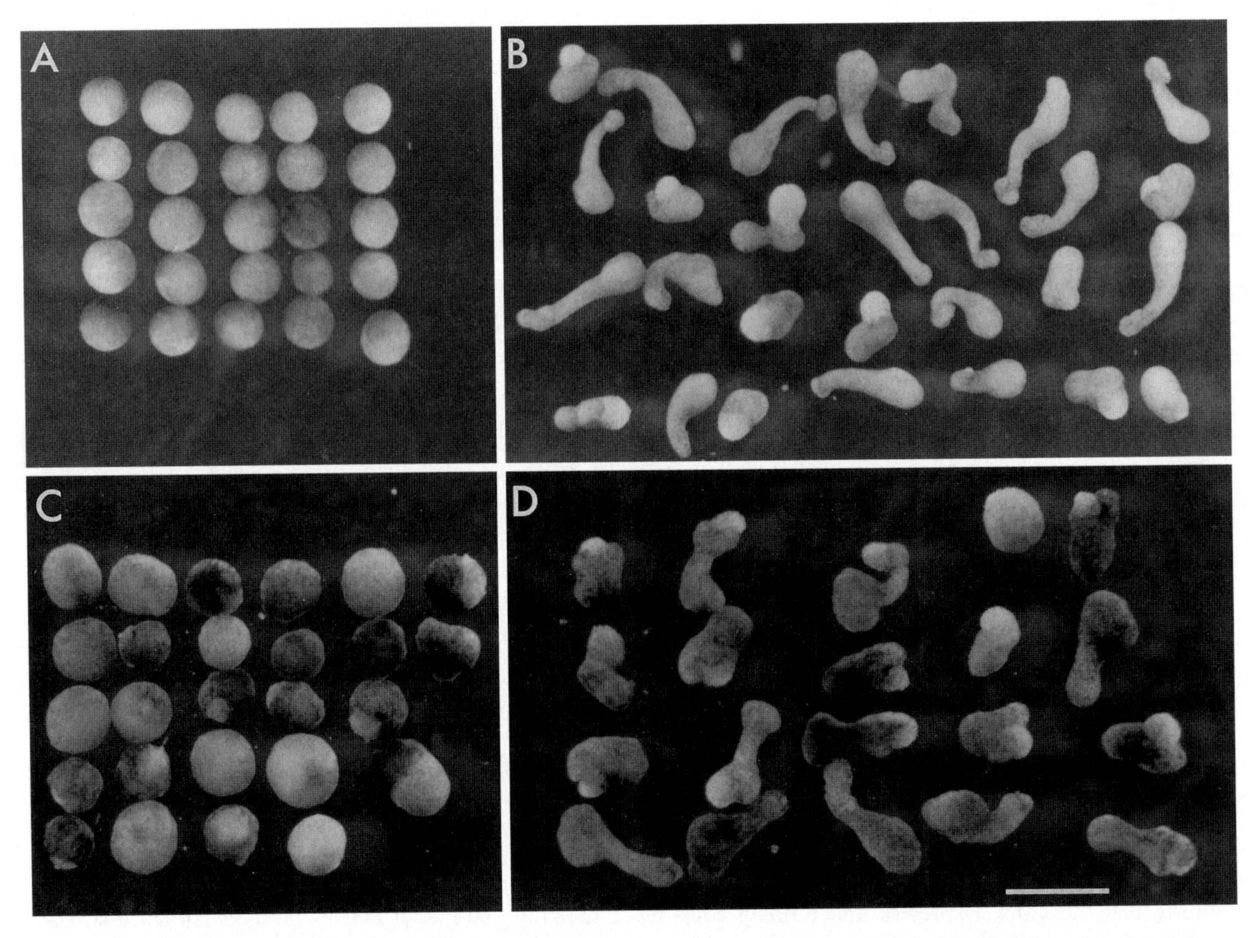

Fig. 2. BMP-4 inhibits elongation in activin-treated animal cap explants. Animal poles were excised at stage 8 and cultured in the presence (B, C, D) or absence (A) of activin until stage 18. In embryos that had been injected with 1. 5 ng of XBMP-4 RNA at the one cell stage, activin-induced elongation was severely inhibited (C), but was unaffected by the injection of antisense RNA (D). These experiments were carried out in collaboration with Dr Leslie Dale (Birmingham University, UK; see Dale et al., 1992) with whose kind permission this figure is included.

overexpression of intracellular responses to induction rather than through the extracellular signals (see below).

Cell spreading and migration

The other type of cell behaviour observed during gastrulation is cell migration. In the intact *Xenopus* embryo, this is most easily studied in the earliest-involuting cells, which go on to form head mesoderm. These cells do not undergo convergent extension and may be important in ensuring that exogastrulation does not occur (see above). In vitro, cell spreading may be studied by dissociating prospective mesodermal cells and seeding them onto fibronectin (FN)-coated tissue-culture plates. Mesodermal cells from all regions of the marginal zone spread and migrate on this substratum, as will prospective endodermal cells, but cells derived from the animal pole region do not adhere, and remain as loosely attached spheres (Nakatsuji, 1986; Winklbauer, 1988). Prospective ectodermal cells can, however, be induced to spread and migrate on fibronectin if they are treated with activin (Smith et al., 1990; see Fig. 3). This is a rapid response, and one that can be induced in single cells; it does not, therefore, require a 'community effect' (Gurdon, 1988).

Recently, we have gone on to investigate whether other mesoderm-inducing factors also cause animal pole cells to spread and migrate on fibronectin. Our first experiments used bFGF, in response to which factor animal pole cells attach strongly to the FN substratum, but do not spread or migrate. This was slightly surprising, because fibroblast growth factor has been suggested to induce mesoderm of ventral character (Slack et al., 1987; Green et al., 1990), and cells derived from the ventral marginal zone of the early gastrula spread on FN in a manner very similar to that of cells from the dorsal marginal zone (J. E. H., unpublished observations). It remains possible that other factors induce cell spreading in the ventral marginal zone, and we are now investigating this by treating cells simultaneously with FGF and activin, and by observing the behaviour of cells derived from animal pole regions that have received injections of *Xwnt-8* or XBMP-4 mRNA and are cultured in the presence, or absence, of activin and FGF. Preliminary results indicate that FGF inhibits activin-induced spreading on FN, and it is possible that this provides a way of regulating cell migration during gastrulation.

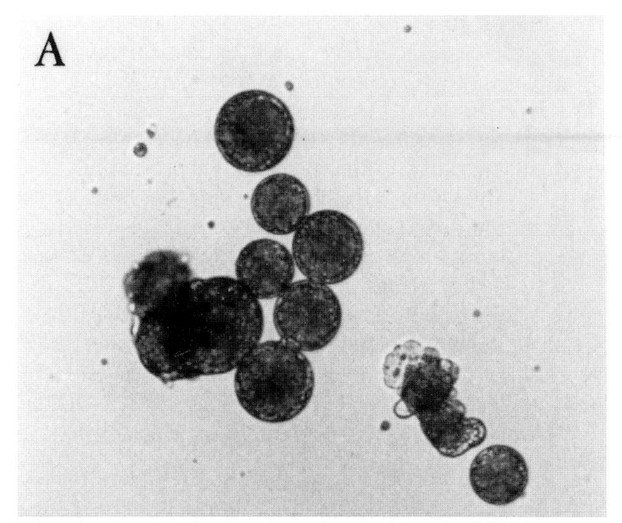

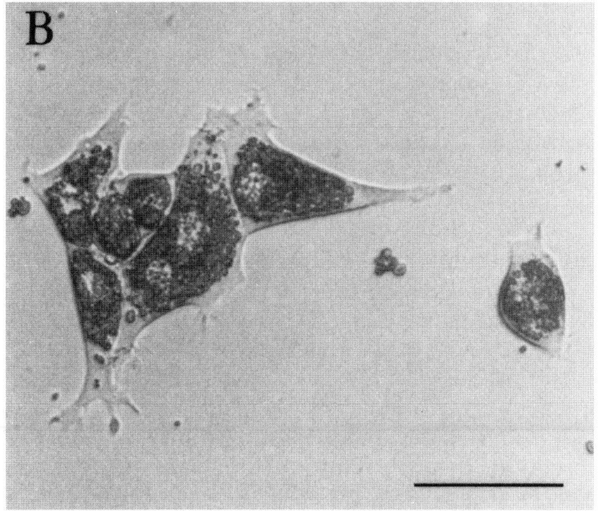

Fig. 3. Mesoderm induction causes cells to spread and migrate on fibronectin. Uninduced animal cap cells do not spread on a fibronectin-coated substratum (A), whereas those treated with activin do (B). Scale bar is 25 μm.

The cellular basis of gastrulation

Accurate and detailed descriptions of cell behaviour during gastrulation have been made by Keller and colleagues (see chapter by Keller in this volume), and it might appear that studying the gastrulation-like movements seen in response to mesoderm-inducing factors would have little to add to these. As with the other aspects of gastrulation discussed in this paper, however, the advantage of studying gastrulation in response to defined factors is that the situation is simplified: a narrower range of cell behaviours will occur, and the existence of one type of behaviour will not confuse analysis of another.

Convergent extension

Keller has suggested that the dramatic extension produced by convergent extension movements can be achieved through relatively short-range cell intercalation. We have investigated this question in activin-induced convergent extension of animal caps by microinjecting a lineage tracer into *Xenopus* embryos at the two-cell stage such that half the embryo becomes labelled (DeSimone et al., 1991). Such embryos were allowed to develop to the mid-blastula stage, when animal pole regions were dissected and cultured in the presence or absence of activin. Histological analysis of such caps immediately after dissection showed that little mingling of labelled and unlabelled cells had occurred during early development, in agreement with the results of Wetts and Fraser (1989). We therefore went on to examine cell mixing in explants allowed to develop until the equivalent of stage 12. 5, by which time animal caps exposed to activin had undergone considerable elongation, while controls remained spherical.

In both induced and uninduced explants, blastomeres mixed with each other only to a very limited extent. In both, a few individual cells could be found away from the main group, but overall the impression, in agreement with Keller, is that convergent extension does not involve long-range cell migration but a short-range, and perhaps directed, exchange of neighbours. Keller and Hardin (1987) suggested that such exchange of neighbours might occur through 'jostling' of adjacent cells, and this idea is supported by the appearance of induced and uninduced cells in the scanning electron microscope. At the equivalent of the early gastrula stage, induced and uninduced explants appear similar, with rounded cells at the surface of the explant that give the impression of being motile. By the late gastrula stage, however, the external surface of uninduced animal caps is very smooth, with the cells forming a flat epithelial-like sheet; the induced explants, by contrast, still have a motile appearance.

Cell spreading and migration

We have made little progress in the analysis of the cellular activities involved in spreading and migration, although it seems likely that these will involve major changes in the cytoskeleton and in proteins associated with the cytoskeleton. Rather, we have concentrated on changes in expression of cell adhesion molecules such as the integrins.

The molecular basis of gastrulation

To understand the molecular basis of gastrulation, it is necessary to discover the sequence of events between the receipt of a mesoderm-inducing signal such as activin and the onset of gastrulation behaviour. One way of going about this is to follow the events in the order in which they occur. This approach is being followed already in an effort to understand the patterning of the *Xenopus* mesoderm, and receptors both for FGF and activin have been demonstrated in the early embryo (Gillespie et al., 1989; Musci et al., 1990; Friesel and Dawid, 1991; Amaya et al., 1991; Kondo et al., 1991; Mathews et al., 1992), progress has been made in understanding second messenger pathways (Maslanski et al., 1992), and several mesoderm-specific genes such as *Xhox3* (Ruiz i Altaba and Melton, 1989) Xenopus *Brachyury* (Smith et al., 1991) and *goosecoid* (Cho et al., 1991) have been found. The next step in this approach is to ask whether overexpression of, for example, *Brachyury* can lead to mesoderm-like movements in animal pole cells, and such work is in progress (V. T. Cunliffe, J. E. H. and J. C. S., unpublished work).

The other approach is complementary to that outlined above. A very rapid response to mesoderm induction is the acquisition of the ability of cells to spread and migrate on FN. If the molecular events responsible for this change in behaviour can be elucidated, we can in principle work backwards from them towards the initial inductive event. One class of molecules likely to be involved is the integrin family. Indeed, there is already good evidence that FN-integrin interactions play an important role in amphibian gastrulation. This comes from work on *Pleurodeles*, which has shown that involution and migration are blocked by injection of anti-fibronectin (FN) antibodies into the blastocoel of the embryo (Boucaut et al., 1984), by intrablastocoelic injection of peptides corresponding to the cell binding site of FN (Boucaut et al., 1985) and by intracellular injection of an antibody targeted to the cytoplasmic domain of integrin β_1 (Darribère et al., 1990). Intrablastocoelic injection of antibodies raised against the extracellular domain of integrin β_1 also inhibits gastrulation by blocking formation of the matrix over which the cells migrate. Thus it is likely that the migration element of gastrulation is, at least in *Pleurodeles*, dependent on the interaction of FN with integrin β_1. Recently, we have investigated whether the same is true in *Xenopus*, and, if so, whether mesoderm-inducing factors such as activin influence the expression of integrins in *Xenopus*.

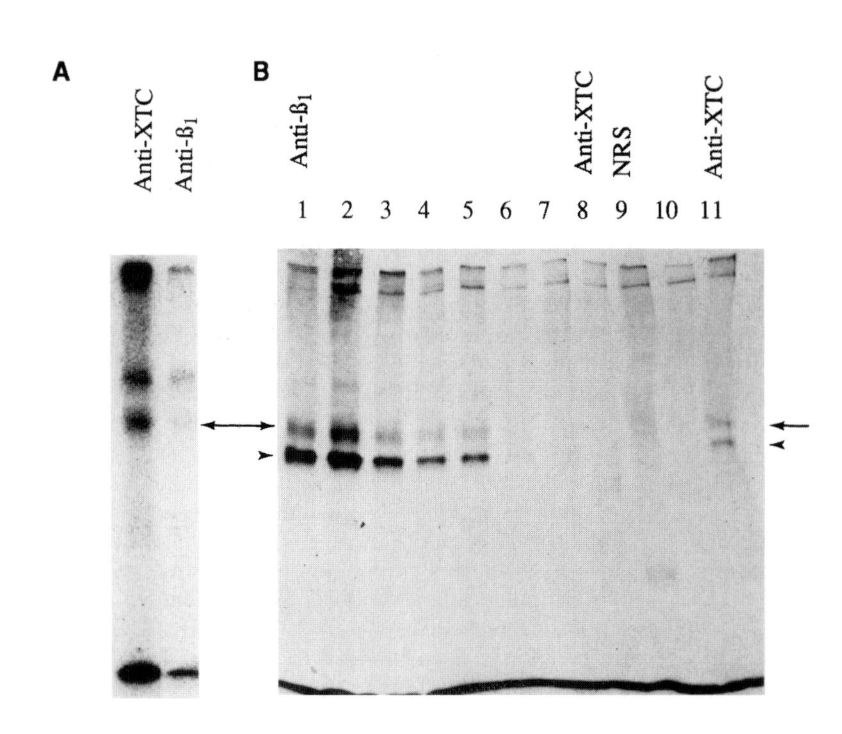

Fig. 4. An antiserum raised against XTC cells recognizes only integrin β_1. Proteins immunoprecipitated with an anti-XTC cell antiserum (Anti-XTC) or an anti-integrin β_1 antiserum (Anti-β_1) from surface- (A) or metabolically labelled (B) neurula extracts show the same electrophoretic mobility in SDS gels. The mature form (arrows) appears on the surface but the immature form (arrowheads) does not. Further evidence that the two sera recognize the same molecule comes from preclearing experiments (B). Sequential precipitations (lanes 1-7) with the anti-β_1 serum from metabolically labelled neurula extracts can remove all the target molecule from this solution; subsequent immunoprecipitation with the anti-XTC cell serum yields nothing (lane 8). Preclearing precipitations with normal rabbit serum (precipitations 1 and 7 are shown here; lanes 9 and 10) do not remove the target molecule for the anti-XTC serum (lane 11).

Our initial experiments used an anti-integrin β_1 antiserum (Marcantonio and Hynes, 1988) and a fibronectin affinity column to identify FN-binding proteins on the cell surfaces of mid-gastrula stage *Xenopus* embryos (Howard et al., 1992). This work indicated that the major fibronectin-binding protein on the surface of *Xenopus* gastrula cells contains integrin β_1, and we therefore went on to analyze the expression of this molecule during gastrula and later stages; previous studies had demonstrated the presence of integrin β_1 mRNA and protein up to and during blastula stages (DeSimone and Hynes, 1988; Smith et al., 1990; Gawantka et al., 1992). Immunoprecipitation from surface-labelled gastrula extracts revealed very little integrin β_1, as previously reported (Krotoski and Bronner-Fraser, 1990; Gawantka et al., 1992), although expression was much higher by neurula stages. This expression pattern was confirmed by immunoprecipitation of integrin β_1 from metabolically labelled cells, which revealed a 105×10^3 M_r polypeptide at all stages, representing the precursor form of integrin β_1, while mature species of M_r 116-120$\times10^3$ were only clearly detectable after neurulation. Immunocytochemical studies performed in our laboratory and elsewhere (Krotoski and Bronner-Fraser, 1990) reflect this pattern of synthesis, with low levels of expression of integrin β_1 - so low as to be undetectable with many antisera - until notochord and somite formation.

The low levels of expression of the mature functional form of integrin β_1 cast some doubt on its role in gastrulation in *Xenopus*. The anti-integrin β_1 antiserum used for the expression studies was raised against a C-terminal peptide (Marcantonio and Hynes, 1988), but intracellular microinjection of this antibody did not disrupt function (J. E. H., unpublished observations). We therefore raised antisera against the extracellular domain of the mature molecule using XTC cells, which we knew to express integrin β_1 at high levels, as an immunogen. The resulting sera were indeed targeted to the external portion of integrin β_1 (Fig. 4), and were used to demonstrate that the spreading of newly induced mesodermal cells is dependent on integrin β_1 (Fig. 5; Howard et al., 1992). This is true not only for activin-induced animal cap cells but also for cells excised from the dorsal marginal zone, providing us with further evidence that activin closely mimics the action of the endogenous inducer. Furthermore, as in *Pleurodeles*, gastrulation is disrupted if this anti-integrin β_1 serum is injected into the blastocoel, apparently because the matrix of the blastocoel roof is disrupted (Fig. 6; Howard et al., 1992). These data do not contradict the work of Keller et al. (1985), which shows that gastrulation in *Xenopus* can occur even if the blastocoel roof is removed; rather, they suggest that an overlying matrix is required only in the very early stages of involution and that the coordinated force of convergent extension is then sufficient to complete the process in the absence of further directional cues. In *Pleurodeles*, involuting cells migrate more independently, convergent extension is therefore necessarily less significant, and cells require continued directional cues.

These results indicate that integrin β_1 is involved in gastrulation in *Xenopus*, and to understand gastrulation it will be necessary to understand how the function of integrin β_1 is regulated. One approach to this question is to use mesoderm-inducing factors such as activin, which, as we describe above, causes animal pole cells to undergo both convergent extension and cell migration, the latter being dependent on integrin β_1. An understanding of how the function of this molecule is regulated by activin in this context could therefore provide insight into the intracellular events which initiate gastrulation. In our attempts to investigate this, we have, unfortunately, been unable to show any increase in expression of integrin β1 in response to activin, either at the level of rate of synthesis (Fig. 7; Smith et al., 1990) or of cell-surface expression (J. E. H., unpublished observation). This suggests that the mechanism underlying this very rapid response is a more subtle modulation of integrin function. This might involve a change in lipid environment (Conforti et al., 1990; Hermanowski-Vosatka et al., 1992), an altered interaction with the cytoskeleton, or a different dimerization partner (Dedhar, 1990). This last modification cannot be studied until α-specific antisera become available in *Xenopus*. In future work, we intend to address these problems.

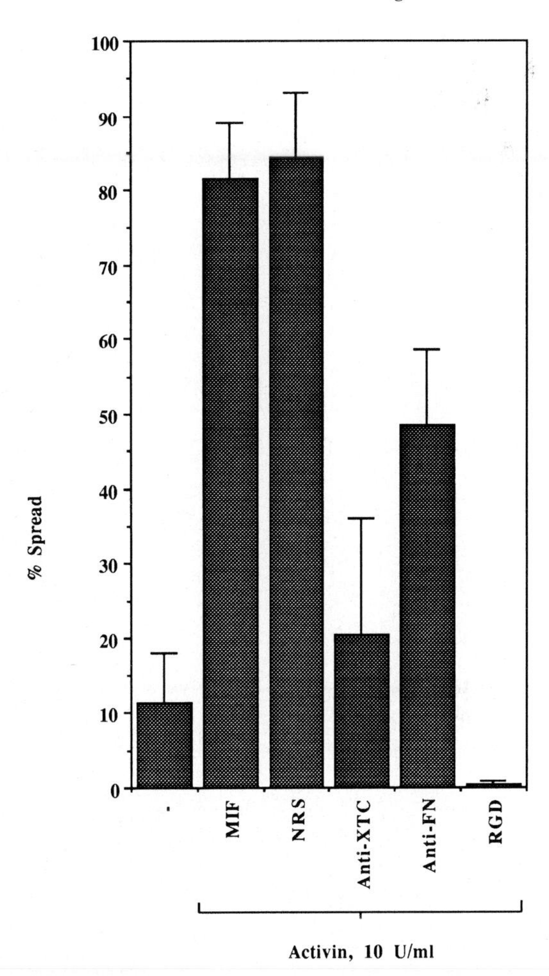

Fig. 5. Anti-XTC antisera block the spreading of activin-induced *Xenopus* blastomeres. Cells were treated with activin, or left untreated, and plated in the presence of the indicated antisera. The percentages of cells that had spread were assessed 1-2 hours later. The treatments are: – , no activin; remaining cases were all treated with activin; MIF, no antiserum; NRS, normal rabbit serum; Anti-XTC, Anti-XTC antiserum; Anti-FN, anti-fibronectin; RGD, 5 mM GRGDTP.

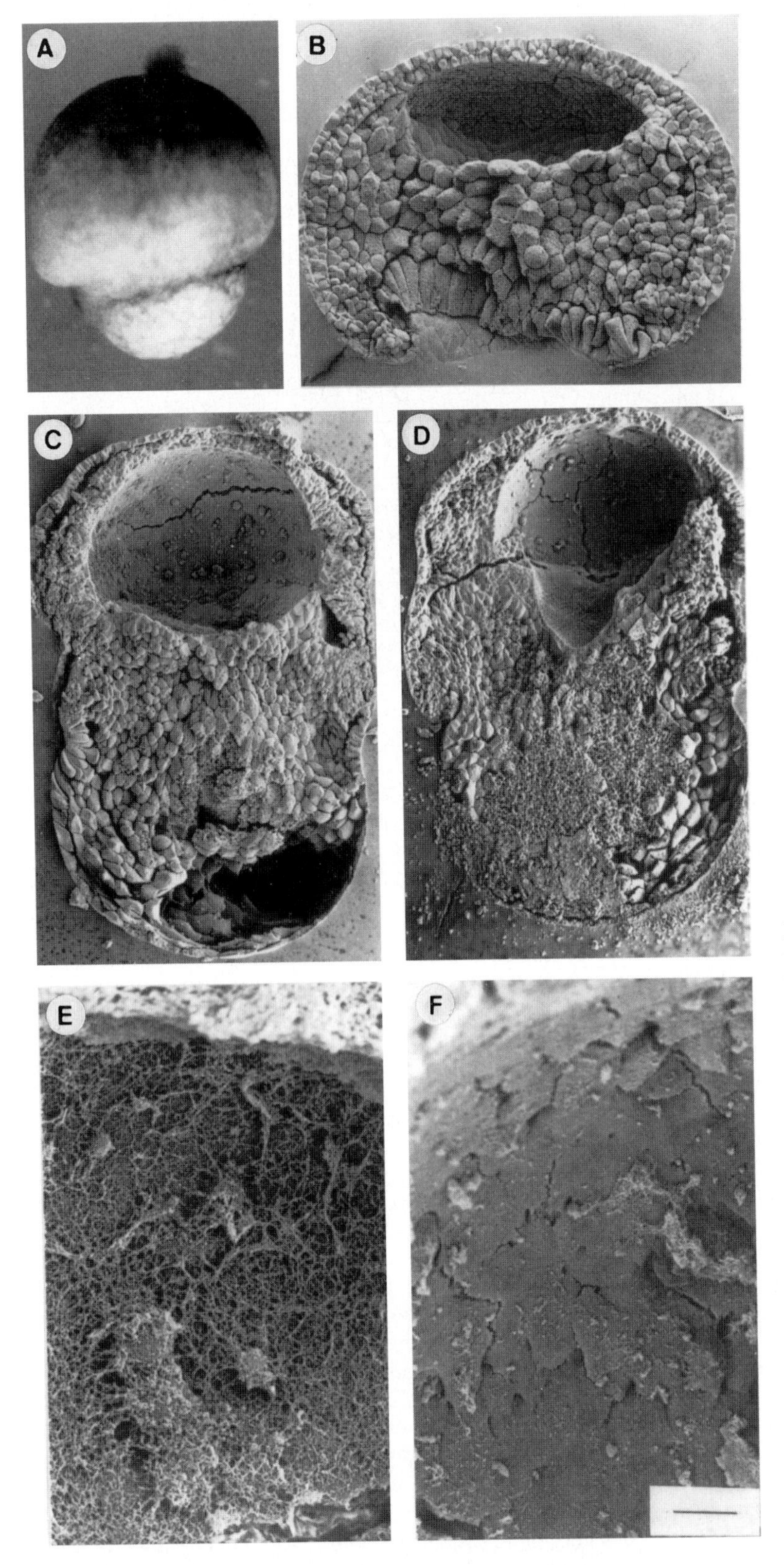

Fig. 6. Fibronectin-integrin interactions are required for gastrulation to occur in *Xenopus*. Involution in untreated embryos (B) normally begins at stage 10 and continues until the ingressing cells on the dorsal side (left in all pictures) have passed the centre of the blastocoel roof. Injection of anti-fibronectin (A,C,E) and anti-XTC cell (D, F) antisera into the blastocoel of stage 9 embryos severely restricts involution of mesodermal cells. In the presence of anti-FN antibodies in the blastocoel, the blastopore lip forms normally (A; stage 10. 5 embryo), but involution does not occur (A, C). In these scanning electron micrographs, the embryos were fixed at stage 10. 5 (B) or 12 (C-F) and cleaved as in Hirst and Howard (1992). Scanning electron micrographs of the blastocoel roof of embryos receiving injections of anti-XTC antiserum suggest that this disruption is caused by loss of the extracellular matrix (ECM) (F). In embryos injected with anti-fibronectin antibody the ECM is intact (E), and disruption is assumed to result from an inability of involuting cells to interact with FN. Bar represents 200 μm (B, C, D), 20 μm (E, F), 300 μm (A). Micrographs were kindly produced by Liz Hirst.

Conclusions

Gastrulation is a fundamental problem in developmental biology and, to arrive at an understanding of the process, it will be necessary to combine the techniques of experimental embryology, cell biology and molecular biology. In this paper, we have described the use of mesoderm-inducing factors such as activin A in the study of gastrulation. The ability of factors such as these to induce naive cells to undergo gastrulation movements provides a powerful tool

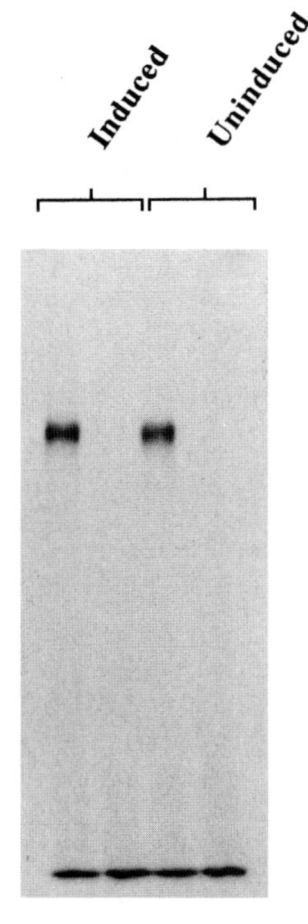

Fig. 7. Synthesis of integrin β_1 is not significantly affected by activin treatment. Animal pole explants were incubated in the absence (lanes 1 and 2) or presence (lanes 3 and 4) of activin, together with [^{35}S]methionine. Levels of integrin β_1 synthesized during the labelling period were assessed by immunoprecipitation with the anti integrin β_1 antiserum (lanes 1 and 3) or normal rabbit serum (lanes 2 and 4).

for coming to understand what makes cells behave in particular ways at particular times.

References

Amaya, E., Musci, T. J. and Kirschner, M. W. (1991). Expression of a dominant negative mutant of the FGF receptor disrupts mesoderm formation in Xenopus embryos. *Cell* **66**, 257-270.

Black, S. D. (1989). Experimental reversal of the normal dorsal-ventral timing of blastopore formation does not reverse axis polarity in *Xenopus laevis* embryos. *Dev. Biol.* **134**, 376-381.

Boucaut, J. -C., Darribère, T., Boulekbache, H. and Thiery, J. -P. (1984). Prevention of gastrulation but not neurulation by antibodies to fibronectin in amphibian embryos. *Nature* **307**, 364-367.

Boucaut, J. -C., Darribère, T., Li, S. D., Boulekbache, H., Yamada, K. M., and Thiery, J. -P. (1985). Evidence for the role of fibronectin in amphibian gastrulation. *J. Embryol. Exp. Morph.* **89 Supplement**, 211-227.

Chakrabarti, A., Matthews, G., Colman, A. and Dale, L. (1992). Secretory and inductive properties of *Drosophila wingless* (Dwnt-1) protein in *Xenopus* oocytes and embryos. *Development*, in press.

Cho, K. W. Y., Blumberg, B., Steinbeisser, H. and De Robertis, E. M. (1991). Molecular nature of Spemann's organizer: the role of the Xenopus homeobox gene *goosecoid*. *Cell* **67**, 1111-1120.

Christian, J. L., Olsen, D. J. and Moon, R. T. (1992). *Xwnt-8* modifies the character of mesoderm induced by bFGF in isolated *Xenopus* ectoderm. *EMBO J.* **11**, 33-41.

Conforti, G., Zanetti, A., Pasquali, R. I., Quaglino, D. J., Neyroz, P. and Dejana, E. (1990). Modulation of vitronectin receptor binding by membrane lipid composition. *J. Biol. Chem.* **265**, 4011-4019.

Cooke, J. (1973). Morphogenesis and regulation in spite of continued mitotic inhibition in *Xenopus* embryos. *Nature* **242**, 55-57.

Cooke, J. (1989). Mesoderm-inducing factors and Spemann's organiser phenomenon in amphibian development. *Development* **107**, 229-241.

Cooke, J. and Smith, J. C. (1987). The midblastula cell cycle transition and the character of mesoderm in the u.v.-induced nonaxial *Xenopus* development. *Development* **99**, 197-210.

Cooke, J. and Smith, J. C. (1989). Gastrulation and larval pattern in *Xenopus* after blastocoelic injection of a *Xenopus*-derived inducing factor: experiments testing models for the normal organization of mesoderm. *Dev. Biol.* **131**, 383-400.

Cooke, J. and Smith, J. C. (1990). Measurement of developmental time by cells of early embryos. *Cell* **60**, 891-894.

Dale, L., Howes, G., Price, B. M. J. and Smith, J. C. (1992). Bone morphogenetic protein 4: a ventralizing factor in early *Xenopus* development. *Development* **115**, in press.

Darribère, T., Guida, K., Larjava, H., Johnson, K. E., Yamada, K. M., Thiery, J. -P. and Boucaut, J. -C. (1990). In vivo analyses of integrin β_1 subunit function in fibronectin matrix assembly. *J. Cell Biol.* **110,** 1813-1823.

Dawid, I. B. and Sargent, T. D. (1990). The role of growth factors in embryonic induction in amphibians. *Curr. Top. Dev. Biol.* **24**, 31-55.

Dedhar, S. (1990). Integrins and tumor invasion. *BioEssays* **12**, 583-590.

DeSimone, D. W. and Hynes, R. O. (1988). Xenopus laevis integrins. Structural conservation and evolutionary divergence of integrin beta subunits. *J Biol. Chem.* **263**, 5333-5340.

DeSimone, D. W., Smith, J. C., Howard, J. E., Ransom, D. G. and Symes, K. (1991). The expression of fibronectins and integrins during mesodermal induction and gastrulation in *Xenopus*. In *Gastrulation: Movements, Patterns and Molecules* (ed. R. Keller, W. Clark and F. Griffin), pp185-198.

Friesel, R. and Dawid, I. B. (1991). cDNA cloning and developmental expression of fibroblast growth factor receptors from *Xenopus laevis*. *Mol. Cell Biol.* **11**, 2481-2488.

Gawantka, V., Ellinger-Ziegelbauer, H. and Hausen, P. (1992). β_1-integrin is a maternal protein that is inserted into all newly formed plasma membranes during early *Xenopus* embryogenesis. *Development* **115**, 595-606.

Gerhart, J. C., Danilchik, M., Doniach, T., Roberts, S., Rowning, B. and Stewart, R. (1989). Cortical rotation of the *Xenopus egg*: consequences for the anteroposterior pattern of embryonic dorsal development. *Development* **1989 Supplement**, 37-51.

Gillespie, L. L., Paterno, G. D. and Slack, J. M. W. (1989). Analysis of competence: receptors for fibroblast growth factor in early *Xenopus* embryos. *Development* **106**, 203-208.

Green, J. B. A., Howes, G., Symes, K., Cooke, J. and Smith, J. C. (1990). The biological effects of XTC-MIF: quantitative comparison with *Xenopus* bFGF. *Development* **108**, 173-183.

Green, J. B. A. and Smith, J. C. (1990). Graded changes in dose of a *Xenopus* activin A homologue elicit stepwise transitions in embryonic cell fate. *Nature* **347**, 391-394.

Gurdon, J. B. (1988). A community effect in animal development. *Nature* **336**, 772-774.

Hermanowski-Vosatka, A., Van Strijp, J. A. G., Swiggard, W. J. and Wright, S. D. (1992). Integrin modulating factor-1: A lipid that alters the function of leukocyte integrins. *Cell* **68**, 341-342.

Hirst, E. M. A., and Howard, J. E. (1992). A simple technique to control the position and orientation of dry fracture planes for studying surfaces hidden within bulk tissue by scanning electron microscopy. *J. Microscopy*. In press.

Howard, J. E., Hirst, E. M. and Smith, J. C. (1992). Are b_1 integrins involved in *Xenopus* gastrulation? *Mech. Dev.* In press.

Jones, C. M., Lyons, K. M., Lapan, P. M., Wright, C. V. E. and Hogan, B. M. L. (1992). DVR-4 (Bone Morphogenetic Protein-4) as a posterior-ventralizing factor in Xenopus mesoderm induction. *Development* **115**, 639-648.

Keller, R. (1991). Early embryonic development of Xenopus laevis. *Methods in Cell Biology* **36**, 61-113.

Keller, R. E. and Danilchik, M. (1988). Regional expression, pattern and timing of convergence and extension during gastrulation of *Xenopus laevis*. *Development* **103**, 193-210.

Keller, R. E., Danilchik, M., Gimlich, R. and Shih, J. (1985). The function of convergent extension during gastrulation of *Xenopus laevis*. *J. Embryol. Exp. Morph.* **89 Supplement**, 185-209.

Keller, R. E. and Hardin, J. (1987). Cell behaviour during active cell rearrangement: evidence and speculations. *J. Cell Sci Suppl.* **8**, 369-393.

Kimelman, D. and Kirschner, M. (1987). Synergistic induction of mesoderm by FGF and TGF-beta and the identification of an mRNA coding for FGF in the early Xenopus embryo. *Cell* **51**, 869-877.

Kondo, M., Tashiro, K., Fujii, G., Asano, M., Miyoshi, R., Yamada, R., Muramatsu, M. and Shiokawa, K. (1991). Activin receptor mRNA is expressed early in *Xenopus* embryogenesis and the level of the expression affects the body axis formation. *Biochem. Biophys. Res. Comm.* **181**, 684-690.

Köster, M., Plessow, S., Clement, J. H., Lorenz, A., Tiedemann, H. and Knöchel, W. (1991). Bone morphogenetic protein 4 (BMP-4), a member of the TGF-b family, in early embryos of *Xenopus laevis*: analysis of mesoderm-inducing activity. *Mech. Dev.* **33**, 191-199.

Krotoski, D. and Bronner-Fraser, M. (1990). Distribution of integrins and their ligands in the trunk of *Xenopus laevis* during neural crest cell migration. *J. Exp. Zool.* **253**, 139-150.

McMahon, A. P. and Moon, R. T. (1989). Ectopic expression of the proto-oncogene *int-1* in Xenopus embryos leads to duplication of the embryonic axis. *Cell* **58**, 1075-1084.

Marcantonio, E. E. and Hynes, R. O. (1988). Antibodies to the conserved cytoplasmic domain of the integrin b_1 subunit react with proteins in vertebrates, invertebrates and fungi. *J. Cell Biol.* **106**, 1765-1772.

Maslanski, J. A., Leshko, L. and Busa, W. B. (1992). Lithium-sensitive production of inositol phosphates during amphibian embryonic mesoderm induction. *Science* **256**, 243-245.

Mathews, L. S., Vale, W. W. and Kintner, C. R. (1992). Cloning of a second type of activin receptor and functional characterization in *Xenopus* embryos. *Science* **255**, 1702-1705.

Musci, T. J., Amaya, E. and Kirschner, M. W. (1990). Regulation of the fibroblast growth factor receptor in early *Xenopus* embryos. *Proc. Natn. Acad. Sci. USA* **87**, 8365-8369.

Nakatsuji, N. (1986). Presumptive mesoderm cells from *Xenopus laevis* gastrulae attach to and migrate on substrata coated with fibronectin or laminin. *J. Cell Sci.* **86**, 109-118.

Ruiz i Altaba, A. and Melton, D. A. (1989). Interaction between peptide growth factors and homeobox genes in the establishment of antero-posterior polarity in frog embryos. *Nature* **341**, 33-38.

Satoh, N. (1982). Timing mechanisms in early embryonic development. *Differentiation* **22**, 156-163.

Slack, J. M. W., Darlington, B. G., Heath, J. K. and Godsave, S. F. (1987). Mesoderm induction in early *Xenopus* embryos by heparin-binding growth factors. *Nature* **326**, 197-200.

Smith, J. C. (1989). Mesoderm induction and mesoderm-inducing factors in early amphibian development. *Development* **105**, 665-677.

Smith, J. C. (1992). A *wnt*er's tale. *Current Biology* **2**, 177-179.

Smith, J. C., Price, B. M. J., Green, J. B. A., Weigel, D. and Herrmann, B. G. (1991). Expression of a Xenopus homolog of *Brachyury* (*T*) is an immediate-early response to mesoderm induction. *Cell* **67**, 79-87.

Smith, J. C., Symes, K., Hynes, R. O. and DeSimone, D. (1990). Mesoderm induction and the control of gastrulation in *Xenopus laevis*: the roles of fibronectin and integrins. *Development* **108**, 229-238.

Smith, W. C. and Harland, R. M. (1991). Injected Xwnt-8 RNA acts early in Xenopus embryos to promote formation of a vegetal dorsalizing center. *Cell* **67**, 753-765.

Sokol, S., Christian, J. L., Moon, R. T. and Melton, D. A. (1991). Injected Wnt RNA induces a complete body axis in Xenopus embryos. *Cell* **67**, 741-752.

Symes, K. and Smith, J. C. (1987). Gastrulation movements provide an early marker of mesoderm induction in *Xenopus laevis*. *Development* **101**, 339-349.

Wetts, R. and Fraser, S. E. (1989). Slow intermixing of cells during *Xenopus* embryogenesis contributes to the consistency of the fate map. *Development* **105**, 9-15.

Whitman, M. and Melton, D. A. (1989). Growth factors in early embryogenesis. *Ann. Rev. Cell Biol.* **5**, 93-117.

Winklbauer, R. (1988). Differential interaction of *Xenopus* embryonic cells with fibronectin *in vitro*. *Dev. Biol.* **130**, 175-183.

Winklbauer, R. (1990). Mesodermal cell migration during *Xenopus* gastrulation. *Dev. Biol.* **142**, 155-168.

Winklbauer, R. and Selchow, A. (1992). Motile behavior and protrusive activity of migratory mesoderm cells from the Xenopus gastrula. *Dev. Biol.* **150**, 335-351.

Yasuda, G. K. and Schubiger, G. (1992). Temporal regulation in the early embryo: is MBT too good to be true? *Trends in Genetics* **8**, 124-127.

Development 1992 Supplement, 137-142 (1992)
Printed in Great Britain © The Company of Biologists Limited 1992

Muscle gene activation in *Xenopus* requires intercellular communication during gastrula as well as blastula stages

J. B. GURDON, K. KAO, K. KATO and N. D. HOPWOOD

Wellcome CRC Institute, Tennis Court Road, Cambridge CB2 1QR, UK

Summary

In *Xenopus* an early morphological marker of mesodermal induction is the elongation of the mesoderm at the early gastrula stage (Symes and Smith, 1987). We show here that the elongation of equatorial (marginal) tissue is dependent on protein synthesis in a mid blastula, but has become independent of it by the late blastula stage. In animal caps induced to become mesoderm, the time when protein synthesis is required for subsequent elongation immediately follows the time of induction, and is not related to developmental stage. For elongation, intercellular communication during the blastula stage is of primary importance.

Current experiments involving cell transplantation indicate a need for further cell:cell interactions during gastrulation, and therefore after the vegetal-animal induction during blastula stages. These secondary cell interactions are believed to take place among cells that have already received a vegetal induction, and may facilitate some of the later intracellular events known to accompany muscle gene activation.

Key words: mesoderm, induction, *Xenopus,* intercellular communication, gastrula, blastula, protein synthesis, muscle genes.

Introduction

Mesoderm-forming induction in *Xenopus* is the subject of intense investigation at present, particularly in respect of growth factor-like molecules, which can cause cultured animal pole cells to undergo various types and degrees of mesodermal differentiation (Green and Smith, 1991; Jessel and Melton, 1992). Although many morphological responses of mesoderm are not observed till neurula or later stages in development, for example the formation of notochord and muscle, and the appearance of blood cells, it is generally believed that the initial receipt of inductive signals specifying the entire mesodermal lineage takes place during cleavage and is complete by the late blastula stage. Three reasons for this are the following. First, ectoderm (animal cap) tissue of *Xenopus* has lost its competence to form mesoderm by the early gastrula stage, when induced experimentally (Nakamura et al., 1970; Gurdon et al., 1985a). Second, animal cells are capable of responding to inductive signals at about stage $6\frac{1}{2}$ (48 cell stage), as shown by placing early blastula animal cells in contact with vegetal tissue at the end of its inductive life (Jones and Woodland, 1987). It is therefore believed that a signal from vegetal cells starts to be received by animal cells at about the 48 cell stage and continues for about 6 hours, by which time embryos have reached the early gastrula stage. Third, the earliest responses to mesoderm induction include the transcriptional activation of the gene *Mix1*, which reaches its maximal expression by stage 10 and declines thereafter (Rosa, 1989). The transcription of several other mesoderm response genes is initiated before gastrulation. Among these are *XMyoD* (Hopwood et al., 1989; Harvey, 1991; Scales et al., 1990; Frank and Harland, 1991; Rupp and Weintraub, 1991), *Xbra* (Smith et al., 1991), and *Xgoosecoid* (Cho et al., 1991).

In this article, we first describe results concerning the timing of the inductive signal which elicits an early mesoderm-specific response, namely tissue elongation. This is an in vitro assay for the convergent extension movements of gastrulation (review by Keller, 1991). We then comment on the results of current cell transplantation experiments, which lead to the conclusion that some kinds of cell communication take place during gastrulation, and therefore after receipt of the initial inductive signal from vegetal cells.We think that these cell:cell interactions during gastrulation may be required for some subsequent mesodermal responses such as muscle gene activation.

Gastrula elongation and the inhibition of protein synthesis

Previous work (Cascio and Gurdon, 1987) showed that muscle gene activation following mesodermal induction is dependent on protein synthesis during late blastula and early gastrula stages; the reversible suppression of protein synthesis by cycloheximide at these stages prevented the initiation of cardiac and cytoskeletal actin gene transcription. We have now extended this analysis by determining the time when protein synthesis is required for cells to undergo the elongation movement known to be a characteristic of gastrulation and a very early response to mesoderm induction (Symes and Smith, 1987; Smith et al., 1990). Although changes in single cell motility on fibronectin is an earlier response to induction (Smith et al., 1990), we have chosen to use in vitro tissue elongation as a more convenient indication of induction.

Either cycloheximide or puromycin can suppress protein synthesis when added to the medium in which parts of embryos are cultured (Fig. 1). Furthermore the inhibition is reversible, sooner after puromycin than cycloheximide. The in vitro elongation of embryo explants, which represents the normal gastrulation movements of mesoderm cells, is prevented by these same inhibitors in a dose-dependent way and is correlated with the extent of protein synthesis inhibition (Fig. 2).

To determine the stage during blastula formation at which subsequent elongation is sensitive to protein synthesis, we incubated vegetally depleted embryos in inhibitors, starting at various mid or late blastula stages. The result is that these explants are completely prevented from elongating when protein synthesis is inhibited starting at stage 8, but elongate normally if inhibition is commenced at stage 9 or later (Fig. 3).

We next asked whether the time when protein synthesis is required for subsequent elongation is related to a particular stage in development (stage 8) or to the time elapsed from receipt of an inductive signal. This was tested by exposing animal caps at stage 7 to XTC-cell inducing medium, and then treating them with puromycin at various times after that. Elongation was scored as described in Table 1 when control embryos had reached stage 13; those vegetally depleted embryos that showed elongation did so during the time when whole control embryos were gastrulating. The result (Table 1) is that animal caps did not elongate when treated with puromycin 1-2 hours after induction by XTC medium, but did so normally when puromycin was added at stage 9 or later, that is 4 or more hours after induction. However, when animal caps were exposed to XTC medium at stage 9 and then at once to puromycin, elongation was inhibited (Table 1). In conclusion, subsequent elongation requires protein synthesis immediately after induction, irrespective of the stage when induction takes place. This aspect of timing may be contrasted with that previously described for muscle gene activation (Gurdon et al., 1985a) and for elongation itself (Symes and Smith, 1987), which always take place at a particular developmental stage, independently of the time of induction, namely at the mid and early gastrula stages respectively.

We have discussed elongation as an example of an early mesoderm-specific response to vegetal induction. The animal cap experiments outlined above show that the vegetal signal leading to mesoderm induction is immediately followed by a period of protein synthesis which is required for subsequent elongation. For comparison, protein synthesis is required for the initiation of transcription of *XMyoD* (Harvey, 1991), but not for that of *Mix1* (Rosa, 1989), *Xbrachyury* (Smith et al., 1991) or *goosecoid* (Cho et al., 1991). It is important to appreciate that most of the work on early amphibian induction involves experimental combinations of animal and vegetal tissue or animal caps treated with growth factor-like substances, and that these do not necessarily reflect the timing of induction in normal development. On the other hand, the direct analysis of equatorial cells indicates the normal timing of events in those cells that contribute the great majority of the mesoderm. We find that the time when equatorial cells of a blastula normally synthesize protein required for future elongation is during stage 8. This could mean that these cells received a vegetal induction shortly before stage 8, or that they inherited cytoplasmic substances from the egg with the same effect (Gurdon et al., 1985b). In either case, it seems that equatorial cells have completed their receipt of, and immediate response to, vegetal signal before gastrulation. This agrees with the time of transcription of early response genes, as mentioned above. It will be interesting to see whether the time of protein synthesis sensitivity for *Mix1*, *goosecoid*, etc. is the same for normal equatorial cells as for in vitro induced animal pole cells.

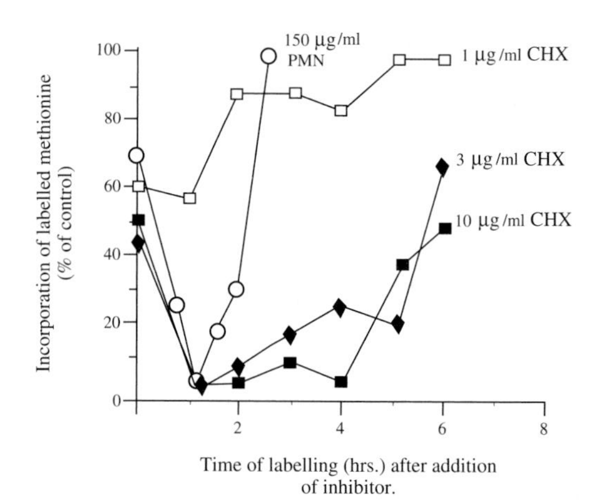

Fig. 1. Cycloheximide (CHX) and puromycin (PMN) reduce protein synthesis reversibly. Stage 8 embryos whose vegetal region had been removed to increase exposure to the medium were incubated at 23°C in various concentrations of CHX (1,3 and 10 μg/ml) or PMN (150-200 μg/ml) for one hour, and this was followed by extensive washing in MBS. By removing the vegetal one third of embryos, the animal and equatorial two thirds remained "open" and readily accessible to inhibitor and washes. Radioactively labelled methionine was added to the incubation medium of embryos for one hour at hourly intervals at the beginning of and following CHX addition (horizontal axis). The embryos were then frozen to determine the level of incorporation of label into acid-insoluble material (vertical axis). Each point represents the data collected for at least 4 embryos taken from two different batches of embryos. Values were normalized to one embryo before they were compared to the controls.

Muscle gene activation requires cell interactions during gastrulation

When a vegetal induction has been received by cells during blastula stages, it might be thought that further events, lead-

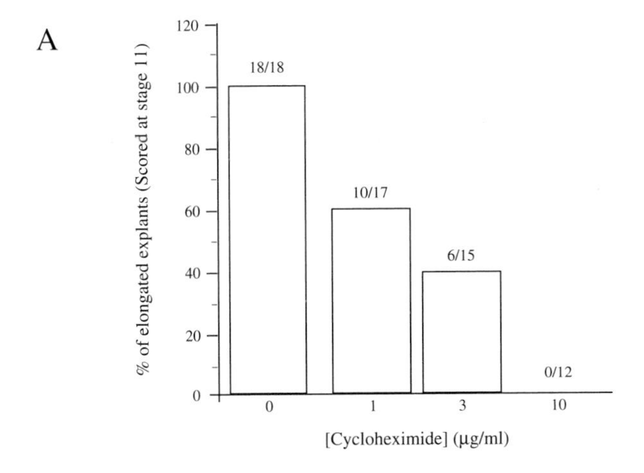

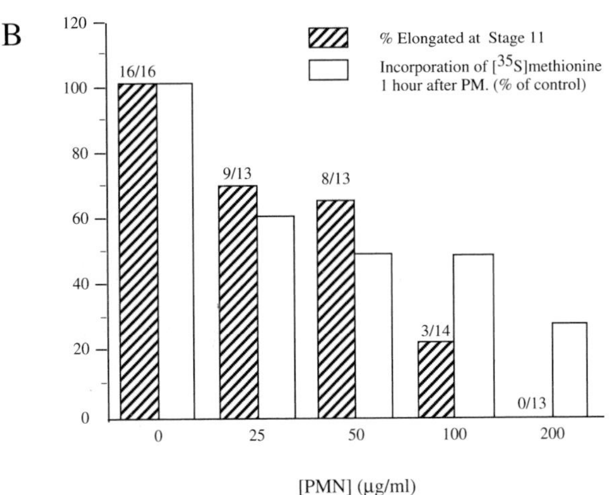

Fig. 2. Cycloheximide and puromycin prevent elongation. Embryos had their vegetal region removed at stage 8, were allowed to heal in MBS for half an hour, and were then exposed to various concentrations of CHX (A) or PMN (B). Also shown in B is the level of incorporation of [^{35}S]methionine after PMN treatment. At stage 11, the embryos were scored for elongation which was considered to have taken place when an embryo's longest dimension was at least twice its shortest diameter. Any elongation observed always took place at the normal time during gastrulation.

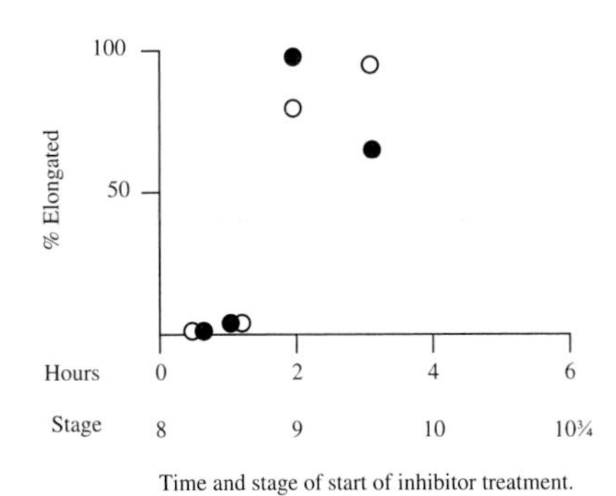

Fig. 3. The inhibition of elongation by protein synthesis inhibitors is stage-dependent. Embryos had their vegetal region removed at stage 8 and were treated with 5 μg/ml of CHX (open circles) or 200 μg/ml PMN (solid circles) for one hour starting at the times and stages indicated. Each point represents the start of a period of inhibition of protein synthesis lasting for four hours (CHX) or for two hours (PMN).

ing to muscle gene activation, are solely intracellular. The first indication that this is not the case and that further cell interactions are required came from experiments in which isolated cells were cultured with or without calcium (Gurdon et al., 1984). Late blastula cells, which had received their vegetal inductive signal, were dissociated, cultured as loose cells in Ca-free medium, and then reaggregated by Ca^{2+} addition at various times. It turned out that Ca^{2+} addition and cell reaggregation are absolutely necessary, during the early gastrula stage, for muscle genes to be subsequently transcribed, but not for the elongation or gastrulation type of cell movement (Smith et al., 1990), nor for the activation of cytokeratin genes, which takes place independently of any cell interactions (Sargent et al., 1986). Another indication that mesoderm-forming cell interactions continue during gastrulation comes from the grafting experiments of Smith et al. (1985). These authors found that ventral equatorial tissue, which normally forms blood cells and mesenchyme, will form muscle if cultured in association with dorsal equatorial tissue. They suggest that this represents a normal "dorsalization" signal taking place at stage 10, and perhaps later. All of these results imply that some interaction must take place among cells during gastrulation, and that this is different from the inductive interaction which precedes it.

Further indications that cell interactions during gastrulation are required for the full activation of muscle genes come from cell transplantation experiments. In some of these, single future muscle cells have been transplanted to a non-muscle region of other embryos. Kato et al. (unpublished) find that only when a cell is taken from a late gastrula, can it continue its differentiation as a muscle cell; if taken from earlier stages, similar cells fail to differentiate as muscle. The conclusion is that, during gastrulation, future muscle cells normally undergo some process that commits them to stable muscle gene activation. These results extend the single cell transplantation experiments of Heasman et al. (1984) by using muscle-specific gene markers with which to recognize differentiation.

Another design of cell transplant experiment using future muscle cells indicates the importance of cell interactions during gastrulation. Cells at the mid-gastrula stage are placed in sandwiches composed of blastula ectoderm, either as single cells or as reaggregates, and then cultured overnight before being tested for muscle gene expression. This is observed only in the reaggregated cells and not in single cells, a result that we interpret as indicating the need for an interaction among cells that have already embarked on a mesodermal pathway of differentiation due to events prior to gastrulation. Since the only difference between the single and reaggregated cells is the nature of their neighbours, we consider that these results point to the importance of cell interactions, during gastrulation, for muscle gene expression. These results (in preparation) extend the concept of a community effect, originally described for blastula cells in vegetal tissue sandwiches (Gurdon, 1988, 1989), to events in normal muscle cell differentiation. These

Table 1. *Elongation of animal caps requires protein synthesis immediately after XTC treatment, independently of developmental stage*

Stage of XTC treatment	Initiation of PMN treatment (Hrs after XTC)	Stage of PMN treatment	Elongation score No. (%) +	++	+++	Total
7	-	-	0	0	12(100)	12
7	1	8	0	0	0	10
7	2	$8\frac{1}{2}$	6(60)	0	0	10
7	3	9	5(42)	3(25)	0	12
7	4	$9\frac{1}{2}$	0	7(58)	4(33)	12
9	1	$9\frac{1}{2}$	0	0	0	10

XTC treatment was for 1 hour.
Puromycin (PMN) treatment was at 150 µg/ml for 30 minutes.
Elongation was scored by the ratio, at stage 11, of an embryo's longest dimension to its shortest diameter: +, 1.5-2.0; ++, 2.0-3.0; +++, greater than 3.0.

experiments and results differ in two respects from those of Godsave and Slack (1989, 1991) who described muscle cell differentiation in the progeny of single cells. First, the in vitro cultures used by these authors were initiated as single mid-blastula (stage 8) cells, each of which divided to form daughters, and hence a small group of cells; our experiments were initiated with stage 11 cells which divided little if at all. Secondly, Godsave and Slack cultured cells on a fibronectin- and laminin-coated substratum, and added gamma-globulin to the medium, procedures that might influence cell differentiation. In our experiments, implanted cells were surrounded by a normal ectodermal environment with which they are normally in contact.

Molecular events associated with the initiation of muscle actin gene transcription

For several years, our group has been analyzing the promoter of the *Xenopus* cardiac actin gene, which is strongly expressed in skeletal muscle, and, together with skeletal actin, is the first structural muscle protein gene to be transcribed in development. This gene is first expressed at the mid-gastrula stage (Mohun et al., 1984; Cascio and Gurdon, 1986), and we may ask whether any of the events associated with this might result from cell interactions during gastrulation.

Three regions of the cardiac actin promoter have been shown by deletion analysis to be of major importance in initiating its transcription during gastrulation (see review by Gurdon et al., 1992). One is an upstream GC-rich region about which very little is currently known. Another is a CArG box sequence [CC(6A or T)GG], of which there are four copies, but only the most 3′ of these centered at −85 is of key importance. Several proteins have been identified that bind to this sequence; most of these are present from fertilization onwards, and, subject to the possibility of secondary modifications, are available to bind to the actin promoter before gastrulation. These factors are reviewed by Mohun et al. (this volume).

The third significant part of the cardiac actin promoter is the M region (Taylor et al., 1991), which contains three copies of the E-box motif (CANNTG), to which proteins of the MyoD family can bind. In *Xenopus*, *XMyoD* and *XMyf5* genes are activated early in development, their mRNAs accumulating rapidly during early gastrulation, about two hours before the first appearance of transcripts of the cardiac actin gene (Hopwood et al., 1989, 1991; for review, which refers also to the work of other laboratories, see Gurdon et al., 1992). By the end of gastrulation, these mRNAs are restricted to the developing myotomes, in which the cardiac actin gene is also specifically expressed (Hopwood et al., 1989, 1991). XMyoD protein, which is known to accumulate only in myotomal nuclei (Hopwood et al., 1992), and XMyf5 protein, can bind specifically to the M region (Taylor et al., 1991). XMyoD and XMyf5 are therefore likely to be key factors in determining the specificity of cardiac actin gene transcription.

Further support for this view is provided by ectopic expression experiments, in which synthetic XMyoD or XMyf5 mRNA was microinjected into early embryos. Either kind of mRNA was able to activate transcription of the cardiac actin gene in animal cap ectoderm, which would not normally express it (Hopwood and Gurdon, 1990; Hopwood et al., 1991). Furthermore, we have recently found a threshold effect for the stable activation of muscle-specific gene expression in animal cap cells (Hopwood et al., unpublished). Below a threshold dose of XMyoD mRNA, transcription of the cardiac actin gene is only transient, but above it expression is sustained and later muscle markers are also activated. These results indicate that MyoD family members are sufficient to activate muscle-specific gene expression in normal embryonic cells, and that this activation might normally be stabilized by a threshold mechanism. However, parallel staining of XMyoD mRNA-injected animal caps and normal, uninjected future muscle cells using an anti-XMyoD monoclonal antibody showed that some injected animal cap cells contain, until the end of gastrulation, more XMyoD protein than the early muscle cells, without subsequently expressing muscle markers stably. For this reason, it is likely that other muscle-specific factors, not themselves activated by XMyoD, are required for normal muscle-specific gene activation. XMyf5 could have a significant additive, though not strongly synergistic effect (Hopwood et al., 1991), but there may be a role for other factors yet to be discovered.

For the purposes of the present article, we may point out that it is during neurulation (stages 13-18), that above-threshold animal caps have a normal myotomal concentration of XMyoD protein, and just subthreshold animal caps do not; before this stage, subthreshold animal caps have a higher concentration of XMyoD protein than normal myotome cells. One interpretation of this observation is that neurulation is the stage when cells sense, and respond to, the threshold concentration of XMyoD protein. It is also possible that the threshold concentration of XMyoD protein is sensed by animal cap cells at an earlier stage in development, when, it must also be supposed, they lack some other factor(s) needed by animal (but not future myotomal) cells to cooperate with XMyoD.

In summary, a considerable amount of information exists about genes and factors concerned with the initiation of muscle actin gene transcription. Although the association of factors with genes must involve intracellular processes, it is quite possible that the regulation of these events, such as the association of regulatory proteins with each other or with the muscle actin gene promoter, may be influenced by cell:cell interactions which take place during gastrulation.

Conclusion

The mesoderm of *Xenopus* embryos is believed to be formed as a result of a cell:cell interaction between inducing vegetal and responding animal cells. This interaction is completed by the end of the blastula stage. Since many of the responding genes, such as those that encode structural muscle proteins, do not start transcription at a significant rate until mid gastrulation or later, several events, yet to be identified, must take place between receipt of the vegetal inductive signal and the transcription of response genes. We have summarized here some of the reasons, which are still preliminary, for believing that further cell:cell interactions take place during gastrulation, and have an important role in connecting mesoderm induction with stable gene activation. These interactions might help to regulate some of the intracellular events involved in muscle gene activation, such as the association of myogenic proteins with a muscle gene promoter.

We thank J. C. Smith for a sample of XTC cell extract, and the Cancer Research Campaign for support of this work. The first author is also a member of Cambridge University Zoology Department.

References

Cascio, S. and Gurdon, J. B. (1986). The timing and specificity of actin gene activation in early *Xenopus* development. In *Molecular Approaches to Developmental Biology, UCLA Symposia on Molecular and Cellular Biology* 51. (ed. R. A. Firtel and E. H. Davidson). pp. 195-204. New York: Alan R. Liss, Inc.

Cascio, S. and Gurdon, J. B. (1987). The initiation of new gene transcription during *Xenopus* gastrulation requires immediately preceding protein synthesis. *Development* **100**, 297-305.

Cho, K. W. Y., Blumberg, B., Steinbeisser, H. and De Robertis, E. M. (1991). Molecular nature of Spemann's organizer: the role of the *Xenopus* homeobox gene *goosecoid*. *Cell* **67**, 1111-1120.

Frank, D. and Harland, R. M. (1991). Transient expression of XMyoD in non-somitic mesoderm of *Xenopus* gastrulae. *Development* **113**, 1387-1393.

Godsave, S. F. and Slack, J. M. W. (1989). Clonal analysis of mesoderm induction in *Xenopus laevis*. *Dev. Biol.* **134**, 486-490.

Godsave, S. F. and Slack, J. M. W. (1991). Single cell analysis of mesoderm formation in the *Xenopus* embryo. *Development* **111**, 523-530.

Green, J. B. A. and Smith, J. C. (1991). Growth factors as morphogens: do gradients and thresholds establish the body plan? *Trends in Genet.* **7**, 245-250.

Gurdon, J. B. (1988). A community effect in animal development. *Nature* **336**, 772-774.

Gurdon, J. B. (1989). From egg to embryo: the initiation of cell differentiation in amphibia. *Proc. Roy. Soc. Lond. B.* **237**, 11-25.

Gurdon, J. B., Brennan, S., Fairman, S. and Mohun, T. J. (1984). Transcription of muscle-specific actin genes in early *Xenopus* development: nuclear transplantation and cell dissociation. *Cell* **38**, 691-700.

Gurdon, J. B., Fairman, S., Mohun, T. J. and Brennan, S. (1985a). The activation of muscle-specific actin genes in *Xenopus* development by an induction between animal and vegetal cells of a blastula. *Cell* **41**, 913-922.

Gurdon, J. B., Mohun, T. J., Fairman, S. and Brennan, S. (1985b). All components required for the eventual activation of muscle-specific actin genes are localized in the subequatorial region of an uncleaved Amphibian egg. *Proc. Nat. Acad. Sci. USA* **82**, 139-142.

Gurdon, J. B., Hopwood, N. D. and Taylor, M. V. (1992). Myogenesis in *Xenopus* development. *Seminars in Developmental Biology*, **3**, 255-266.

Harvey, R. P. (1991). Widespread expression of MyoD genes in *Xenopus* embryos is amplified in presumptive muscle as a delayed response to mesoderm induction. *Proc. Nat. Acad. Sci. USA* **88**, 9198-9202.

Heasman, J., Snape, A., Smith, J. and Wylie, C. C. (1984). Fates and states of determination of single vegetal pole blastomeres of *Xenopus laevis*. *Cell* **37**, 185-194.

Hopwood, N. D., Pluck, A. and Gurdon, J. B. (1989). MyoD expression in the forming somites is an early response to mesoderm induction in *Xenopus* embryos. *EMBO J.* **8**, 3409-3417.

Hopwood, N. D., Pluck, A. and Gurdon, J. B. (1991). *Xenopus* Myf-5 marks early muscle cells and can activate muscle genes ectopically in early embryos. *Development* **111**, 551-560.

Hopwood, N. D., Pluck, A., Gurdon, J. B. and Dilworth, S. M. (1992). Expression of XMyoD protein in early *Xenopus laevis* embryos. *Development* **114**, 31-38.

Jessel, T. M. and Melton, D. A. (1992). Diffusible factors in Vertebrate embryonic induction. *Cell* **68**, 257-270.

Jones, E. A. and Woodland, H. R. (1987). The development of animal cap cells in *Xenopus*: a measure of the start of animal cap competence to form mesoderm. *Development* **101**, 557-563.

Keller, R. E. (1991). Early embryonic development of *Xenopus laevis*. In Xenopus laevis *Practical Uses in Cell and Molecular Biology*. (ed. B. Kay and B. Teng). *Methods in Cell Biol.* **36**, 61-113.

Mohun, T. J., Brennan, S., Dathan, N., Fairman, S. and Gurdon, J. B. (1984). Cell type-specific activation of actin genes in the early amphibian embryo. *Nature* **311**, 716-721.

Nakamura, O., Takasaki, H. and Ishihara, M. (1970). Formation of the organizer from combinations of presumptive ectoderm and endoderm. *Proc. Jap. Acad.* **47**, 313- 318.

Rosa, F. M. (1989). Mix1, a homeobox mRNA inducible by mesoderm inducers, is expressed in the presumptive endodermal cells of *Xenopus* embryos. *Cell* **57**, 967-974.

Rupp, R. A. W. and Weintraub, H. (1991). Ubiquitous MyoD transcription at the midblastula transition precedes induction-dependent MyoD expression in presumptive mesoderm of *X. laevis*. *Cell* **65**, 927-937.

Sargent, T. D., Jamrich, M. and Dawid, I. B. (1986). Cell interactions and the control of gene activity during early development of *Xenopus laevis*. *Dev. Biol.* **114**, 238-246.

Scales, J., Olson, E. and Perry, M. (1990). Two distinct *Xenopus* genes with homology to MyoD1 are expressed before somite formation in early embryogenesis. *Mol. Cell Biol.* **10**, 1516-1524.

Smith, J. C., Dale, L. and Slack, J. M. W. (1985). Cell lineage labels and region-specific markers in the analysis of inductive interactions. *J. Embryol. exp. Morph.* **89 Supplement**, 317-331.

Smith, J. C., Symes, K., Hynes, R. O. and DeSimone, D. (1990).

Mesoderm induction and the control of gastrulation in *Xenopus laevis*: the roles of fibronectin and integrins. *Development* **108**, 229-238.

Smith, J. C., Price, B. M. J., Green, J. B. A., Weigel, D. and Hermann, G. (1991). Expression of a *Xenopus* homologue of Brachyury (T) is an immediate early response to mesoderm induction. *Cell* **67**, 79-87.

Symes, K. and Smith, J. C. (1987). Gastrulation movements provide an early marker of mesoderm induction in *Xenopus laevis*. *Development* **101**, 339-349.

Taylor, M. V., Gurdon, J. B., Hopwood, N. D., Towers, N. and Mohun, T. J. (1991). *Xenopus* embryos contain a somite specific MyoD-like protein which binds to a promoter site required for muscle actin expression. *Genes Dev.* **5**, 1149-1160.

Development 1992 Supplement, 143-149 (1992)

Specification of the body plan during *Xenopus* gastrulation: dorsoventral and anteroposterior patterning of the mesoderm

J. M. W. SLACK, H. V. ISAACS, G. E. JOHNSON, L. A. LETTICE, D. TANNAHILL and J. THOMPSON

Imperial Cancer Research Fund Developmental Biology Unit, Department of Zoology, University of Oxford, South Parks Road, Oxford OX1 3PS, UK

Summary

Although the mesoderm itself is induced at the blastula stage, its subdivision mainly occurs in response to further inductive signals during gastrulation. In the late blastula, most of the mesoderm has a ventral-type commitment except for the small organizer region which extends about 30° on each side of the dorsal midline. During gastrulation, dorsal convergence movements bring the cells of the lateroventral marginal zone up near the dorsal midline and into the range of the dorsalizing signal emitted by the organizer. This dorsalizing signal operates throughout gastrulation, can cross a Nuclepore membrane, and is not mimicked by lithium, FGFs or activin.

Anteroposterior specification also takes place during gastrulation and is probably controlled by a dominant region at the posterior end of the forming axis.

We have studied the expression patterns in *Xenopus* of three members of the FGF family: bFGF, int-2 and a newly discovered species, eFGF. These all have mesoderm inducing activity on isolated animal caps, but are likely also to be involved with the later interactions. RNAase protections and in situ hybridizations show that the int-2 and eFGF mRNAs are concentrated at the posterior end, while bFGF is expressed as a posterior to anterior gradient from tailbud to head.

Studies of embryos in which bFGF is overexpressed from synthetic mRNA show that biological activity is far greater when a functional signal sequence is provided. This suggests that int-2 and eFGF, which possess signal sequences, are better candidates for inducing factors in vivo than is bFGF.

Key words: *Xenopus*, gastrulation, mesoderm induction, dorsoventral specification, anteroposterior specification, fibroblast growth factors, activins.

Introduction

Gastrulation is a time of extensive morphogenetic movements and in the vertebrate embryo it is also a time of extensive regional specification. The formation of the *Xenopus* body plan starts with cortical rotation in the egg and mesoderm induction in the blastula, but the main events of anteroposterior and dorsoventral specification both occur during gastrulation (Fig.1). So, in contrast to *Drosophila*, where the principal territories of the body plan are set up before gastrulation, in the vertebrates we have to understand the inductive interactions which bring about specification in the context of the simultaneous cell and tissue movements.

The main advance in the understanding of early *Xenopus* development in recent years has been the identification of a number of inducing factors belonging to the FGF, activin and wnt families. The FGFs and activins were first identified as mesoderm inducing factors, but we now think it is likely that they also have a role in the later interactions. In this paper we shall briefly review the embryology of dorsoventral and anteroposterior specification in *Xenopus* and consider which factors are candidates for which of the biological functions under consideration.

Dorsoventral specification

The initial step in dorsoventral (DV) specification is the cortical rotation which occurs following fertilization (reviewed by Gerhart et al., 1989). This is in some way necessary for the establishment of a "DV" centre, also called a "Nieuwkoop centre", in the dorsovegetal quadrant. During the blastula stages, the DV centre induces a small territory on its animal side to become the *organizer*, while around the remainder of the equatorial circumference a signal is emitted from the vegetal cells that induces a ring of "ventral-type" mesoderm from the equatorial part of the animal hemisphere (Gimlich, 1986; Dale and Slack 1987b; Stewart and Gerhart, 1990). Eggs that were irradiated with ultraviolet light shortly after fertilization do not undergo the cortical rotation and form no DV centre. They gastrulate in a radially symmetrical manner with the whole circumference of the marginal zone behaving like the ventral half of a normal embryo. Since they form abundant blood and loose mesenchyme around their circumference it seems that their ventral mesoderm inducing signal functions normally and is probably present around the whole circumference also in normal embryos. Although the regional specificity of mesoderm induction was originally thought to be vested entirely

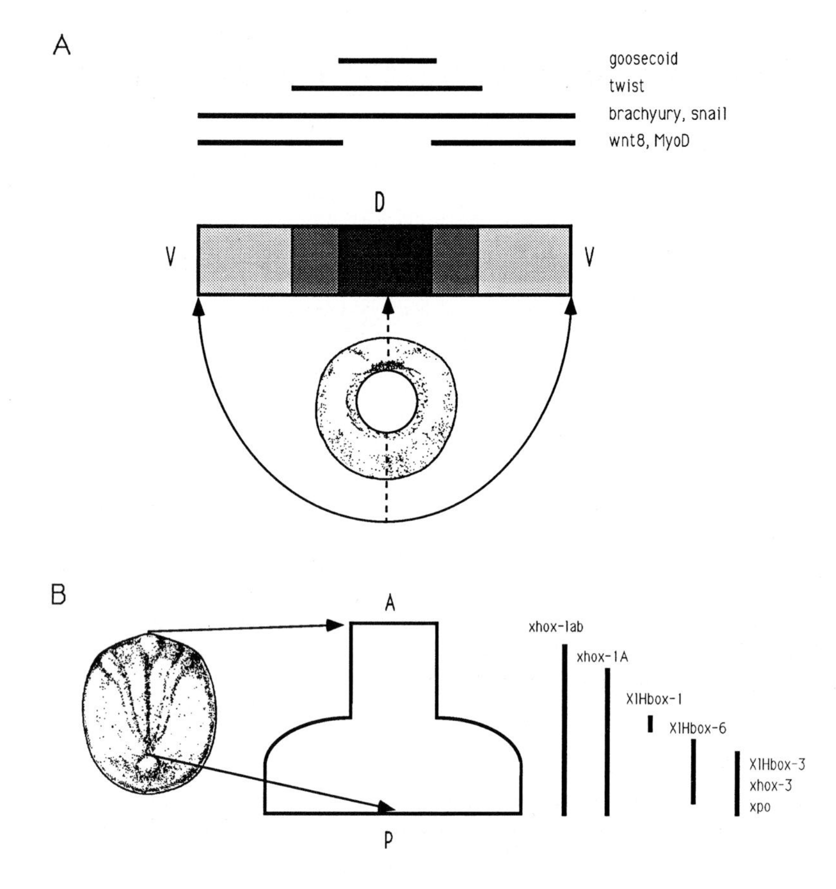

Fig. 1. Regionalisation of the mesoderm in the *Xenopus* gastrula. (A) Expression domains during gastrulation of genes with different dorsoventral domains. (B) Expression domains in the neurula of genes with different anteroposterior domains. Sample references to the expression patterns are as follows:
goosecoid: Cho et al. (1991a)
twist: Hopwood et al. (1989a)
brachyury: Smith et al. (1991)
snail: Sargent and Bennett (1990)
wnt-8: Christian et al. (1991); Smith and Harland (1991)
MyoD: Hopwood et al. (1989b); Frank and Harland (1991)
xhox-lab: Sive and Cheng (1991)
xhox-1A: Harvey et al. (1986)
XlHbox-1: Carrasco and Malacinski (1987); Oliver et al. (1988)
XlHbox-6:Sharpe et al. (1987)
XlHbox-3 (Xhox-36): Condie and Harland (1987)
xhox-3: Ruiz i Altaba et al. (1991)
xpo: Sato and Sargent (1991)

in the signal, it has been shown recently that there is also a difference of competence, especially of the region just above the equator on the dorsal side, which is more prone to become organizer than the rest of the animal hemisphere (Sokol and Melton, 1991).

So by the middle blastula stage the specification map of the embryos shows a small (60-90°) region forming notochord and a large (270-300°) region forming loose mesenchyme and blood cells. A specification map is compiled by explanting small pieces of tissue and allowing them to self differentiate in culture. It is a measure of the commitment achieved by the tissue up to the time of explantation. By contrast a fate map of an embryo is compiled by labelling each region in situ and allowing the intact embryo to develop further, and it shows what each region will become in normal undisturbed development. The fate map of the cleavage and blastula stages is not the same as the specification map. It shows that a substantial proportion of the axial tissues, for example about 60% of the myotomal muscle, is derived from the ventral half (Dale and Slack, 1987a). This means that much of the mesoderm which initially has a ventral character must be promoted to axial status at some later stage.

The available evidence suggests that this process, which we call *dorsalization*, occurs during gastrulation. Juxtaposition of dorsal and ventral tissues from early gastrulae has long been known to cause dorsalization of the ventral component, resulting in the formation of muscle masses and pronephric tubules instead of loose mesenchyme and blood cells (Slack and Forman, 1980; Dale and Slack, 1987b; Fig. 2). We have now shown, using equivalent combinations made at different stages throughout gastrulation, that dorsalization can occur at high frequency between stage 10 and stage 12. We have also carried out heterochronic combinations and it may be deduced from these that the dorsal midline tissue continues to emit a dorsalizing signal until after closure of the blastopore, while the competence of the ventral tissue to respond falls sharply after stage 12.

These experiments show that dorsalization *can* occur up until the end of gastrulation, but does it actually do so? The best evidence that it does (apart from the comparison between fate and specification maps referred to above) is obtained by looking at the relative sizes of axis and blood forming territories in normal embryos compared to embryos in which gastrulation movements have been inhibited. One way of inhibiting gastrulation movements is by injection with suramin (Gerhart et al., 1989) although other methods lead to the same result. In suramin treated embryos, not only is there a truncation at the anterior end, but in the trunk region the axis is much smaller and the blood forming territory much larger than usual (Fig.3). Since a treatment at the beginning of gastrulation can prevent dorsalization, the interaction must occur after this stage. We also know that competence of ventral tissue to become dorsalized is lost after stage 12 (see above) so it follows that dorsalization must normally be occurring during gastrulation. The movements of gastrulation involve massive dorsal convergence of marginal zone cells, which bring about three quarters of the mesoderm up into the axial region. This means that the signal may only need to have a very short range, of a few cell diameters, and it is presumably those cells that come close to the dorsal midline that receive the signal and

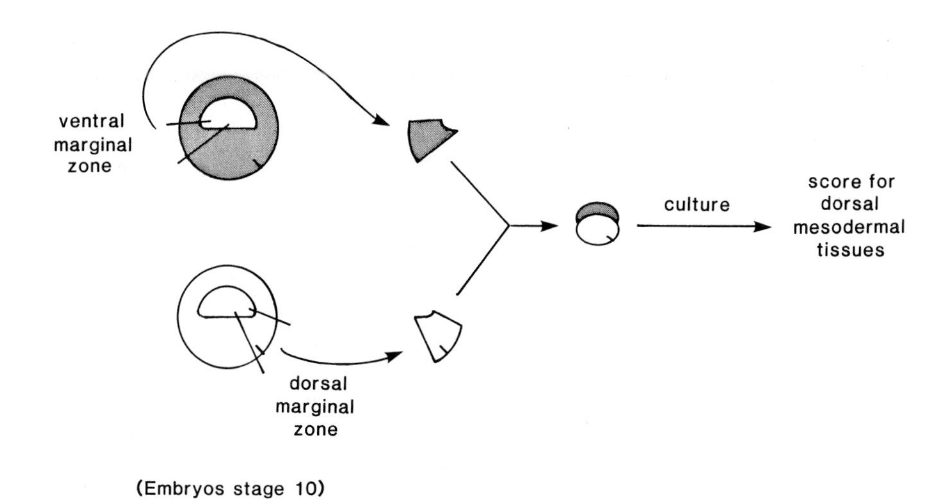

Fig. 2. Design for experiments on dorsalization of the mesoderm: FDA-labelled ventral marginal zone explants are combined with unlabelled dorsal explants and cultured for two days. The formation of labelled muscle blocks or pronephric tubules is indicative of dorsalization.

become dorsalized, while those that remain in the ventrolateral region do not see the signal and remain ventral in character.

We have recently examined the dorsalizing signal using a transfilter apparatus of the type described by Slack (1991). This shows that the signal can be transmitted across a liquid gap, but with lower efficiency than the mesoderm inducing signals. We have also tested a variety of cytokines, including activin A and FGFs, on gastrula stage ventral explants to see if any of them can provoke dorsalization, but none has done so, even when they have successfully brought about mesoderm induction in blastula stage animal caps treated simultaneously. Lithium ions can dorsalize ventral explants from the middle blastula as previously reported (Slack et al., 1988), but can no longer do so by the late blastula or gastrula stages. At present therefore we have no idea about the nature of the dorsalizing signal, except that it is unlikely to be a member of the FGF or activin classes.

Anteroposterior specification

There is a high degree of cellular intercalation occurring on the dorsal side during gastrulation (Wilson and Keller, 1991). Because of this any small group of cells in the dorsal lip region will become stretched out and scrambled by the cell mixing and so it seems unlikely that any anteroposterior levels could be specified before gastrulation. The evidence we have from several lines of work suggests that anteroposterior specification and gastrulation go hand in hand (reviewed by Gerhart, 1989; Slack and Tannahill, 1992).

In general, experiments in which anterior and posterior tissues are juxtaposed seem to result in the anterior member becoming posteriorized while the posterior member remains unchanged (reviewed by Slack and Tannahill, 1992). Much of the data underlying this statement are quite old, the experiments being performed on urodeles and without adequate labels to distinguish graft from host cells. Recently we have repeated some of these on axolotls using FDA-labelled grafts and the indication so far is that the results are indeed correct. For example if an early dorsal lip is grafted into the position of a late dorsal lip, it becomes integrated into the axial structures of the posterior trunk and tail (Fig.4). On the other hand, a late lip grafted into an early gastrula does not populate the head. It still populates the trunk and tail, and the head is left severely malformed. The idea of posterior dominance is consistent with the widespread belief that the genes of the Antennapedia-like homeobox clusters (HOX genes) are coding factors for different anteroposterior levels, since these genes are activated in a serial threshold arrangement, all being on at the posterior end and each territory in the posterior to anterior direction being specified by the loss of one more gene product. As to the actual mechanism of generation of a sequence of posterior to anterior states, two possibilities have been the subject of recent informal discussions, which can be called for short the "timing" model and the "signalling" model.

In the timing model, the early dorsal lip is seen as possessing an anterior specification and of acquiring a progressively more posterior character with time. For example some substance, M, could accumulate in the lip region and as its concentration rises so more and more posterior genes would be activated. In those cohorts of cells that involuted away from the lip, the altered environment of the embryo interior would stop this accumulation and "freeze" the tissue at that level of posterior specification achieved by the time of leaving the lip. The end result will be a series of territories arranged from anterior to posterior in positions which have an ordered sequence of anterior to posterior states of specification.

In the signalling model, the lip is seen as permanently posterior in character. It emits a morphogen, M, that forms a posterior to anterior gradient across the involuted tissue. Each of the AP coding genes is turned on at a differrent threshold concentration. Cohorts of cells respond to this signal as a function of distance and hence become progressively more anterior in character as they invaginate away from the lip. Their states do not become irreversibly determined until the end of gastrulation.

In both models we are obliged to assume that something changes on invagination as a result of exposure to the internal environment, which might mean exposure to the blastocoelic fluid, or contact with the blastocoel roof, or both. The main difference between the models is that in the second the anteroposterior level remains labile in the dorsal mesoderm for some time after invagination, whereas in the former it becomes fixed straight away. We feel that more embryological work is required to define this process more closely and that this is an essential adjunct to the molecular work to be described below.

"Mesoderm inducing" factors

When the first pure substances were shown to have mesoderm inducing activity, it was confidently expected that they would indeed be performing this task in vivo. However it has not yet been possible to prove that either bFGF or the activins definitely have a role in this process. No mRNA has been detected for activin A or B in early stage *Xenopus* embryos (Thomsen et al., 1990), although some activin-like protein has been found (Asashima et al., 1991). bFGF is expressed as mRNA and protein in the early embryo (Kimelman et al., 1988; Slack and Isaacs, 1989) but we have concluded after a series of overexpression experiments that it cannot be secreted from cells (Thompson and Slack, 1992). In these experiments, synthetic mRNA for bFGF is injected into fertilized eggs, they are allowed to develop to the blastula stage and then animal caps are explanted (Fig.5). Despite the synthesis of large amounts of bFGF protein, and its concentration in cell nuclei, only a limited degree of autoinduction is found in such caps. A similar study by Kimelman and Maas (1992) also showed only limited activity from large doses of RNA in comparable "ventral type" caps. However the biological activity goes up by over 100-fold if the bFGF is provided with a signal sequence from the immunoglobulin gene, or if another member of the family, which does possess a signal sequence, such as human kFGF, is used (Thompson and Slack, 1992).

In an attempt to test directly the role of activins and bFGF, a series of transfilter experiments were carried out in which the inducing and responding tissues were separated by an assembly of membranes about 100 μm wide (Slack, 1991). A high frequency of control inductions was obtained although these are of a ventral character, with a variable content of muscle, so it may be that the DV signal cannot cross the liquid gap. Inclusion in the liquid gap of high concentrations of follistatin, a naturally occurring inhibitor of activin, or of neutralizing antibody to bFGF, failed to inhibit the transfilter inductions. This suggests that these substances are not the factors released from the vegetal cells, although they may still have a role at a subsequent stage of mesoderm induction within the responding tissue.

Because bFGF seemed not to be secreted from the vegetal cells, and indeed the overexpression experiments suggest that it cannot be secreted at all, we turned our attention away from bFGF and towards a search, in *Xenopus*, for other members of the FGF family, as will be described below.

If the activins and bFGF are not normally secreted from vegetal cells, why do they show activity when applied to animal caps? The explanation probably lies in the fact that the receptors are maternally coded and are present in an active form on the cell surfaces of the early stages (Gillespie et al., 1989; Musci et al., 1990; Friesel and Dawid,

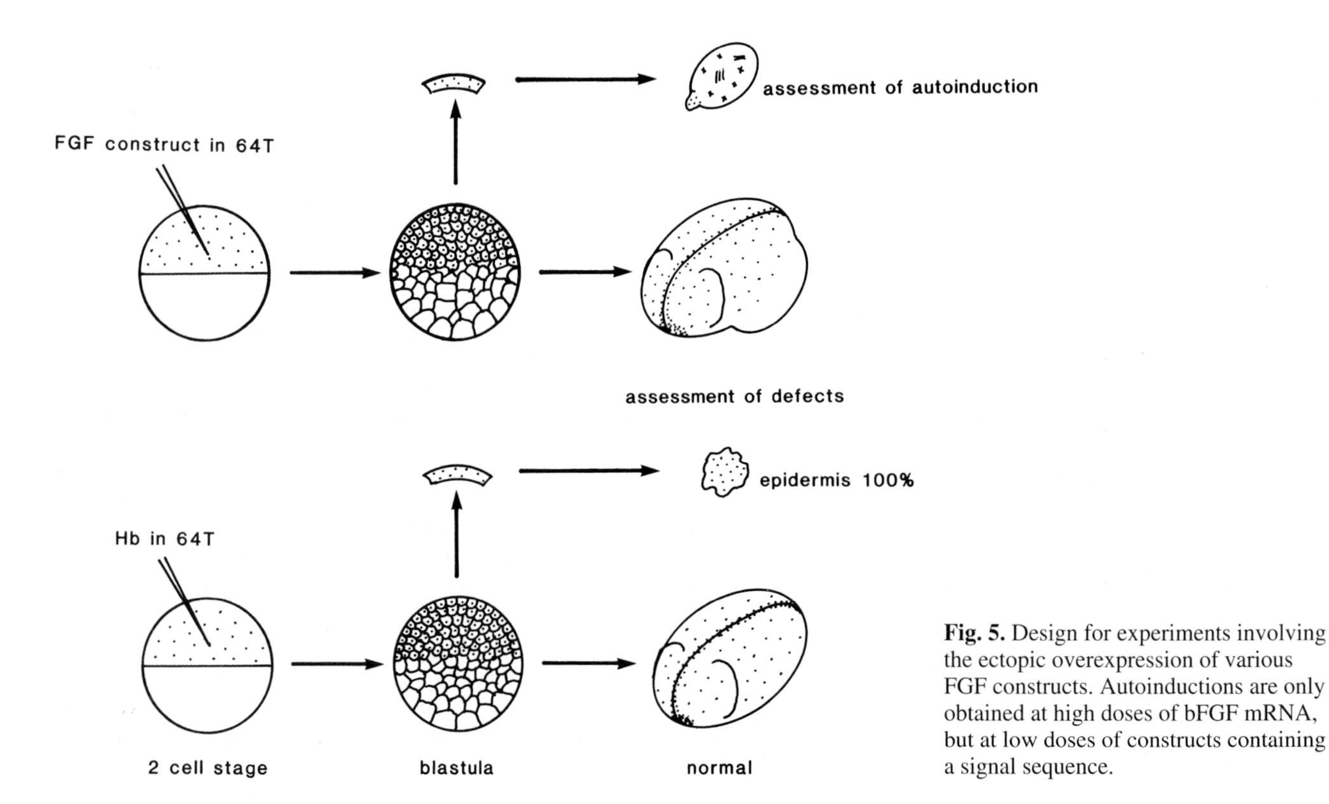

Fig. 5. Design for experiments involving the ectopic overexpression of various FGF constructs. Autoinductions are only obtained at high doses of bFGF mRNA, but at low doses of constructs containing a signal sequence.

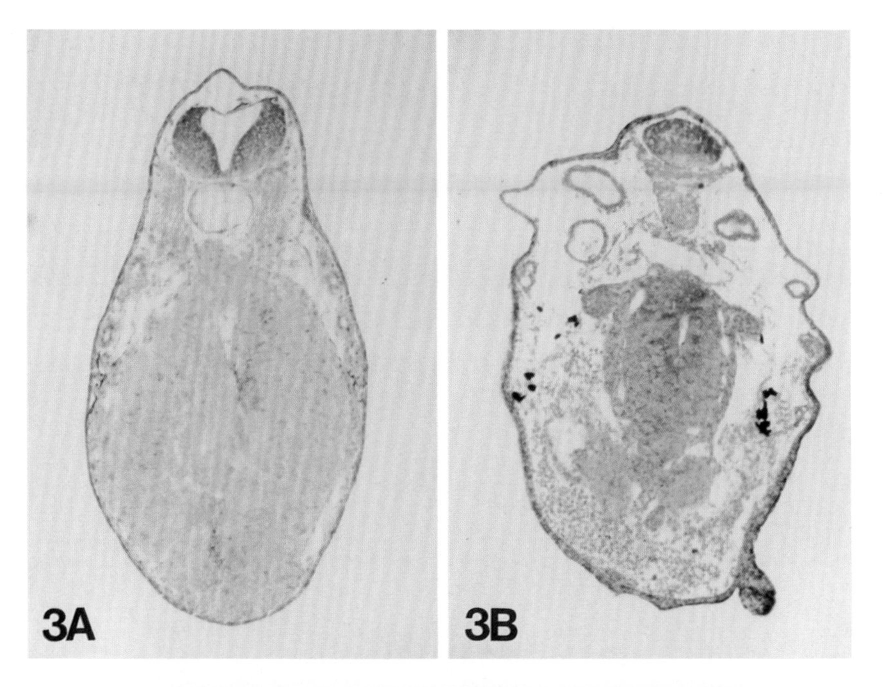

Fig. 3. Effect on dorsoventral proportions of blocking cell movements during gastrulation. (A) Normal embryo, TS through the level of the pronephros. (B) An embryo injected with suramin at the early gastrula stage. The proportion of blood cells relative to axial mesoderm is very considerably increased.

Fig. 4. Effect of grafting the dorsal lip of an early gastrula to the dorsal lip of a late gastrula. This experiment was performed on axolotl embryos and the graft was FDA labelled. The graft becomes integrated into the tissues of the posterior axis.

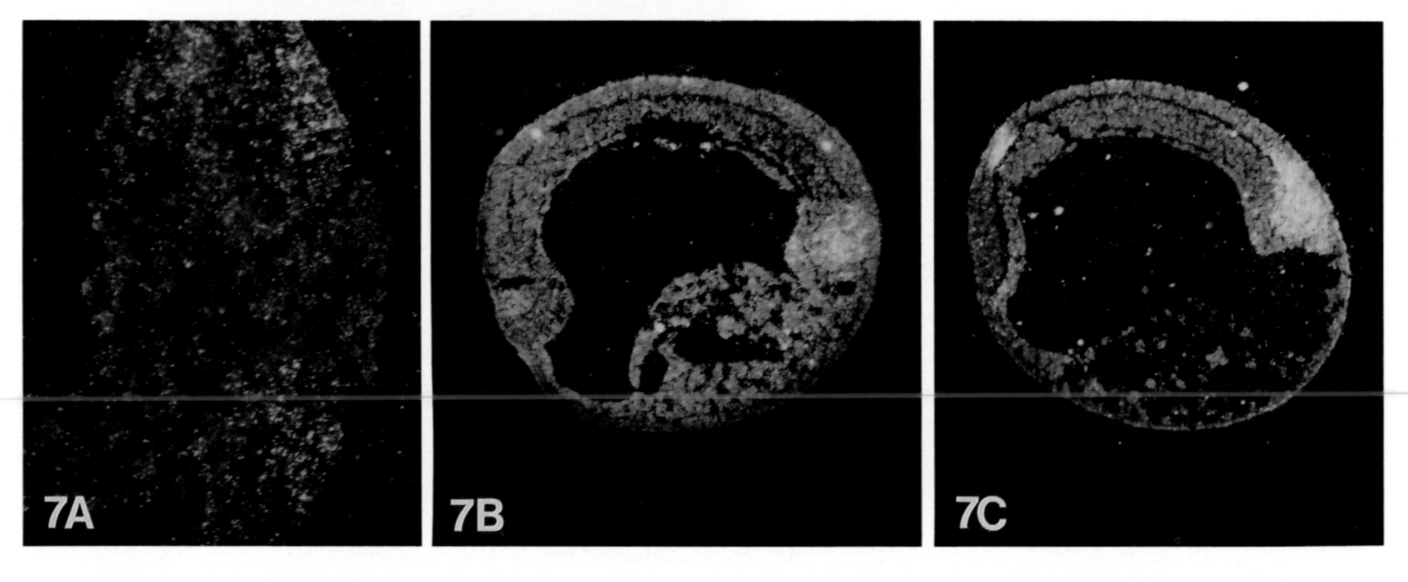

Fig. 7. In situ hybridization of endogenous mesoderm inducing factors. Anterior lies to the left. (A) eFGF in the middle gastrula stage showing activity in the mesoderm near the blastopore. (B) eFGF in the early neurula, parasagittal section, showing activity in the posterior mesoderm. (C) int-2 in the early neurula showing activity in the posterior mesoderm and in the anterior of the neural plate.

1991; Kondo et al., 1991). We know that the number of second messages in the signal transduction pathways is quite limited and so if a stimulus is received by blastula cells that elevates (say) tyrosine kinase activity then a whole cascade of protein phosphorylations will be started which will push the cells down the mesoderm pathway (Whitman and Melton, 1989; Gillespie et al., 1992). The FGF receptors are all tyrosine kinases and the activin receptors are SerThr kinases, so stimulation by the two classes of factor is likely to bring about distinct but overlapping patterns of protein phosphorylation. This may account for distinct but overlapping biological response, briefly summarised by saying that activin gives dorsal type inductions while FGF gives ventral type inductions (Smith, 1989; Green et al., 1990).

There is now reasonable agreement on the criteria that need to be satisfied to identify the true mesoderm inducing factor or factors. Firstly they must be expressed at the right stage and in sufficient quantity. Secondly the purified protein must show the expected biological activity. Thirdly inhibition of the factor should cause inhibition of the process in vivo. These have not yet been satisfied for bFGF or activin, although are closely approached for eFGF (see below). Recently it has been shown that synthetic mRNA from the *wnt-8* gene will mimic the DV signal if injected into vegetal blastomeres (Sokol et al., 1991; Smith and Harland, 1991). The *wnt-8* gene itself is not expressed until gastrulation, and then in the ventrolateral part of the marginal zone, so it cannot be regarded as a credible candidate itself. Even if there are maternally coded *wnt* mRNAs with similar activity, more experiments would obviously need to be done to satisfy the conditions listed. A similar list of criteria would hold for any putative inductive signal involved in later events.

Expression patterns of active factors

Regardless of the exact role that activins and FGFs ultimately turn out to have with regard to mesoderm induction, they are quite likely also to be involved in later inductive interactions necessary to establish the body plan, such as those described above. In order to make a reasonable guess about their functions the first step is to establish the developmental expression pattern of each factor. We have in our laboratory attempted to do this using RNAase protections and in situ hybridizations to detect mRNA. We have studied four factors. Firstly there is bFGF, the prototype member of the FGF family amd the one first shown to be a mesoderm inducing factor (Slack et al., 1987). Then there is int-2, originally discovered as an insertion site for murine mammary tumour virus and later shown to have mesoderm inducing activity (Paterno et al., 1989). Thirdly, there is eFGF, which was cloned in our laboratory as part of a search for potentially secretable FGFs (Isaacs et al., 1992). It is a molecule about equally similar to human kFGF and FGF-6, it has a signal sequence and protein expressed in bacteria has mesoderm inducing activity. Fourthly we have also included activin B, originally cloned by Thomsen et al. (1990), which is very similar in biological activity to activin A (Smith et al., 1990).

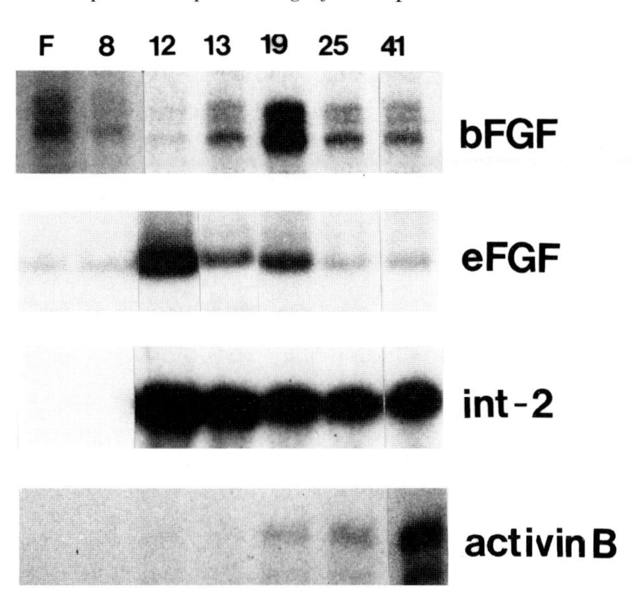

Fig. 6. Developmental time courses of four endogenous mesoderm inducing factors: bFGF, eFGF, int-2 and activin B. The figure shows parallel RNAase protections at different stages. The RNA loading was 20 µg for the FGFs and 40 µg for the activin B.

In whole embryos bFGF and eFGF are both expressed maternally although at low levels. Int-2 and eFGF (zygotic) come on in the early gastrula, bFGF (zygotic) in the early neurula, and activin B in the late neurula (Fig.6). Dissections of blastulae show that neither bFGF nor eFGF are localized at this stage. In the gastrula, both eFGF and int-2 are expressed first in the blastopore lip region (Fig.7). They remain on in the blastopore lip but not in the invaginated mesoderm as gastrulation proceeds, suggesting that they must be turned off in the mesoderm that has migrated in away from the lip. Int-2 also comes on as a patch in the prospective mid/hindbrain region of the forming neural plate, and we have shown that this is an early response to neural induction (Tannahill et al., 1992). In the neurula and tailbud stages both factors show a sharp restriction to the extreme posterior of the mesoderm, later becoming the tailbud.

The zygotic expression of bFGF and activin B both commence after the end of gastrulation. bFGF is expressed preferentially in the posterior but much less sharply than eFGF and int-2, so there is a gradient from the tail to the head end (Fig.8A). Activin B on the other hand is initially more abundant at the head end and later is expressed also in the posterior (Fig.8B).

Evidence for a role in AP or DV patterning

Several studies in the last few years have implicated the FGFs and activins in anteroposterior specification. However there have been some problems of interpretation because the experiments do not involve direct respecification of gastrula tissues. Ruiz i Altaba and Melton (1989a) have examined the behaviour of animal caps, induced with FGF or activin, and then implanted into early gastrulae by the "Ein-

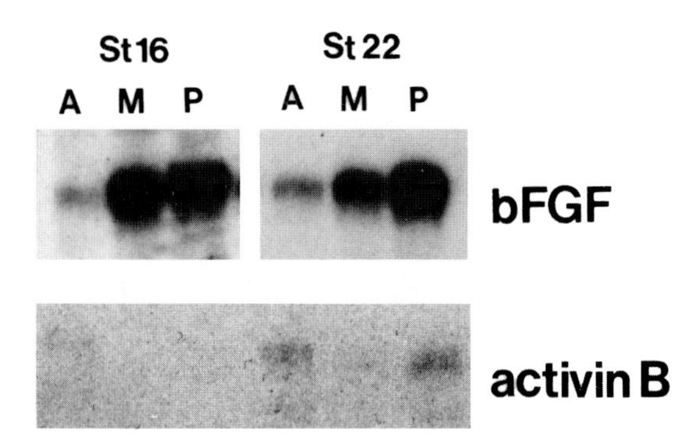

Fig. 8. Anterior-posterior distribution of messenger RNA for bFGF and activin B. RNAase protections are shown of anterior, middle and posterior thirds from the stages indicated. 20 μg of RNA was used for each lane.

steckung" procedure. Activin-treated caps tend to induce heads while FGF-treated caps tend to induce tails. We have always been puzzled by the fact that isolated FGF-treated caps develop into ventral structures, wheras posterior ones were obtained in these Einsteck experiments. It now seems that the difference lies in the exposure to additional signals after implantation. We have found that when ventral marginal zone (VMZ) explants are implanted into embryos they cause the formation of extra tails, although in isolation, like FGF induced caps, they form ventral tissues. Ruiz i Altaba and Melton have also studied a gene called *Xhox-3*, which is an *eve* type homeobox gene expressed in the posterior mesoderm (Ruiz i Altaba and Melton, 1989b; Ruiz i Altaba et al., 1991). It is turned on to a higher degree in animal caps by FGF than by activin, overexpression causes reduction of the head, and injection of an antibody causes, in a proportion of cases, defects in the posterior.

Cho and De Robertis (1990) have investigated the activation of HOX cluster genes in animal caps treated with activin or FGF. *XlHbox1*, normally expressed in the anterior trunk region, is preferentially activated by activin, and XlHbox6, normally expressed in the mid- and hind-trunk region, is preferentially activated by FGF. In a further study, Cho et al. (1991a) found that overexpression of *XlHbox6* alone was sufficient to cause supernumerary tail formation after subsequent Einsteckung procedure. It could even override the head forming effect of activin induction.

In both these sets of experiments the assay is rather indirect since the formation of axial structures in Einsteckung experiments involves participation by both the graft and the host tissue, and requires some further inductive signals from the host. However for both *XlHbox6* and *Xhox-3* there is a *prima facie* case that the genes are involved in anteroposterior specification. In neither case does overexpression in isolated caps result in mesoderm formation so they seem to be controlling positional coding rather than tissue type. Both are activated by FGF but we do not know how direct this relationship is, and it is possible that there are other intervening genes to be discovered.

An experiment that directly addresses the role of the FGF family was performed by Amaya et al. (1991). They made a version of the FGF receptor lacking the cytoplasmic domain. This forms unproductive dimers with the endogenous receptor and prevents response to exogenous FGF. Overexpression of this construct in intact embryos should theoretically lead to failure of those processes that depend on any member of the FGF family. Although a detailed study of the morphology of the affected embryos has not yet been published, it seems that they have less mesoderm than expected, the dorsoventral arrangement of tissues is deranged, and they may lack posterior parts of the axis.

If we put together the expression patterns reported above with the functional experiments reviewed in the present section, it does look rather probable that, in addition to their involvement in mesoderm induction, an important role of the FGF family is concerned with anteroposterior specification. One possibility is that the FGFs are posterior morphogens and that a gradient is set up by secretion from the blastopore lip region. *XlHbox6* and *Xhox3* and probably other genes as well would then be turned on above a certain concentration threshold and they would be responsible for activating the appropriate terminal differentiation genes for the trunk and tail. Another possibility is that the FGF type factors are permissive for the maintenance of an uncommitted state, and that once cells leave the blastopore lip region, or later the tailbud, they become specified by reference to other signals. In either case it is likely that we shall have to contend with a certain redundancy of function since at least eFGF and int-2, which have overlapping biological activities, are expressed in a very similar tight zone in the posterior.

References

Amaya, E., Musci, T. J. and Kirschner, M. W. (1991). Expression of a dominant negative mutant of the FGF receptor disrupts mesoderm formation in Xenopus embryos. *Cell* **66**, 257-270.

Asashima, M., Nakano, H., Uchiyama, H., Sugino, H., Nakamura, T., Eto, Y., Ejima, D., Nishimatsu, S., Ueno, N. and Kinoshita, K. (1991). Presence of activin (erythroid differentiation factor) in unfertilized eggs and blastulae of Xenopus laevis. *Proc. Natl. Acad. Sci. USA* **88**, 6511-6514.

Carrasco A. E. and Malacinsk, J. M. (1987). Localization of Xenopus homeobox gene transcripts during embryogenesis and in the adult nervous system. *Dev. Biol.* **121**, 69-81.

Cho, K. W. Y., Blumberg, B., Steinbeisser, H. and de Robertis, E. M. (1991a). Molecular nature of Spemann's organizer: the role of the Xenopus homeobox gene goosecoid. *Cell* **67**, 1111-1120.

Cho, K. W. and De Robertis, E. M. (1990). Differential activation of Xenopus homeo box genes by mesoderm-inducing growth factors and retinoic acid. *Genes Dev.* **4**, 1910-6.

Cho, K. W., Morita, E. A., Wright, C. V. and De Robertis, E. M. (1991b). Overexpression of a homeodomain protein confers axis-forming activity to uncommitted Xenopus embryonic cells. *Cell* **65**, 55-64.

Christian, J. L., McMahon, J. A., McMahon, A. P. and Moon, R. A. (1991). *Xwnt-8*, a *Xenopus Wnt-1/int-1* related gene responsive to mesoderm inducing growth factors may play a role in ventral mesodermal patterning during embryogenesis. *Development* **111**, 1045-1055.

Condie, B. G. and Harland, R. M. (1987). Posterior expression of a homeobox gene in early *Xenopus* embryos. *Development* **101**, 93-105.

Dale, L. and Slack, J. M. W. (1987a). Fate map for the 32 cell stage of *Xenopus laevis*. *Development* **99**, 527-551.

Dale, L. and Slack, J. M. W. (1987b). Regional specification within the mesoderm of early embryos of *Xenopus laevis*. *Development* **100**, 279-295.

Frank, D. and Harland, R. M. (1991). Transient expression of XMyoD in non-somitic mesoderm of *Xenopus* gastrulae. *Development* **113**, 1387-1393.

Friesel, R. and Dawid, I. B. (1991). cDNA cloning and developmental expression of fibroblast growth factor receptors from *Xenopus laevis*. *Mol. Cell. Biol.* **11**, 2481-2488.

Gerhart, J., Danilchik, M., Doniach, T., Roberts, S., Rowning, B., and Stewart, R. (1989) Cortical rotation of the *Xenopus* egg: consequences for the anteroposterior pattern of embryonic dorsal development. *Development* **107 Supplement** 37-51.

Gillespie, L. L., Paterno, G. D. and Slack, J. M. W. (1989). Analysis of competence: Receptors for fibroblast growth factor in early *Xenopus* embryos. *Development* **106**, 203-208.

Gillespie, L. L., Paterno, G. D., Mahadevan, L. C. and Slack, J. M. W. (1992). Intracellular signalling pathways involved in mesoderm induction by FGF. *Mech. Dev.* in press.

Gimlich, R. L. (1986). Acquisition of developmental autonomy in the equatorial region of the Xenopus embryo. *Dev. Biol.* **115**, 340-352.

Green, J. B. A., Howes, G., Symes, K., Cooke, J. and Smith, J. C. (1990). The biological effects of XTC-MIF: quantitative comparison with *Xenopus* bFGF. *Development* **108**, 173-183.

Harvey, R. P., Tabin, C. J. and Melton, D. A. (1986). Embryonic expression and nuclear localization of Xenopus homeobox (Xhox) gene products. *EMBO J.* **5**, 1237-1244.

Hopwood, N. D., Pluck, A. and Gurdon J. B. (1989a). A Xenopus mRNA related to Drosophila twist is expressed in response to induction in the mesoderm and the neural crest. *Cell*, **59**, 893-903.

Hopwood, N. D., Pluck, A. and Gurdon J. B. (1989b). Myo D expression in the forming somites is an early response to mesoderm induction in Xenopus embryos. *EMBO J.* **8**, 3409-3417.

Isaacs, H. V., Tannahill, D. and Slack, J. M. W. (1992). Expression of a novel FGF in the *Xenopus* embryo. A new candidate inducing factor for mesoderm formation and anteroposterior specification. *Development* **114**, 711-720.

Kimelman, D., Abraham, J. A., Haaparanta, T., Palisi, T. M. and Kirschner, M. W. (1988). The presence of fibroblast growth factor in the frog egg: its role as a natural mesoderm inducer. *Science* **242**, 1053-1056.

Kimelman, D. and Maas, A. (1992). Induction of dorsal and ventral mesoderm by ectopically expressed *Xenopus* basic fibroblast growth factor. *Development*, **114**, 261-269.

Kondo, M., Tashiro, K. Fujii, K., Asano, M. Miyoshi, R., Yamada, R., Muramatsu, . and Shiokawa, K. (1991). Activin receptor mRNA is expressed early in Xenopus embryogenesis and the level of the expression affects body axis formation. *Biochem. Biophys. Res. Comm.* **181**, 684-690.

Musci, T. J., Amaya, E. and Kirschner, M. W. (1990). Regulation of the fibroblast growth factor receptor in early Xenopus embryos. *Proc. Natl. Acad. Sci. USA* **87**, 8365-8369.

Oliver, G., Wright, C. V., Hardwicke, J. and De Robertis, E. M. (1988). Differential antero-posterior expression of two proteins encoded by a homeobox gene in Xenopus and mouse embryos. *EMBO J.* **7**, 3199-3209.

Paterno G. D., Gillespie, L. L., Dixon, M. S., Slack J. M. W. and Heath, J. K. (1989). Mesoderm inducing properties of int-2 and kFGF: two oncogene encoded growth factors related to FGF. *Development* **106**, 79-83.

Ruiz i Altaba, A. and Melton, D. A. (1989a). Bimodal and graded expression of the *Xenopus* homeobox gene Xhox3 during embryonic development. *Development* **106**, 173-83.

Ruiz i Altaba, A. and Melton, D. A. (1989b). Interaction between peptide growth factors and homoeobox genes in the establishment of antero-posterior polarity in frog embryos. *Nature* **341**, 33-38.

Ruiz i Altaba, A., Choi, T. and Melton, D. A. (1991). Expression of the xhox3 homeobox protein in Xenopus embryos: blocking its early function suggests the requirement of xhox3 for normal posterior development. *Dev. Growth Diffn.* **33**, 651-669.

Sargent, M. G. and Bennett, M. F. (1990). Identification in *Xenopus* of a structural homologue of the *Drosophila* gene *snail*. *Development*, **109**, 967-973.

Sato, S. M. and Sargent, T. D. (1991). Localized and inducible expression of *Xenopus* posterior (Xpo), a novel gene active in early frog embryos, encoding a protein with a CCHC finger domain. *Development* **112**, 747-753.

Sive, H. L. and Cheng, P. F. (1991). Retinoic acid perturbs the expression of xhox.lab genes and alters mesodermal determination in Xenopus laevis. *Genes and Dev.* **5**, 1321-1332.

Sharpe, C. R., Fritz, A., De Robertis, E. M. and Gurdon, J. B. (1987). A homeobox-containing marker of posterior neural differentiation shows the importance of predetermination in neural induction. *Cell* **50**, 749-758.

Slack, J. M. W. (1991). The nature of the mesoderm inducing signal in *Xenopus*: a transfilter induction study. *Development* **113**, 661-671.

Slack, J. M. W., Darlington, B. G., Heath, J. K. and Godsave, S. F. (1987). Mesoderm induction in early Xenopus embryos by heparin-binding growth factors. *Nature* **326**, 197-200.

Slack, J. M. W. and Forman, D. (1980). An interaction between dorsal and ventral regions of the marginal zone in early amphibian embryos. *J. Embryol. Exp. Morph.* **56**, 283-299.

Slack, J. M. W. and Isaacs, H. V. (1989). Presence of basic fibroblast growth factor in the early *Xenopus* embryo. *Development* **105**, 147-154.

Slack, J. M. W., Isaacs, H. V., and Darlington, B. G. (1988). Inductive effects of fibroblast growth factor and lithium ion on *Xenopus* blastula ectoderm. *Development* **103**, 581-590.

Slack, J. M. W. and Tannahill, D. (1992). Mechanism of anteroposterior axis specification in vertebrates. Lessons from the amphibians. *Development* **114**, 285-302.

Smith, J. C. (1989). Mesoderm induction and mesoderm-inducing factors in early amphibian development. *Development* **105**, 665-677.

Smith, J. C., Price, B. M. J., Van Nimmen, K. and Huylebroek, D. (1990). Identification of a potent Xenopus mesoderm inducing factor as a homologue of activin A. *Nature* **345**, 729-731.

Smith, J. C., Price, B. M. J., Green, J. B. A., Weigel, D. and Herrman, B. G. (1991). Expression of a Xenopus homolog of Brachyury (T) is an immediate early response to mesoderm induction. *Cell* **67**, 79-87.

Smith, W. C. and Harland, R. M. (1991). Injected X-Wnt 8 RNA acts early in Xenopus embryos to promote formation of a vegetal dorsalizing centre. *Cell* **67**, 753-766.

Sokol, S., Christian, J. L., Moon, R. T. and Melton, D. A. (1991). Injected Wnt RNA induces a complete body axis in Xenopus embryos. *Cell* **67**, 741-752.

Sokol, S. and Melton, D. A. (1991). Pre-existent pattern in Xenopus animal pole cells revealed by induction with activin. *Nature* **351**, 409-411.

Stewart, R. M. and Gerhart, J. C. (1990). The anterior extent of dorsal development of the *Xenopus* embryonic axis depends on the quantity of organizer in the late blastula. *Development* **109**, 363-372.

Tannahill, D. Isaacs, H. V., Close, M. J., Peters, G. and Slack, J. M. W. (1992). Developmental expression of the Xenopus int-2 (FGF-3) gene: activation by mesodermal and neural induction. *Development* **115**, 695-702.

Thomsen, G., Woolf, T., Whitman, M., Sokol, S., Vaughan, J., Vale, W. and Melton, D. A. (1990). Activins are expressed early in Xenopus embryogenesis and can induce axial mesoderm and anterior structures. *Cell* **63**, 485-493.

Thompson, J. and Slack, J. M. W. (1992). Overexpression of fibroblast growth factors in Xenopus embryos. *Mech. Dev.* (in press).

Whitman, M. and Melton, D. A. (1989). Induction of mesoderm by a viral oncogene in early Xenopus embryos. *Science* **244**, 803-806.

Wilson, P. and Keller, R. (1991). Cell rearrangement during gastrulation of *Xenopus*: direct observation of cultured explants. *Development* **112**, 289-300.

Development 1992 Supplement, 151-156 (1992)
Printed in Great Britain © The Company of Biologists Limited 1992

Relationships between mesoderm induction and the embryonic axes in chick and frog embryos

CLAUDIO D. STERN, YOHKO HATADA, MARK A. J. SELLECK* and KATE G. STOREY

Department of Human Anatomy, South Parks Road, Oxford OX1 3QX, UK

*Present address: Developmental Biology Center, University of California at Irvine, Irvine, California 92717, USA

Summary

The hypoblast is generally thought to be responsible for inducing the mesoderm in the chick embryo because the primitive streak, and subsequently the embryonic axis, form according to the orientation of the hypoblast. However, some cells become specified as embryonic mesoderm very late in development, towards the end of the gastrulation period and long after the hypoblast has left the embryonic region. We argue that induction of embryonic mesoderm and of the embryonic axis are different and separable events, both in amniotes and in amphibians. We also consider the relationships between the dorsoventral and anteroposterior axes in both groups of vertebrates.

Key words: chick embryo, amphibian embryo, pattern formation, mesoderm induction, embryonic axis, Hensen's node, stem cells.

Introduction

Establishment of the basic body plan in vertebrate embryos depends on two distinct, but interrelated processes. One is gastrulation, whereby the three germ layers and embryonic axes become established. The other is the commitment of cells to mesoderm and the subsequent differentiation of these cells into various mesodermal cell types. In Amphibians, the mesoderm of the early embryo is determined by an inductive interaction between vegetal cells (a transient population of yolky cells; see Hadorn, 1970) and the ectoderm of the marginal zone (Nieuwkoop, 1969, 1985; see Green and Smith, 1991; Slack et al., 1992; Smith and Howard, 1992 for reviews). Certain peptide growth factors are able to replace the vegetal cells in in vitro assays (the 'animal cap assay'; see Green and Smith, 1991): when an isolated piece of animal cap ectoderm is treated with appropriate concentrations of certain members of the FGF or TGFβ families of growth factors, mesodermal cell types differentiate. The higher the concentration of growth factor, the more 'dorsal/axial' the cell type formed. The highest concentrations of activin are able to generate notochord and cells that have organising activity. At these high concentrations, blastopore-specific genes like *Brachyury* (*XBra*) (Smith et al., 1991), *goosecoid* (Cho et al., 1991; Blum et al., 1992; De Robertis et al., 1992), *XFKH1* (Dirksen and Jamrich, 1992) and perhaps *Xlim1* (Taira et al., 1992) are expressed (see Smith and Howard, 1992). Lower concentrations of activin produce muscle and induce α-actin expression. FGF-related growth factors also induce mesoderm in a concentration-dependent manner: low concentrations give blood, mesenchyme and endothelium, higher concentrations give muscle and induce expression of α-actin (see Slack and Tannahill, 1992). These results are generally interpreted to mean that the vegetal cells produce growth factors related to the FGF and TGFβ families, which induce the marginal zone ectoderm cells to become mesodermal. In support of this hypothesis, Asashima et al. (1991) have recently described the presence of maternally derived activin-related activities in the egg and early embryo, and Isaacs et al. (1992) have reported the presence of a member of the FGF family (XeFGF) in the early embryo.

In the chick embryo, mesoderm is thought to arise as a result of a similar interaction, between the hypoblast and the overlying epiblast (Waddington, 1933, repeated and confirmed by Azar and Eyal-Giladi, 1981). When the hypoblast (consisting of yolky entodermal cells that do not contribute to the embryo proper) is rotated with respect to the overlying epiblast, the primitive streak forms according to the orientation of the hypoblast. However, as in the frog (Sokol and Melton, 1991), the epiblast of the chick also has its own polarity: if the hypoblast is dissociated into single cells, and then reaggregated into a sheet before being combined with intact epiblast, a primitive streak forms according to the original orientation of the epiblast (Mitrani and Eyal-Giladi, 1981).

Mitrani and his colleagues have argued that the hypoblast can be replaced by activin, similar to the case in *Xenopus*. If the centre of the embryo is deprived of both hypoblast and marginal zone and incubated in the presence of activin, a normal embryonic axis, containing notochord and somites, develops (Mitrani and Shimoni, 1990; Mitrani et al., 1990b). Mitrani et al. (1990b) conclude that the hypoblast may be the endogenous source of activin in the embryo and that it may secrete this factor in a graded way, with the highest concentrations being emitted at its posterior midline. However, although it is clear from these experiments and others (e.g. Mitrani et al., 1990a; Cooke and Wong, 1991) that the chick epiblast **can** respond to activin and FGF, there is as yet no direct evidence that these molecules can act as true mesoderm-inducing factors on the chick epiblast (see Slack, 1991). One reason for this is that mesodermal differentiation of chick epiblast cells cannot yet be assessed independently from formation of an embryonic axis.

Induction of the mesoderm and its dorso-ventral subdivision still continue throughout gastrulation

In *Xenopus*, it has been suggested that mesodermal cell diversity is generated prior to the establishment of an overt body plan. Different mesoderm types are segregated, such that there is a transition from dorsal mesoderm (e.g. notochord) to ventral (blood, endothelium) across the marginal zone. The 'three signal model' (Slack et al., 1984), proposed to explain both the induction of mesoderm and its subdivision into different dorsoventral cell types, suggests that a ventral vegetal (VV) signal (perhaps FGF) instructs ventral marginal zone cells to become mesodermal. A second, dorsal vegetal (DV) signal, emanating from a small group of cells (the 'Nieuwkoop centre') in the most dorsal part of the vegetal hemisphere, instructs the most dorsal marginal zone cells (the site of the future dorsal lip of the blastopore) to become 'Spemann organizer' cells. These cells would emit an organizing (O) signal, which subdivides the marginal zone mesodermal belt into different dorsoventral cell types, according to their distance from the Spemann organizer. Later in development, the Spemann organizer cells emit neural inducing signal(s), instructing the neighbouring animal cap ectoderm to become neural.

The competence of *Xenopus* animal caps to respond both to vegetal cells and to mesoderm-inducing factors like FGF and activin declines at the beginning of gastrulation (stage 9-10; see Gurdon, 1987; Green and Smith, 1991; Slack and Tannahill, 1992). However, at least some individual cells in the amphibian dorsal lip (Delarue et al., 1992) and chick Hensen's node (Selleck and Stern, 1991) still give rise to progeny that are located in both ectoderm and mesendoderm at the end of gastrulation. Clearly, these cells cannot have been induced to form mesoderm before gastrulation. Therefore, some mechanisms inducing mesoderm must still operate at the end of the gastrula stage, even though the competence of cells to respond to known mesoderm-inducing factors has all but disappeared by this stage.

The mesoderm also retains its ability to be regionalized into different dorsoventral cell types at least until the start of neurulation. Single cells can contribute to notochord and somites at this stage (Selleck and Stern, 1991), and cells located at the posterior end of the paraxial mesoderm can contribute both to somites and to more lateral/ventral mesoderm (mesonephros, endothelium, blood; Stern et al., 1988). Grafts of Hensen's node from a late primitive streak stage quail embryo into the lateral part of a similarly staged chick host embryo can produce paraxial (somite) and lateral mesoderm from the host, although the degree to which somites form depends on distance from the host axis (Hornbruch et al., 1979). Whether these somites form from the lateral plate of the host or from newly induced mesoderm remains to be established, but these experiments show that the competence of the mesoderm to become subdivided into dorsoventral regions has not completely disappeared before the end of gastrulation. Taken together, these conclusions suggest that, at least in the chick, commitment of mesoderm cells is still occurring at the start of neurulation. By this time, the hypoblast has been displaced into extraembryonic regions and is therefore unlikely to be responsible for mesoderm induction or for its regionalisation at these stages.

Thus, mesoderm induction seems to occur over a protracted period of development, in more than one step, and more than one signal must be involved.

Anteroposterior patterning of the mesoderm: evidence for stem cells in Hensen's node

As well as producing a diversity of mesodermal cell types, generation of the basic body plan requires the axes of the embryo to become established. We have seen that in fate maps of Hensen's node, some cells contribute progeny to notochord only, some to somite only and others to both notochord and somite (Selleck and Stern, 1991; see above). But single cell lineage analysis in Hensen's node also revealed an otherwise unsuspected spatial organisation of the mesodermal descendants of the marked cells. Injection of the fluorescent lineage tracer lysine-rhodamine-dextran (LRD) into a single cell in the node generates several clusters of labelled cells, regularly spaced along the length of the notochord or somitic mesoderm (Selleck and Stern, 1991, 1992a, b; Fig. 1). The spacing between clusters differs in the two tissues: in the notochord, clusters are separated by about 1.5-2 somite lengths, whilst in the somitic mesoderm the distance appears to be about 5-7 somites. The results have been interpreted as indicating that the node contains a population of multipotent cells with stem cell properties, which give rise to founder cells with more restricted fates (viz. notochord or somite; Selleck and Stern, 1992b; Fig. 2). The founder cells also have stem cell properties; to account for the differences in spacing between adjacent clusters, the rate of cell division is proposed to be faster in notochord founder cells than in somitic precursors.

The distance of 5-7 somites between adjacent clusters in the somitic mesoderm agrees well with the findings of Primmett and colleagues (Primmett et al., 1988, 1989; Stern et al., 1988). They found that heat shock generates periodic anomalies in the somitic mesoderm, with a spacing of about 6-7 somites, and that this periodicity correlates directly with the rate of cell division of somite precursor cells. Indeed, somite precursors in the segmental plate mesoderm divide about every 10 hours, which is the time taken for about 7 somites to form (Primmett et al., 1989). These experiments suggest that the anteroposterior axis of the mesoderm becomes regionalised in the notochord and somite precur-

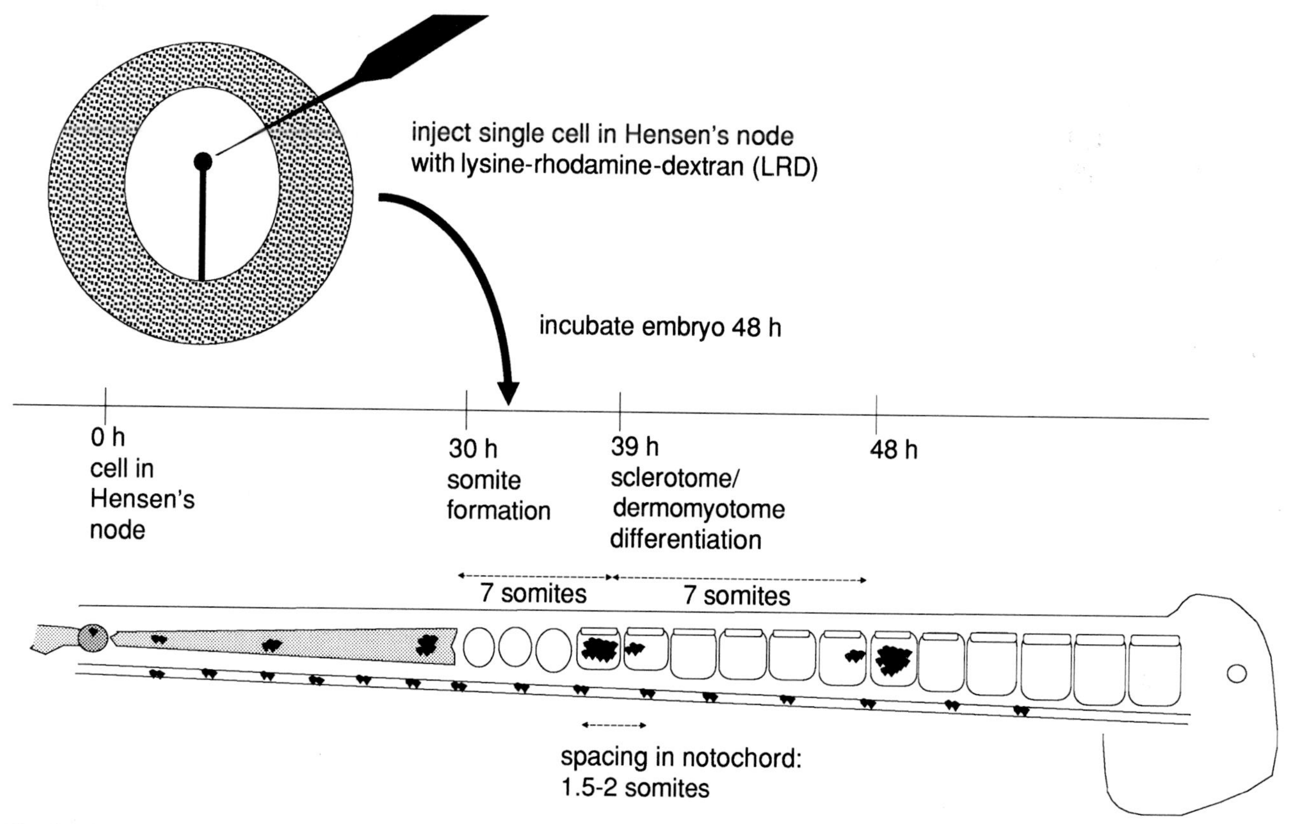

Fig. 1. Diagram summarising periodic clusters of cells revealed by injection of lysine-rhodamine-dextran (LRD) into a single cell in Hensen's node, based on the results of Selleck and Stern (1991, 1992a,b and unpublished observations). The spacing between clusters of labelled cells in the somitic mesoderm is about 6 somites; this is correlated with a time scale following the fate of the somitic descendants of the injected cell over 2 days after leaving Hensen's node. In the notochord, the clusters of descendants of the injected cell are 1.5-2 somite-lengths apart.

sor cells in the node, in association with the cell division cycle of these progenitor cells.

Relationships between the anteroposterior and dorsoventral axes

Several experiments have revealed effects of mesoderm-inducing factors on development of the anteroposterior axis of the early embryo. For example, when dominant-negative mutants are constructed for the FGF receptor in *Xenopus* (Amaya et al., 1991), not only is the ventral mesoderm affected, but also the posterior part of the embryo is deficient or fails to develop. For this reason, as well as the finding that activin-treated explants placed in the blastocoele of a host embryo ('Einsteckung' assay) can generate head structures whilst FGF-treated explants only generate tails (Ruiz i Altaba and Melton, 1989; Sokol and Melton, 1991; Slack and Tannahill, 1992), these results suggest that members of the FGF family are posterior inducers as well as ventral inducers. Microinjection of mRNA encoding goosecoid (Cho et al., 1991; De Robertis et al., 1992) or members of the *wnt* family of proto-oncogenes (Smith and Harland, 1991; Sokol et al., 1991) into ventral blastomeres can also generate an ectopic axis.

Thus, there appears to be a correlation between the ability of a substance to induce dorsal mesoderm with its ability to generate head structures. Therefore, inducing factors are often referred to as 'ventral/posterior' or 'dorsal/anterior' inducers (see Slack and Tannahill, 1992). But if the same factors are responsible for posterior and ventral induction or for anterior and dorsal induction, how do these two axes become separate in development?

Origin of posterior structures

Given the apparent relationship between anterior and dorsal, and posterior and ventral, how do structures such as the notochord (dorsal) of the tail (posterior) become established? In diagrams of the three signal model, the embryo is often shown in mid-sagittal section (e.g. Slack et al., 1984; see also Slack and Tannahill, 1992), with the ventral region being thought of also as posterior. However, fate maps of the early amphibian embryo (e.g. Hadorn, 1970; Nieuwkoop et al., 1985; see also Keller et al., 1992) show the presumptive tail bud region just above the equator, about 90° away from the prospective ventral part of the embryo (Fig. 3).

Classical fate maps of the chick blastoderm before the appearance of the primitive streak (e.g. Rudnick, 1935; Pasteels, 1940; Waddington, 1956; Balinsky, 1975) seem to place the presumptive tail at an equivalent, lateral position close to the marginal zone (Fig. 3). Prior to and during the early stages of primitive streak formation, 'Polonnaise'-like movements of the epiblast make the left and right tail pri-

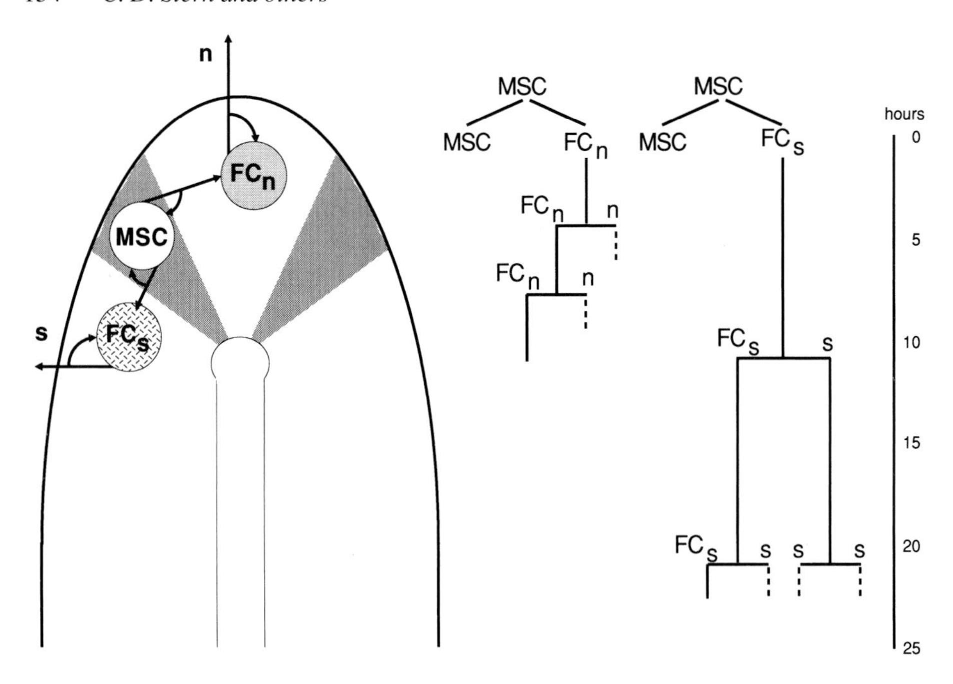

Fig. 2. A model of the organization of stem cells and founder cells in Hensen's node of the chick embryo at stage 3+. The median, anterior quadrant of the node contributes cells to the notochord but not to the somites, the regions lateral to the primitive pit populate the medial halves of the somites, and the region between the previous two (shaded) contributes to both notochord and somites (Selleck and Stern, 1991). Based on these results and those summarised in Fig. 2, the following scheme is proposed (see Selleck and Stern, 1992b): the region that contributes cells to both notochord and somites (shaded) contains multipotent stem cells (MSC) which renew themselves and give rise to founder cells that contribute progeny to either the notochord (FC_n) or the somite (FC_s). Each of these founder cells also have stem cell properties, since they can also renew themselves. The same model is displayed as a lineage tree on the right. n, notochord cells; s, medial somite cells. In the lineage tree, the length of the vertical lines is proportional to the length of the cell division cycle of the cell shown at the top of the line.

mordia converge towards the posterior margin, whilst the cells that originally lay posteriorly in the marginal zone shift forwards (see Stern, 1990). Some of these cells end up in Hensen's node and subsequently contribute to the prechordal plate, definitive (gut) endoderm and chordamesoderm (see Selleck and Stern, 1991). Although the classical fate maps mentioned above were produced mostly without good lineage markers and before a good staging system was available for pre-primitive streak stages of chick development (Eyal-Giladi and Kochav, 1976), they appear to be remarkably accurate; recent studies in our laboratory (Selleck and Stern, 1991, 1992a; YH and CDS, in preparation) confirm their conclusions.

Such fate maps of both chick and amphibian embryos, indicate that: (a) presumptive head structures and prospective dorsal mesoderm are located close to each other, in the region of Hensen's node or the dorsal lip of the blastopore of the gastrula-stage embryo; (b) ventral mesoderm originates from a region which in the amphibian appears to be located ventrally in the blastula; in the chick, the ventral mesoderm seems to come from cells situated in the more central epiblast, away from the marginal zone (see Stern and Canning, 1990; Stern, 1992); (c) posterior ventral structures come from a region about 90° away from both the prospective dorsal lip and ventral marginal zone in the amphibian blastula, and from a marginal region about 90° away from the posterior margin of the chick blastoderm. But this still leaves us with the question: where are the progenitors of the dorsal structures of the trunk and tail?

If, as we have discussed above, the node contains notochord and somite precursor cells with stem cell properties, then the posterior notochord and somites will be derived from progenitors common to more anterior notochord and somites at the late primitive streak stage. Somehow, the descendants of these progenitors must acquire their anteroposterior positional information **after** this stage. One possibility is that such positional information is dependent on the number of cell divisions undergone by stem cells before each of their descendants leaves the node region. For example, the progenitor cells might become posteriorised by exposure to some substance, like retinoic acid, present locally within the node such that, the longer the time spent in the node, the more posterior the character of their descendants.

One further piece of evidence supports this conclusion. When Hensen's nodes of increasing age are grafted into the area opaca of a competent host embryo, the anteroposterior extent of the structures formed from the host depends on the age of the node (see Storey et al., 1992; Kintner and Dodd, 1991; for amphibians, see Spemann and Mangold, 1924; Nieuwkoop et al., 1985; Hemmati Brivanlou et al., 1990; Sharpe, 1990). The older the node, the more posterior the structures that develop. However, host-derived structures only form if the transplanted node comes from a donor embryo younger than the definitive streak stage. If the node is older than this, the structures formed are derived from self-differentiation of the grafted node; nevertheless, they express the most posterior markers as if the grafted node were able to pattern itself as far as the tail. This conclusion is consistent with the idea that patterning posterior to the otic vesicle (the level at which Hensen's node appears to be located at the definitive streak stage; see Rudnick, 1935; Balinsky, 1975) is related to the length of time spent by progenitor cells within the node region.

Origin of the definitive (gut) endoderm, neural induction and regionalisation

During the early stages of chick gastrulation, cells destined

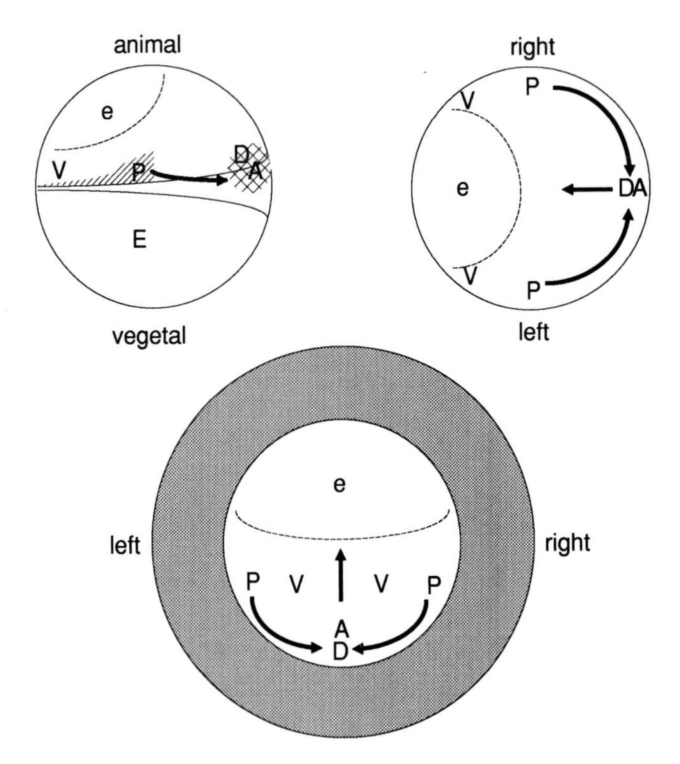

Fig. 3. Presumptive dorsoventral and anteroposterior axes in the fate maps of an amphibian late blastula (upper diagrams) and chick blastoderm at about stage XI (lower diagram). The amphibian map on the left shows the surface of the blastula viewed from its left side, and the map on the right shows a similar blastula observed from the animal pole. The chick blastoderm is viewed from the epiblast side. D, dorsal; V, ventral; A, anterior; P, posterior; E, yolky vegetal 'entoderm' (amphibian); e, surface ectoderm. The arrows show the direction of some of the main cell movements. In the upper left diagram, the presumptive tail region is shown as a shaded area and the presumptive head is shown by cross-hatching. In the chick diagram, the area opaca is shaded.

to form the tail bud in amphibians and birds converge from the lateral margins to the future posterior margin, as we have discussed above. At this time, presumptive definitive endoderm cells become located close to precursors of the notochord (Rudnick, 1935; Pasteels, 1940; Waddington, 1956; Rosenquist, 1966, 1971, 1983; Nicolet, 1970, 1971; Balinsky, 1975; Selleck and Stern, 1991; YH and CDS, in preparation). The regions giving rise to each of these tissues all seem to meet at Hensen's node. The endoderm cells have been suggested to be responsible for neuralization of the overlying ectoderm (see Dias and Schoenwolf, 1990; Storey et al., 1992), whilst the notochord and/or somitic precursors may be responsible for its regionalization (Storey et al., 1992).

Conclusions

The above discussion suggests that in the tail and posterior trunk regions, the dorsal and paraxial mesoderm components are derived from late descendants of chordal and somitic founder cells in the node region, and which were located in the midline of the mid-blastula stage embryo. Intermediate and lateral mesodermal components of the tail, on the other hand, are recruited by the regressing primitive streak from cells that have migrated towards the midline during the early stages of gastrulation.

This brings us back to the original question: what is the relationship between induction of the mesoderm and specification of the embryonic axes? Examination of fate maps and analysis of other experimental findings can help us to separate dorsal from anterior, ventral from posterior, and all of the above from mesoderm induction. But we still have to address the questions of when each of these axes is specified, and whether the role of the hypoblast in the chick and of the vegetal tissue of the frog is mainly to induce mesoderm, to set up dorsoventral pattern or to specify the axes of the embryo. Acknowledgement of the identity and location of these three embryonic dimensions could help us to understand better the role of the so-called mesoderm-inducing factors in early vertebrate development.

Several discussions with Jeremy Green contributed greatly to the development of some of our ideas about the origin of the tail in the frog. We are also grateful to Marianne Bronner-Fraser for her helpful comments on the manuscript.

References

Amaya, E., Musci, T. J. and Kirschner, M. W. (1991). Expression of a dominant negative mutant of the FGF receptor disrupts mesoderm formation in Xenopus embryos. *Cell* **66**, 256-270.

Asashima, M., Nakano, H., Uchiyama, H., Sugino, H., Nakamura, T., Eto, Y., Ejima, D., Nishimatsu, S.-I., Ueno, N. and Kinoshita, K. (1991). Presence of activin (erythroid differentiation factor) in unfertilized eggs and blastulae of Xenopus laevis. *Proc. natn. Acad. Sci. USA* **88**, 6511-6514.

Azar, Y. and Eyal-Giladi, H. (1981). Interaction of epiblast and hypoblast in the formation of the primitive streak and the embryonic axis in the chick, as revealed by hypoblast-rotation experiments. *J. Embryol. Exp. Morph.* **61**, 133-144.

Balinsky, B. I. (1975). *An Introduction to Embryology* (4th ed.) Philadelphia: W.B. Saunders

Blum, M., Gaunt, S. J., Cho, K. W. Y., Steinbesser, H., Bittner, D. and De Robertis, E. M. (1992). On the role of the mouse homeobox gene goosecoid during gastrulation. *Cell* (in press).

Cho, K. W. Y., Blumberg, B., Steinbesser, H. and De Robertis, E. M. (1991). Molecular nature of Spemann's organizer: the role of the Xenopus homeobox gene goosecoid in gastrulation. *Cell* **67**, 1111-1120.

Cooke, J. and Wong, A. (1991). Growth-factor-related proteins that are inducers in early amphibian development may mediate similar steps in amniote (bird) embryogenesis. *Development* **111**, 197-212.

Delarue, M., Sanchez, S., Johnson, K. E., Darribère, T. and Boucaut, J.-C. (1992). A fate map of superficial and deep circumblastoporal cells in the early gastrula of *Pleurodeles waltl*. *Development* **114**, 135-146.

De Robertis, E. M., Blum, M., Niehrs, C. and Steinbesser, H. (1992). *Goosecoid* and the organizer. *Development* **1992 Supplement**, 167-171.

Dias, M. and Schoenwolf, G. C. (1990). Formation of ectopic neurepithelium in chick blastoderms: age-related capacities for induction and self-differentiation following transplantation of quail Hensen's nodes. *Anat. Rec.* **229**, 437-448.

Dirksen, M. L. and Jamrich, M. (1992). A novel, activin-inducible, blastopore lip-specific gene of Xenopus laevis contains a fork head DNA-binding domain. *Genes Dev.* **6**, 599-608.

Eyal-Giladi, H. and Kochav, S. (1976). From cleavage to primitive streak formation: a complementary normal table and a new look at the first stages of the development of the chick. I. General morphology. *Dev. Biol.* **49**, 321-337.

Green, J. B. A. and Smith, J. C. (1991). Growth factors as morphogens: do gradients and thresholds establish body plan? *Trends Genet.* **7**, 245-250.

Gurdon, J. B. (1987). Embryonic induction: molecular prospects. *Development* **99**, 285-306.

Hadorn, E. (1970). *Experimentelle Entwicklungsforschung*. Berlin: Springer Verlag.

Hemmati Brivanlou, A., Stewart, R. M. and Harland, R. M. (1990). Region-specific neural induction of an engrailed protein by anterior notochord in *Xenopus*. *Science* **250**, 800-802.

Hornbruch, A., Summerbell, D. and Wolpert, L. (1979). Somite formation in the early chick embryo following grafts of Hensen's node. *J. Embryol. Exp. Morph.* **51**, 51-62.

Isaacs, H. V., Tannahill, D. and Slack, J. M. W. (1992). Expression of a novel FGF in the *Xenopus* embryo. A new candidate inducing factor for mesoderm formation and anteroposterior specification. *Development* **114**, 711-720.

Keller, R., Shih, J. and Domingo, C. (1992). The patterning and functioning of protrusive activity during convergence and extension of the *Xenopus* organiser. *Development* **1992 Supplement**, 81-91.

Kintner, C. R. and Dodd, J. (1991). Hensen's node induces neural tissue in *Xenopus* ectoderm. Implications for the action of the organizer in neural induction. *Development* **113**, 1495-1505.

Mitrani, E. and Eyal-Giladi, H. (1981). Hypoblastic cells can form a disk inducing an embryonic axis in chick epiblast. *Nature* **289**, 800-802.

Mitrani, E. and Shimoni, Y. (1990). Induction by soluble factors of organized axial structures in chick epiblasts. *Science* **247**, 1092-1094.

Mitrani, E., Gruenbaum, Y., Shohat, H. and Ziv, T. (1990a). Fibroblast growth factor during mesoderm induction in the early chick embryo. *Development* **109**, 387-393.

Mitrani, E. Ziv, T., Thomsen, G., Shimoni, Y., Melton, D. A. and Bril, A. (1990b). Activin can induce the formation of axial structures and is expressed in the hypoblast of the chick. *Cell* **63**, 495-501.

Nicolet, G. (1970). Analyse autoradiographique de la localization des différentes ébauches presomptives dans la ligne primitive de l'embryon de poulet. *J. Embryol. Exp. Morph.* **23**, 79-108.

Nicolet, G. (1971). Avian gastrulation. *Adv. Morphogen.* **9**, 231-262.

Nieuwkoop, P. D. (1969). The formation of mesoderm in Urodelean amphibians. I. Induction by the endoderm. *Wilhelm Roux Arch. EntwMech. Organ.* **162**, 341-373.

Nieuwkoop, P. D. (1985). Inductive interactions in early amphibian development and their general nature. *J. Embryol. Exp. Morph.* **89 Supplement**, 333-347.

Nieuwkoop, P. D., Johnen, A. G. and Albers, B. (1985). *The Epigenetic Nature of Early Chordate Development. Inductive Interaction and Competence*. Cambridge Univ. Press.

Pasteels, J. (1940). Un aperçu comparatif de la gastrulation chez les chordés. *Biol. Rev.* **15**, 59-106.

Primmett, D. R. N., Stern, C. D. and Keynes, R. J. (1988). Heat-shock causes repeated segmental anomalies in the chick embryo. *Development* **104**, 331-339.

Primmett, D. R. N., Norris, W. E., Carlson, G. J., Keynes, R. J. and Stern, C. D. (1989). Periodic segmental anomalies induced by heat-shock in the chick embryo are associated with the cell cycle. *Development* **105**, 119-130.

Rosenquist, G. C. (1966). A radioautographic study of labeled grafts in the chick blastoderm development from primitive streak stages to stage 12. *Contrib. Embryol. Carnegie Inst. Wash.* **38**, 71-110.

Rosenquist, G. C. (1971). The location of the pregut endoderm in the chick embryo at the primitive streak stage as determined by radioautographic mapping. *Devl Biol.* **26**, 323-335.

Rosenquist, G. C. (1983). The chorda center in Hensen's node of the chick embryo. *Anat. Rec.* **207**, 349-355.

Rudnick, D. (1935). Regional restriction of potencies in the chick during embryogenesis. *J. Exp. Zool.* **71**, 83-99.

Ruiz i Altaba, A. and Melton, D. A. (1989). Interaction between peptide growth factors and homeobox genes in the establishment of antero-posterior polarity in frog embryos. *Nature* **341**, 33-38

Selleck, M. A. J. and Stern, C. D. (1991). Fate mapping and cell lineage analysis of Hensen's node in the chick embryo. *Development* **112**, 615-626.

Selleck, M. A. J. and Stern, C. D. (1992a). Commitment of mesoderm cells in Hensen's node of the chick embryo to notochord and somite. *Development* **114**, 403-415.

Selleck, M. A. J. and Stern, C. D. (1992b). Evidence for stem cells in the mesoderm of Hensen's node and their role in embryonic pattern formation. **In** *Development of Embryonic Mesoderm*. (ed. J. W. Lash, R. Bellairs and E. J. Sanders). New York: Plenum Press (in press).

Sharpe, C. R. (1990). Regional neural induction in *Xenopus laevis*. *BioEssays* **12**, 591-596.

Slack, J. M. W., Dale, L. and Smith, J. C. (1984). Analysis of embryonic induction by using cell lineage markers. *Phil. Trans. Roy. Soc.* **B 307**, 331-336.

Slack, J. M. W., Isaacs, H. V., Johnson, G. E., Lettice, L. A., Tannahill, D. and Thompson, J. (1992). Specification of the body plan during *Xenopus* gastrulation: dorsoventral and anteroposterior patterning of the mesoderm. *Development* **1992 Supplement**, 143-149.

Slack, J. M. W. and Tannahill, D. (1992). Mechanism of anteroposterior axis specification in vertebrates: lessons from the amphibians. *Development* **114**, 285-302.

Slack, J. M. W. (1991). Molecule of the moment. *Nature* **349**, 17-18.

Smith, J. C. and Howard, J. E. (1992). Mesoderm-inducing factors and the control of gastrulation. *Development* **1992 Supplement**, 127-136.

Smith, J. C., Price, B. M. J., Green, J. B. A., Weigel, D. and Herrmann, B. G. (1991). Expression of a Xenopus homolog of Brachyury (T) is an immediate-early response to mesoderm induction. *Cell* **67**, 79-87.

Smith, W. C. and Harland, R. M. (1991). Injected XWnt-8 RNA acts early in Xenopus embryos to promote formation of a vegetal dorsalizing center. *Cell* **67**, 753-765.

Sokol, S. and Melton, D. A. (1991). Pre-existent pattern in Xenopus animal pole cells revealed by induction with activin. *Nature* **351**, 409-411.

Sokol, S., Christian, J. L., Moon, R. T. and Melton, D. A. (1991). Injected Wnt RNA induces a complete body axis in Xenopus embryos. *Cell* **67**, 741-752.

Spemann, H. and Mangold, H. (1924). Über Induktion von Embryonanlagen durch Implantation artfremder Organisatoren. *Wilhelm Roux Arch. EntwMech. Organ.* **100**, 599-638.

Stern, C. D. (1990). The marginal zone and its contribution to the hypoblast and primitive streak of the chick embryo. *Development* **109**, 667-682.

Stern, C. D. (1992). Mesoderm induction and development of the embryonic axis in amniotes. *Trends Genet.* **8**, 158-163.

Stern, C. D. and Canning, D. R. (1990). Origin of cells giving rise to mesoderm and endoderm in chick embryo. *Nature* **343**, 273-275.

Stern, C. D., Fraser, S. E., Keynes, R. J. and Primmett, D. R. N. (1988). A cell lineage analysis of segmentation in the chick embryo. *Development* **104 Supplement**, 231-244.

Storey, K. G., Crossley, J. M., De Robertis, E. M., Norris, W. E. and Stern, C. D. (1992). Neural induction and regionalisation in the chick embryo. *Development* **114**, 729-741.

Taira, M., Jamrich, M., Good, P. J. and Dawid, I. B. (1992). The LIM domain-containing homeobox gene Xlim-1 is expressed specifically in the organizer region of Xenopus gastrula embryos. *Genes Dev.* **6**, 356-366.

Waddington, C. H. (1933). Induction by the endoderm in birds. *Wilhelm Roux Arch. EntwMech. Organ.* **128**, 502-521.

Waddington, C. H. (1956). *Principles of Embryology*. London: Allen and Unwin.

Development 1992 Supplement, 157-165 (1992)
Printed in Great Britain © The Company of Biologists Limited 1992

Brachyury - a gene affecting mouse gastrulation and early organogenesis

R. S. P. BEDDINGTON, P. RASHBASS and V. WILSON

Centre for Genome Research, King's Buildings, West Mains Road, Edinburgh EH9 3JQ, UK

Summary

Mouse embryos that are homozygous for the *Brachyury* (*T*) deletion die at mid-gestation. They have prominent defects in the notochord, the allantois and the primitive streak. Expression of the *T* gene commences at the onset of gastrulation and is restricted to the primitive streak, mesoderm emerging from the streak, the head process and the notochord. Genetic evidence has suggested that there may be an increasing demand for *T* gene function along the rostrocaudal axis. Experiments reported here indicate that this may not be the case. Instead, the gradient in severity of the *T* defect may be caused by defective mesoderm cell movements, which result in a progressive accumulation of mesoderm cells near the primitive streak.

Embryonic stem (ES) cells which are homozygous for the *T* deletion have been isolated and their differentiation in vitro and in vivo compared with that of heterozygous and wild-type ES cell lines. In +/+ ↔ *T/T* ES cell chimeras the *Brachyury* phenotype is not rescued by the presence of wild-type cells and high level chimeras show most of the features characteristic of intact *T/T* mutants. A few offspring from blastocysts injected with *T/T* ES cells have been born, several of which had greatly reduced or abnormal tails. However, little or no ES cell contribution was detectable in these animals, either as coat colour pigmentation or by isozyme analysis. Inspection of potential +/+ ↔ *T/T* ES cell chimeras on the 11th or 12th day of gestation, stages later than that at which intact *T/T* mutants die, revealed the presence of chimeras with caudal defects. These chimeras displayed a gradient of ES cell colonisation along the rostrocaudal axis with increased colonisation of caudal regions. In addition, the extent of chimerism in ectodermal tissues (which do not invaginate during gastrulation) tended to be higher than that in mesodermal tissues (which are derived from cells invaginating through the primitive streak). These results suggest that nascent mesoderm cells lacking the *T* gene are compromised in their ability to move away from the primitive streak. This indicates that one function of the *T* gene may be to regulate cell adhesion or cell motility properties in mesoderm cells. Wild-type cells in +/+ ↔ *T/T* chimeras appear to move normally to populate trunk and head mesoderm, suggesting that the reduced motility in *T/T* cells is a cell autonomous defect.

Key words: *Brachyury*, mouse gastrulation, embryonic stem cells.

Introduction

At present the mouse is pre-eminent amongst vertebrate experimental organisms as a source of developmental mutants. For a start, there is a relatively large repertoire of spontaneous or physically induced developmental mutants (Lyon and Searle, 1989), although the catalogue of homozygous embryonic lethals is somewhat biased towards genes that are semi-dominant, and have relatively obvious heterozygous phenotypes in viable offspring. This list of developmental mutants is now rapidly being augmented by mutations generated by transgenesis, and in particular by mutations created in embryonic stem (ES) cells (Evans and Kaufman, 1981; Martin, 1981). ES cells provide an opportunity not only for generating random mutations by transgene insertion (some of which may affect development) but also for disrupting previously identified genes, whose genomic sequence is known and which have been shown to be expressed during particular stages of embryogenesis (see Reith and Bernstein, 1991). As a result of this new technology mouse developmental biology is entering a new era of extensive genetic analysis. However, in all cases an intact mutant embryo, while identifying genes necessary for normal development, often provides only gross information regarding the developmental consequences of gene malfunction: the primary perturbation may be obscured by subsequent defective tissue interactions producing a complex phenotype (see Beddington et al., 1991).

In this paper we will describe experiments aimed at further resolving the developmental effects of a well-known mouse mutant, *Brachyury* (*T*). The first part will be devoted to a brief review of previous work describing the morphological, genetic and molecular basis of the *Brachyury* phenotype. This will provide both a context for our own work and also illustrate some of the strengths and weaknesses of descriptive studies, most of which were performed before it was possible to recognise homozygous mutants prior to the inception of abnormalities. The data that we present comprise an analysis of the behaviour of *T/T* ES cells in chimeric embryos throughout gestation, and demonstrate that the dynamic behaviour of mutant cells intermixed with wild-type cells can reveal subtle alterations in morphogenetic movements.

Morphological features of *Brachyury*

Brachyury was first recognised almost 70 years ago by Dobrovolskaïa-Zavadskaïa (Dobrovolskaïa-Zavadskaïa, 1927) because heterozygous animals have short, and often slightly kinked, tails. Subsequently, it was found that homozygous embryos die at mid-gestation, about 10.5 day post coitum (dpc), and have distinctive caudal abnormalities. The allantois, which should form a major component of the chorioallantoic placenta, fails to extend and traverse the exocoelom (Gluecksohn-Schoenheimer, 1944). Consequently, the embryo is denied a placental connection and is deprived of adequate nutritive supply. This is probably the physiological cause of embryonic death. However, embryonic pattern posterior to the forelimb region is also disturbed. Somites posterior to the seventh pair of somites are absent or abnormal, the neural folds fuse but the neural tube is severely kinked in the caudal region and the surface ectoderm tends to form large fluid filled blisters. Central features of *T/T* embryos are the apparent absence of a notochord and profound thickening of the primitive streak (Chesley, 1935; Gluecksohn-Schoenheimer, 1938; Gruneberg, 1958). In addition, the node at the extreme anterior of the primitive streak, the normal origin of the notochord, is less distinct than in wild-type embryos (Fujimoto and Yanagisawa, 1983). Careful descriptive studies over the last 65 years, together with a wealth of evidence implicating the notochord in the patterning of the neural tube and possibly the somites (e.g. Clarke et al., 1991; Hemmati-Bravanlou et al., 1990; Kitchin, 1949; Placzek et al., 1990; Smith and Schoenwolf, 1989; van Straaten et al., 1985; Yamada et al., 1991), have pointed to the defective notochord being a prime cause of many of the embryonic abnormalities. There is some debate as to whether the notochord fails to form altogether or in fact does delaminate, but subsequently degenerates or fuses with the adjacent gut or ventral neural tube, thereby becoming unrecognisable in histological sections (Chesley, 1935; Gruneberg, 1958; Spiegelman, 1976; Yanagisawa, 1990). Immunocytochemical staining of 10 dpc *T/T* embryos using an antibody raised against cellular retinol binding protein (Maden et al., 1990), which is present in the notochord, suggest that a notochord may be present in caudal regions (P. Rashbass, V. Wilson, M. Maden and R. Beddington, in preparation). Certainly, the absence of notochord is an improbable explanation for the failure of allantois to differentiate normally and the allantoic defects are more likely to result from abnormal deployment of cells emerging from the primitive streak, or inappropriate differentiation of cells once incorporated into the allantoic bud.

The observation that the mesoderm:ectoderm ratio is elevated in the caudal 15% of 8 dpc putative *T/T* embryos but reduced compared to wild-type embryos in the region immediately anterior to the primitive streak (Yanagisawa et al., 1981) would support the notion that morphogenetic movements are abnormal during gastrulation. The mesoderm:ectoderm ratio is normal in the anterior half of the embryo (Yanagisawa et al., 1981) and there is no significant difference in either mitotic index (Yanagisawa et al., 1981) or the incidence of [^{3}H]thymidine labelling (Yanagisawa and Fujimoto, 1977b) in different axial regions. Furthermore, no increase in cell death was noted in the posterior region underlying the primitive streak. This argues that it is migration of mesoderm away from the streak which is compromised in the latter stages of gastrulation. Direct measurements of active mesoderm migration on extracellular matrix in vitro reveal that *T/T* cells from 8-9 dpc embryos have a slightly but significantly reduced migration rate (e.g. 8 dpc, 39.4 ± 11 μm h^{-1}) compared to wild-type mesoderm (8 dpc, 52.8±22.6 μm h^{-1}) (Hashimoto et al., 1987).

In the heterozygote the tail is short and often kinked. Again the notochord in the caudal region (usually confined to the tail but sometimes extending as far forwards as the cloaca) is abnormal during embryonic development. It may be branched, improperly separated from the hindgut or neural tube and often has a prominent central lumen. From studies of both homozygotes and heterozygotes Grüneberg concludes that "a common denominator for all the abnormalities of the notochord may be a change in surface properties" (Gruneberg, 1958).

That the surface of mutant cells may be altered has been tested directly by comparing the ability of wild-type or *T/T* mutant cells to form aggregates (Yanagisawa and Fujimoto, 1977a). Cells from the trunk, head and forelimb bud were disaggregated and the diameter of aggregates formed in suspension culture measured. Mutant cells, from any one of these regions, consistently formed smaller aggregates and this implies a difference in cell surface adhesive properties. However, the true relevance of these data is not clear since cells were isolated from both affected and unaffected tissues. Glycosyltransferase activity has also been shown to be reduced in *T/T* mutants (Shur, 1982) and abnomalities of the extracellular matrix have been described (Jacobs-Cohen et al., 1983).

Histogenetic potential of *T/T* cells

Several experiments have shown that *T/T* cells are capable of differentiating into a wide variety of mature differentiated tissues. Initially, Ephrussi explanted mutant embryonic tissues in vitro and showed that these could survive beyond the time of embryonic death and could differentiate into an array of cell types comparable to those formed by wild-type embryos (Ephrussi, 1935). Subsequently, ectopic transfer of posterior regions recovered from 8.5 - 9.5 dpc *T/T* embryos showed that a diverse mixture of mature differentiated tissues, representative of derivatives of all three germ layers, could develop in the resulting experimental teratomas (Bennett et al., 1977). There was no predisposition for *T/T* embryos to give rise to teratocarcinomas containing undifferentiated embryonal carcinoma (EC) cells indicating that the epiblast, the progenitor of EC cells (Diwan and Stevens, 1976), matured at an equivalent rate to wild-type embryos. When anterior and posterior regions

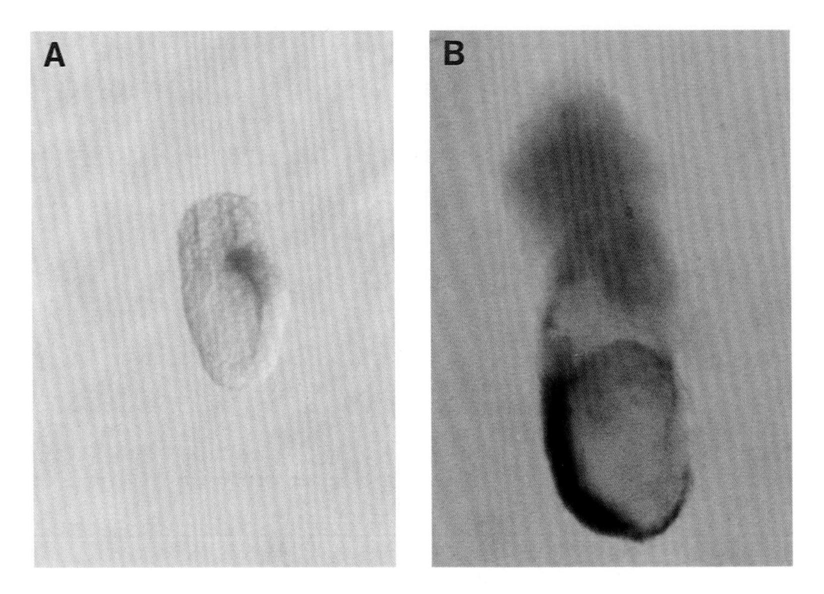

Fig. 1. (A) Whole-mount in situ hybridisation to *T* mRNA in early primitive-streak stage wild-type embryos. The *T* gene is expressed in the primitive streak region from the onset of gastrulation. (B) Whole-mount in situ hybridisation to *T* mRNA in late primitive-streak stage wild-type embryos. (Photographs reproduced by permission of B. Rosen.)

were compared a decrease in the frequency, but not the complete absence, of bone and cartilage was detected in tumours formed from posterior regions (Fujimoto and Yanagisawa, 1979). Taken together, these results suggest that the embryonic defects seen in *T/T* conceptuses stem from an organizational failing rather than an inability to differentiate into specific tissues. However, it should be noted that neither notochord nor allantoic differentiation can be recognised by histological inspection of experimental teratomas.

Genetics of *Brachyury*

The original *Brachyury* mutation (*T*) has been shown to be a large deletion (160-200 kb). Two other mutants have been described which exhibit an identical phenotype. One, T^{2J}, is also a large deletion (81-110 kb) and the other T^{ktl} (Justice and Bode, 1990) was induced by ethylnitrosourea and likely to be a point mutation or small deletion. No other gene has been identified in the *Brachyury* deletion and if T^{ktl} is indeed a point mutation, then this argues that the observed phenotype stems from the absence of a single gene.

There are three further *T* alleles that have similar phenotypes to each other but which differ from that of the original *Brachyury* mutant. In T^{Wis} (Shedlovsky et al., 1988) there is an insertion of a retroviral-like element in the seventh exon (Herrmann et al., 1990) and in T^c (Searle, 1966) there is a 19 bp deletion in the last exon (Herrmann et al., 1990). $T^{c\text{-}2H}$ has a frameshift mutation affecting the same region of the carboxy terminus as the T^c deletion (B. G. Herrmann, personal communication). What is interesting about these mutations is that they present a more severe phenotype than deletion of the *T* gene. In both heterozygotes and homozygotes abnormalities occur at a more rostral axial level (Herrmann, 1991; Searle, 1966). Thus heterozygotes usually have no tail and homozygotes show no sign of somites. The boundary of embryonic defects in the homozygotes is shifted rostrally to the cervical region.

Evidence that T^c is an antimorph comes from dosage studies. *T* alleles that delete the *T* gene can be complemented by the *T* locus duplication t^{wLub2} (MacMurray and Shin, 1988; Winking and Silver, 1974). However, T^c, while not being independent of wild-type gene copy number, is only partially complemented by this duplication, which indicates that the T^c product serves to antagonise wild-type activity (MacMurray and Shin, 1988). In addition it has been shown that T^c/T embryos have a less severe phenotype than T^c/T^c (Searle, 1966). However, since *T* is a deletion this amelioration of the T^c effect cannot be due to residual *T* activity. It is more likely that the T^c gene product acts like a dominant negative mutation. If the *T* gene product is only active in association with a second gene product (Herrmann, 1991, 1992; Lyon and Meredith, 1964), perhaps as a dimer, the T^c protein may interfere with this association or affect the biological activity of resulting protein complexes.

Whatever the mode of action of such an antimorph there emerges a compelling gradient of phenotype where the more severe alleles (T^c and T^{Wis}) affect more rostral levels in both homozygotes and heterozygotes than the weaker deletion mutants (*T* and T^{2J}). A similar phenomenon has been observed when comparing the severity of tail defects in *T/+* and *T/t* embryos (Yanagisawa, 1990). On the face of it, this indicates that there is an increased requirement for *T* activity as one moves caudally along the axis, the cranial region being independent of *T* activity but tail formation requiring high levels. As the anteroposterior axis is laid down sequentially in the mouse this can be viewed as an increased requirement for *T* activity with time rather than with distance. In other words, late gastrulation and tailbud differentiation have higher demands for *T* activity than does early gastrulation. Alternatively, the level of *T* activity may be constant but the consequences of defective gastrulation accumulate with time. Thus, a more severe effect on gastrulation will culminate in abnormalities earlier, and therefore more rostrally, than less severe disruption. The chimeric analysis presented in this paper (see below) is consistent with a progressive accumulation of defective cells caudally rather than an increased demand for *T* activity.

The nature of the *T* gene and its expression pattern

The sequence of the cloned mouse *T* gene does not immediately reveal the nature of the protein product (Herrmann et al., 1990). The sequence has an open reading frame of 436 amino acids and shows limited homology to *MyoD1* (Willison, 1990). Antibodies raised against the *Brachyury* homologue in zebrafish demonstrate that the protein is localised to the nucleus (see Herrmann, 1992). Therefore, at present, all available data are consistent with the T protein being a transcription factor.

The expression pattern of the *T* gene in mouse embryos is largely consistent with the observed pattern of abnormalities seen in mutants (Herrmann, 1991; Wilkinson et al., 1990). Furthermore, the expression of *T*, or its homologue, in zebrafish (see Herrmann, 1992), *Xenopus*, and mouse are directly comparable with respect both to embryonic stage and position. In the mouse it is first expressed in the primitive streak at the onset of gastrulation (Fig. 1A), and the *Xenopus* homologue of *T* has been shown to be induced, in the absence of protein synthesis, by the mesoderm inducing activity of peptide growth factors (Smith et al., 1991). Expression continues in the primitive streak throughout gastrulation and can be detected in ectoderm adjacent to the streak and nascent mesoderm underlying the streak (Herrmann, 1991; Fig. 1B). However, expression in the mesoderm disappears as the cells move away from the streak and assume their lateral, paraxial or extraembryonic positions. Only the head process and notochord continue to express high levels of *T*. Interestingly, the allantois, with the possible exception of a very early basal component (A. McMahon and J. McMahon, unpublished data) does not express *T*. Thus, the expression pattern of the gene supports the notion that its primary sites of action are the primitive steak and the notochord, but does little to explain the defects in allantois development.

In T^{Wis}/T^{Wis} embryos, the mutant *T* gene is expressed normally during the early stages of gastrulation (Herrmann,

1991). However, expression declines rapidly at about 8 dpc, expression being lost first in the head process and notochord precursor and anterior part of the primitive streak and finally at the posterior end of the streak. This loss of expression, particularly in the primitive streak, cannot be accounted for solely by cell death but suggests instead that the normal pattern of *T* gene expression is dependent, either directly or indirectly, on normal T protein activity (Herrmann, 1992).

Chimeric analysis of mutant development

All the studies on *Brachyury* to date suggest a strong correlation between expression of the *T* gene, whose product is probably a transcription factor, and the normal differentiation or survival of axial mesoderm: the head process and notochord. Expression in the primitive streak may affect cell survival but it also has an influence on the ability of nascent mesoderm to move away from the streak, although this may only be true for the latter stages of gastrulation since the cranial region forms normally. The defects in the allantois may be a consequence of this abnormal migration or they may indicate that for normal allantoic differentiation *T* expression is required as cells pass through the streak.

Mixing wild-type and mutant cells together in a chimera allows the cell autonomous function of a gene to be assessed. Furthermore, it may allow the analysis of mutant cell behaviour in embryos surviving beyond the stage at which intact mutant embryos die. In mouse, chimerism in all tissues can only be achieved by the addition of cells to the preimplantation embryo. However, there are no morphological criteria for identifying *T/T* preimplantation embryos. Cloning of the *T* gene makes genetic characterisation theoretically possible, but would involve laborious polymerase chain reaction assays on biopsies of individual embryos (Handyside et al., 1990). The alternative, of isolating and genetically characterising *T/T* ES cells is appealing for several reasons. First of all, once estabished, such lines provide a continuous source of mutant cells whose development can be monitored either in vivo or in vitro. Secondly, the availability of ES cells null for the *T* gene presents an ideal substrate for genetic manipulation of *T* expression or of genes acting downstream of it.

Isolation of ES cell lines

We have isolated and genetically characterised several *T/T*, *T/+* and +/+ ES cell lines from blastocysts derived from heterozygous BTBR *T/+* matings (Rashbass et al., 1991). These cells are all homozygous for the *glucose phosphate isomerase-1a* gene (*Gpi-1^a*). These lines have similar morphological characteristics and growth rates in vitro. The only minor difference observed is that *T/T* embryoid bodies take longer to disaggregate in trypsin than do heterozygous or wild-type lines. Following inoculation under the testis capsule all lines, regardless of genotype, form teratocarcinomas. Inspection of the differentiated tissues present in these tumours does not show any consistent, qualitative differences, although one *T/T* cell line (BTBR6) failed to give rise to bone and cartilage and a second null line (BTBR10) gave a lower incidence of these tissues (Table 1). This is reminiscent of the results from ectopic transfer of embryonic fractions (Bennett et al., 1977; Fujimoto and Yanagisawa, 1979).

9th and 10th day of gestation chimeras

When introduced into wild-type embryos, which are *Gpi-1^b/Gpi-1^b*, a significant difference was observed between *T/T* and *T/+* lines (Beddington et al., 1991; Rashbass et al., 1991). Heterozygous lines formed normal chimeric conceptuses at 8.5-9.5 dpc. On the other hand, two independent *T/T* cell lines formed predominantly abnormal chimeras at midgestation (Rashbass et al., 1991; V. Wilson and R. Beddington, unpublished observations). Typically, these *T/T* ↔+/+ chimeras have a phenotype which mimics that of the intact *T/T* mutant (Table 2). Chimeric embryos with an ES cell contribution greater than 70%, as judged by ES cell glucose phosphate isomerase (GPI) isozyme activity, appear almost indistinguishable from intact mutants. Embryos with a lower but detectable ES cell contribution almost invariably have an abnormal allantois, some thickening of the primitive streak and sectioned material shows regions of necrotic or absent notochord (Rashbass et al., 1991). The extent of neural tube disruption and the number of somites formed is more variable in embryos that are less than 70% chimeric. In the absence of a single cell marker, which distinguishes all mutant cells from wild-type cells, it is impossible to be dogmatic about the cell autonomous nature of these effects but the cell death seen in the notochord and beneath the primitive streak suggest that *T* may have a cell autonomous function in these tissues. Furthermore, chimerism in the allantois invariably results in defective development of this extraembryonic tissue and the severity of the abnormalities correlates with the degree of chimerism. This too could be a cell autonomous effect if there is a requirement for allantoic precursors to express *T* as they invaginate through the streak.

Table 1. *Histogenetic potential of* T/T, T/+ *and* +/+ *ES cells in teratomas*

Cell line	No. teratomas analysed	EC cells*	Neural tissue	Epithelial cysts	Skin	Cartilage	Bone	Striated muscle
BTBR6 & BTBR10 (*T/T*)	10	9 (90)	10 (100)	10 (100)	2 (20)	4 (40)	3 (30)	3 (30)
BTBR1.3 (*T/+*)	2	2 (100)	2 (100)	2 (100)	2 (100)	2 (100)	1 (50)	1 (50)
BTBR4 (+/+)	3	2 (66.7)	3 (100)	3 (100)	1 (33.3)	2 (66.7)	2 (66.7)	2 (66.7)

*Cell types are scored according to morphological criteria in histologically stained tumour sections.
Figures in parentheses are the percentage of the total number analysed.

Table 2. *Phenotype and frequency of* T/T ↔ +/+ *and* T/+ ↔ +/+ *ES cell chimeras recovered at 8.5-9.5 days*

Cell line	No. implantation sites containing embryos	No. normal	No. normal chimeras (% of normal embryos)	No. abnormal	No. abnormal chimeras (% of abnormal embryos)
BTBR6 & BTBR10 (*T/T*)	50	36	1 (2.8)	14	14 (100)
BTBR1.3 (*T/+*)	8	8	5 (62.5)	0	0 (0)

Table 3. *Phenotype and levels of chimerism of liveborn offspring derived from blastocysts injected with* T/T *or* T/+ *ES cells*

Cell line	No. liveborn/ blastocysts transferred (%)	No. abnormal tail (% of liveborn)	No. abnormal chimeric* (% of abn. tail)	No. normal tail (% of liveborn)	No. normal chimeric (% of normal)
BTBR6 & BTBR10 (*T/T*)	23/49 (46.9)	8 (34.8)	1† (12.5)	15 (65.2)	0 (0)
BTBR1.3 & BTBR7 (*T/+*)	32/52 (61.5)	2 (6.3)	2 (100)	30 (93.7)	6 (20)

*Chimerism is judged by coat colour except as indicated.
†Trace only by GPI analysis.

Liveborn and 11th and 12th day of gestation chimeras

This analysis has now been extended to examine what happens to *T/T* cells in low level chimeras later in gestation (V. Wilson, P. Rashbass and R. Beddington, unpublished data). Initially, blastocysts injected with *T/T* or *T/+* ES cells were reimplanted in pseudopregnant recipients and allowed to develop to term. Table 3 shows the frequency of liveborn young and the incidence and level of chimerism. As expected, *T/+* ES cells could contribute to viable liveborn chimeras. This had already been witnessed in preimplantation embryo aggregation chimeras where the liveborn *T/+* ↔ +/+ chimeras were genotyped retrospectively according to tail length and transmission of the mutant *T* allele to offspring (Bennett, 1978). Blastocysts injected with *T/T* ES cells gave rise to liveborn young but no coat colour chimerism was evident in those pups that survived. Five of these surviving offspring had short or deformed tails. A further 3 neonates, which died shortly after birth, had short, curly or absent tails. Surprisingly, only a trace of GPI 1A activity was found in the tail remnant of one of these animals (Table 3). Nonetheless, the striking tail abnormalities strongly suggested that these 8 offspring had been chimeric at some stage of their development. However, despite the lack of coat colour chimerism, it was possible that all 8 might simply be extremely low level chimeras (less than 2% contribution) which could not be detected by the GPI assay. Alternatively, they might once have contained detectable levels of ES cell contribution which had been selected against during the latter part of gestation. To resolve this question we examined embryos injected with *T/T* or *T/+* ES cells on the 11th and 12th days of gestation.

Table 4 shows the incidence of normal and abnormal chimeras recovered on the 11th or 12th day of gestation. In general, *T/+* ↔+/+ chimeras either appeared grossly normal or displayed a high incidence of mild tail defects, such as kinking or slight irregularity of the tail tip. The allantois had fused with the chorion and as expected there were no visible abnormalities in the cranial or trunk region of the embryo reminiscent of the homozygous phenotype. In contrast, *T/T* ↔ +/+ chimeras show a range of defects extending from a severe phenotype, similar to dying intact homozygous mutants, to localised defects in the allantois or distal tail region. The tail defects included truncation, branching and the appearance of blood filled sacs at the caudal extreme of the tail. Normal and abnormal embryos were subdivided into different axial regions and the extent of chimerism measured following GPI gel electrophoresis.

Embryos classified as normal were not chimeric, with the exception of a single embryo, which proved to have an ES contribution of approximately 10%. On the other hand, over 85% of the abnormal embryos contained *T/T* ES cell descendants (Table 4). A very interesting axial pattern of chimerism was evident (Fig. 2). In all chimeras there was an increase from head to tail in the *T/T* cell contribution. Indeed, in some embryos a contribution was only evident in caudal structures (Fig. 2) or in the allantois (data not shown). Enzymatic separation (Beddington, 1987; Levak-Svajger and Svajger, 1971) of forelimb bud and hindlimb bud regions of some abnormal embryos into separate neural tube and paraxial and lateral mesoderm fractions also showed a consistent trend. The neural tissue was almost invariably more chimeric than mesodermal tissues in *T/T* ↔+/+ (Fig. 3). This was not the case in *T/+* ↔+/+

Table 4. *Phenotype and frequency of chimerism in 9.5-11.5 day embryos recovered from blastocysts injected with* T/T *or* T/+ *ES cells*

Cell line	No. implantation sites containing embryos	No. normal	No. normal chimeric	No. abnormal	No. abnormal chimeric
BTBR10	76	51	1	25	21
BTBR6 (*T/T*)	23	5	0	18	16
Total	99	56	1 (1.8%)	43	37 (86.1%)
BTBR7 (*T/+*)	34	21	3 (14.3%)	13	10 (76.9%)

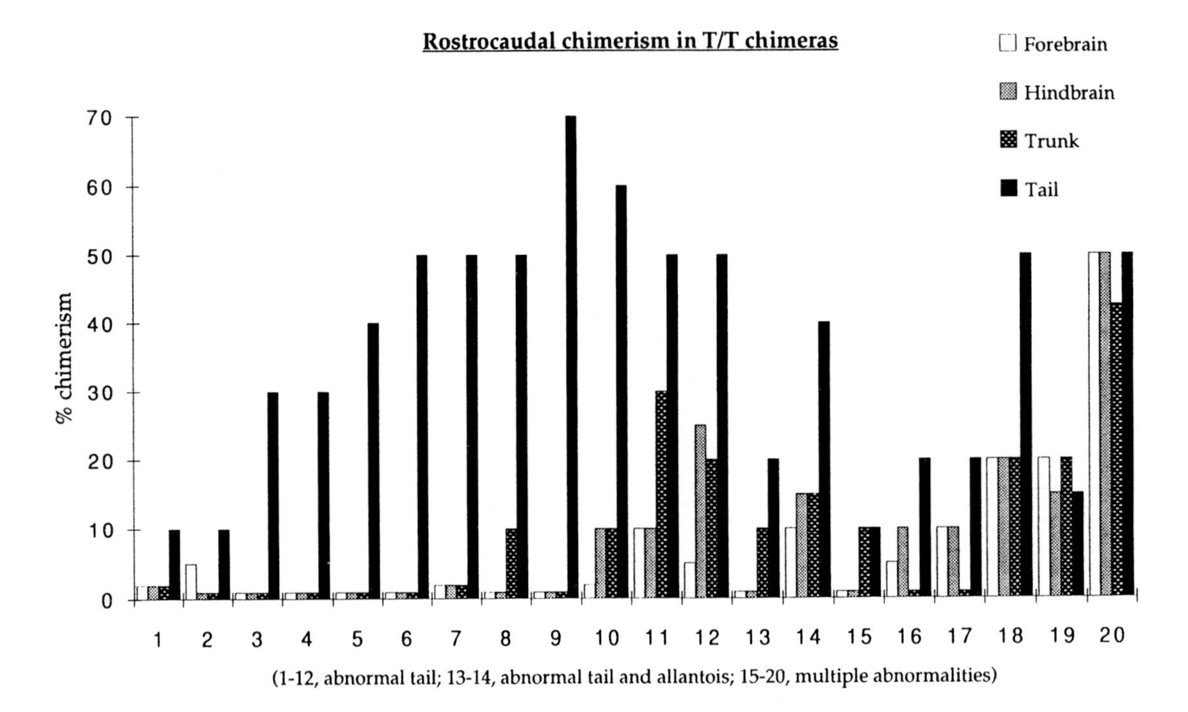

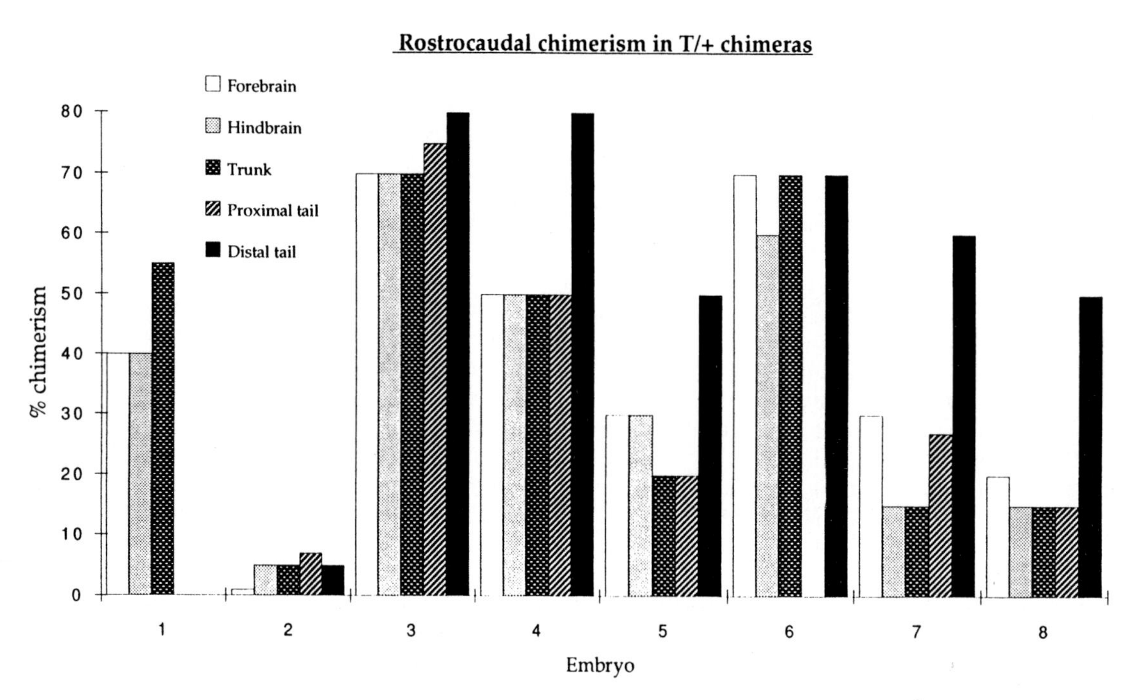

Fig. 2. Histograms depicting the percentage of ES cell contribution along the rostrocaudal axis in 11th and 12th day *T/T* ↔+/+ and *T/+* ↔+/+ chimeras. In *T/T* ↔+/+ chimeras there is a pronounced skew in the contribution of *T/T* ES cells towards the caudal end of the embryo. This is still apparent in *T/+*↔+/+ chimeras but less extreme. Samples in which ES cell contribution was undetectable are assigned an arbitrary value of 1% chimerism, to distinguish them from samples which were not analysed. The latter are assigned a value of zero.

chimeras where there was a more or less equivalent level of chimerism in the neural tube and mesoderm fractions.

Conclusions

There are three possible explanations for this rostrocaudal gradient of *T/T* ES cell colonisation. Embryos earlier in development could have been equally chimeric throughout, but there was subsequently strong selection against *T/T* cells in rostral regions. This would seem unlikely since the gene is not expressed in the head and no defects are apparent in this region, which might be expected if there were extensive cell death. Certainly, no such cell death has been

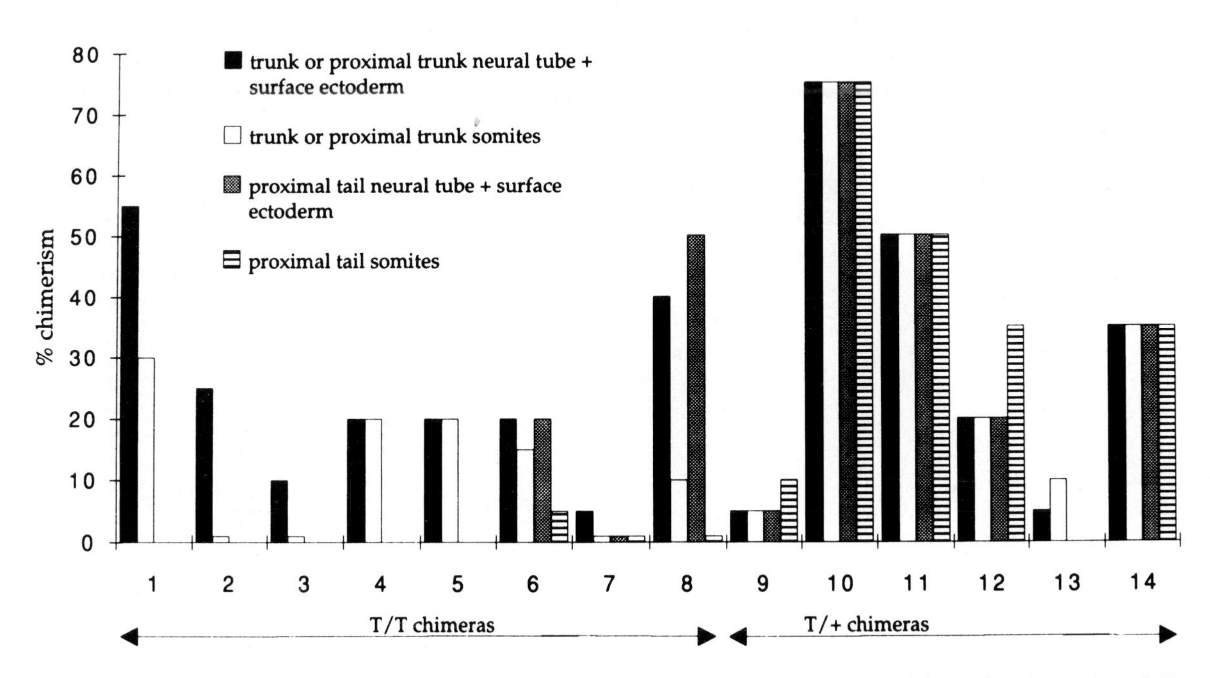

Fig. 3. Histogram comparing the contribution of ES cells to neurectoderm as opposed to mesodermal tissues in *T/T* ↔+/+ and *T/+* ↔+/+ chimeras. In *T/T*↔+/+ chimeras the neurectoderm tends to be more chimeric than the mesoderm but this is not the case in *T/+* ↔+/+ chimeras.

observed in intact *T/T* mutants. Moreover, high level chimeras contain significant populations of *T/T* cells in the cranial region (e.g. Embryo 20, Fig. 2). The second possibility is that *T/T* ES cells assume a very abnormal position in the epiblast before or during gastrulation, compared to the random distribution observed with wild-type ES cells (R.Beddington, unpublished observations; Suemori et al., 1990), which leads to this preferred caudal fate. Again the high level chimeras make this improbable because *T/T* cells can be distributed throughout the axis. Furthermore, inspection of separated germ layer derivatives (Fig. 3) would argue that *T/T* cells were present in anterior regions of the epiblast which give rise to neurectodermal tissues (Beddington, 1981, 1982; Lawson et al., 1991). The third, and most plausible, explanation is that *T/T* cells accumulate at the caudal end of the embryo because once invaginated during gastrulation they fail to migrate away from the streak efficiently. Consequently, *T/T* mesodermal cells emerging from the streak will tend to end up in the cloacal region (the conventional anatomical boundary between primitive streak and tailbud derived embryonic tissues). They will also make a relatively higher contribution to the tail bud when it begins to form at the caudal end of the embryo during the early forelimb bud stage. This will lead to an elevated level of *T/T* chimerism in the tail compared to more rostral levels of the embryo. Proliferation of the mesodermal cells in the tail bud is responsible for elongation of the tail (Tam, 1984). If *T/T* cells are compromised in their ability to move away from this growth zone then they will further accumulate at the distal tip of the growing tail. The restriction of tail abnormalities to this region in *T/T* ↔+/+ chimeras endorses this interpretation.

A prediction that follows the hypothesis that *T/T* cells fail to migrate away from the streak would be that the level of chimerism in the neurectoderm, a tissue that never invaginates, should always be higher than the level of chimerism in the paraxial and lateral mesoderm (Fig. 4). This prediction is borne out in *T/T* ↔+/+ chimeras when compared to *T/+* ↔+/+ embryos (Fig. 3). This bias against *T/T* cells colonising mesodermal tissues rostral to the primitive streak also implies that this impairment of movement is a cell autonomous effect, rather than being caused by extracellular matrix defects, since wild-type cells can evidently move away from the streak normally. Therefore, we would predict that a gene or genes activated by *T* expression would code for intracellular or cell surface components involved in active movement or in cell adhesion.

A possible model for the mode of action of the *T* gene can be made. Assuming, on the basis of the dominant negative mutant phenotypes (see above), that *T* acts as part of a transcription factor dimer or complex (Herrmann, 1992) then one function of this complex would be to regulate a gene, or genes, whose products are cell autonomous components of adhesion or migration processes. In T^{Wis}/T^{Wis} or T^c/T^c embryos the abnormal T protein can still form a dimer or protein complex but this binding of the mutant T protein results in an inactive complex. Consequently, a subset of adhesion or motility proteins would be altered in the cells beneath the streak, thereby seriously compromising their morphogenetic movements. In *T/T* embryos, the absence of T protein does not affect the availability of the other component(s) of the transcriptional regulator and there is enough residual activity in a transcription complex lacking T protein to produce low level transcription of downstream

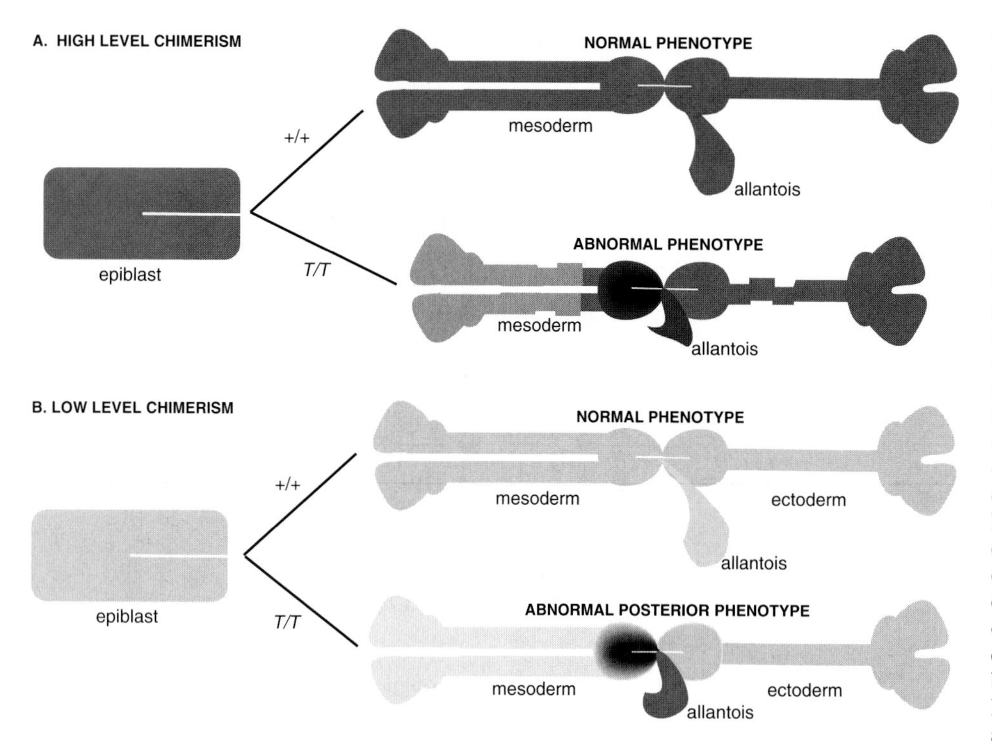

Fig. 4. Diagram depicting the distribution of *T/T* ES cells in high level and low level chimeras compared to the distribution of wild-type cells in chimeras. The epiblast, from which the fetus and extraembryonic mesoderm are derived, is shown on the left and increasing levels of chimerism represented by darker shades of grey. Stereotyped 9th day embryos are shown on the right with the mesodermal constituent on one side (left) and the ectodermal component on the other (right). The allantois is drawn emerging from the caudal extreme of the embryo. In both high level (A) and low level (B) chimeras made with wild-type ES cells, contribution is usually equivalent throughout the length of the axis, and the levels of chimerism do not tend to vary between different tissues. High level *T/T* ↔ +/+ chimeras are abnormal and exhibit many of the classical features of intact *T/T* mutants. There is an accumulation of *T/T* ES cells in the caudal region of the embryo and the mesoderm is less chimeric than the neurectoderm. In low level *T/T* ↔ +/+ chimeras the embryos are more normal in the trunk region but develop abnormalities in the tail. They develop to a later stage and show a pronounced accumulation of *T/T* cells caudally at the expense of *T/T* cells contributing to more rostral mesoderm.

genes. As a result, morphogenetic movements of the mesoderm are still affected but less severely than in T^{Wis}/T^{Wis} or T^c/T^c. In the chimera where *T/T* cells are in competition with wild-type ones these motility defects result in accumulation of *T/T* cells at the caudal end of the streak or tail.

This scenario does not fully explain either the defects in the notochord or the abnormalities in the allantois. The sustained expression of *T* in the notochord after gastrulation has ceased indicates that the gene product may have additional function in this tissue and the cell death observed in the notochord of *T/T* ↔+/+ chimeras (Rashbass et al., 1991) would be consistent with the T protein being required for cell survival. The effects on the allantois are more mysterious. Analysis of in vitro chimeras, formed by grafting [^{3}H]thymidine-labelled cells into the primitive streak of late-streak or early-somite stage embryos, indicate that the allantois forms as a self-contained population of cells in less than 24 hours (Tam and Beddington, 1987). Chimeras created at the late-streak stage contain labelled cells in the allantois whereas those grafted at the early-somite stage do not. It is possible that accumulation of *T/T* cells at the caudal end of the embryo physically blocks the recruitment of cells into the allantoic bud. However, the degree of chimerism expected to produce such an efficient blockage might be expected to result in concomitant tail chimerism. This is not always the case; some *T/T* ↔+/+ chimeras, albeit a low percentage, are abnormal only in the allantois and contain detectable numbers of *T/T* cells only in this tissue. A more compelling argument against simple physical blockage is that in chimeras there does not appear to be an under representation of *T/T* cells in the allantois. Moreover, the effect must be more complex than a simple 'log-jam' effect because there is a characteristic alteration in the morphology and behaviour of allantoic cells. They show a pronounced preference to spread over the amnion rather than cohere and traverse the exocoelom. Thus, the absence of *T* expression results in the altered phenotype of cells which even in wild-type embryos are not expressing the gene. It is possible, therefore, that *T* expression is an obligatory component of the initial specification of normal allantoic tissue within the streak region, or that it is an essential early step in the subsequent differentiation of allantoic precursors.

This work was supported by the Agriculture and Food Research Council and the Imperial Cancer Research Fund. We would like to thank Linda Manson and Louise Anderson for their valuable technical assistance.

References

Beddington, R. S. P. (1981). An autoradiographic analysis of the potency of embryonic ectoderm in the 8th day postimplantation mouse embryo. *J. Embryol. exp. Morph.* **64**, 87-104.

Beddington, R. S. P. (1982). An autoradiographic analysis of tissue potency in different regions of the embryonic ectoderm during gastrulation in the mouse. *J. Embryol. exp. Morph.* **69**, 265-285.

Beddington, R. S. P. (1987). Isolation, culture and manipulation of post-implantation mouse embryos. In *Mammalian Development: A practical Approach.* (Ed. M. Monk). Oxford: IRL Press.

Beddington, R. S. P., Püschel, A. W. and Rashbass, P. (1991). Chimeras to study gene function in mesodermal tissues during gastrulation and early organogenesis. In *Postimplantation Development in the Mouse. Ciba Found. Symp.* **165**, p.61-74.

Bennett, D. (1978). Rescue of a lethal *T/t* locus genotype by chimerism with normal embryos. *Nature* **272**, 539.

Bennett, D., Artzt, K., Magnuson, T. and Spiegelman, M. (1977). Developmental interactions studied with experimental teratomas derived from mutants at the *T/t* locus in the mouse. In *Cell Interactions in Differentiation*. (Ed. M. Kaarkinen-Jaaskelainen, L. Saxen and L. Weiss), pp. 389-398. New York: Academic Press.

Chesley, P. (1935). Development of the short-tailed mutant in the house mouse. *J. exp. Zool.* **70**, 429-459.

Clarke, J. D. W., Holder, N., Soffe, S. R. and Storm-Mathison, J. (1991). Neuroanatomical and functional analysis of neural tube formation in notochordless *Xenopus* embryos; laterality of the ventral spinal cord is lost. *Development* **112**, 499-516.

Diwan, S. B. and Stevens, L. C. (1976). Development of teratomas from ectoderm of mouse egg cylinders. *J. Natl. Cancer Inst.* **57**, 937-942.

Dobrovolskaïa-Zavadskaïa, N. (1927). Sur la mortification spontanée de la queue chez la souris nouveau-née et sur l'existence d'un caractère heriditaire "non-viable". *C. R. Soc. Biol.* **97**, 114-116.

Ephrussi, B. (1935). The behaviour in vitro of tissues from lethal embryos. *J. Exp. Zool.* **70**, 197-204.

Evans, M. J. and Kaufman, M. H. (1981). Establishment in culture of pluripotential cells from mouse embryos. *Nature* **292**, 154-155.

Fujimoto, H. and Yanagisawa, K. O. (1979). Effects of the *T* mutation on histogenesis of the mouse embryo under the testis capsule. *J. Embryol. exp. Morph.* **50**, 21-30.

Fujimoto, H. and Yanagisawa, K. O. (1983). Defects in the archenteron of mouse embryos homozygous for the *T*-mutation. *Differentiation* **25**, 44-47.

Gluecksohn-Schoenheimer, S. (1944). The development of normal and homozygous *brachy* (*T/T*) mouse embryos in the extraembryonic coelom of the chick. *Proc. Natl. Acad. Sci. USA* **30**, 134-140.

Gluecksohn-Schoenheimer, S. (1938). The development of two tailless mutants in the house mouse. *Genetics* **23**, 573-584.

Gruneberg, H. (1958). Genetical studies on the skeleton of the mouse. XXIII. The development of *Brachyury* and *Anury*. *J. Embryol. exp. Morph.* **6**, 424-443.

Handyside, A. H., Kontogianni, E. H., Hardy, K. and Winston, R. M. L. (1990). Pregnancies from biopsied human preimplantation embryos sexed by Y-specific DNA amplification. *Nature* **344**, 768-770.

Hashimoto, K., Fujimoto, H. and Nakatsuji, N. (1987). An ECM substratum allows mouse mesodermal cells isolated from the primitive streak to exhibit motility similar to that inside the embryo and reveals a deficiency in the *T/T* mutant cells. *Development* **100**, 587-598.

Hemmati-Bravanlou, A., Stewart, R. M. and Harland, R. M. (1990). Region-specific neural induction of an engrailed protein by anterior notochord in *Xenopus*. *Science* **250**, 800-802.

Herrmann, B. G. (1991). Expression pattern of the *Brachyury* gene in whole-mount T^{Wis}/T^{Wis} mutant embryos. *Development* **113**, 913-917.

Herrmann, B. G. (1992). Action of the *Brachyury* gene in mouse embryogenesis. In *Postimplantation Development in the Mouse. Ciba Found. Symp.* **165**, 78-86.

Herrmann, B. G., Labeit, S., Poustka, A., King, T. R. and Lehrach, H. (1990). Cloning of the *T* gene required in mesoderm formation in the mouse. *Nature* **343**, 617-622.

Jacobs-Cohen, R. J., Spiegelman, M. and Bennett, D. (1983). Abnormalities of cells and extracellular matrix of *T/T* embryos. *Differentiation* **25**, 48-55.

Justice, M. J. and Bode, V. C. (1990). ENU-induced allele of *Brachyury* (T^{kt1}) exhibits a developmental lethal phenotype similar to the original Brachyury (*T*) mutation. *J. exp. Zool.* **254**, 286-295.

Kitchin, I. C. (1949). The effects of notochordectomy in *Ambystoma mexicanum*. *J. exp. Zool.* **112**, 393-415.

Lawson, K. A., Meneses, J. J. and Pedersen, R. A. (1991). Clonal analysis of epiblast during germ layer formation in the mouse embryo. *Development* **113**, 891-911.

Levak-Svajger, B. and Svajger, A. (1971). Differentiation of endodermal tissues in homografts of primitive ectoderm from two-layered rat embryonic shields. *Experientia* **27**, 683-684.

Lyon, M. E. and Searle, A. G. (1989). *Genetic Variants and Strains of the Laboratory Mouse*. 2nd ed. pp. 876. Oxford: Oxford University Press.

Lyon, M. F. and Meredith, R. (1964). The nature of *t* alleles in the mouse. II. Genetic analysis of an unusual mutant allele and its derivatives. *Heredity* **19**, 313-325.

MacMurray, A. and Shin, H.-S. (1988). The antimorphic nature of the T^c allele at the mouse locus T. *Genetics* **120**, 545-550.

Maden, M., Ong, D. E. and Chytil, F. (1990). Retinoid binding protein distribution in the developing mammalian nervous system. *Development* **109**, 75-80.

Martin, G. R. (1981). Isolation of a pluripotent cell line from early mouse embryos cultured in medium conditioned by teratocarcinoma stem cells. *Proc. Natl. Acad. Sci., USA* **78**, 7634-7638.

Placzek, M., Tessier-Lavigne, M., Yamada, T., Jessell, T. and Dodd, J. (1990). Mesodermal control of neural cell identity: floor plate induction by the notochord. *Science* **250**, 985-988.

Rashbass, P., Cooke, L. A. Herrmann, B. G. and Beddington, R. S. P. (1991). A cell autonomous function of *Brachyury* in *T/T* embryonic stem cell chimeras. *Nature* **353**, 348-350.

Reith, A. D. and Bernstein, A. (1991). Molecular basis of mouse developmental mutants. *Genes Dev.* **5**, 1115-1123.

Searle, A. G. (1966). *Curtailed*, a new dominant T-allele in the house mouse. *Genet. Res.* **7**, 86-95.

Shedlovsky, A., King, T. R. and Dove, W. F. (1988). Saturation germ line mutagenesis of the murine *t* region including a lethal allele at the *quaking* locus. *Proc. Natl. Acad. Sci. USA* **85**, 180-184.

Shur, B. D. (1982). Cell surface glycosyltransferase activities during normal and mutant (*T/T*) mesenchyme migration. *Dev. Biol.* **91**, 149-162.

Smith, J. C., Price, B. M. J., Green, J. B. A., Weigel, D. and Herrmann, B. G. (1991). Expression of the *Xenopus* homolog of *Brachyury* (*T*) is an immediate-early response to mesoderm induction. *Cell* **67**, 79-87.

Smith, J. L. and Schoenwolf, G. C. (1989). Notochordal induction of cell wedging in the chick neural plate and its role in neural tube formation. *J. exp. Zool.* **250**, 49-62.

Spiegelman, M. (1976). Electron microscopy of cell associations in T-locus mutants. In *Embryogenesis in Mammals*. (Ed. K. Elliott and M. O'Connor) pp. 199-226. Amsterdam: Elsevier/North Holland.

Suemori, H., Kadokawa, Y., Goto, K., Araki, I., Kondoh, H. and Nakatsuji, N. (1990). A mouse embryonic stem cell line showing pluriptotency of differentiation in early embryos and ubiquitous beta-galactosidase expression. *Cell Differ. Dev.* **29**, 181-186.

Tam, P. P. L. (1984). The histogenetic capacity of tissues in the caudal end of the embryonic axis of the mouse. *J. Embryol. exp. Morph.* **82**, 253-266.

Tam, P. P. L., and Beddington, R. S. P. (1987). The formation of mesodermal tissues in the mouse embryo during gastrulation and early organogenesis. *Development* **99**, 109-126.

van Straaten, H. W. M., Thors, F., Wiertz-Hoessels, E. J. L., Hekking, J. W. M. and Drukker, J. (1985). Effect of a notochord implant on the early morphogenesis of the neural tube and neuroblasts: histometrical and histological results. *Dev. Biol.* **110**, 247-254.

Wilkinson, D., Bhatt, S. and Herrmann, B. G. (1990). Expression pattern of the mouse *T* gene and its role in mesoderm formation. *Nature* **343**, 657-659.

Willison, K. (1990). The mouse *Brachyury* gene and mesoderm formation. *Trends Genet.* 104-105.

Winking, H. and Silver, L. M. (1974). Characterization of a recombinant mouse *t* haplotype that expresses a dominant lethal maternal effect. *Genetics* **108**, 1013-1020.

Yamada, T., Placzek, M., Tanaka, H., Dodd, J. and Jessell, T. M. (1991). Control of cell pattern in the developing nervous system: Polarizing activity of the floor plate and notochord. *Cell* **64**, 635-647.

Yanagisawa, K. O. (1990). Does the *T* gene determine the anteroposterior axis of a mouse embryo? *Jpn. J. Genet.* **65**, 287-297.

Yanagisawa, K. O. and Fujimoto, H. (1977a). Differences in rotation mediated aggregation between wild-type and homozygous Brachyury (T) cells. *J. Embryol. exp. Morph.* **40**, 277-283.

Yanagisawa, K. O. and Fujimoto, H. (1977b). Viability and metabolic activity of homozygous *Brachyury* (*T*) embryos. *J. Embryol. Exp. Morph.* **40**, 271-276.

Yanagisawa, K. O., Fujimoto, H. and Urushihara, H. (1981). Effects of the *Brachyury* (*T*) mutation on morphogenetic movement in the mouse embryo. *Dev. Biol.* **87**, 242-248.

Development 1992 Supplement, 167-171 (1992)
Printed in Great Britain

goosecoid and the organizer

EDDY M. DE ROBERTIS, MARTIN BLUM, CHRISTOF NIEHRS and HERBERT STEINBEISSER

Molecular Biology Institute and Department of Biological Chemistry, University of California, Los Angeles, CA 90024-1737, USA

Summary

The molecular nature of Spemann's organizer phenomenon has long attracted the attention of embryologists. *goosecoid* is a homeobox gene with a DNA-binding specificity similar to that of *Drosophila bicoid*. *Xenopus goosecoid* is expressed on the dorsal side of the embryo before the dorsal lip is formed. Cells expressing *goosecoid* are fated to become pharyngeal endoderm, head mesoderm and notochord. Transplantation of *goosecoid* mRNA to the ventral side of *Xenopus* embryos by microinjection mimics the properties of Spemann's organizer, leading to the formation of twinned body axes. *goosecoid* is activated by dorsal inducers and not affected by ventral inducers. In the mouse, *goosecoid* is expressed in the anterior tip of the primitive streak. The availability of two early markers, *goosecoid* and *Brachyury*, opens the way for the comparative analysis of the vertebrate gastrula. The results suggest that the *goosecoid* homeodomain protein is an integral component of the biochemical pathway leading to Spemann's organizer phenomenon.

Key words: *goosecoid*, Spemann's organizer, gastrulation, *Drosophila*, *Xenopus*.

Introduction

It is now generally agreed that development results from a series of cell-cell interaction events. The experiment that contributed more than any other to this view was that of Hans Spemann and Hilde Mangold (1924), showing that a small fragment of the gastrula, the dorsal lip, had the ability, after transplantation, to form a twinned body axis in the opposite (ventral) side of a host embryo. The transplanted tissue contributed only a small part of this secondary axis, and thus was able to recruit, or organize, cells into complex anatomical structures. The quest to understand the inductive properties of the organizer still provides great impetus to experimental embryologists (reviewed by Spemann 1938; Nakamura and Toivonen, 1978; Hamburger, 1988; Marx, 1991).

In the present paper, we will discuss how microinjection of purified *goosecoid* homeobox-containing mRNA can mimic most of the properties of Spemann's organizer.

Homeobox genes expressed in the organizer

Homeobox genes encode DNA-binding proteins, which frequently are involved in the specification of positional information in the embryo (reviewed by Gehring 1987; Kessel and Gruss, 1990; De Robertis et al., 1991; McGinnis and Krumlauf, 1992). It is therefore of interest that several such genes are expressed in the dorsal lip region of the *Xenopus* embryo. By screening a cDNA library derived from manually dissected *Xenopus* dorsal lips, four different homeobox genes were isolated by Blumberg et al. (1991). Three were related to genes isolated previously (*X. caudal* 1 and 2, *X. labial*), but the most abundant clone (23 out of 30 isolates) contained a novel type of homeobox. Because its homeobox contained similarities to the *Drosophila* genes *gooseberry* in the initial ⅔ of the homeodomain and to *bicoid* in the DNA recognition helix, this gene was christened *goosecoid*. Additional genes expressed in the dorsal lip have been isolated by other groups from *Xenopus* embryo cDNA libraries by screening with homologous probes. *X-LIM-1* contains a homeobox as well as a conserved cysteine-rich domain (Taira et al., 1992), and while *XFKH*-1 lacks a homeobox, it has sequence similarities to the DNA-binding domain of the *Drosophila fork-head* homeotic gene (Dirksen and Jamrich, 1992).

The expression of *goosecoid* in the *Xenopus* gastrula

All the genes mentioned above are surely important in building the body axis, but we will deal here only with *goosecoid,* which is particularly interesting because it has some degree of functional similarity to the *Drosophila* anterior morphogen *bicoid* (Nüsslein-Volhard, 1991). In an *in vitro* assay a *goosecoid* recombinant protein was shown to bind target sequences with a DNA-binding specificity similar to that of *Drosophila bicoid* on target sequences derived from the promoter of the gap gene *hunchback* (Blumberg et al., 1991). In addition, *goosecoid* is expressed very early in *Xenopus* embryos. As can be seen in Fig. 1B, *goosecoid* expression is detectable by whole-mount in situ hybridization as a crescent on the dorsal marginal zone of the

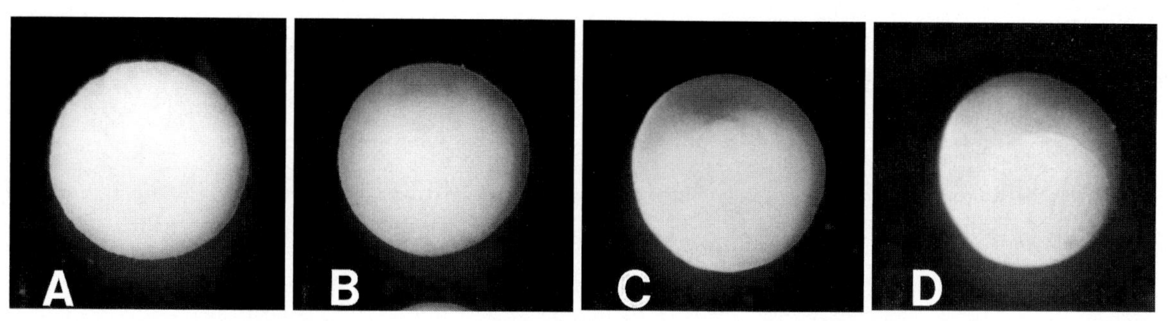

Fig. 1. Time course of *goosecoid* expression in *Xenopus* embryos. Whole-mount *in situ* hybridization (Harland, 1991) at (A) stage 8, mid blastula; (B) stage 9, late blastula one hour before the onset of gastrulation; (C) stage 10, early gastrula in which the dorsal lip is formed; (D) stage 11, mid gastrula with a circular blastopore. Note that the patch of goosecoid expression in the marginal zone is present before gastrulation starts (B), and that it invaginates into the interior of the embryo during gastrulation (D). Unfortunately the black and white reproduction of this figure, which was originally in color, does not have enough contrast.

embryo at least one hour before the start of gastrulation. At the start of gastrulation (Fig. 1C) *goosecoid* is most intense in an arc of about 60° just above the dorsal lip (Cho et al., 1991). This corresponds to the region where Spemann's organizer, as defined by transplantation experiments, is located (Cooke, 1972; Gerhart et al., 1989).

In sagittal sections, *goosecoid* is seen to be expressed in the internal layer of the dorsal lip region. As shown in Fig. 2, cells close to the incipient indentation of the dorsal lip (indicated by an arrow) contain mostly nuclear transcripts, while cells located further away from the dorsal lip contain increasing amounts of cytoplasmic *goosecoid* RNA. The cells that express *goosecoid* correspond to the dorsal-most invaginating cells. Their normal fate is to become pharyngeal endoderm, head endoderm and notochord (Keller, 1976, 1991).

Microinjection of *goosecoid* mRNA gives rise to a twinned body axis

When full-length *goosecoid* mRNA is microinjected into the ventral side of the four-cell embryo (Cho et al., 1991), the formation of a second dorsal lip, and subsequently of a twinned body axis, ensues (Fig. 3). Some of these axes are complete, containing head structures such as eyes, cement gland and hatching glands. More frequently, only trunk structures containing massive amounts of notochord are formed (Cho et al., 1991). These structures are made at the expense of the tail, so that the resulting tadpoles are considerably shortened.

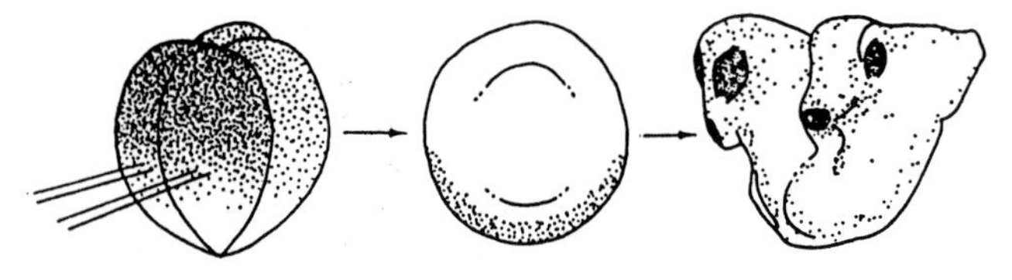

Fig. 3. Diagram depicting the effects of injecting *goosecoid* mRNA into the two ventral blastomeres of a 4-cell *Xenopus* embryo (i.e., opposite to the site where this gene will normally be expressed). A second dorsal lip is formed, which leads to the formation of a twinned body axis. After experiments of Cho et al. (1991), redrawn by Koichiro Shiokawa.

Blastomeres microinjected with *goosecoid* are able to recruit neighboring uninjected cells into the secondary axes (C. Niehrs, K.W.Y. Cho and E. De Robertis, unpublished results), as occurs in Spemann's organizer phenomenon. Thus, the *goosecoid* homeobox protein has non-cell autonomous effects, a property that has been noted for some *Drosophila* homeobox genes during midgut development (Immerglück et al., 1990; Reuter and Scott, 1990).

The conclusion from these microinjection experiments is that transplantation of a single gene product, *goosecoid* mRNA, can mimic Spemann's organizer phenomenon.

How is *goosecoid* activated?

Our understanding of mesoderm induction has progressed greatly in the past few years. The pioneering work of Peter Nieuwkoop showed that mesoderm induction results from signals emanating from the vegetal cells (reviewed by Nieuwkoop, 1973). Furthermore, ventral and dorsal vegetal cells differ in their inductive power when placed in contact with pluripotent animal cap cells. As indicated in Fig. 4, ventral induction leads to the formation of tissues such as blood, mesothelium (coelom) and lateral plate mesoderm. Addition of FGF to animal cap fragments mimics this ventral induction (Slack et al., 1987; Slack, this volume). Dorsal and vegetal blastomeres induce notochord and muscle when conjugated with animal cap cells (Boterenbrood and Nieuwkoop, 1973). Growth factors of the activin type (Smith et al., 1989; Green and Smith, 1990) and of the *Wnt* type (McMahon and Moon, 1989a,b) can mimic this dorsal induction.

The nature of the molecules that cause mesoderm induction in vivo is not known (Slack, 1991), but this is a very active area of research at present. Interesting recent developments suggest that (1) the microinjection of activin mRNA into single blastomeres can produce secondary axes (Thomsen et al., 1990) and (2) that injection of *Wnt*-8 mRNA into vegetal blastomeres can produce extensive secondary axes, including complete rescue of UV-treated embryos, presumably via induction of Spemann organizer function (Sokol et al., 1991; Smith and Harland, 1991). Because *Wnt*-8 is normally localized in the ventral side of the embryo, it is considered unlikely to be the endogenous

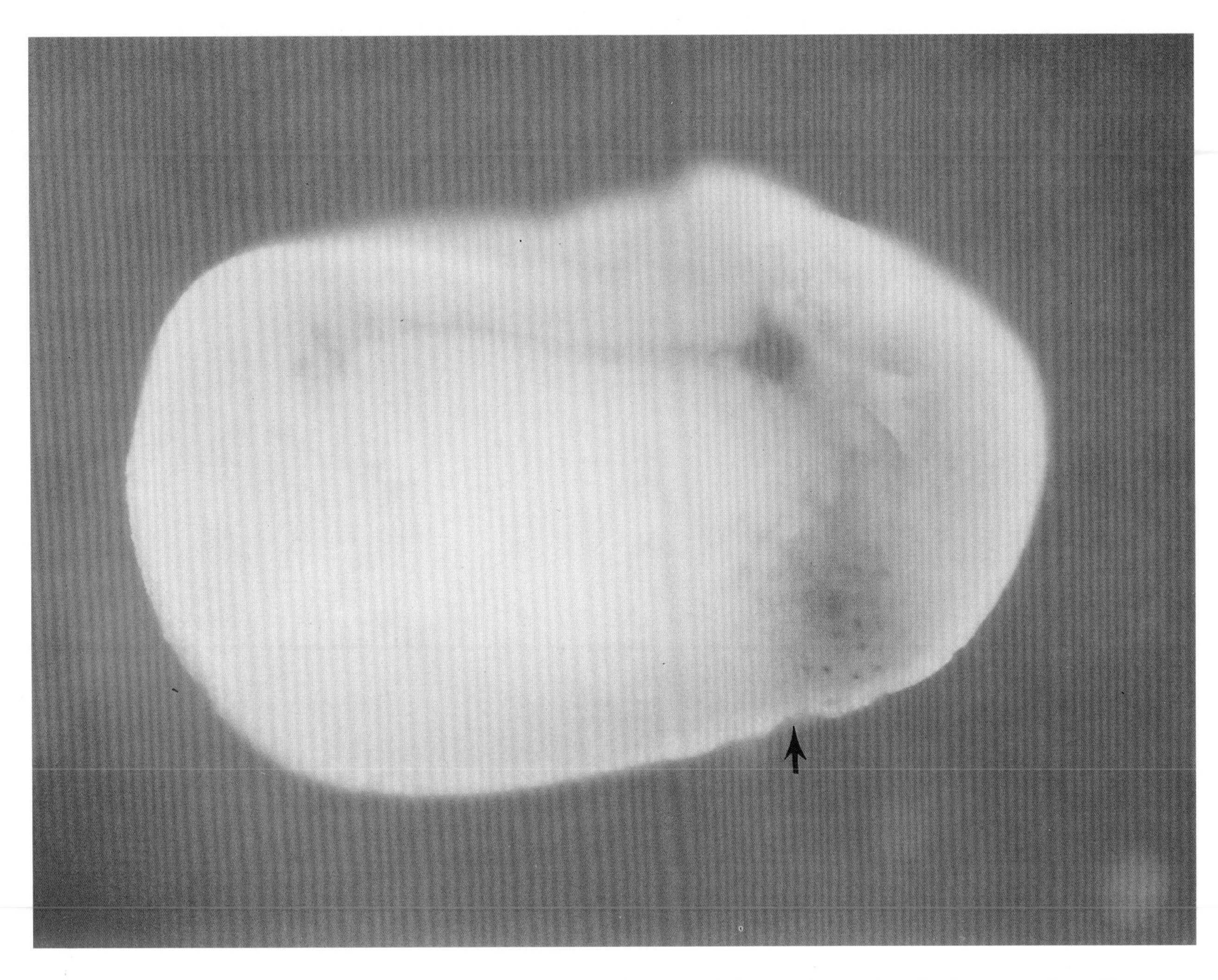

Fig. 2. *goosecoid* expression at the start of gastrulation, sagittal section through the organizer region of a *Xenopus* embryo. The arrow indicates the incipient dorsal blastopore lip. Note that *goosecoid* RNA is nuclear in the proximity of the dorsal lip (suggesting that the gene is first transcribed at this site) and cytoplasmic as cells move into deeper layers of the marginal zone. The fate of *goosecoid*-expressing cells is to become pharyngeal endoderm, head mesoderm and notochord.

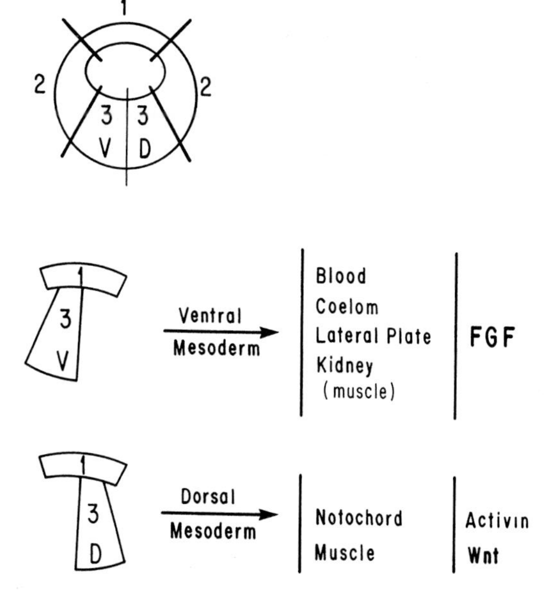

Fig. 4. Mesoderm induction can be of a ventral or dorsal nature. Explanted animal caps conjugated and cultured with either ventral or dorsal vegetal blastomeres form different types of mesoderm (after experiments by Boterenbrood and Nieuwkoop, 1973). These inductions can be mimicked by treating animal cap cells with FGF (ventral mesoderm) or with activin or *Wnt* (dorsal signals).

dorsal inducer, but a similar substance, as yet uncloned, could be secreted by dorsal and vegetal blastomeres.

The current model, summarized in Fig. 5, is therefore that there exists a radial signal (or signals) that induces a ring of cells in the overlying marginal zone to become ventral mesoderm. On the dorsal side, an additional signal (or signals) is released from the vegetal cells (Boterenbrood and Nieuwkoop, 1973; Gimlich and Gerhart, 1984), originating in a region that has been designated the Nieuwkoop center (Gerhart et al., 1989). The signal from the Nieuwkoop center acts upon the overlying marginal zone cells and induces Spemann's organizer tissue. *goosecoid* is expressed in the latter region.

As expected, *goosecoid* is induced by activin but not by the ventral inducer FGF (Cho et al., 1991). The effect of *Wnt*-type factors is presently under investigation. *goosecoid* is a primary response gene to activin induction, i.e., it's transcripts will accumulate even in the absence of protein synthesis. We have previously argued that this result would place *goosecoid* high up in the hierarchy of genetic events leading to axis formation (Cho et al., 1991). This interpretation should now be reassessed in view of recent findings that a great many genes are primary targets for activin in *Xenopus* animal cap explants. These include *Mix-1* (an endoderm-specific homeobox gene isolated by screening for activin-induced transcripts; Rosa, 1989), *Myo D* (Rupp and Weintraub, 1991), *Brachyury* (a gene expressed as a ring in the marginal zone of the *Xenopus* embryo as well as in the organizer region; Smith et al. (1991) and known to be required for notochord and posterior axis formation in the mouse, Rasbash et al., 1991), *X-LIM-1* (Taira et al., 1992) and *XFKH1* (Dirksen and Jamrich, 1992). In fact, transcripts for many of these genes are present at low levels already at the time when the animal cap are excised at the mid blastula stage. This applies to *Myo D* (Rupp and Weintraub, 1991), *X-LIM-1* (Taira et al., 1992), *goosecoid* (Cho et al., 1991) and presumably other inducible genes. Low levels of *goosecoid* mRNA are present in the *Xenopus* unfertilized egg and early cleavage stages (H. Steinbeisser, unpublished observations) but went unnoticed in our initial study (Blumberg et al., 1991). These maternal transcripts are not detectable by whole-mount in situ hybridization (Fig. 1A).

In order to dissect the hierarchy of these genes in axis formation, if indeed one exists, it will be necessary to carry out loss-of-function studies. For example, one would like to know whether *goosecoid* is expressed in a *Brachyury* mutant, and vice versa. Such studies should be possible in the mouse embryo.

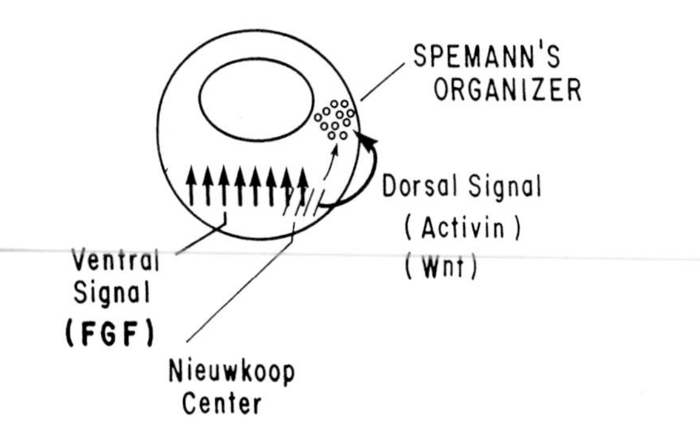

Fig. 5. A current view of mesoderm induction in *Xenopus* (see Gerhart et al., 1989). A ventral signal is released radially by the vegetal cells (e.g., FGF), inducing the entire marginal zone to become mesoderm. On the dorsal side an additional signal (perhaps activin- or *wnt*-like) is released by vegetal cells of the Nieuwkoop center, inducing Spemann organizer tissue in the overlying marginal zone cells.

goosecoid in mouse gastrulation

As is well known, all vertebrate embryos appear very similar to each other at mid-embryogenesis (the so-called phylotypic stage of the vertebrate embryo, Wolpert, 1991; Feduccia and McCrady, 1991). On the other hand, embryos from the various vertebrate classes appear to differ greatly at the gastrula stage. For example, in teleosts the main gastrulation movement is epiboly, in which the embryo proper envelops the yolk mass; in amphibians, which in general have holoblastic cleavage, the main morphogenetic movement is the invagination of the endomesoderm through the circular blastopore; while in birds and mammals (anmiotes) the main morphogenetic movement is the delamination of the future endodermal and mesodermal cells through the linear primitive streak. One of the main lessons that we have learned in the past few years, in particular from the extraordinary conservation of the workings of *Hox* genes (reviewed by De Robertis et al., 1990; Kessel and Gruss, 1990; McGinnis and Krumlauf, 1992), is that while embryogenesis may appear to differ greatly, the molecular mechanisms involved are universal. Southern blot analysis indicates that all classes of vertebrates contain a *goosecoid* homologue (Blum et al., 1992). Thus the *goosecoid* marker

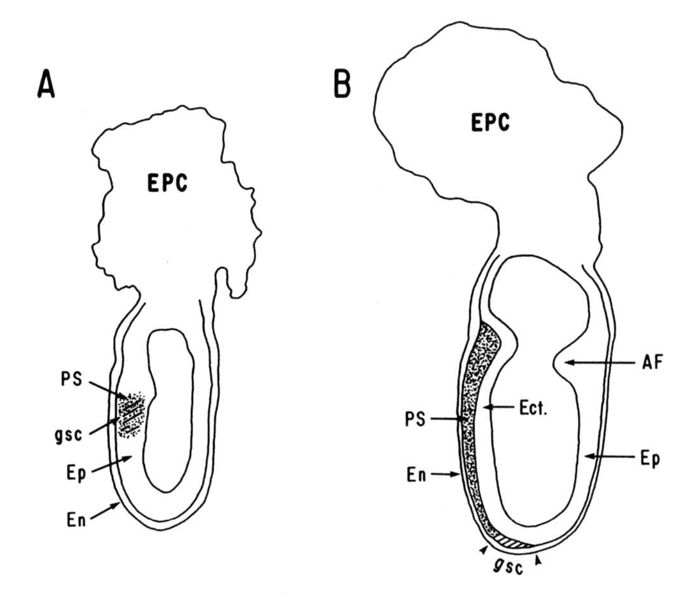

Fig. 6. Expression of *goosecoid* in the mouse gastrula. (A) At 6.4 days *goosecoid* (gsc) mRNA is expressed as a patch on the side of the epiblast, coinciding with the site at which the epithelial-mesenchymal transition that initiates primitive streak formation takes place. (B) At $6^3/_4$ days the region of *goosecoid* expression is located in the mesoderm at the anterior end of the primitive streak, located at the tip of the egg cylinder (gsc). This region corresponds to the future head process from which the definitive endoderm, head mesoderm, and notochord are derived. The *goosecoid* signal is represented by hatched bars; the primitive streak is stippled. The epiblast (Ep), embryonic endoderm (En), primitive streak mesoderm (PS), ectoderm (Ect), amniotic fold (AF) and ectoplacental cone (EPC) are indicated. After studies by Blum et al., 1992.

provides an opportunity for a comparative analysis of the organizer region in gastrulation.

In mammals, gastrulation occurs at the time in which the embryo, due to its small size and uterine implantation, is most inaccessible to study. Mouse *goosecoid* has been cloned, and analysis of its expression provides a useful marker for dorsal development. The mouse epiblast has the shape of a cup, called the egg cylinder (see Lawson, 1991). At 6.4 days post-coitum mouse *goosecoid* mRNA can be seen on the side of the egg cylinder, in the cells of the epiblast that begin the epithelial-mesenchymal transition that marks the start of primitive streak formation (Fig. 6A). As the primitive streak elongates, the *goosecoid*-expressing cells are seen to be located at the anterior end of the advancing mesoderm (Blum et al., 1992). The direction of these movements is in agreement with the mouse gastrula fate map (Lawson, 1991). By $6^3/_4$ days the *goosecoid*-expressing cells are located at the anterior-most tip of the primitive streak, as indicated in Fig. 6B. This region is the precursor of the head process, which gives rise to notochord, head mesoderm and the definitive endoderm. Shortly thereafter *goosecoid* mRNA becomes undetectable (Blum et al., 1992), and does not reappear until 3½ days later (Gaunt et al., 1992).

The distribution of *goosecoid* mRNA is in agreement with transplantation experiments of mouse gastrula fragments into *Xenopus* embryos. Head organizer activity was found at the tip of the egg cylinder at about $6^3/_4$ days, while the base of the egg cylinder, which contains most of the primitive streak, induces tail or no structures at all (Blum et al., 1992).

The availability of two markers, *goosecoid*, which marks the organizer and invaginating prechordal plate (Cho et al., 1991; Blum et al., 1992), and *Brachyury*, which marks the notochord as well as the marginal zone in *Xenopus* and the primitive streak in the mouse (Smith et al., 1991; Herrmann, 1991), now permits the identification of homologous regions in these two gastrulae. Together with the transplantation experiments described above, the molecular studies are consistent with the anterior end of the mouse primitive streak (Fig. 6B) corresponding to the *Xenopus* dorsal lip and the more posterior primitive streak to the lateral and ventral blastopore lip regions.

Conclusions

The isolation of *goosecoid* provides a molecular marker that should permit identification of the equivalent of the organizer region in a number of species in which manipulative experiments are not possible. Microinjection of a purified molecule, *goosecoid* mRNA, into ventral blastomeres mimics Spemann's organizer phenomenon. This gene is induced in the marginal zone by dorsal inducers which in turn mimic the signal released by the Nieuwkoop center. While surely many more components remain to be discovered, at present the results seem to suggest that the *goosecoid* DNA-binding protein is part of the following pathway:

Nieuwkoop center cells
↓
release of dorsal intercellular signal/s
↓
induction of dorsal marginal zone cells
↓
synthesis of *goosecoid* DNA-binding protein
↓
Spemann's organizer activity

We thank Larry Tabata for help with the illustrations and Anatalia Cuellar and Diana Mihaylova for technical assistance. M. B. is a DAAD/NATO postdoctoral fellow, C. N. an ACS (California division) senior fellow, and H. S. a DFG postdoctoral fellow. Our work is supported by grants of the NIH (HD21502-07) and of the HFSPO.

References

Blum, M., Gaunt, S. J., Cho, K. W. Y., Steinbeisser, H., Blumberg, B., Bittner, D. and De Robertis, E. M. (1992). Gastrulation in the mouse: the role of the homeobox gene *goosecoid*. *Cell* **69**, 1097-1106.

Blumberg, B., Wright, C. V. E., De Robertis, E. M. and Cho, K. W. Y. (1991). Organizer-specific homeobox genes in *Xenopus laevis* embryos. *Science* **253**, 194-196.

Boterenbrood, E. C. and Nieuwkoop, P. D. (1973). The formation of the mesoderm in urodelean amphibians. V. Its regional induction by the endoderm. *Wilhelm Roux Arch. Dev. Biol.* **173**, 319-332.

Cho, K. W. Y., Blumberg, B., Steinbeisser, H. and De Robertis, E. M.

(1991). Molecular nature of Spemann's organizer: the role of the *Xenopus* homeobox gene *goosecoid. Cell* **67**, 1111-1120.

Cooke, J. (1972). Properties of the primary organization field in the embryo of *Xenopus laevis*. II. Positional information for axial organization in embryos with two head organizers. *J.Embryol. Exp. Morph.* **28**, 27-46.

De Robertis, E. M., Oliver, G. and Wright, C. V. E. (1990). Homeobox genes and the vertebrate body plan. *Scientific American* **263**, 46-52.

De Robertis, E. M., Morita, E. A. and Cho, K. W. Y. (1991). Gradient fields and homeobox genes. *Development* **112**, 669-678.

Dirksen, M. L. and Jamrich, M. (1992). A novel, activin-inducible, blastopore lip-specific gene of *Xenopus laevis* contains a *fork head* DNA-binding domain. *Genes Dev.* **6**, 599-608.

Feduccia, A. and McCrady, E. (1991). *Torrey's Morphogenesis of the Vertebrates*, Fifth Edition. New York: John Wiley.

Gaunt, S. J., Blum, M. and De Robertis, E. M. (1992). Expression of the mouse *goosecoid* gene during mid-embryogenesis may mark mesenchymal cell lineages in the developing head, limbs and body wall. *Development*, in press.

Gehring, W. J. (1987). Homeo boxes in the study of development. *Science* **236**, 1245-1252.

Gerhart, J., Danilchik, M., Doniach, T., Roberts, S., Rowning, B. and Stewart, R. (1989). Cortical rotation of the *Xenopus* egg: consequences for the anteroposterior pattern of embryonic dorsal development. *Development* **107 Supplement**, 37-51.

Gimlich, R. L. and Gerhart, J. C. (1984). Early cellular interactions promote embryonic axis formation in *Xenopus laevis*. *Dev. Biol.* **104**, 117-130.

Green, J. B. A. and Smith, J. C. (1990). Graded changes in dose of a *Xenopus* activin A homologue elicit stepwise transitions in embryonic cell fate. *Nature* **347**, 391-394.

Hamburger, V. (1988). *The Heritage of Experimental Embryology*. Oxford: Oxford University Press.

Harland, R. M. (1991). In situ hybridization; an improved whole mount method for *Xenopus* embryos. *Meth. in Cell Biol.* **36**, in press.

Herrmann, B. G. (1991). Expression pattern of the *Brachyury* gene in whole-mount T^{wis}/T^{wis} mutant embryos. *Development* **113**, 913-917.

Immerglück, K., Lawrence, P. A. and Bienz, M. (1990). Induction across germ layers in *Drosophila* mediated by a genetic cascade. *Cell* **62**, 261-268.

Keller, R. E. (1976). Vital dye mapping of the gastrula and neurula of *Xenopus laevis*. II. Prospective areas and morphogenic movements of the deep layer. *Dev. Biol.* **51**, 118-137.

Keller, R. E. (1991). Early embryonic development of *Xenopus laevis*. *Meth. in Cell Biol.* **36**, 61-113.

Kessel, M. and Gruss, P. (1990). Murine developmental control genes. *Science* **249**, 374-379.

Lawson, K. A., Meneses, J. J. and Pedersen, R. A. (1991). Clonal analysis of epiblast fate during germ layer formation in the mouse. *Development* **113**, 891-911.

Marx, J. (1991). How embryos tell heads from tails. *Science* **254**, 1586-1588.

McGinnis, W. and Krumlauf, R. (1992). Homeobox genes and axial patterning. *Cell* **68**, 283-302.

McMahon, A. P. and Moon, R. T. (1989a). *int-1* - a proto-oncogene involved in cell signalling. *Development* **107 Supplement**, 161-167.

McMahon, A. P. and Moon, R. T. (1989b). Ectopic expression of the proto-oncogene *int-1* in *Xenopus* embryos leads to a duplication of the embryonic axis. *Cell* **58**, 1075-1084.

Nakamura, O. and Toivonen, S., Eds. (1978). *Organizer - a Milestone of a Half-Century from Spemann.* Amsterdam: Elsevier/North-Holland Press.

Nieuwkoop, P. D. (1973). The "organization center" of the amphibian embryo: its spatial organization and morphogenic action. *Adv.Morphogen.* **10**, 1-39.

Nüsslein-Volhard, C. (1991). Determination of the embryonic axes of *Drosophila. Development* **Supplement 1**, 1-10.

Rasbash, P., Cooke, L. A., Herrmann, B. G. and Beddington, R. S. P. (1991). A cell autonomous function of *Brachyury* in T/T embryonic chimaeras. *Nature* **353**, 348-350.

Reuter, R. and Scott, M. P. (1990). Expression and function of the homeotic genes *Antennapedia* and *Sex combs reduced* in the embryonic midgut of *Drosophila. Development* **109**, 289-303.

Rosa, F. M. (1989). *Mix.1*, a homeobox mRNA inducible by mesoderm inducers, is expressed mostly in the presumptive endodermal cells of Xenopus embryos. *Cell* **57**, 965-974.

Rupp, R. A. and Weintraub, H. (1991). Ubiquitous *Myo D* transcription at the midblastula transition precedes induction-dependent *Myo D* expression in presumptive mesoderm of *X.laevis*. *Cell* **65**, 927-937.

Slack, J. M. W. (1991). The nature of the mesoderm-inducing signal in *Xenopus*: a transfilter induction study. *Development* **113**, 661-669.

Slack, J. M. W., Darlington, B. G., Heath, J. K. and Godsave, S. F. (1987). Mesoderm induction in early *Xenopus* embryos by heparin-binding growth factors. *Nature* **326**, 197-200.

Smith, J. C., Cooke, J., Green, J. B. A., Howes, G., and Symes, K. (1989). Inducing factors and the control of mesodermal pattern in *Xenopus laevis*. *Development* **107**, 149-159.

Smith, J. C., Price, B. M. J., Green, J. B. A., Weigel, D. and Herrmann, B. G. (1991). Expression of a *Xenopus* homolog of *Brachyury* (T) is an immediate-early response to mesoderm-induction. *Cell* **67**, 79-87.

Smith, W. C. and Harland, R. M. (1991). Injected *Xwnt*-8 RNA acts early in *Xenopus* embryos to promote formation of a vegetal dorsalizing center. *Cell* **67**, 753-765.

Sokol, S., Christian, J. C., Moon, R. T. and Melton, D. A. (1991). Injected *Wnt* RNA induces a complete body axis in *Xenopus* embryos. *Cell* **67**, 741-752.

Spemann, H. (1938). *Embryonic Development and Induction* New Haven, Connecticut: Yale University Press.

Spemann, H. and Mangold, H. (1924). Über Induktion von Embryonalanlagen durch Implantation artfremder Organisatoren. *Roux' Arch. f. Entwicklungsmech. Org.* **100**, 599-638.

Taira, M., Jamrich, M., Good, P. J. and Dawid, I. B. (1992). The LIM domain-containing homeobox gene XLIM-1 is expressed specifically in the organizer region of *Xenopus* gastrula embryos. *Genes Dev.* **6**, 356-366.

Thomsen, G., Woolf, T., Whitman, M., Sokol, S., Vaughan, J., Vale, W. and Melton, D. A. (1990). Activins are expressed early in Xenopus embryogenesis and can induce axial mesoderm and anterior structures. *Cell* **63**, 485-493.

Wolpert, L. (1991). *The Triumph of the Embryo*, pp. 183-187. Oxford: Oxford University Press.

Development 1992 Supplement, 173-181 (1992)
Printed in Great Britain

Dorsoventral development of the *Drosophila* embryo is controlled by a cascade of transcriptional regulators

CHRISTINE THISSE and BERNARD THISSE

LGME du CNRS, U184 de l'INSERM, Institut de Chimie Biologique, Faculté de Médecine, 11 rue Humann 67085 Strasbourg Cedex, France

Present address: Institute of Molecular Biology, University of Oregon, Eugene, OR 97403, USA

Summary

Maternal genes involved in dorsoventral (D/V) patterning of the *Drosophila* embryo interact to establish a stable nuclear concentration gradient of the Dorsal protein which acts as the morphogen along this axis. This protein belongs to the *rel* proto-oncogene and NF-KB transcriptional factor family and acts by controlling zygotic gene expression. In the ventral part of the embryo, *dorsal* specifically activates transcription of the gene *twist* and ventrally and laterally *dorsal* represses the expression of *zerknüllt*, a gene involved in the formation of dorsal derivatives. The extent of *dorsal* action is closely related to the affinity and the number of *dorsal* response elements present in these zygotic gene promoters.

***twist* is one of the first zygotic genes necessary for mesoderm formation. It codes for a 'b-HLH' DNA-binding protein which can dimerize and bind to DNA in vitro and to polytene chromosomes in vivo. In addition, in cultured cells *twist* has been shown to be a transcriptional activator. Thus, the first events of embryonic development along the D/V axis are controlled at the transcriptional level.**

Key words: transcriptional regulation, dorsoventral differentiation, *dorsal*, *twist*.

Introduction

Genetic analysis of early development in the *Drosophila* embryo has revealed that embryonic pattern formation depends on both maternal and zygotic genes. The establishment of the anterioposterior (A/P) pattern, along which the body plan and metameric segmentation are defined, requires three sets of maternal genes: the anterior genes for the head and thorax (Frohnhöfer and Nüsslein-Volhard, 1986, 1987), the posterior genes for the abdomen (Lehmann and Nüsslein-Volhard, 1986, 1987), and the terminal gene system for both acron and telson (Schüpbach and Wieschaus, 1986; Klingler et al., 1988).

By contrast the dorsoventral (D/V) axis, along which the embryonic germ layers are defined, requires only one set of genes: the dorsal-ventral system, which include eleven genes of the dorsal group and the gene *cactus* (Anderson, 1987; Roth et al., 1989; for review see Nüsslein-Volhard, 1991; St Johnston and Nüsslein-Volhard, 1992). Each gene in the dorsal group displays a complete dorsalization as the lack-of-function phenotype: only elements that normally derive from the dorsalmost region of the egg are formed, while ventral and lateral elements are lacking (Nüsslein-Volhard, 1979). The twelfth gene, *cactus*, shows partial ventralization as a lack-of-function phenotype: pattern elements normally derived from the dorsal and dorsolateral regions are absent in mutant embryos, while ventral and ventrolateral elements are formed along the entire D/V axis (Schüpbach and Wieschaus, 1989; Roth et al., 1989).

These twelve genes act in a complex way to establish the spatial coordinates of the D/V axis (for review see St Johnston and Nüsslein-Volhard, 1992). Analysis of the phenotype observed in mutants belonging to the D/V system initially suggested that the position along this axis is defined by the local concentration of a morphogen (Nüsslein-Volhard, 1979). Several lines of evidence suggest the *dorsal* (*dl*) gene to be at the end of this cascade of regulation, making it the best candidate for the gene encoding this morphogen: first, *dorsal* is the only mutation that produces a dorsalized phenotype in double mutants with loss-of-function ventralizing *cactus* alleles (Roth et al., 1989, 1991). Thus, *dorsal*, but none of the other dorsal-group genes functions downstream of the *cactus* gene. Second, in transplantation experiments, a localized rescuing activity can be found only in the case of *dorsal*, and this localization only appears at the syncytial blastoderm stage (Santamaria and Nüsslein-Volhard, 1983). Finally, only in the case of *dorsal*, loss-of-function mutations show a dominant (dl^D) effect (Nüsslein-Volhard et al., 1980); that is at 29°C, *dl*/+ females lay eggs that do not develop mesoderm. Thus, this germ layer which corresponds to the most ventral part of the embryo requires a higher level of *dorsal* activity than the lateral and dorsal regions.

Molecular analysis has shown how the product of the

dorsal gene acts as a morphogen for the D/V axis, and how the Dorsal product is able to define spatial coordinates along this axis. The *dorsal* gene was cloned (Steward et al.,1984), its RNA and protein were shown to be synthesized during oogenesis and to be uniformly distributed in the cytoplasm of the egg (Steward et al., 1985; Roth et al., 1989; Rushlow et al., 1989; Steward, 1989). After the ninth cleavage division, the Dorsal protein becomes highly concentrated in the nuclei on the ventral side of the embryo (Roth et al., 1989; Rushlow et al., 1989; Steward, 1989). Laterally, the nuclear and cytoplasmic concentration of Dorsal are approximately equal, and dorsally, the Dorsal protein is excluded from the nuclei and remains in the cytoplasm. Thus, positional information along the D/V axis is defined by a gradient of concentration of Dorsal protein in cell nuclei.

The nucleotide sequence of *dorsal* (Steward, 1987) showed that it encodes a protein with strong sequence similarities to the Rel family proteins, which includes the proto-oncogene c-*rel* and NF-KB, a nearly ubiquitous eukaryotic transcription factor (Ghosh et al., 1990; Kieran et al., 1990; Bours et al., 1990; Nolan et al., 1991). Proteins of this family have several common characteristics (Fig. 1; Gilmore, 1991). They share a highly conserved 300 amino-acid domain located in their amino-terminal part and each Rel protein has its own unique carboxy-terminal half. Rel family proteins can form protein complexes with other family members and other unrelated cellular proteins (such as inhibitor IKB). Proteins in the Rel family appear to be regulated by subcellular localization: they are sequestered in an inactive form in the cytoplasm of cells and become active by translocation into the nucleus. Most of the Rel proteins that have been described are able both to activate and to repress transcription. These proteins have gene activation and cytoplasmic anchoring function within their carboxy-terminal domain. The highly related amino-terminal domain of Rel proteins contains several different functional motifs: a region involved in the formation of homodimers and heterodimers (Ghosh et al., 1990; Kieran et al., 1990; Nolan et al., 1991), a DNA-binding region (Ghosh et al., 1990; Kieran et al., 1990; Nolan et al., 1991). An IKB binding region (Ghosh et al., 1990), a region that inhibits the carboxy-terminal gene activation domain (Bull et al., 1990; Richardson and Gilmore, 1991), and a stretch of basic amino-acids that functions as nuclear translocation signal were also identified (Gilmore and Temin, 1988; Capobianco et al., 1990). A serine residue is located within a consensus recognition sequence for phosphorylation by protein kinase A and is positioned approximately 20 amino-acids amino-terminal to the nuclear targeting signal.

As *dorsal* belongs to the Rel family, it should be a DNA-binding protein and thus would be predicted to act as a transcriptional activator and/or repressor.

Zygotic genes affecting the D/V pattern

Mutations in several zygotic loci affecting the D/V pattern have been identified (Nüsslein-Volhard et al., 1984; Wieschaus et al., 1984; Jürgens et al., 1984). Generally these loci affect only one of the three D/V pattern elements (Fig. 2): dorsal ectoderm (Irish and Gelbart 1987; Rushlow et al., 1987a), ventral ectoderm (Mayer and Nüsslein-Volhard, 1988) and mesoderm (Simpson, 1983).

With such genetic analysis, seven loci required for the specification of dorsal structures have been identified: *decapentaplegic* (*dpp*), *zerknüllt* (*zen*), *screw* (*scw*), *tolloid* (*tld*), *shrew* (*srw*), *twisted gastrulation* (*tsg*) and *short gastrulation* (*sog*). Mutations in these loci cause a general loss of amnioserosa, dorsal ectoderm and dorsolaterally derived structures of the acron and telson (Anderson, 1987; Rushlow and Arora, 1990). Accompanying this loss of dorsal structures is an expansion of ventrolateral pattern elements. Two other zygotic loci, *twist* (*twi*) and *snail* (*sna*), are required for the formation of the ventralmost part of the embryo, the mesoderm. In such mutant embryos, the cells on the ventral side do not invaginate and no ventral furrow is formed (Simpson, 1983; Leptin and Grünewald, 1990) and the resulting embryos lack all derivatives of the mesoderm. This phenotype is very similar to the dl^D phenotype, the weakest phenotype of *dorsal*.

Some of these genes have been cloned and studies of their pattern of expression in wild-type and mutant embryos strongly support the hypothesis that the expression of these zygotic genes depends directly on the *dorsal* protein. In early wild-type embryos, *dpp* and *zen* are both expressed in the dorsalmost 40% of the blastoderm embryo. Their transcripts extend also around both anterior and posterior poles and label some ventral cells (St Johnston and Gelbart, 1987; Rushlow et al., 1987b). Ventrally, transcripts of *twi* and *sna* are first detected in a single continuous stripe, comprising the ventralmost 20% of the embryo, and extending up to and around both poles (Thisse et al., 1987, 1988; Alberga et al., 1991). For all of these zygotic genes, after this initial phase each of these pattern sharpens into a

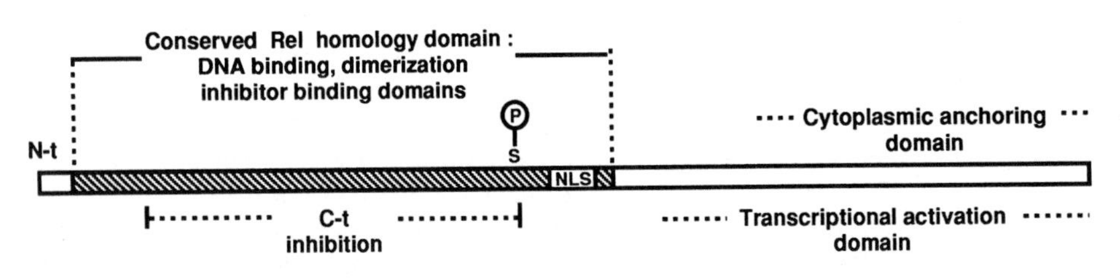

Fig. 1. Summary of structural information concerning the Rel family proteins. The domain of similarities (hatched box) includes a region important for DNA binding, dimerization, inhibitor binding (e. g. IKB), nuclear localization signal (NLS), inhibition of the unique carboxy terminal gene activation domain (Ct inhibition) and a consensus site for phosphorylation by protein kinase A (S-P). The amino-terminal domains vary in length. The carboxy-terminal domain contains transcriptional activation and cytoplasmic anchoring domains (modified from Gilmore, 1991).

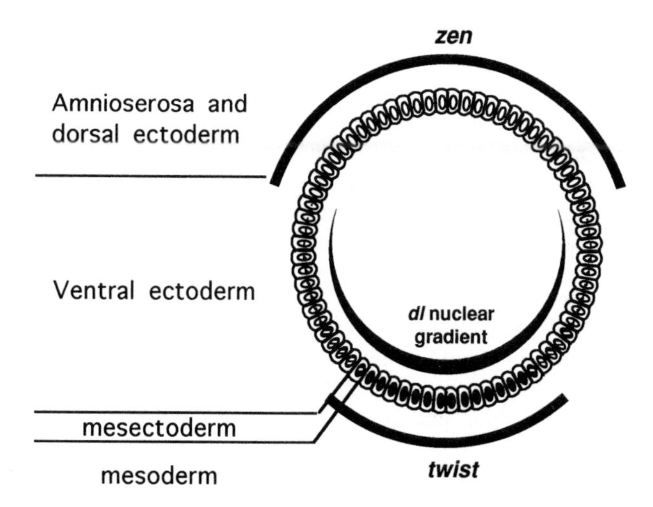

Fig. 2. Dorsoventral fate map of the blastoderm embryo in cross section and expression territories of Dorsal, Twi and Zen proteins. From dorsal to ventral: the amnioserosa and dorsal ectoderm, ventral ectoderm and mesoderm. Cells that connect mesoderm to ectoderm, and express Twi, form the mesectoderm. These cells are involved later in nervous system development. Dorsal protein is observed as a gradient of nuclear concentration with a maximum in the ventral region. In this region, its concentration is sufficient to allow the transcription of the *twi* gene. Due to a high affinity for binding sites on the *zen* promoter, Dorsal is also able to repress *zen* expression in ventral and lateral regions, thus restricting *zen* expression to the dorsal ectoderm and the amnioserosa (see details in the text).

refined pattern which is a derivative of the initial pattern (Ray et al., 1991).

In loss-of-function dorsal group mutations, in which Dorsal protein is excluded from the nuclei at all positions around the D/V axis, the dorsal zygotic genes *zen* and *dpp* are expressed everywhere, while the ventral genes *twi* and *sna* are not expressed anywhere (Roth et al., 1989; St Johnston and Nüsslein-Volhard, 1992). In such mutant embryos, all cells adopt a dorsal fate. Conversely, in the strongest ventralizing mutants, Dorsal protein localizes to all of the nuclei. In these mutants, *twi* and *sna* are expressed all around the circumference of the embryo while *zen* and *dpp* expression is repressed. Finally, in mutant combinations that produce a lateralized phenotype, Dorsal protein is evenly distributed between the nuclei and the cytoplasm and neither the dorsal nor ventral zygotic genes are expressed in the embryo (excepted for the polar regions).

These observations suggest strongly that the nuclear concentration of Dorsal determines the dorsoventral pattern by controlling the expression of the zygotic genes: high nuclear concentration of Dorsal seems to initiate the expression of the *twi* and *sna* genes, which are responsible for the differentiation of the mesoderm in ventral regions, and repress expression of dorsal zygotic genes *dpp* and *zen*, restricting the expression to dorsal regions where they are responsible for the differentiation of dorsal ectodermal derivatives.

Molecular analysis of the Dorsal protein

Regulatory elements of two putative target genes for the Dorsal protein, *zen* and *twi* have been extensively analyzed at the molecular level. For the *zen* gene, a region responsible for its ventral repression was localized between −1.4 and −1.1 kb (Doyle et al., 1989; Ip et al., 1991). This distal regulatory element has the property of a silencer element and can act over a distance to repress ventral expression of a heterologous promoter (Doyle et al., 1989; Ip et al., 1991). As a high nuclear concentration of the Dorsal protein seems to be responsible for the ventral repression of *zen*, this silencer element in the *zen* promoter is a good candidate for carrying Dorsal response elements (DREs). Using gel shift assay experiments, with a bacterially expressed Dorsal protein, in vitro experiments revealed that it is able to bind to the DNA of this particular region (Ip et al., 1991). Three of the four Dorsal binding sites characterized in that study are located within the limits of the ventral repressor element. These observations correlate well with the expected repressor effect of the Dorsal protein on *zen* gene expression. The conserved sequence motif recognized by the Dorsal protein on the *zen* promoter is: G G G several As C C (Ip et al., 1991). These binding sites are closely related to the consensus recognition site of the NF-KB transcriptional factor.

While *dorsal* represses *zen* gene activity, it activates the *twi* gene. The first evidence of a regulation of *twi* transcription by *dorsal* came from northern blot experiments: *twi* RNA is not detected in embryos derived from eggs laid by mutant *dorsal* females (Thisse et al., 1987). Subsequently, an extensive analysis of the regulatory sequences of the *twi* gene was performed by using in vivo P-mediated rescue experiments. These studies revealed different regulatory regions: one negative region between −7 kb and −3 kb; two positive quantitative regions, one located between −3 kb and −0.8 kb, the second one located between +3.2 kb and +6.2 kb. As no *twi* RNA is found in embryos lacking the Dorsal protein, and as 0.8 kb is suffficient to allow transcription of the *twi* gene, these observations suggest that within the 0.8 kb minimal promoter, there are Dorsal response elements (Thisse et al., 1991).

Using different combinations of *twi* 5′ flanking sequences fused to a *lacZ* reporter gene and expressed in P-transformed embryos, two regions containing DREs were defined: the first one, or proximal region, was localized in the minimal promoter between −0.18 kb and −0.4 kb; the second one, or distal region, was localized between −0.8 kb and −1.4 kb (Thisse et al., 1991; Jiang et al., 1991; Pan et al., 1991). Deletion of this distal element causes a reduction of the *twi* pattern of expression. Expression is lost at both poles and is restricted to the ventralmost part of the embryo (Thisse et al., 1991., and Fig. 3). In addition, as judged from in situ hybridization, a substantial narrowing in the lateral limits of *twi* expression was observed for constructs lacking the distal element (Jiang et al., 1991; Pan et al., 1991). These two observations demonstrate that both proximal and distal regions are required for a wild-type expression pattern of the *twi* gene. Nevertheless, a β-galactosidase construct carrying the 3′ *twi* quantitative enhancer does not reveal the ventral restriction observed by in situ hybridization experiments, as mesectodermal cells (the more dorsally *twi*-expressing cells in wild-type embryos) are clearly labelled during gastrulation (Thisse et al., 1991 and Fig. 3). The ventral restriction observed by in situ hybridization corresponds to a reduction of the level, but not to a complete loss of *twi* expression. On the contrary, the expression in the poles of the embryos is never seen with such a truncated promoter even if the 3′ regulatory sequences are present. Evidently, the control of *twi* expression in the poles is different from its control in the ventral part of the embryo. As was recently proposed (Ray et al., 1991), the terminal gene system could act through or with *dorsal* to affect expression of the *twi* gene at the poles.

Cotransfection assays in cultured Schneider 2 cells demonstrated that the Dorsal protein specifically activates expression from the *twi* promoter (Thisse et al., 1991). These experiments defined two activation regions which correspond to activation regions previously determined in phenotypic rescue and β-galactosidase expression experiments. Some interactions between regions of activation on the *twi* promoter and sequences localized on the vector used were also observed. This result suggests that cooperative interactions between different DREs may play an important role in the regulation of *twi* by *dorsal*.

In vitro studies were performed in order to define sequences of the *twi* promoter bound by the Dorsal protein. Gel shift and DNAase I foot printing assays, using a bacterially expressed Dorsal protein identified binding sites in the proximal and distal activation regions (Thisse et al., 1991; Jiang et al., 1991; Pan et al., 1991; see also Fig. 4). Two major binding sites were identified in the proximal region. The sequence motifs, GAGAAAACCC and GGGAAAATGC are closely related to the consensus sequence of Dorsal binding sites described for the *zen* promoter. Surprisingly, other Dorsal binding sites, present on the *twi* promoter, and described by three independent laboratories are different. Such differences may be due to different bacterially expressed Dorsal protein extracts. We used a full-length Dorsal protein which was solubilized in urea and then progressively renaturated (Thisse et al., 1991). Two other laboratories used a truncated Dorsal pro-

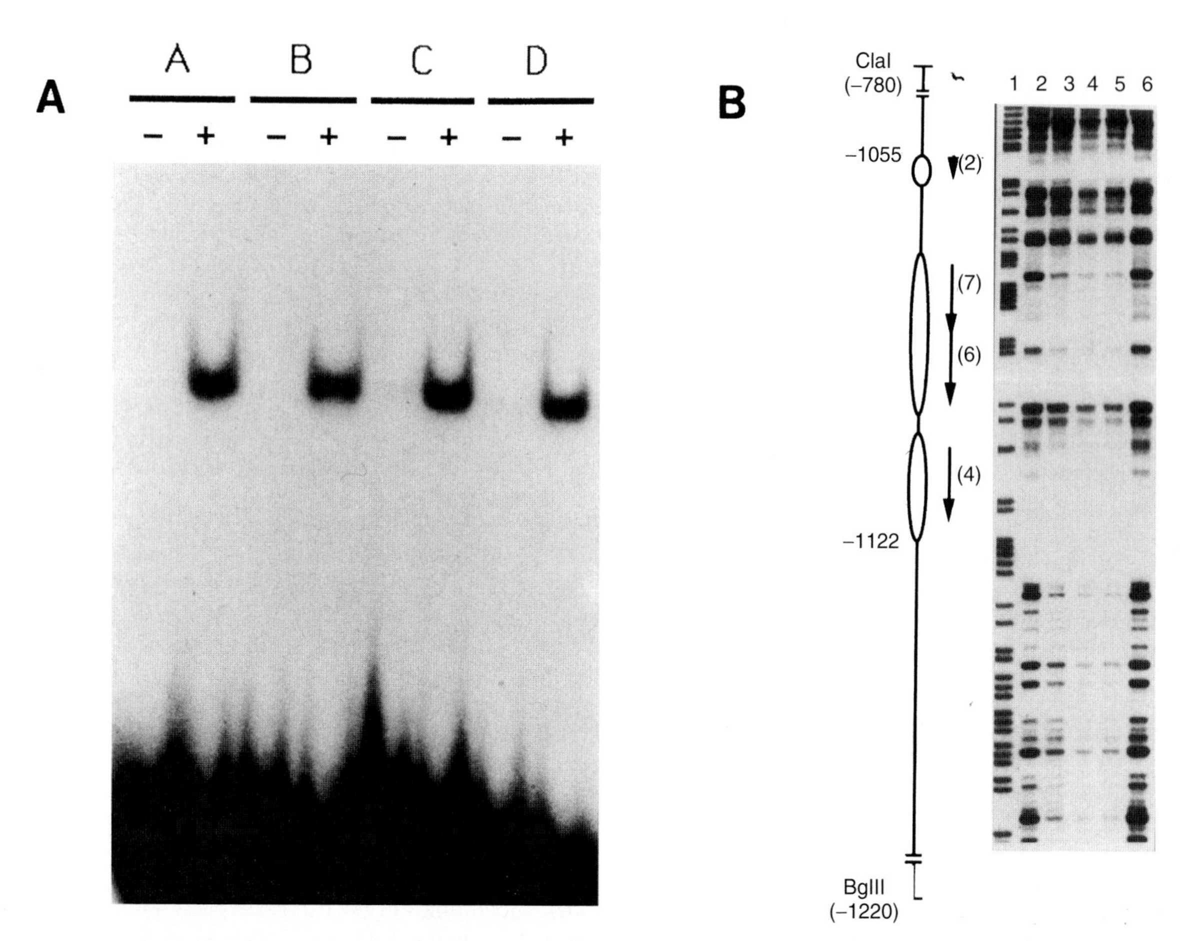

Fig. 4. (A) Gel retardation assays showing the binding of the bacterially synthesized Dorsal protein on *twist* promoter and pUC motifs. 2 μg of protein of an extract from IPTG-induced *E. coli*, were used with different ^{32}P-labeled probes in gel retardation assays. (A) Oligonucleotide corresponding to one site of the proximal region of activation by *dorsal*. (B and C) Oligonucleotides corresponding to motifs of the distal region. (D) Oligonucleotide corresponding to the most conserved motif in the pUC vector. +: extract from *E. coli* expressing *dorsal*. −: extract from IPTG induced *E. coli* containing the vector without insert (Thisse et al., 1991). (B) Footprint reaction using the bacterially induced full-length Dorsal protein. Example showing the DNAase I footprinting analysis of coding strand of region A. Lane 1: G+A product of chemical sequencing reactions. Lanes 2 and 6: DNAase I digestion of naked DNA. Lanes 3-5: pattern of protection with respectively 4, 8 and 16 μg of bacterial Dorsal protein. The diagram to the left of the autoradiogram represents the binding of the protein. The vertical arrows indicate the location and orientation of the conserved motifs (Thisse et al., 1991).

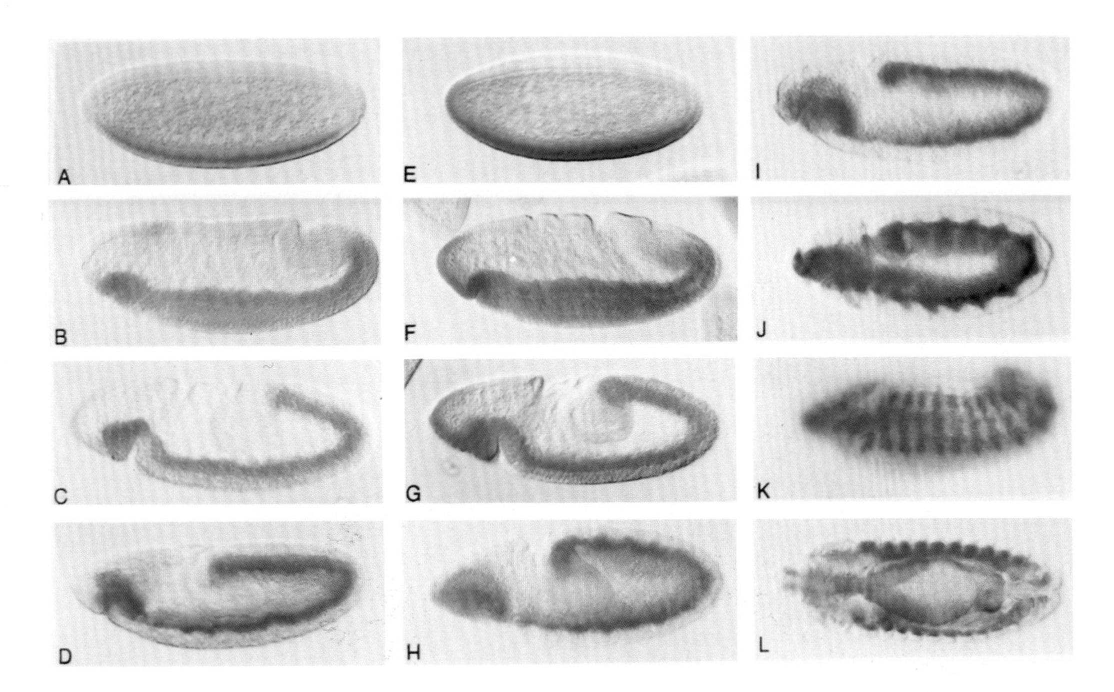

Fig. 3. Distribution of β-galactosidase expression in whole-mount embryos of different *twi*-β-galactosidase lines. Embryos of *twi*-β-galactosidase lines (with the 3′ region of *twi*) were stained using anti-β-galactosidase antibodies. Whole-mount preparations were photographed using Nomarski optics. (A-D) Embryos of the 0.8 kb *twi*-β-galactosidase lines lacking *twi* expression at their poles. (E-L) Embryos of 1.4 kb *twi*-β-galactosidase lines lacking *twi* expression at posterior pole during blastoderm stage due to the low rate of β-galacosidase synthesis (Thisse et al. 1991). (A and E) Cellular blastoderm embryos (stage 5), (B and F) gastrulating embryos (late stage 6), (C and G) embryos at germband extension (stage 7), (D and H) embryos at extended germband stage (stage 9), (I) embryo at stage 10, (J) embryo at stage 11, (K-L) embryos at stage 14. L is a dorsal view. Note for I to L that all the embryonic tissues involved in the larval muscle formation are labelled. Orientation of the embryos in A to K are lateral views, L is a top view. Staging is according to Campos-Ortega and Hartenstein (1985).

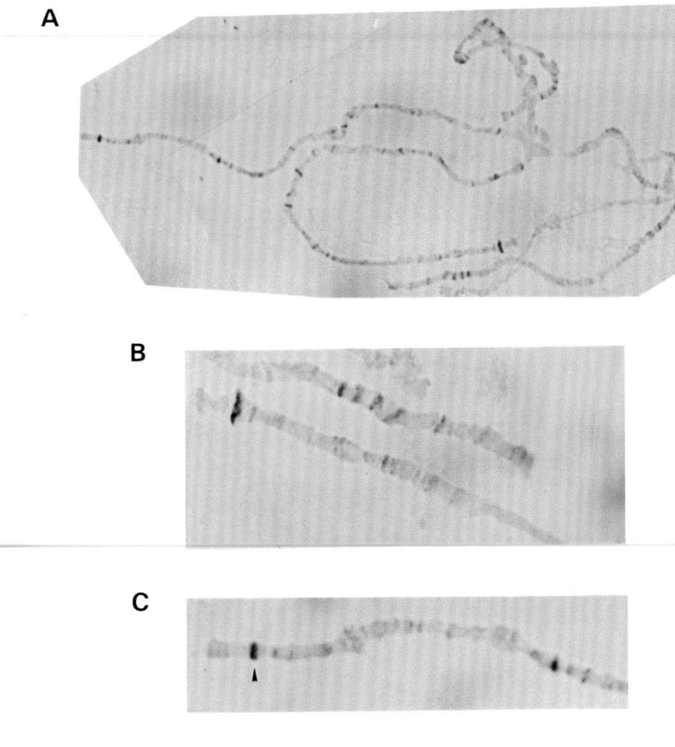

Fig. 6. Localization of Twi binding sites on polytene chromosomes of salivary glands of third instar larvae. Twi binding sites have been identified by immunocytochemical reaction using anti-Twi antisera. Chromosomes are stained with hematoxylin. (A) Entire nucleus showing chromosomes stained both with hematoxylin and anti-Twi antibodies. About 60 major binding sites have been identified. (B) Enlargment showing the tips of the right arm (top) and the left arm (bottom) of the third chromosome. (C) Enlargment showing the tip of the right arm of the second chromosome. One major band is located in 60B (arrow), which corresponds to the β*3 tubulin* locus.

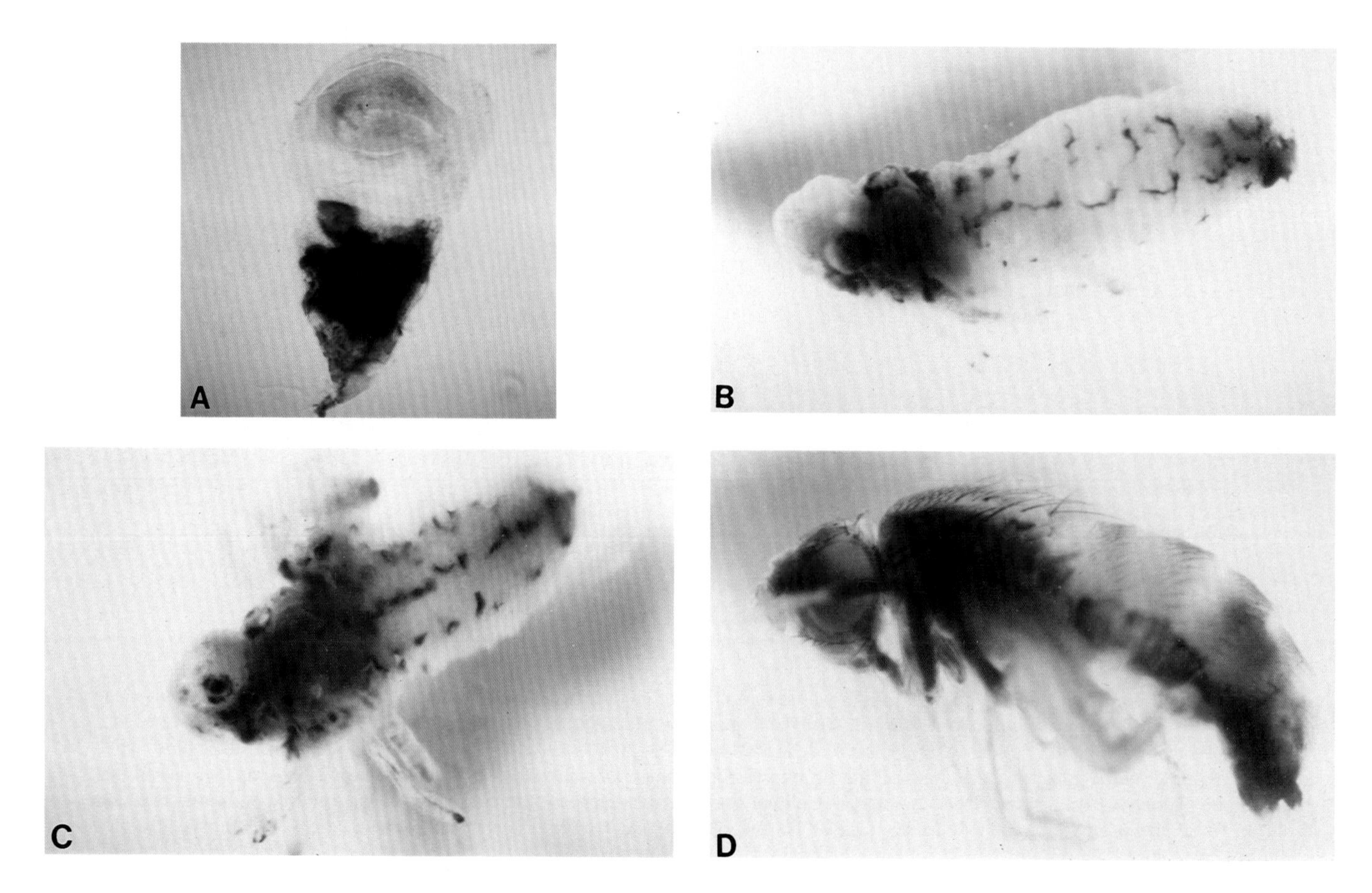

Fig. 8. Late distribution of β-galactosidase in *twi-lacZ* transgenic lines. β-galactosidase expression was revealed by X-Gal staining on whole-mount pupae and adult or dissected imaginal discs. (A) Wing imaginal disc: X-Gal staining is observed in adepithelial cells located in the thoracic part of the disc. These cells are involved in the formation of muscles necessary for flight. (B, C) β-galactosidase expression in pupae: staining is observed in all cells involved in adult muscle development. (D) β-galactosidase expression in an adult just after emergence: expression of β-galactosidase is revealed in all the adult muscles (low staining observed for abdominal dorsal muscles probably result from less diffusion of the X-Gal in this experiment).

tein (Jiang et al., 1991; Pan et al., 1991). Denaturation or partial deletion of this protein may modify some characteristics in the binding properties. Possibly, all these different sites would correspond in vivo to binding sites of native Dorsal protein. They could be low affinity sites and then cooperative mechanisms would increase the specificity of Dorsal action.

In addition, physiologically, the affinity of Dorsal for the *twi* promoter must be low, as *twi* transcription is initiated by Dorsal only in the ventral region of the embryo, where its nuclear concentration is maximal. In fact, binding affinity of Dorsal protein for *twi* and *zen* motifs have been compared in vitro. The affinity of Dorsal for *twi* motifs was found to be five times lower than the affinity of Dorsal for *zen* motifs (Thisse et al., 1991; Jiang et al., 1991; see also Fig. 5). This result correlates well with data concerning the spatial expression territories of these two genes, that is the minimum nuclear concentration of Dorsal able to activate *twi* is clearly higher than the maximum nuclear concentration of Dorsal able to repress *zen* (Fig. 2). Recently it was observed that the same Dorsal binding site can mediate either activation or repression depending upon its context within the promoter (Courey et al., 1992).

Thus, as for the anterior system for which the transmission of the maternal information to the zygote is mediated by the transcriptional activation of zygotic genes such as *hunchback* by the Bicoid protein (Driever and Nüsslein-Volhard, 1989), the transmission of the maternal information to the zygote along the D/V axis is mediated by transcriptional control of zygotic gene expression by the maternal gene product Dorsal.

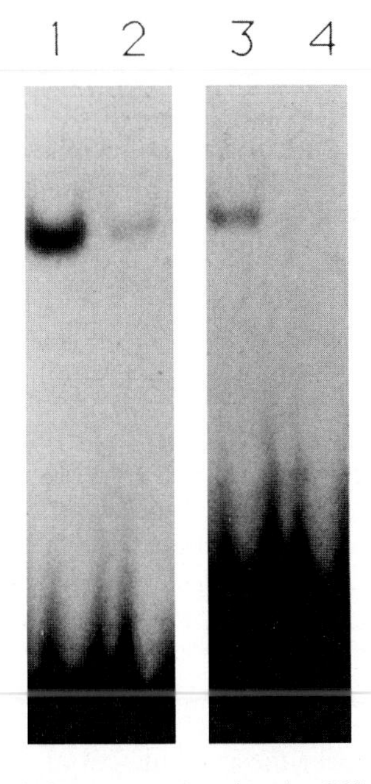

Fig. 5. Gel retardation assays showing the difference of binding affinity of Dorsal for *twist* and *zen* motifs. 0.5 μg (1 and 3) or 0.2 μg (2 and 4) of protein from IPTG induced *E. coli* were used with different ^{32}P-labeled probes in gel retardation assays. 1 and 2: gel retardation assay with a synthetic oligonucleotide corresponding to sequences −1534 to −1558 of the *zen* promoter ('strong' Dorsal binding sites described by Ip et al., (1991)), 3 and 4: gel retardation assay with the oligonucleotide B (see Fig. 4) from the distal region of activation by Dorsal on the *twi* promoter (Thisse et al., 1991).

Function of the zygotic genes

Four zygotic genes involved in D/V patterning have been extensively analyzed. The gene *zen* is necessary for formation of dorsal structures and it encodes a homeodomain-containing protein that binds to DNA (Rushlow et al., 1987a) and thus probably acts by regulating expression of downstream zygotic genes. The *dpp* gene is required for the formation of dorsal ectoderm and encodes a protein sharing strong sequence similarities with TGFβ (Gelbart, 1989), but its function in establishing the dorsal epidermis is not clearly understood. Two zygotic genes, involved in the formation of ventral structures, *sna* and *twi* encode putative DNA-binding proteins. The protein product of *sna* is a member of the zinc finger family (Boulay et al., 1987), and therefore is probably involved in transcriptional regulation. There is no direct molecular evidence at this time that the Sna protein regulates transcription, but the expression pattern of early embryonic markers in *sna* mutant embryos suggests that *sna* functions as a ventral repressor of laterally expressed genes, such as *single minded* (Leptin, 1991). Finally, *twi* encodes a protein containing a b-HLH domain, a domain described in a growing number of proteins involved in cellular differentiation and regulation of transcription (Murre et al., 1989). The b-HLH domain consists of a basic DNA-binding motif and helix-loop-helix dimerization motif (Murre et al., 1989). Presence of this domain in the Twi protein suggests that *twi* encodes a DNA-binding protein that would be able to dimerize and may act by regulating the expression of downstream mesodermal genes. Molecular arguments favor these predictions.

First, the Twi protein is able to homodimerize in vitro. Our finding based on glutaraldehyde cross-linking data, using a bacterially expressed Twi protein, is that one Twi monomer can complex with another and form a dimer. Under these experimental conditions, on SDS-PAGE, we observed a band that did not exist in a control extract with un-crosslinked Twi protein (data not shown). The apparent relative molecular mass corresponding to this band is twice that of monomeric Twi . This band is recognized by anti-Twi antibodies, strongly suggesting that it corresponds to the covalently linked Twi homodimer.

Secondly, we looked at the ability of the Twi protein to bind to DNA. We used one of the avantages of *Drosophila melanogaster*, that is the existence of polytene chromosomes in salivary glands of third instar larvae. The *twi* gene was cloned in an expression vector under the control of the promoter of the *sgs3* gene (salivary gland secretion protein 3) specific to salivary glands of third instar larvae (a gift of M. Martin). Transgenic flies were obtained that express *twi* in their salivary glands. By using anti-Twi antibodies, we detected Twi protein in the nuclei of cells of these salivary glands and localized Twi binding sites on polytene chromosomes (Fig. 6). Sixty major binding sites were observed. One of these sites was in 60B near the distal extremity of the right arm of chromosome 2 (Fig. 6B). This site is at or near the β*3 tubulin* locus. This gene encodes a mesodermal variant of the β tubulin (Gash et al., 1989) that is not expressed in *twi* mutant embryos (Leiss et al., 1988), suggesting that *twi* is necessary for expression of this mesodermal gene, thus making it a potential target for the Twi protein. Unfortunately, complementary experiments using both cotransfection in cultured Schneider 2 cells and binding studies of the Twi protein on polytene chromosomes of transformed β*3 tubulin* lines carrying all the regulatory sequences of the β*3 tubulin* gene (a gift from R. Renkawitz-Pohl) reveal that Twi is not able to bind specifically to β*3 tubulin* regulatory sequences.

Such methods allowing the characterization of targeted genes have been previously described for some DNA-binding proteins, for example: *zeste* (Pirrotta et al., 1988), *polycomb* (Zink and Paro, 1989), *serendipity* β (Payre et al., 1990) and *polyhomeotic* (DeCamillis et al., 1992). All these genes are expressed in salivary glands of third instar larvae in normal development, but that is not the case for *twi*. In addition, all Twi binding sites were detected in the interband regions of chromosomes and thus this binding seems to be highly dependent of the chromatin structure. An absence of correlation between cytologically mapped binding sites on third instar larvae polytene chromosomes and in vitro DNA-binding specificities has also been described (Payre and Vincent, 1991). These authors proposed that only a subset of recognition sites for a given DNA-binding protein is occupied in a specific tissue at a given development stage. If this is true, this in vivo approach may not allow characterization of target genes, in the case of a gene like *twi* that is normally expressed during early embryonic development. Nevertheless, by using this method we have demonstrated that Twi is able to bind to specific sites on chromosomes.

In order to study the function of *twi*, we tested its ability to regulate its own transcription. In embryos mutant for strong *twi* alleles, *twi* RNA is initially transcribed normally, but during cellularization, the RNA begins to fade and finally disappears at the beginning of gastrulation. Thus, the early *twi* transcription pattern is established but its expression cannot be maintained in absence of the *twi* function (Leptin, 1991), consistent with the notion that *twi* expression is initiated by maternal genes, but maintained by an auto-regulatory function. Using cotransfection assays in cultured Schneider 2 cells we studied whether Twi protein could transactivate a CAT reporter gene under the control of the *twi* regulatory sequences. No transactivation of the reporter gene was detected, showing that Twi homodimer is not able to regulate its own expression. Surprisingly, in control experiments we observed a strong transactivation of the CAT gene from the pBLCAT2 vector we used (Fig. 7). Then looking in the pBLCAT2 vector sequence we found some CANNTG motifs. This motif, also called the E box, was previously described as a consensus binding site for other b-HLH proteins (Kingston, 1989). Three of these E boxes are localized upstream of the tk promoter of the pBLCAT2 vector in positions –72, –341 and –364. Deletion of a short restriction fragment upstream of the tk promoter eliminates the two distal E boxes and abolishes completely the transactivation by *twi* of the CAT gene (Fig. 7). This result strongly suggests that E boxes present in the

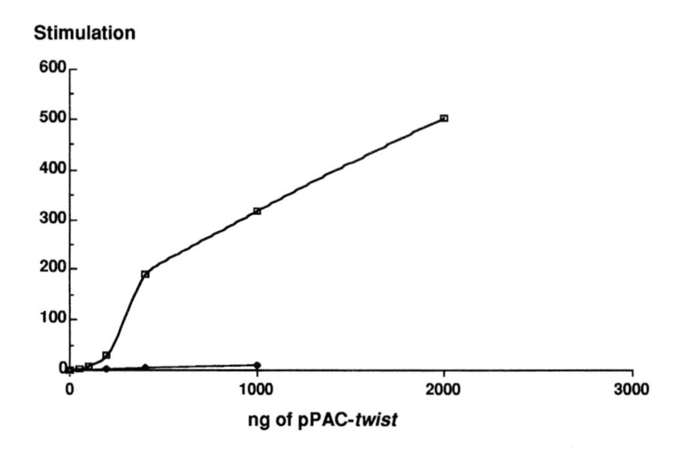

Fig. 7. Activation of CAT gene transcription by the Twi protein on the pBLCAT2 and pBLCAT2Δ vectors. 5 μg of pBLCAT2 (upper curve) and pBLCAT2Δ (lower curve) were cotransfected with the indicated amounts of pPAC-*twi*, a vector expressing *twi* under the control of the *actin 5C* promoter. The expression of β-galactosidase from the pACH vector used as control (Thisse et al., 1991) is not affected by the Twi protein.

region we deleted are involved in the transcriptional activation by Twi and that Twi is able to bind such sequences. This last hypothesis was tested in vitro using gel retardation assays, with a bacterially expressed Twi protein and an oligonucleotide corresponding to the pBLCAT2 sequence carrying the two distal E boxes. A specific retarded band was observed demonstrating that Twi is able to bind E boxes in vitro (data not shown).

In conclusion, the *twi* gene encodes a b-HLH protein, that homodimerizes in vitro and binds to chromosomes in vivo. It is also able to bind to DNA and it recognizes E box motifs and acts as a transcriptional activator.

Physiological targets of the Twi protein have yet to be well characterized but several genes are good candidates. For example, *tinman* (previously *msh*-2), is expressed in the early mesoderm between the end of the blastoderm stage and germ band elongation but is not expressed in *twi* mutant embryos (Bodmer et al., 1990). The second candidate is PS2α. This gene is expressed in cells of the presumptive mesoderm from the blastoderm stage until germ band retraction (Bogaert et al., 1987). PS2α is detected at the surface of basal cells of visceral myoblasts and is involved in the attachment of mesodermal cells to the ectoderm. No expression of this gene is detected in mutant *twi* embryos (Ingham, 1988; Leptin, 1991). Present and future work will determine whether *twi* is able to activate the transcription of these genes.

Phenotypic analysis of *twi* mutants showed that its activity is required for the invagination of the mesodermal layer at gastrulation (see the article by Leptin et al., in this issue). Molecular studies revealed that *twi* expression and probably function extends after these early steps of embryonic development. In fact, the *twi* product is detected in mesodermal territories until its differentiation in splanchnopleuric and somatopleuric mesoderm derivatives. After completion of germ band shortening, *twi* expression remains in a small residual population of cells of each segment (Thisse et al., 1988; Bate et al., 1991). This population of cells proliferates during larval life. During the third instar larval stage, *twi* expression is detected in all imaginal discs (Bate et al., 1991; Broadie and Bate, 1991; see also Fig. 8A) and in some cells of each abdominal segment (Bate et al., 1991; Currie and Bate 1991; Broadie and Bate, 1991; and personal observations). For example, in a wing disc, *twi* expression is observed in adepithelial cells which are presumptive myoblasts of the adult musculature. Observation of *twi* expression during larval and pupal life and cell ablation experiments showed that cells with persistent *twi* expression are the embryonic precursors of adult muscles (Bate et al., 1991; Broadie and Bate, 1991). Using the advantage of the great stability of ß-galactosidase, it is possible to follow muscle formation in transgenic *twi-lacZ* lines, from precursor cells, during larval life, to the completely differentiated muscles of the adult (Fig. 8).

The Twi protein is observed in muscle precursors during early pupal stage, then disappears when muscle differentiation markers appear (Bate et al., 1991). Thus *twi* expression is only observed in undifferentiated cells, and *twi* may act by retaining these cells in an embryonic state prior to the onset of differentiation. This late function of *twi* in the muscle pathway is necessarily different from its function during mesodermal invagination at the gastrulation stage and, of course, the kind of genes regulated by *twi* during these two different processes must be different. We postulate that these two functions are mediated by two different sets of protein complexes. As Twi is a b-HLH protein and is able to form homodimers, it may also form heterodimer(s) with other uncharacterized b-HLH proteins. Such heterodimers may have new binding specificities and thus regulate new sets of genes. The early function of *twi* would be mediated by one oligomeric form (homodimeric or heterodimeric in a complex with b-HLH protein expressed at blastoderm stage) and for its later function, under an another oligomeric form, as a complex with other uncharacterized b-HLH protein expressed during late development and eventually with a more restricted tissue specificity.

Concluding remarks

During these last few years, one of the most interesting questions concerning the early development of the *Drosophila* embryo, namely, how the information coming from the mother is transmitted to the zygote, has been resolved by a combination of genetic and molecular analyses. First, genetic studies allowed the characterization of genes involved in this process; then the study of gene hierarchies allowed the classification of genes into a regulatory cascade resulting in the storage in the egg of molecular positional informations. Four localized maternal signals define the basic organization and polarity of the two major embryonic axes. These signals specify cell states and provide a prepattern of development. The anterior-posterior prepattern is formed by the spatially regulated transcription of the gap genes (St Johnston and Nüsslein-Volhard, 1992). For the D/V axis, the coordinate system results from a gradient of nuclear concentration of the Dorsal protein. This protein acts as a transcriptional activator or repressor depending of the target promoters. These two properties of activation and repression carried out by the same protein allows the division of the D/V prepattern into three regions: dorsally, the dorsal ectoderm expressing only dorsal zygotic genes; ventrally, the mesoderm, expressing only genes involved in ventral structures; and laterally, the ventral ectoderm that expresses neither dorsal nor ventral zygotic genes.

Interestingly, many of the zygotic genes regulated by the morphogen of the D/V axis encode putative transcriptional factors that are able to amplify and refine the early maternal signal. This succession of transcriptional regulation would probably continue during the next downstream steps as several DNA-binding proteins expected to act as transcriptional factors are putative targets for genes such as *twi* and *sna* (for example, *tinman* coding for a homeodomain-containing protein may be a target for *twi*).

In fact, function of early zygotic genes (immediately downstream of the maternal signal) like *twi* or *sna*, is probably complex. These genes act during the first step of embryonic development, that is just when the first movements occur in the embryo. As the time between the initiation of their expression (during the blastoderm stage) and gastrulation is too short to allow a complex cascade of transcriptional regulation, these genes must act directly on this process. This observation suggests that genes coding for proteins involved in cell shape and/or cell movements during ventral furrow formation would be early targets for *twi* and *sna*. Later, *twi* and *sna* would act via the activation and/or repression of other transcriptional factors.

An additional level of regulation would be a consequence of the properties of homo and heterodimerization of b-HLH proteins. These proteins could form heterodimers with other b-HLH proteins expressed at different stages of embryogenesis and with a possible tissue specificity. *twi*, in such heterodimers might have some new binding properties, allowing the transcriptional control of a new set of genes. Such mechanisms could be involved in the later function of *twi* in the myoblastic pathway.

Thus, with only a very small number of initial genes that are directly regulated by the maternal information, it would be possible to control in time and in space a large number of genes whose expression will be responsible for progressive cell differentiation and development of the embryo.

We are grateful to P. Chambon for his continued interest in this project. We also would like to thank J. Postlethwait for comments and help on this manuscript. We wish to thank R. Renkawitz-Pohl for providing β3-*tubulin*-β-galactosidase lines, M. Martin for the sgs3 promoter and S. Noselli for his protocol of revelation of proteins on polytene chromosomes with specific antibodies. This work was supported by funds from the Association pour la Recherche sur le Cancer, the Ligue Nationale Française Contre le Cancer, the Institut National de la Santé et de la Recherche Médicale and the Centre National de la Recherche Scientifique.

References

Alberga, A., Boulay, J.-L., Kempe, E., Dennefeld, C. and Haenlin, M. (1991). The *snail* gene required for mesoderm formation in *Drosophila* is expressed dynamically in derivatives of all three germ layers. *Development* **111**, 983-992

Anderson, K. V. (1987). Dorsal-ventral embryonic pattern genes of *Drosophila. Trends in Genet.* **3**, 91-97.

Bate, M., Rushton, E. and Currie, D. A. (1991). Cells with persistent *twist* expression are the embryonic precursor of adult muscle in *Drosophila. Development* **113**, 79-89.

Bodmer, R., Jan, L. Y. and Jan, Y. N. (1990). A new homeobox-containing gene, *msh-2*, is transiently expressed early during mesoderm formation. *Development* **110**, 791-804.

Bogaert, T., Brown, N. and Wilcox, M. (1987). The *Drosophila* PS2 antigen is an invertebrate integrin that, like the fibronectin receptor becomes localized to muscle attachments. *Cell* **51**, 929-940.

Broadie, K. S. and Bate, M. (1991). The development of adult abdominal muscle in *Drosophila*: ablation of identified muscle precursor cells. *Development* **113**, 103-118.

Boulay, J-L., Dennenfeld, C. and Alberga, A. (1987). The *Drosophila* developmental gene *snail* encodes a protein with nucleic acid binding fingers. *Nature* **330**, 392-395.

Bours, V., Villalobos, J., Burd, P. R., Kelly, K. and Siebenlist, V. (1990). Cloning of a mitogen-inducible gene encoding a KB DNA binding protein with homolgy to the *rel* oncogene and to cell cycle motifs. *Nature* **348**, 76-79.

Bull, P., Morly, K. L., Hoekstra, M. F., Hunter, T. and Vermam, I. M. (1990). The mouse c-*rel* protein has an N-terminal regulatory domain and a C-terminal transactivation domain. *Mol. Cell Biol.* **10**, 5473-5485.

Campos-Ortega, J. A. and Hartenstein, V. (1985). *The Embryonic Development of* Drosophila melanogaster. Berlin: Springer Verlag.

Capobianco, A. J., Simons, D. L. and Gilmore, T. D. (1990). Cloning and expression of a chicken c-*rel* cDNA - Unlike P59v-*rel*, P68c-*rel* is a cytoplamic protein in chicken fibroblast. *Oncogene* **5**, 257-265.

Courey, A., Pan, D., Huang, J., Shirokawa, J., Schwyter, D. and Dubnicoff, T. (1992). Molecular basis of transcriptional activation and repression by the *dorsal* morphogen. *Abstract Volume of the 33rd Annual Drosophila Research Conference, Wyndham Franklin Plaza, Philadelphia, Pennsylvania* p 31.

Currie, D. A. and Bate, M. (1991). The development of adult abdominal muscles in *Drosophila*: myoblasts express *twist* and are associated with nerves. *Development* **113**, 91-102.

DeCamillis, M., Cheng, N., Pierre, D. and Brock, H. W. (1992). The *polyhomeotic* gene of *Drosophila* encodes a chromatin protein that shares polytene chromosome-binding sites with *polycomb. Genes Dev.* **6**, 223-232.

Doyle, H. J., Kraut, R. and Levine, M. (1989). Spatial regulation of *zerknüllt*: a dorsal-ventral patterning gene in *Drosophila. Genes Dev.* **3**, 1518-1533.

Driever, W. and Nüsslein-Volhard, C. (1989). The *bicoid* protein is a positive regulator of *hunchback* transcription in the early *Drosophila* embryo. *Nature* **337**, 138-143.

Frohnhöfer, H. G. and Nüsslein-Volhard, C. (1986). Organisation of anterior pattern in the *Drosophila* embryo by the maternal gene *bicoid. Nature* **324**, 120-125.

Frohnhöfer, H. G. and Nüsslein-Volhard, C. (1987). Maternal genes required for the anterior localization of *bicoid* activity in embryo of Drosophila. *Genes Dev.* **1**, 880-890.

Gash, A., Hinz, U. and Renkawitz-Pohl, R. (1989). Intron and upstream sequences regulate expression of the *Drosophila ß3-tubulin* gene in the visceral and somatic musculature, respectively. *Dev. Biol.* **86**, 3215-3218.

Gelbart, W. M. (1989). The *decapentaplegic* gene: a TGF-ß homologue controlling pattern formation in *Drosophila. Development* **Supplement 104**, 65-74.

Gilmore, T. D. (1991). Malignant transformation by mutant Rel protein. *Trends in Genet.* **7**, 318-322

Gilmore, T. D. and Temin, H. M. (1988). V-*rel* oncoproteins in the nucleus and the cytoplasm transform chicken spleen cells. *J. Virol.* **62**, 703-714.

Ghosh, S., Gilford, A. M., Riviere, L. R., Tempst, P., Nolan, G. P. and Baltimore, D. (1990). Cloning of the p50 DNA binding subunit of NF-KB: homology to *rel* and *dorsal. Cell* **62**, 1019-1029.

Ingham, P. W. (1988). The molecular genetics of embryonic pattern formation in *Drosophila. Nature* **335**, 25-34.

Ip, Y. T., Kraut, R., Levine, M. and Rushlow, C. A. (1991). The *dorsal* morphogen is a sequence-specific DNA-binding protein that interacts with a long-range repression element in Drosophila. *Cell* **64**, 439-446.

Irish, V. F. and Gelbart, W. M. (1987). The *decapentaplegic* gene is required for dorsal-ventral patterning of the *Drosophila* embryo. *Genes Dev.* **1**, 868-879.

Jiang, J., Kosman, D., Ip, Y. T. and Levine, M. (1991). The *dorsal* morphogene gradient regulates the mesoderm determinant *twist* in early *Drosophila* embryo. *Genes Dev.* **5**, 1881-1891.

Jürgens, G., Wieschaus, E., Nüsslein-Volhard, C. and Kluding, H. (1984). Mutations affecting the pattern of the larval cuticule in *Drosophila melanogaster*. II. Zygotic loci on the third chromosome. *Roux Arch. Dev. Biol.* **193**, 283-295.

Kieran, M., Blank, V., Logeat, F., Vandekerckhove, J., Lottspeich, F., Le Bail, O., Urban, M. B., Kourilsky, P., Baeuerle, P. A. and Israël A. (1990). The DNA binding subunit of NF-KB is identical to factor KBF1 and homologous to the *rel* oncogene product. *Cell* **62**, 1007-1018.

Kingston, R. E. (1989). Transcription control and differentiation: the HLH family, c-*myc* and C/EBP. *Current Opinion in Cell Biol.* **1**, 1081-1087.

Klinger, M., Endélyi, M., Szabad, J. and Nüsslein-Volhard, C. (1988). Function of *torso* in determining the terminal anlagen of the *Drosophila* embryo. *Nature* **335**, 275-277.

Lehmann, R. and Nüsslein-Volhard, C. (1986). Abdominal segmentation, pole cell formation, and embryonic polarity require the localized activity of *oskar*, a maternal gene in *Drosophila. Cell* **47**, 141-152.

Lehmann, R. and Nüsslein-Volhard, C. (1987). Involvement of the *pumilio* gene in the transport of an abdominal signal in the *Drosophila* embryo. *Nature* **329**, 167-170.

Leiss, D., Hinz, U., Gasch, A., Mertz, R. and Renkawitz-Pohl, R. (1988). ß3-*tubulin* expression characterize the differentiating mesodermal germ layer during *Drosophila* embryogenesis. *Development* **104**, 525-531.

Leptin, M. and Grunewald, B. (1990). Cell shape changes during gastrulation in *Drosophila. Development* **110**, 73-84.

Leptin, M. (1991). *twist* and *snail* as positive and negative regulators during *Drosophila* mesoderm development. *Genes Dev.* **5**, 1568-1576.

Mayer, U. and Nusslein-Volhard, C. (1988). A group of genes required for pattern formation in the ventral ectoderm of the *Drosophila* embryo. *Genes Dev.* **2**, 1496-1511.

Nolan, G. P., Ghosh, S., Lou, H. C., Tempst, P. and Baltimore, D. (1991). DNA binding and IKB inhibition of the cloned p65 subunit of NF-KB and rel related polypeptide. *Cell* **64**, 961-969.

Murre, C., Schonleber-Mc Caw, P. and Baltimore, D. (1989). A new DNA binding and dimerization motif in immunoglobulin enhancer binding, *daughterless*, *MyoD* and *myc* proteins. *Cell* **56**, 777-783.

Nüsslein-Volhard, C. (1979). Maternal effect mutations that alter the spatial coordinates of the embryo of *Drosophila melanogaster.* In *Determination of Spatial Organisation,* (ed. G. Subtelney and I.R. Kœnigsberg) pp 185-211. New York: Academic Press.

Nüsslein-Volhard, C., Lohs-Schardin, M., Sander, K. and Cremer, C. (1980). A dorso-ventral shift of embryonic primordia in a new maternal-effect mutant of *Drosophila. Nature* **283**, 474-476.

Nüsslein-Volhard, C., Wieschaus, E. and Kluding, H. (1984). Mutations affecting the pattern of the larval cuticule in *Drosophila melanogaster.* I. zygotic loci on the second chromosome. *Roux's Arch. Dev. Biol.* **193**, 267-282.

Nüsslein-Volhard, C. (1991). Determination of embryonic axes of *Drosophila. Development* **Supplement 1**, 1-10.

Pan, D., Huang, J-D. and Courey, A. J. (1991). Functional analyses of the *Drosophila twist* promoter reveals a *dorsal* binding ventral activator region. *Genes Dev.* **5**, 1892-1901.

Payre, F., Noselli, S., Lefrere, V. and Vincent, A. (1990). The closely related Drosophila *sry*-β and *sry*-δ zinc finger proteins show differential embryonic expression and distinct pattern of binding sites on polytene chromosomes *Development* **110**, 141-149.

Payre, F. and Vincent, A. (1991). Genomic targets of the *serendipity* β and δ zinc finger proteins and their respective DNA recognition sites. *EMBO J.* **10**, 2533-2541.

Pirrotta, V., Bickel, S. and Makani, C. (1988). Developmental expression of the *Drosophila zeste* protein on polytene chromosomes. *Genes Dev.* **2**, 1839-1850.

Ray, R. P., Arora, K., Nüsslein-Volhard, C. and Gelbart, W. M. (1991). The control of cell fate along the doral-ventral axis of the *Drosophila* embryo. *Development* **113**, 35-54.

Richardson, P. and Gilmore, T. D. (1991). V-*rel* is an inactive member of the Rel family of transcriptional activator proteins. *J. Virol.* **65**, 3122-3130.

Roth, S., Stein, D. and Nüsslein-Volhard, C. (1989). A gradient of nuclear localization of the *dorsal* protein determines dorso-ventral pattern in the *Drosophila* embryo. *Cell* **59**, 1189-1202.

Roth, S., Hiromi, Y., Godt, D. and Nüsslein-Volhard, C. (1991). *cactus*, a maternal gene required for proper formation of the dorso-ventral morphogen gradient in *Drosophila* embryos. *Development* **112**, 371-388.

Rushlow, C., Frasch, M., Doyle, H. and Levine, M. (1987a). Maternal regulation of *zerknüllt*: a homeobox gene controlling differenciation of dorsal tissues in *Drosophila*. *Nature* **330**, 583-586.

Rushlow, C., Frasch, M., Doyle, H. and Levine, M. (1987b). Molecular characterization of the *zerknüllt* region of the *Antennapedia* gene complex in *Drosophila*. *Genes Dev.* **1**, 1268-1279.

Rushlow, C., Han, K., Manley, J. L. and Levine, M. (1989). The graded distribution of the *dorsal* morphogen is initiated by selective nuclear transport in *Drosophila*. *Cell* **59**, 1165-1177.

Rushlow, C. and Arora, K. (1990). Dorsal-ventral polarity and pattern formation in the Drosophila embryo. *Seminar in Cell Biol.* **1**, 137-149.

Santamaria, P. and Nüsslein-Volhard, C. (1983). Partial rescue of *dorsal*, a maternal effect mutation affecting the dorso-ventral pattern of the *Drosophila* embryo, by the injection of wild-type cytoplasm. *EMBO J.* **2**, 1695-1699.

Schüpbach, T. and Wieschaus, E. (1986). Maternal-effect mutations altering the anterior-posterior pattern of *Drosophila* embryo. *Roux's Arch. Dev. Biol.* **195**, 302-317.

Schüpbach, T. and Wieschaus, E. (1989). Female sterile mutation on the second chromosome of *Drosophila melanogaster*. I. Maternal effect mutations. *Genetics* **121**, 101-117.

Simpson, P. (1983). Maternal zygotic gene interactions during formation of the dorsoventral pattern in *Drosophila* embryos. *Genetics* **105**, 615-632.

St Johnston, R. D. and Gelbart, W. M. (1987). *Decapentaplegic* transcripts are localized along the dorsal-ventral axes of the *Drosophila* embryos. *EMBO J.* **6**, 2785-2791.

St Johnston, R. D. and Nüsslein-Volhard, C. (1992). The origin of pattern and polarity in the *Drosophila* embryo. *Cell* **68**, 201-219.

Steward, R., McNally, F. J. and Schedl, P. (1984). Isolation of the *dorsal* locus in *Drosophila*. *Nature* **311**, 262-265.

Steward, R., Ambrosio, L. and Schedl, P. (1985). Expression of the *dorsal* gene. *Cold Spring Harbor Symp. Quant. Biol.* **50**, 223-228.

Steward, R. (1987). *Dorsal*, an embryonic polarity gene in *Drosophila*, is homologous to the vertebrate proto-oncogene, c-*rel*. *Science* **238**, 692-694.

Steward, R. (1989). Relocalization of the *dorsal* protein from the cytoplasm to the nucleus correlates with its function. *Cell* **59**, 1179-1188.

Thisse, B., Stoetzel, C., El Messal, M. and Perrin-Schmitt, F. (1987). Genes of the *Drosophila* maternal dorsal group control the specific expression of the zygotic gene *twist* in presumptive mesodermal cells. *Genes Dev.* **1**, 709-715.

Thisse, B., Stoetzel, C., Gorostiza-Thisse, C. and Perrin-Schmitt, F. (1988). Sequence of the *twist* gene and nuclear localization of its protein in endomesodermal cells of early *Drosophila* embryos. *EMBO J.* **7**, 2175-2183.

Thisse, C., Perrin-Schmitt, F., Stoetzel, C. and Thisse, B. (1991). Sequence specific transactivation of the *Drosophila twist* gene by the *dorsal* gene product. *Cell* **65**, 1191-1201.

Wieschaus, E., Nüsslein-Volhard, C and Kluding, J. (1984). Mutations affecting the pattern of the larval cuticule in Drosophila melanogaster. I. zygotic loci on the second chromosome. *Wilhelm Roux's Arch. Devl Biol.* **193**, 296-307.

Zink, B. and Paro, R. (1989). In vivo binding pattern of trans-regulator of homeotic genes in *Drosophila melanogaster*. *Nature* **337**, 468-471.

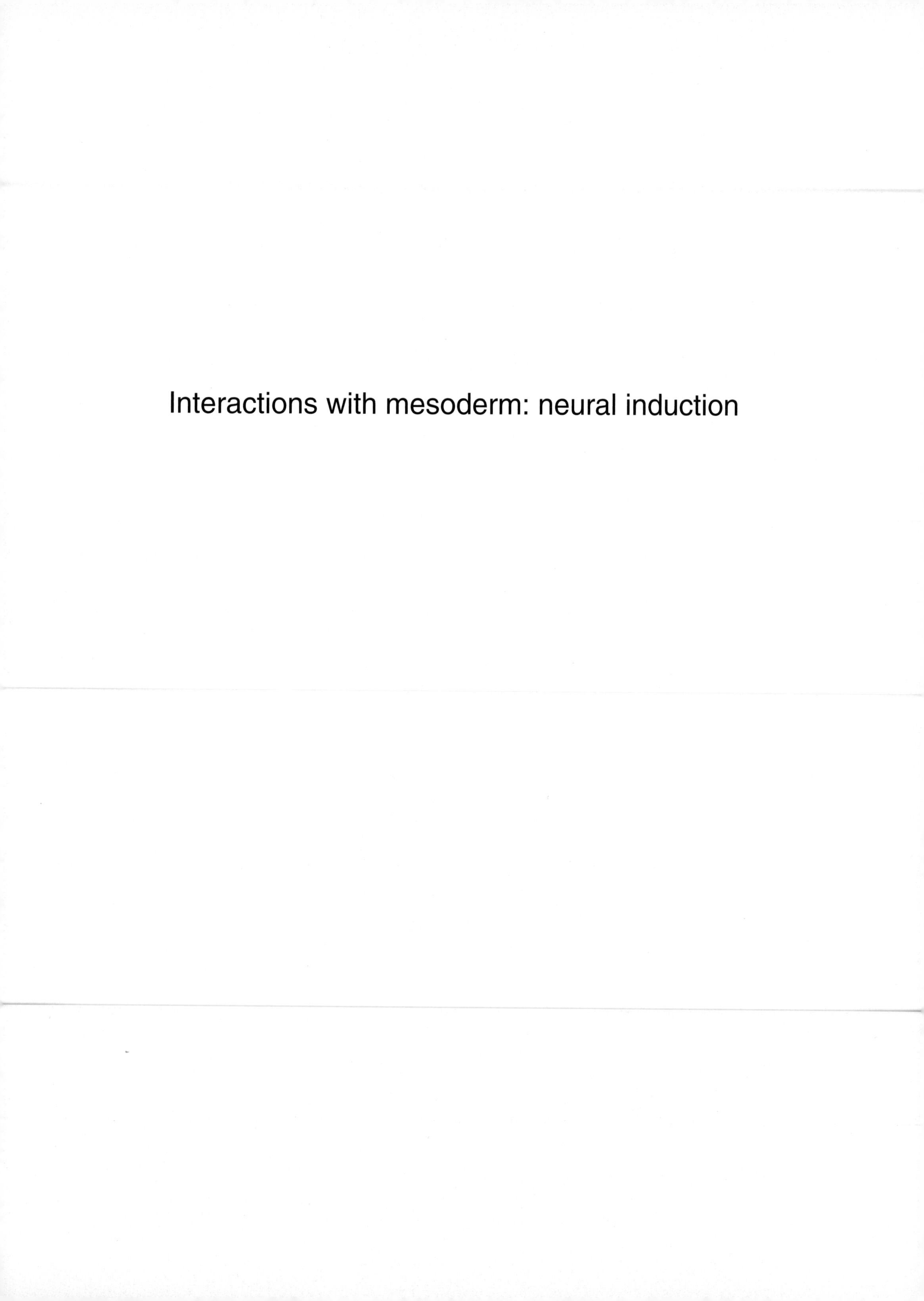

Interactions with mesoderm: neural induction

Development 1992 Supplement, 183-193 (1992)
Printed in Great Britain © The Company of Biologists Limited 1992

Induction of anteroposterior neural pattern in *Xenopus* by planar signals

TABITHA DONIACH*

Department of Molecular and Cell Biology, 301 Life Sciences Addition, University of California, Berkeley, CA 94720, USA

*Present address: Department of Obstetrics, Gynecology and Reproductive Sciences, Box 0556, University of California, San Francisco, CA 94143, USA

Summary

Neural pattern in vertebrates has been thought to be induced in dorsal ectoderm by 'vertical' signals from underlying, patterned dorsal mesoderm. In the frog *Xenopus laevis*, it has recently been found that general neural differentiation and some pattern can be induced by 'planar' signals, i.e. those passing through the single plane formed by dorsal mesoderm and ectoderm, without the need for vertical interactions. Results in this paper, using the frog *Xenopus laevis,* indicate that four position-specific neural markers (the homeobox genes *engrailed-2(en-2), XlHbox1* and *XlHbox6* and the zinc-finger gene *Krox-20*) are expressed in planar explants of dorsal mesoderm and ectoderm ('Keller explants'), in the same anteroposterior order as that in intact embryos. These genes are expressed regardless of convergent extension of the neurectoderm, and in the absence of head mesoderm. In addition, *en-2* and *XlHbox1* are not expressed in ectoderm when mesoderm is absent, but they and *XlHbox6* are expressed in naïve, ventral ectoderm which has had only planar contact with dorsal mesoderm. *en-2* expression can be induced ectopically, in ectoderm far anterior to the region normally fated to express it, suggesting that a prepattern is not required to determine where it is expressed. Finally, the mesoderm in planar explants expresses *en-2* and *XlHbox1* in an appropriate regional manner, indicating that A-P pattern in the mesoderm does not require vertical contact with ectoderm. Overall, these results indicate that anteroposterior neural pattern can be induced in ectoderm soley by planar signals from the mesoderm. Models for the induction of anteroposterior neural pattern by planar and vertical signals are discussed.

Key words: *Xenopus laevis*, *engrailed-2*, *XlHbox1*, *XlHbox6*, A-P patterning, Keller explants, neural induction.

Introduction

The concept that neural development is induced in the ectoderm by the dorsal mesoderm originated from the famous 'organizer' transplantation experiment performed by Hilde Mangold in 1921 (Spemann and Mangold, 1924). In this experiment, the dorsal lip of the blastopore of an amphibian (urodele) early gastrula was found to induce a secondary nervous system in ventral ectoderm of another gastrula. The dorsal lip tissue, composed of dorsal mesoderm, was named the organizer because the secondary axis displayed large-scale anteroposterior and dorsoventral organization. At the time, Spemann proposed two routes by which the inductive signals from the dorsal lip could reach the ectoderm: the planar route, in which signals pass within the continuous plane of tissue (also referred to as tangential or horizontal induction), or the vertical one, in which they pass to the overlying ectoderm after the dorsal lip tissue has involuted (Spemann, 1938).

Support for vertical induction

In 1933, Holtfreter found evidence in favor of the vertical route of induction and against the planar route, observing that urodele exogastrulae lack any histological evidence of neural differentiation. In these abnormal embryos, the mesoderm and endoderm do not involute, but move outwards during gastrulation, thereby precluding vertical interactions while maintaining planar ones. Vertical induction received further support when Otto Mangold (1933) found that dorsal mesoderm (the 'chordamesoderm', or presumptive notochord mesoderm) from early urodele neurulae could induce neural development in ventral ectoderm when inserted into the blastocoel of an early gastrula. Moreover, Mangold discovered that different regions of chordamesoderm along the anteroposterior (A-P) axis generally induced neural tissue of an equivalent A-P level. This led to the model that the chordamesoderm itself contains A-P pattern information in the form of regionalized inducers which induce a parallel pattern in the overlying ectoderm. Many workers have since found that chordamesoderm from different A-P regions of the late gastrula and early neurula can induce different A-P levels of neural differentiation in ectoderm that has been wrapped around it (for examples see Ter Horst, 1948; Sala, 1955; Sharpe and Gurdon, 1990; Hemmati-Brivanlou et al., 1990; Saha and Grainger, 1992). Taken together, the evidence for vertical induction and against planar induction led to common acceptance of vertical induction as the main pathway of neural induction (Spemann, 1938; Hamburger, 1988).

Planar induction revived

It was not until the advent of molecular markers of early neural development that planar neural induction re-emerged as a possibility. Kintner and Melton (1987) made the surprising discovery that *Xenopus laevis* exogastrulae express NCAM, a gene specific to neural tissue at the stages examined. Using in situ hybridization, they found that NCAM RNA was present in the ectoderm next to the evaginated mesoderm. Subsequently, Dixon and Kinter (1989) and Keller et al. (1992b) found that planar explants of dorsal mesoderm and ectoderm ('Keller' explants, see below) express NCAM and another general neural marker NF3, and Savage and Phillips (1989) found that Epi-1, an epidermal marker, is turned off in ectoderm that has had only planar contact with dorsal lip mesoderm. Also, the convergent extension movements associated with the neurectoderm are induced by planar contact with mesoderm (Keller and Danilchik, 1988, Keller et al., 1992b, c). It should be pointed out that this new work has been carried out on *Xenopus* embryos, not on the urodele embryos used by the classical embryologists. It remains to be determined at a molecular level whether planar neural induction occurs in urodeles.

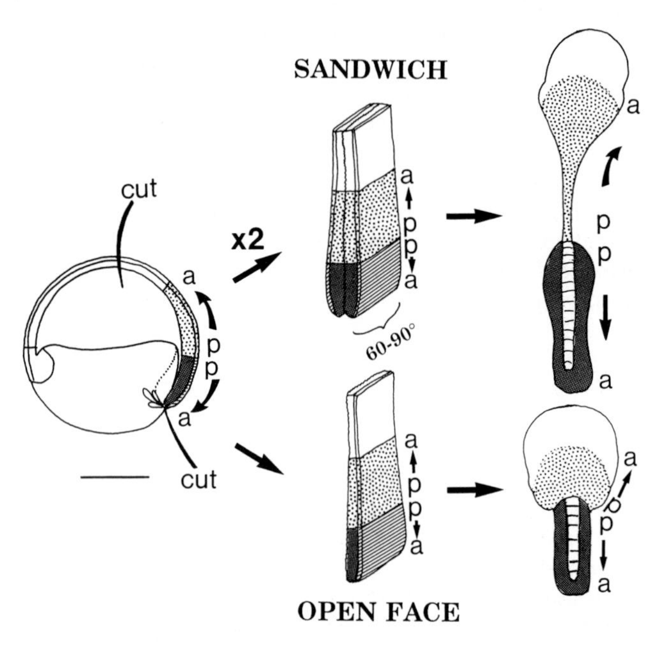

Fig. 1. Regions of early gastrula used to make Keller explants. On left is a stage 10+ embryo in sagittal cross section, dorsal to the right, animal pole up. The shading approximates tissue types predicted by the fate map (Keller, 1975, 1976; Keller et al., 1992a): the animal (upper) hemisphere consists of ectoderm that will give rise to epidermis (white) and neurectoderm (light stippling). Dorsal mesoderm is shown in dark grey and archenteron roof endoderm is striped. Explants were made by cutting out a rectangle of tissue reaching from approximately the animal pole the blastopore, as indicated by the upper and lower 'cuts' respectively, and about 60-90° wide around the equator. The head mesoderm, bounded by the dotted line and the dorsal mesoderm, was removed from explants. The predicted A-P polarity of the mesoderm and ectoderm is indicated. Sandwich explants were made by putting two of these rectangles together with their inner surfaces apposed; these undergo convergent extension (narrowing and elongating) in both the posterior neurectoderm and the meso-endoderm. Open face explants undergo convergent extension only in the meso-endoderm. The layer of endoderm is not shown in the explants depicted on the right. In these, the white column down the center of the mesoderm represents the notochord. Scale bar: 500 μm.

In *Xenopus* at least, these findings support a role for planar signals in the induction of markers of general neural differentiation, but do not indicate whether A-P pattern is established in this neural tissue. A first indication that some neural pattern may be induced by planar signals came when Ruiz i Altaba (1990) found that exogastrulae express the homeobox gene *Xhox-3* in a restricted pattern in the ectoderm. This gene is normally specifically expressed in the midbrain-hindbrain area at the stages studied. However, with a single marker it was not possible to determine if there was A-P polarity to this neural pattern. Also, results from exogastrulae are difficult to interpret because it is hard to be sure that vertical interactions have not occurred during exogastrulation, since the internal movements of the mesoderm cannot be followed.

In this paper, the expression of four position-specific neural markers has been examined in Keller explants to investigate whether planar signals can induce A-P neural pattern. Keller explants are explants of dorsal mesoderm and ectoderm, made before mesodermal involution (and therefore prior to any vertical interactions), and cultured flat throughout gastrulation and neurulation. Extensive experiments have shown that vertical interactions do not occur in these explants (Keller et al., 1992a,b). The results of the present study (see also Doniach et al., 1992) show that the four genes, *engrailed-2* (Hemmati-Brivanlou and Harland, 1989), *Krox-20* (Wilkinson et al., 1989; Bolce et al., 1992), *XlHbox1* (Oliver et al., 1988) and *XlHbox6* (Wright et al., 1990), each expressed in a specific region along the A-P axis of normal embryos, are expressed in Keller explants in the normal A-P order. These and other experiments presented here indicate that planar signals can induce A-P neural pattern. In addition, there is evidence for autonomous A-P pattern in the mesoderm of these explants.

Materials and methods

Culture of embryos and explants

Eggs were fertilized, dejellied and cultured in 33% modified Ringers (MMR) as in Vincent et al. (1986). All embryos used were albinos. Explants were made in Sater's modified Danilchik's medium (SMDM; A. Sater, R. Steinhardt and R. Keller, personal communication), pH 8.1, using an eyebrow hair knife and a hair loop, and cultured in plastic tissue culture dishes in SMDM until controls reached stage 22-27 (as indicated). Culture temperatures varied from 15°C to 23°C, depending on the rate of development that was required. Embryos were staged according to Nieuwkoop and Faber (1967).

Keller explants

Explants were made at stage 10 to 10+, using regions of

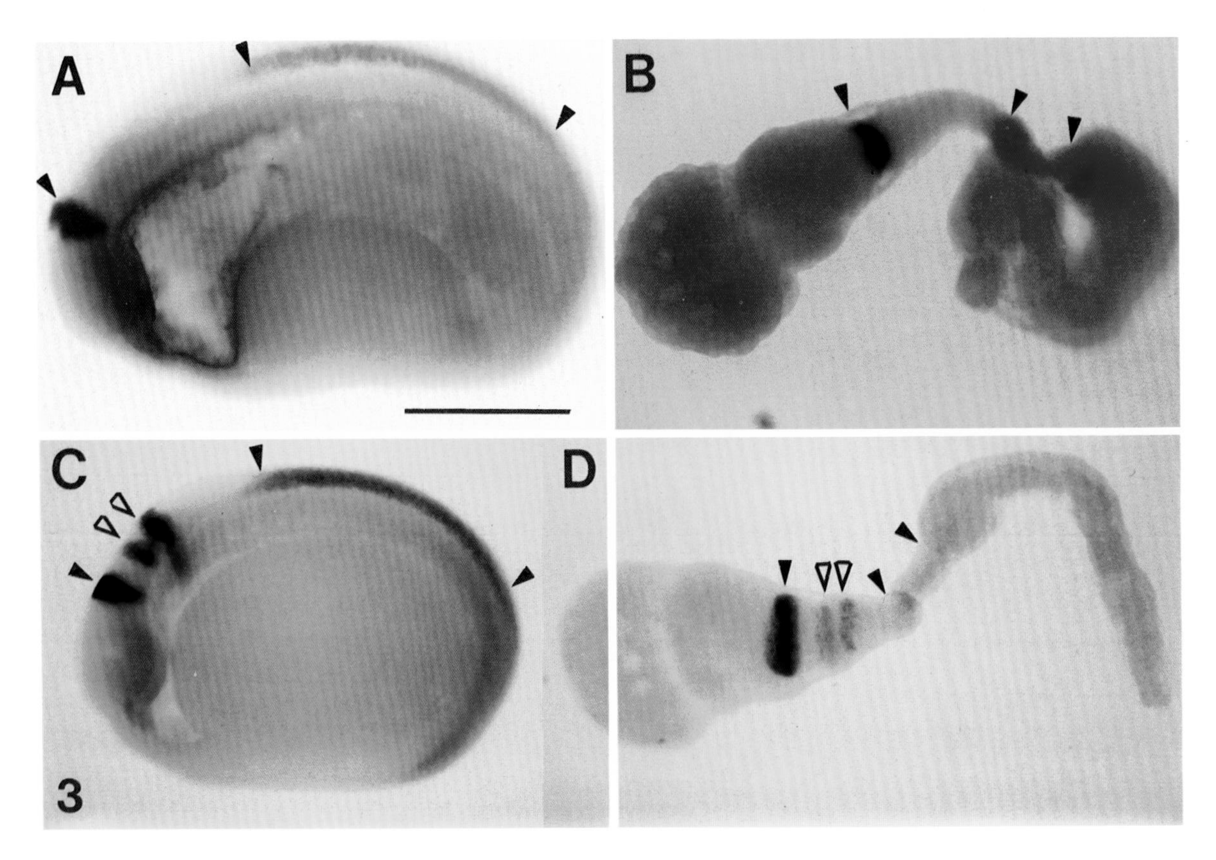

Fig. 3. Double and triple detection of *en-2, Krox-20* and *XlHbox6* by whole-mount in situ hybridization in whole control embryos (left side) and sandwiches (right side). Orientation is as in Fig. 2. Digoxigenin probes for the markers were hybridized simutaneously to determine whether the bands of expression seen in Fig. 2 are spatially separate. (A,B) *en-2* and *XlHbox6* expression in stage 22 control embryo and Keller sandwich, respectively. The RNA signal is indicated with arrowheads. There are two distinct regions of expression; comparison with patterns in Fig. 2 implies that the expression in the flared region of the neurectoderm is *en-2* and that in the neck region is *XlHbox6*. There is additional staining in the anterior mesoderm (not marked); this is *en-2* expression (it is also seen in sandwiches hybridized with *en-2* alone but not with *XlHbox6* alone. In A, there is background staining in the archenteron often seen with this technique (Harland, 1991). In B, the mesodermal portion of the sandwich has curled around; this occurred shortly before fixation at stage 22. (C,D) Expression of *en-2, Krox-20* and *XlHbox6* in control embryo and Keller sandwich, respectively. Filled arrowheads: *en-2* and *XlHbox6*; open arrowheads: *Krox-20*. The relative position of *Krox-20* expression is confirmed in D by the absence of these bands in B (in situ hybridization in D was kindly carried out by Dr. R. Harland). Scale bar 500 µm.

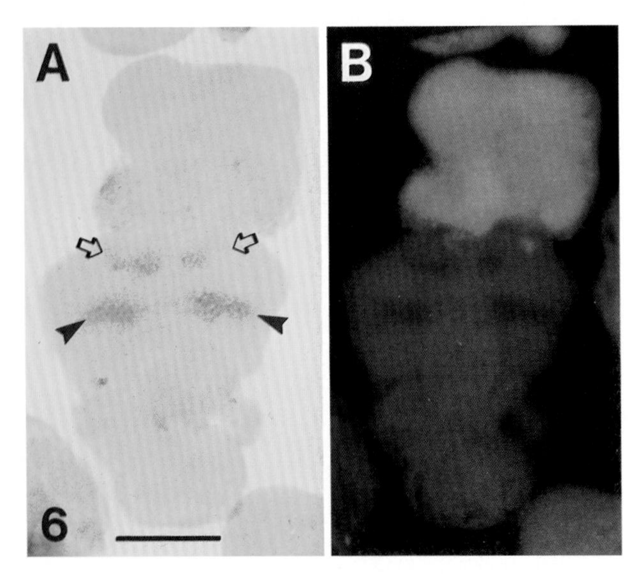

Fig. 6. Ectopic induction of *en-2* expression: Recombinate of FDA-labelled dorsal mesoderm grafted onto the anterior ectoderm of an open-face Keller explant, immunostained for *en-2*. (Orientation: labelled mesoderm: anterior up, posterior down; unlabelled tissue: animal end up, vegetal down, as in Fig. 1) *en-2* expression is in two regions of the unlabelled portion: a 1° region (filled arrows) 2° region (open arrows). Scale bar 250 µm.

the embryo illustrated in Fig. 1. Loose, potentially migratory head mesoderm was gently picked off with an eyebrow hair. Sandwich and open-face explants were then cultured flat under pieces of coverslip resting on silicone vacuum grease (Dow-Corning).

Planar recombinates of mesoderm with ectoderm

In SMDM, the freshly cut edge of a rectangle of dorsal mesoderm from an FDA-labelled (see below) embryo was gently pressed against the freshly cut edge of a piece of ectoderm for a few seconds, and the graft was allowed to heal for 5 minutes. The recombinate was then lightly flattened with a piece of coverslip resting on vacuum grease, cultured in MDM until controls reached stage 22-26, and immunostained as described below.

FDA labelling

Mesoderm donors were lineage labelled by injecting 8 nl of 25 mg/ml fluorescein dextran amine (FDA, a gift of R. Gimlich; Gimlich and Braun,1985) into stage 1 embryos using an air pressure injection system (Tritech Research, Los Angeles) and cultured in the dark throughout development and fixation.

Whole-mount immunocytochemistry and in situ hybridization

Embryos and explants were processed for whole-mount immunocytochemistry (Hemmati-Brivanlou and Harland, 1989), using horseradish peroxidase-tagged 2° antibodies (Bio-Rad or Jackson Immunoresearch), with shortened washes (3×30 minutes) and 0.05% $NiCl_2$ in the diaminobenzidine solution to enhance signal, except when detecting NCAM and 12-101. Whole-mount in situs were done as described in Harland (1991). Digoxigenin-labelled RNA probes were kindly provided by R. Harland and T. Lamb, and the Krox-20 clone was a generous gift of D. Wilkinson. Sequences used as probes are as described in Bolce et al. (1992).

Results

Keller explants

Two types of Keller explants were used (Fig. 1): 'sandwich' explants, in which two sheets of dorsal mesoderm (each covered with a layer of endoderm) and ectoderm are sandwiched together, inner surfaces apposed and 'open-face' explants, in which a single sheet is cultured alone. The involuted head mesoderm (which has not yet come to underly the presumptive neurectoderm) is removed from these explants to avoid potential vertical interactions that might occur, should these highly migratory cells (Winklbauer, 1990; Winklbauer et al., 1991) crawl onto the inner surface of the ectoderm. Keller explants are cultured flat under a coverslip to prevent mesodermal involution. In the absence of such involution, the A-P axes of the mesoderm and ectoderm in these explants point in opposite directions, with the posterior ends of both tissues at their common boundary (Keller, 1975, 1976). In both types of explant, the meso/endodermal portion undergoes essentially normal convergent extension movements (Keller and Danilchik, 1988; Wilson and Keller, 1991), causing elongation along the A-P axis. In sandwiches that are correctly made (i.e. those in which mesoderm has not involuted or migrated between the two sheets of ectoderm (Keller et al., 1992a,b)), the ectodermal portion also converges and extends, with the greatest amount occurring in the presumptive spinal region, producing a long, skinny 'neck'. Progressively less convergent extension occurs in the presumptive hindbrain and midbrain regions, resulting in a gradual widening of the explant towards the anterior end. There is no convergent extension in the ectoderm of open-face explants. Explants that showed irregular or asymmetrical convergent extension were either discarded or not counted in the experiments.

Domains of neural and mesodermal tissue in Keller explants

Dixon and Kintner (1989) showed by RNAase protection assays that there is expression of a general neural marker, NCAM, in Keller sandwiches. When anti-NCAM antibodies (Jacobson and Rutishauser, 1986) are used to stain Keller sandwiches, NCAM protein is detected in the neck region of the ectoderm and more anteriorly, in the flared region beyond the neck (Fig. 2B; Keller et al., 1992b). This defines the neural domain in Keller sandwiches, since NCAM expression is restricted to neural tissue at this stage (Balak et al., 1987; Fig. 2A). NCAM is also expressed in open-face explants (Fig. 2C). Somitic mesoderm, which can be identified using a monoclonal antibody, 12-101 (Kintner and Brockes, 1984), is organized in explants in two blocks flanking the notochord (Fig. 2D-F). The notochord can be identified morphologically, forming a distinctive column of large vacuolated cells.

There is A-P neural pattern in Keller explants

Immunostaining and in situ hybridization were used to localize the expression patterns of three homeobox genes, *en-2, XlHbox1* and *XlHbox6,* and a zinc-finger domain encoding gene, *Krox-20* in Keller explants. In intact embryos, these genes are expressed at specific A-P regions in the developing central nervous system (CNS): *en-2* is expressed at the midbrain-hindbrain junction (Fig. 2G; Hemmati-Brivanlou and Harland, 1989); *Krox-20* in the third and fifth rhombomeres of the hindbrain, with expression in the fifth reaching into the neural crest (Fig. 2P; Wilkinson et al., 1989 and Bolce et al., 1992); *XlHbox1* in a broad band in the anterior spinal region (Fig. 2J; Oliver et al., 1988) and *XlHbox6* throughout the spinal region and tailbud (Fig. 2M; Wright et al., 1990). They are also expressed outside the CNS, as will be described later.

These genes are expressed in Keller sandwiches in the normal A-P order: *en-2* protein is detected in the flared region anterior to the narrow 'spinal region' (Fig. 2H); *Krox-20* RNA in two bands posterior to this, still anterior to the spinal region (Fig.2Q); *XlHbox1* protein at the anterior end of the spinal region (Fig. 2K), and *XlHbox6* protein throughout the length of the spinal region, and extending into the posterior end of the mesoderm (Fig. 2N). This latter region consists of posterior mesoderm and presumptive posterior neurectoderm (Keller et al., 1992a,b). The order of expression can be seen more clearly when two (Fig. 3A,B) and three (Fig. 3C,D) of the markers are detected in

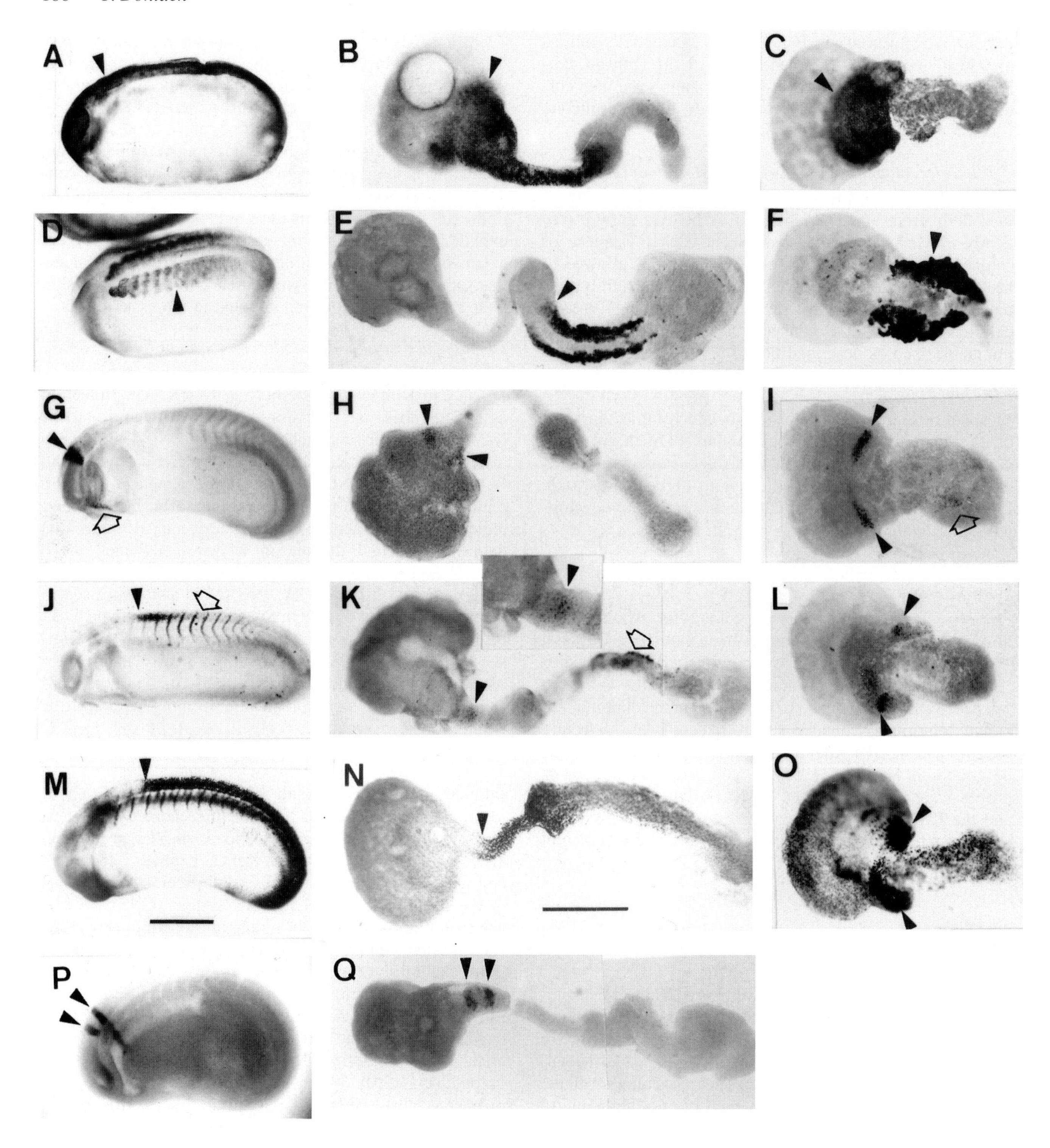

the same sandwich. A majority of the explants expressed these markers (95% for *en-2* (n=39); 64% for *XlHbox1* (n=14); 95% for *Krox-20* (n=22); 91% for *XlHbox6* (n=23)), and all of those that were positive showed expression in the respective patterns described above, although there was some variation in the intensity of the signal. This variation could reflect a difference in amount of expression, or in the sensitivity of the detection method.

A potential problem with Keller sandwiches is that expression of the neural markers could result from mesoderm on one face of the sandwich vertically inducing ectoderm on the other face if the two faces are misaligned. This is avoided in open-face explants. In these, all three homeobox genes are expressed: *en-2* and *XlHbox1* each in bilateral stripes a short distance from the mesoderm (Fig. 2I,L, and *XlHbox6* in larger bilateral regions next to and probably extending into the posterior mesoderm/tailbud region (Fig. 2O; *Krox-20* was not tested). Thus, in the absence of the potential vertical interactions that could occur in sandwiches, these genes are still expressed.

Fig. 2. Expression of neural markers in whole embryos and Keller explants. Left panel: whole embryos, stage 22-26 lateral view, anterior left, dorsal up. Center panel: sandwiches, dorsal view, animal pole left. Right panel: open-face explants, dorsal view, animal pole left. Scale bars 500 μm; scale bar in M applies to examples in the left panel, and that in N, center and right panels. Solid arrowheads point to antigen staining in neurectoderm, open arrows indicate putative mesodermal staining. (A,B,C) NCAM detected with rabbit polyclonal antibodies to the *Xenopus* 180×10^3 M_r isoform (a gift of U. Rutishauser; Jacobson and Rutishauser, 1986), assayed at stage 22. (D,E,F) Muscle detected with 12-101, a monoclonal antibody to a skeletal muscle-specific protein (Kintner and Brockes, 1984; obtained from Developmental Studies Hybridoma Bank), assayed stage 20. (G,H,I) *en-2* protein detected with mouse monoclonal antibody 4D9 (arrowheads, Patel et al., 1989; Hemmati-Brivanlou and Harland, 1989), assayed stage 23-24. Open arrows: *en-2* protein in G, the mandibular arch, and I, the distal mesoderm of an open-face explant. *en-2* expression in sandwiches is consistently lower than in open-face explants. (J,K,L) *XlHbox1* protein detected with affinity purified rabbit polyclonal antibodies to the long form (a gift of C. Wright and E. De Robertis; Oliver et al., 1988). Arrowheads: expression in the CNS, and open arrows: expression in J, trunk somites (striped pattern) and K, somitic mesoderm in a sandwich explant. Somitic mesoderm tends to disintegrate by the stage assayed (stage 26), and is not always present. (K) inset: area of neurectoderm that expresses *XlHbox1* in sandwiches, approx. 2× magnification of K. (M,N,O) *XlHbox6* protein, detected with affinity purified rabbit polyclonal antibodies (arrowheads point to anterior boundary in M and N) (a gift of C. Wright; Wright et al., 1990), assayed at stage 24. Some batches of this antiserum cross react with unidentified nuclear antigens in proximal notochord (unpublished observations) and muscle (Wright et al., 1990), and with an extracellular antigen in the cement gland of intact embryos. (P, Q) *Krox-20* RNA detected by whole-mount in situ hybridization, at stage 21-22 (the probe was kindly made by R. Harland; the probe was a gift of D. Wilkinson). Staining is cytoplasmic. (A-O are reproduced with permission from *Science* journal.)

Head mesoderm does not induce en-2

The involuted head mesoderm is nominally removed when Keller explants are made, but it is possible that some of these highly migratory cells are left behind and could migrate into the ectodermal territory, thereby inducing the expression of the neural markers via vertical interactions. To determine whether head mesoderm is actually capable of inducing *en-2* by vertical contact, the most anterior marker used, head mesoderm (see Fig. 1) and a small amount of attached pharyngeal endoderm were wrapped with competent ectoderm (from stage 10 embryos) and cultured until stage 27. *en-2* was not detected in any of these recombinates (n=22). This implies that if contaminating head mesoderm is present in Keller explants and has vertical contact with the ectoderm, it is unlikely to induce *en-2* expression, and perhaps that of the more posterior markers, although they have not been tested.

Contact with dorsal mesoderm is required for expression of neural markers

The ectodermal expression of these genes could be autonomous, rather than being induced by planar contact with dorsal mesoderm. To test this, dorsal ectoderm was explanted without dorsal mesoderm from early gastrulae

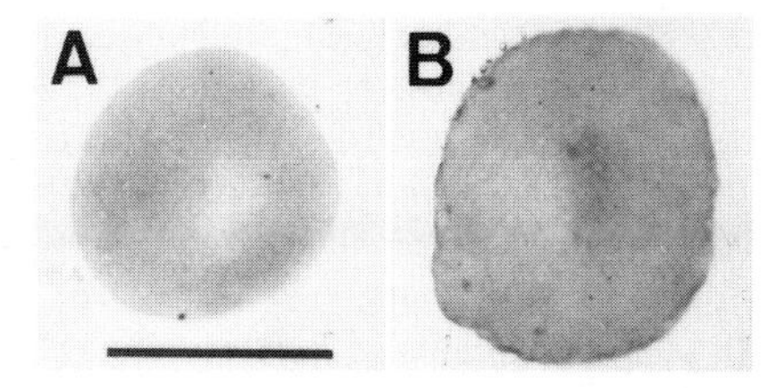

Fig. 4. Isolated dorsal ectoderm, immunostained for (A) *en-2* or (B) *XlHbox1*. Neither marker was detected in these or other explants.

(stage 10+). It does not express *en-2* (0 of 34) or *XlHbox1* (0 of 8; Fig. 4). Therefore, contact with dorsal mesoderm is required for expression of these genes. *XlHbox6* could not be tested because it is expressed in the posterior mesoderm, and *Krox-20* has not yet been tested.

Planar contact is sufficient for neural induction

Is planar contact sufficient to induce these genes in ventral ectoderm, in which they are not normally expressed? Dorsal mesoderm from early gastrulae (stage 10+) that had been labelled with the lineage marker fluorescein dextran amine (FDA; Gimlich and Braun, 1985) was grafted onto the edge of a sheet of unlabelled, stage 10+ ventral ectoderm, creating a planar, open-face recombinate. Normal A-P polarity was preserved by grafting the presumptive posterior

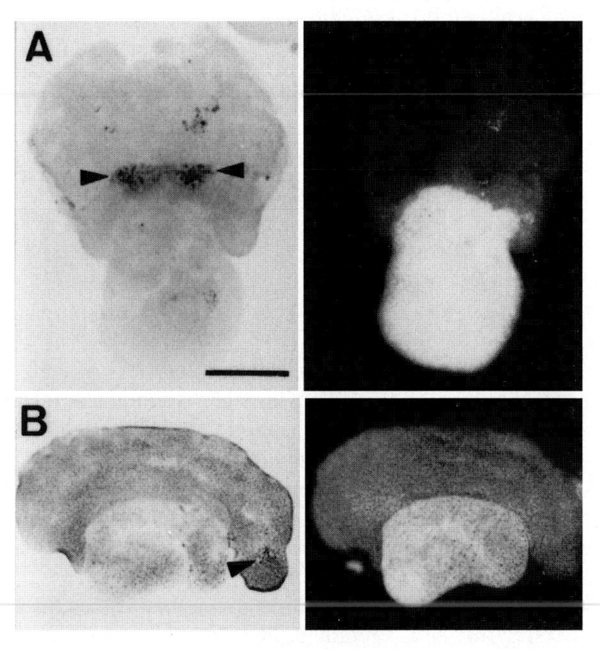

Fig. 5. Induction of *en-2* and *XlHbox1* expression in ventral ectoderm: recombinates of FDA-labelled dorsal mesoderm grafted onto the the edge of unlabelled ventral ectoderm, immunostained for (A) *en-2* (control stage 23) and (B) *XlHbox1* (control stage 26). Left, images viewed with transillumination; right, epifluorescence. Arrowheads point to *en-2* or *XlHbox1*-positive nuclei in the ectoderm. In A, the small patch of staining and fluorescence above the positive nuclei is non-nuclear and is an artifact; also, mesodermal *en-2* expression is visible in the FDA-labelled portion. Scale bar 500μm. (Reproduced with permission from *Science* journal.)

ends of each tissue together. All three homeobox genes were expressed in the unlabelled portion of these recombinates (Fig. 5; 6 of 15 expressed *en-2*, 9 of 18 *XlHbox1*, and 2 of 21 *XlHbox6*). The region of expression, particularly for *en-2*, appears to be closer to the mesoderm than seen in normal open-face explants. These markers were not expressed in control explants of ventral ectoderm alone (0 of 8 expressed *en-2*, 0 of 21 *XlHbox1*, 0 of 13 *XlHbox6*). FDA-labelled mesoderm did not migrate into or mix with the unlabelled ectoderm (Fig. 5A,B), further supporting the notion that contact between mesoderm and ectoderm is exclusively planar. Thus, strictly planar contact is sufficient to induce these genes, even in ventral ectoderm.

Prepattern is not required

The following experiment tests whether *en-2* can be induced ectopically, beyond the latitude predicted by the fate map to express it. A priori, it could be that the A-P expression pattern of genes in the neurectoderm could be dictated by an animal-vegetal prepattern that exists around the whole embryo, including on the ventral side, such that specific genes are only induced in the regions derived from the latitudes fated to express them. To test this, the presumptive posterior edge of FDA-labelled dorsal mesoderm (from stage 10+) was grafted onto the anterior end (near the anterior edge of the presumptive neural plate (Fig. 1)) of an unlabelled open-face explant (also stage 10+). In such recombinates there were two regions of *en-2* expression, one 'primary' region near the mesoderm from the unlabelled portion and a smaller 'secondary' region near the labelled mesoderm, in territory approximately fated to become cement gland or perhaps forebrain (Fig. 6). Keller et al. (1992a) have obtained a similar result using convergent extension as a marker. This demonstrates that a prepattern, in which *en-2* expression is limited to a particular position of the ectoderm along the animal-vegetal axis, does not exist or is at least not required for *en-2* expression.

A-P pattern in the mesoderm

In intact embryos, *en-2*, *XlHbox1* and *XlHbox6* are also expressed in specific regions outside the CNS. *en-2* protein is detected bilaterally in nuclei of loose clusters of cells in the mandibular arch (Fig. 2G; Hemmati-Brivanlou and Harland, 1989; Hemmati-Brivanlou et al., 1991); *XlHbox1* protein is detected in nuclei of somites in the mid-trunk (Fig. 2J) and lateral plate mesoderm of the mid trunk (Oliver et al., 1988); and *XlHbox6*, in nuclei of trunk lateral plate mesoderm and tailbud (posterior) mesoderm (Wright et al., 1990).

In Keller explants, in addition to the expression in the neurectoderm, *en-2* protein is detected in loose, bilateral clusters of nuclei in the anterior (distal) part of the mesoendoderm. In histological sections (not shown), it can be seen that these nuclei are in the deep layer of tissue, suggesting that the respective cells are not from the layer of superficial endoderm (Keller, 1975). The clustered pattern of these nuclei in explants resembles that of those expressing *en-2* in the mandibular arch (Fig. 2I). It has been suggested that the *en-2*-expressing cells in the mandibular arch are derived from the neural crest, so the presence of similar cells in the mesodermal portion of explants was surprising. The location of these cells raised three questions. First, are they derived from the mesoderm, or from neural crest that has migrated from the ectoderm into the mesoderm? Second, if this *en-2* expression is mesodermal, is it autonomous within the anterior mesoderm, or is contact with posterior mesoderm and/or ectoderm required? Third, is the presence of anterior, *en-2* expressing, mesoderm required for the expression of *en-2* in the ectodermal portion of explants, or is posterior mesoderm a sufficient source of inducer?

To answer the above questions, Keller sandwiches were cut into two portions, a 'lower' one consisting of anterior mesoderm, and an 'upper' one consisting of ectoderm and posterior mesoderm, at stage 10+. The two portions were cultured separately but under the same coverslip so as to keep track of each pair of complementary portions from each sandwich. When controls reached stage 26, individual pairs were stained with antibodies to *en-2*. Mesoderm was detected with antibodies to muscle (using mAb 12-101) and notochord (using Tor-70, a monoclonal antibody against the notochordal sheath (Bolce et al., 1992). Only those pairs in which the upper portion also contained a substantial amount of muscle and notochord were scored for *en-2* expression, in order to avoid possible contamination of the lower portion with neural crest from the ectoderm of the upper portion. *en-2* was detected in all of the lower portions (n=9), in loose clusters of nuclei, and it was detected in 3 of 9 of the upper portions of the same pairs, in a band of nuclei characteristic of the neural expression pattern. Therefore, (1) the expression in the lower portions is not neural crest derived, it is mesodermal. This suggests that a substantial portion of the *en-2*-expressing cells in the mandibular arch are mesodermal, although it remains possible that additional ones are neural crest-derived. (2) *en-2* expression in the anterior mesoderm is autonomous, indicating that contact with ectoderm or posterior mesoderm is not required for this regional expression; and (3) the anterior mesoderm, notably that which expresses *en-2*, is not required for expression of *en-2* in the neurectoderm.

XlHbox1 protein is also detected in the nuclei of somites in the middle of the mesodermal portion (Fig. 2K) of Keller sandwiches. Thus, the mesoderm of planar explants exhibits A-P pattern, suggesting that vertical interactions with the ectoderm are not required for patterning of the mesoderm.

Discussion

Position-specific molecular neural markers have been used here to determine whether A-P neural pattern can be induced in planar explants of dorsal mesoderm and ectoderm. All four of the markers (*en-2, Krox-20, XlHbox1,* and *XlHbox6*) are expressed in these explants, and in the normal A-P order. Both of the markers that were tested (*en-2* and *XlHbox1*) are not expressed in ectoderm in the absence of mesoderm, and they and *XlHbox6* are induced in ventral ectoderm when placed in planar contact with dorsal mesoderm. Based on the above results, it is concluded that planar signals alone are sufficient to induce A-P neural pattern. These results corroborate those of Keller et al. (1992b), who found that the convergent extension movements associated with the neurectoderm are induced by planar signals. They are also in agreement with those of Ruiz i Altaba (1990), who found that the anterior neural marker Xhox-3 is expressed in the ectoderm of exogastrulae. In addition, the results indicate that the mesoderm exhibits some A-P pattern of its own in planar explants.

The results are in apparent conflict with the finding that *en-2* protein is often not detected in the ectoderm of *Xenopus* exogastrulae (Hemmati-Brivanlou and Harland, 1989), in which planar interactions presumably occur. It is possible that the ectoderm of extreme exogastrulae is mesodermalized, owing to the absence of the blastocoel, and is therefore no longer competent for neural induction. This needs to be examined. Alternatively, inhibitory interactions may occur in exogastrulae but not in Keller explants. The mixed results obtained with exogastrulae suggest that these abnormal embryos are not a reliable system in which to examine planar induction.

Do Keller explants eliminate vertical interactions?

The conclusions in this study largely depend on evidence that Keller explants truly eliminate vertical interactions. There are four lines of evidence that support this contention. First, extensive experiments by Keller et al. (1992a), in which individual cell movements were traced in Keller sandwiches, have established that full convergent extension of the spinal neurectoderm only occurs in those sandwiches in which there is virtually no mesoderm forming vertical contacts with the ectoderm. When there is vertical contact between ectoderm and mesoderm in Keller explants, convergent extension is reduced and irregular. In this paper, only Keller explants that showed symmetrical and complete convergent extension were considered as usable samples. Second, experiments in this paper indicate that the head mesoderm, which is highly migratory and therefore the most likely to form vertical contacts with ectoderm in Keller explants, is unable to induce *en-2,* the most anterior marker, when placed in vertical contact with animal cap ectoderm. Other experiments (Dixon and Kintner, 1989; T. D. and C. R. Phillips, unpublished) indicate that this tissue is also unable to induce NCAM expression. Thus, even in the event that some vertical interactions occurred between head mesoderm and ectoderm in Keller explants, it is unlikely that expression of the markers studied would result from these interactions. Third, the neural markers were also expressed in open-face explants. These explants avoid the potential problem that exists in sandwiches, in which vertical interactions could occur between ectoderm and mesoderm from different layers of the sandwich. Finally, lineage labelling experiments in this paper show that mesoderm can induce expression of all three of the genes tested without migrating into or beneath the responding ectoderm.

How far can planar signals go?

The four position-specific genes tested are markers for the middle portion of the central nervous system along the A-P axis, from the midbrain-hindbrain junction to at least the anterior spinal cord. At a minimum, the results presented here indicate that the A-P pattern induced by planar signals is complete to this extent. To determine whether full neural pattern from anterior to posterior is induced, it will be necessary to use markers specific to the forebrain and the posterior spinal cord. Although it is not possible to say whether there is any neural pattern anterior to *en-2* in planar explants, it is clear that there is some kind of neural tissue beyond this, because NCAM is detected in the anterior portions of Keller explants beyond *en-2.*

In open-face explants, in which convergent extension does not occur, it is possible to make a rough comparison between the A-P extent of NCAM expression and that of the presumptive neural region at the beginning of gastrulation (Keller, 1975). Both of these cover approximately 50% of the distance between the mesoderm and the animal pole. Thus, the extent of planar induction of neural differentiation in explants is comparable to that in normal embryos. When translated into cell diameters, planar signals would have to travel approximately 50 cell diameters to induce the entire A-P extent of the neurectoderm (R. Keller, personal communication). It is interesting to note that the positional neural markers are expressed in distinct stripes in open face explants, even in the absence of morphogenesis, indicating that morphogenesis itself is not required for patterning (although it presumably contributes to the separation of the stripes). Moreover, the pattern is compressed, much as is the fate map of the early gastrula, implying that the pattern is set up prior to or independently of morphogenesis.

In the experiments in which dorsal mesoderm was grafted onto the edge of ventral ectoderm, the distance between the markers expressed and the mesoderm was apparently shorter than in normal open-face explants. This is analogous to the results of Keller et al. (1992b), in which the region that converged and extended in sandwich explants of equivalent tissues to those above was much shorter than in recombinates using dorsal ectoderm. Keller et al. (1992b) propose that planar signals cannot pass as far in ventral ectoderm as they can in dorsal ectoderm, perhaps reflecting the predisposition of dorsal ectoderm towards neural induction that has been much discussed lately (Sharpe et al., 1987; London et al., 1988; Sokol and Melton, 1991; Otte et al., 1991). The greater distance that planar signals apparently pass in dorsal ectoderm may be a result of such priming by signals prior to neural induction, perhaps going as far back as the cortical rotation during the first cell cycle (London et al., 1988; Gerhart et al., 1989).

A-P pattern in the mesoderm of planar explants

Two of the markers used, *en-2* and *XlHbox1,* are expressed in specific regions outside of the central nervous system (CNS): *en-2* in the mandibular arch and *XlHbox1* in the somites and lateral plate mesoderm of the mid-trunk. Results presented here suggest that at least some of the *en-2*-expressing cells in the mandibular arch are derived from mesoderm anterior to the notochord, rather than from neural crest as has been suggested (Hemmati-Brivanlou et al., 1991). Both *en-2* and *XlHbox1* appear to be expressed in appropriate regions of the mesoderm of planar explants. These results concur with much earlier results of Holtfreter (1933) who observed that the evaginated mesoderm and endoderm of exogastrulae display marked A-P morphology and differentiation. As predicted by the fate map, A-P polarity of the mesoderm is a mirror image to that of the ectoderm. Clearly, the mesoderm does not require vertical contact with ectoderm to develop its A-P pattern. Taken together with the results that *en-2* and *XlHbox1* are also expressed in the ectoderm of planar explants, it appears that these two homeobox genes do not require vertical alignment of the ectoderm and mesoderm to be expressed in the correct position, contrary to recent proposals (DeRobertis et al. 1989; Frohman et al. 1990).

Additional results here indicate that *en-2* expression in the mesoderm does not require contact with ectoderm, at least after the early gastrula stage, since removal of these tissues at the beginning of gastrulation does not prevent *en-2* expression in anterior mesoderm.

Applying vertical induction models to planar induction

The previous models that have been proposed assume neural induction is primarily vertical, and that the dorsal mesoderm provides regionalized pattern information. The simplest model is based on the classical finding that different A-P regions of the dorsal mesoderm from the early neurula can induce corresponding A-P levels of neural tissue (Mangold, 1933). In this model, each level of A-P pattern in the neurectoderm is induced by a specific signal passing vertically from the mesoderm lying directly below it (Fig. 7A). On its own, this model is hard to reconcile with planar induction for two reasons. First, the ectoderm depends on the vertical proximity of each qualitatively different inducing signal to develop a pattern; this is lost when the signal is planar (Fig. 7B). Second, even if it is assumed that there was some A-P information in the ectoderm before neural induction, it is hard to imagine how, for example, an anterior-inducing planar signal originating in the most anterior mesoderm would find its way to its target in the anterior neurectoderm after passing all the way through posterior mesoderm and then posterior neurectoderm.

There have been a variety of gradient models proposed for vertical induction in amphibians. These generally require two inducers originating in the mesoderm, one arranged in an A-P gradient with a high point in the posterior mesoderm, and another that is at a constant level along the A-P axis, or is higher at the anterior end of the mesoderm (Fig. 8A; Saxen, 1989; Nieuwkoop and Albers, 1990). It is tempting to try to adapt these models so that the same signals are used for vertical and planar induction. However, it is difficult to envision how an inducer emanating from the mesoderm could spread through the plane of the ectoderm and end up with an even distribution, let alone a higher amount at the anterior end (Fig. 8B). It is possible that vertical and planar induction use the same inducers, but if so, it seems unlikely that the previous gradient models will apply without additional assumptions.

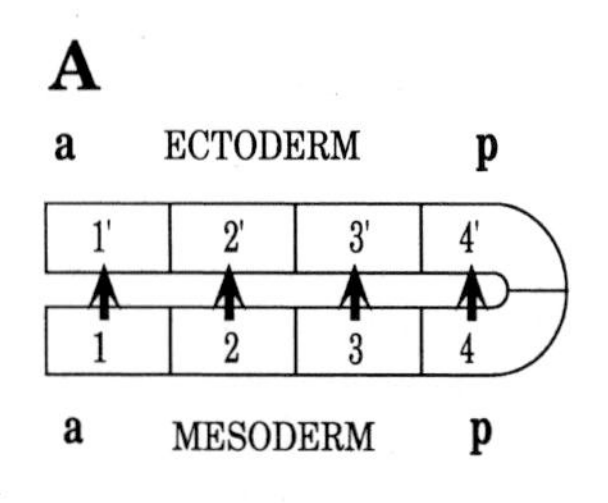

Fig. 7. (A) Schematic representation of a simple model for vertical induction of A-P neural pattern. Involuted mesoderm contains regionalized inducers that induce pattern in ectoderm directly above. (B) When the same inducers use the planar route (indicated by unfolding the sheet of mesoderm and ectoderm shown above), there is no longer any positional information for the ectoderm.

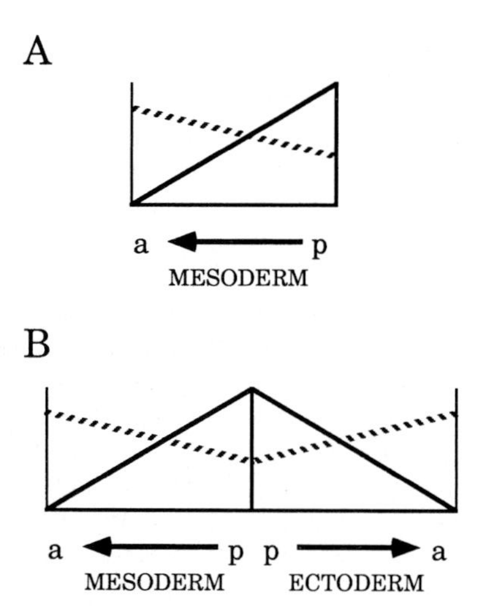

Fig. 8. (A) Simplified distribution of inducers of A-P neural pattern in models proposed by Saxen and Toivonen (Saxen, 1989) or Nieuwkoop (Nieuwkoop and Albers, 1990). *x*-axis, A-P axis; *y* axis, concentration of inducers. In the Saxen-Toivonen model, the dashed line represents the 'neuralizing inducer', which alone induces forebrain development and the black line is the 'mesoderm inducer', which, in combination with neuralizing inducer causes posterior neural development. In the Nieuwkoop model, the dashed line is the 'activator', which induces forebrain, and the black line is 'the transforming agent', which posteriorizes previously activated tissue. (B) Adaptation of the above model for planar induction.

Mechanisms for planar induction of A-P neural pattern

Many mechanisms have been proposed over the years to explain pattern formation. Some will be reconsidered here for planar induction, along with some ways of testing them. There have been several excellent reviews recently about mechanisms of pattern formation (Wolpert, 1989) and neural induction and A-P patterning in amphibians (Jacobson and Sater, 1988; Saxen, 1989; Slack and Tannahill, 1992) that can be referred to for more details.

Concentration gradient

As mentioned above, gradient models have been proposed to explain vertical neural induction. Regardless of whether such gradients operate for vertical induction, it is possible that there is a gradient system for planar induction. Inducer(s) could emanate from the mesoderm, forming a concentration gradient within the plane of ectoderm, with high levels inducing posterior pattern and low levels anterior pattern. A gradient model would be supported if posterior pattern elements are lost when the amount of inducer is reduced (e.g. by cutting off some mesoderm). Results of a similar type have been taken as support for a gradient model for digit patterning in the chick limb bud (Tickle, 1981).

One possible problem with a gradient model is that experiments with *Xenopus* have shown that any reduction in the size of the organizer results in reduction and loss of anterior pattern elements, while those in the posterior are increased (Gerhart et al., 1989). This is the reverse effect of that predicted above for reductions in amount of neural inducer. Not enough is known about how these changes in organizer function or size affect the system that induces neural pattern. However, it seems somewhat contradictory that reduction in the organizer could cause an increase in the amount of neural inducer, thereby leading to an excess of posterior neural pattern.

Prepattern

In recent years molecular evidence has accumulated that there is a bias or predisposition towards neural development in the dorsal but not ventral ectoderm (Sharpe et al., 1987; London et al., 1988; Otte et al., 1991). Taken to an extreme, there could be a precise A-P neural pattern which is laid out cryptically in the ectoderm before induction, and the inductive signal from the mesoderm could be merely permissive for the expression of the pattern. Such a bias, or a prepattern, should it exist, is not strictly required for patterned neural development, as shown in the very first organizer graft (Spemann and Mangold, 1924), in which a secondary nervous system was induced in ventral ectoderm. Results in this paper do not support a strict requirement for a prepattern, either in the dorsal side, or in the latitudes fated for mid-brain, since three of the positional neural markers can be induced in ventral ectoderm, and *en-2* can be induced in ectoderm anterior to its normal position of expression. Thus, if there is a prepattern, it can be overridden by signals from the mesoderm.

Ectodermal competence

It has been proposed that ectodermal competence plays a role in determining the lateral boundaries of the neural plate and the induction of placodes beyond the boundary (such as the lens; Albers, 1987; Servetnick and Grainger, 1991). There are two features to this model: (1) a neural inducer spreads laterally through the neurectoderm (via homeogenetic induction) from the dorsal midline of the neural plate at a certain velocity, and (2) the competence of the ectoderm changes with time, with a series of different responses. Thus, as the signal spreads laterally, medial ectoderm receives the neural inducer early, and develops as CNS; progressively more lateral ectoderm receives the signal progressively later, and responds by becoming neural crest, placode or epidermis, the latter developing as a default after competence to respond to the inducer runs out. An equivalent process could determine A-P pattern in the neurectoderm, if a neural inducer spreads into the ectoderm from the dorsal mesoderm, and if the age of ectoderm when the inducer was received determined the level of A-P neural pattern. This model could be tested using heterochronic planar recombinates, for example dorsal mesoderm and presumptive posterior neurectoderm with younger presumptive anterior neurectoderm. The model would be supported if the posterior pattern is repeated in the presumptive anterior neurectoderm.

Timing gradient

This model is related to a model proposed by Nieuwkoop (1985), in which the duration of vertical contact with involuted dorsal mesoderm determines A-P pattern. To apply this to planar induction, the inducer could instead spread through the plane of the ectoderm from the mesoderm at a given velocity, such that the ectoderm closer to the source (i.e. posterior) is exposed to the inducer for longer than that further away (i.e. anterior). The final amount of time exposed to the inducer could be determined either by a general change in competence or the cessation of the signal. This model is partly supported by the results of Sive et al. (1989). They found that posterior presumptive neurectoderm would express cement gland-specific RNA (cement gland is a tissue normally made beyond the anterior edge of the neural plate), when isolated early in gastrulation, while the same tissue would make NCAM, a neural marker, when isolated later in gastrulation. These results were interpreted in terms of the vertical induction model proposed by Nieuwkoop (1985). However, there is some evidence that cement gland can be induced by planar signals (unpublished results). When applied to planar induction of A-P neural pattern, this model would be supported if progressively longer planar contact between dorsal mesoderm and ectoderm induced progressively more posterior neural markers.

'Phase-shift model'

Two signals could propagate through the ectoderm via homeogenetic induction, each with a different velocity. The time interval between receipt of the signals at a given point in the ectoderm could determine the level of A-P development. A model with these features, referred to as the 'phase-shift' or 'thunderclap' model, was proposed by Goodwin and Cohen (1969). To test this model, recently induced posterior neurectoderm could be assayed for its ability to induce anterior neurectoderm in planar recombinates.

Self organization

In normal neural induction, it is possible that the mesoderm provides minimal pattern information to the ectoderm. Two signals could emanate from the mesoderm, one giving polarity to the ectoderm and the other stimulating general neural development. These signals could reach the ectoderm via either a planar or a vertical path, or both. A-P neural pattern could then self-organize as a result of subsequent cell-cell interactions within the plane of the ectoderm. The ectoderm, of urodeles at least, does have a capacity for some neural self-organization: when exposed to an exogenous neural inducer, it can develop into semi-organized forebrain and eyes. The whole A-P range of neural types, albeit only partly organized, will develop when varying amounts of a second inducer are added (see Saxen, 1989 for review). It is possible that a purified endogenous neural inducer would give a full A-P neural axis, but such an inducer remains to be isolated, in *Xenopus* at least. In the absence of purified inducer, it is difficult to test this model.

However, it may end up as a default model if there is no positive evidence for any of the above models.

Clearly, many experiments need to be done before induction of A-P neural pattern is understood. Most of the experiments suggested above depend on having regional molecular markers, and the existence of a reliable double or triple labelling method in order to distinguish between them. The markers used in this paper are suitable for such experiments, although they are expressed rather close together, especially in the absence of convergent extension. This makes it difficult to explant or recombine tissues fated to express only one of these markers. Markers more widely spaced, such as those specific to the forebrain or posterior spinal cord would be particularly useful for extending studies of A-P patterning.

The role of planar and vertical induction in A-P patterning

The results presented here show that A-P neural pattern can be induced in planar explants, and it is proposed that this occurs in normal development. There has been a lot of previous evidence indicating that vertical signals can also induce neural pattern. The two induction pathways must somehow coordinate in order to achieve a single pattern, whether they use the same or different inducer molecules. It remains to be seen exactly which functions the vertical and planar pathways of induction share, and which are unique. Those functions that overlap may guarantee that the elaborate process of A-P patterning happens correctly every time, as originally proposed by Spemann (1938).

I would like to thank John Gerhart for comments on the manuscript, and for discussions and support; Ray Keller, Amy Sater and John Shih for teaching me how to make explants and for discussions; Carey Phillips for introducing me to planar induction; Jonathan Cooke for discussions; and Cynthia Kenyon for comments on the manuscript. I also thank the Harland lab for help with in situ hybridization, and especially Richard Harland for staining the sample in Fig. 3D, and N. Patel, C. Goodman, C. Wright, E. DeRobertis, P. Kushner and U. Rutishauser for antibodies, D. Wilkinson for the *Krox-20* clone and R. Gimlich for FDA. This work was supported in part by a fellowship from the Jane Coffin Childs Memorial Fund, a training grant from the NIH and funding to J. Gerhart from USPHS grant GM19363.

References

Albers, B. (1987). Competence as the main factor determining the size of the neural plate. *Develop. Growth and Differ.* **29,** 535-545.

Balak, K., Jacobson, M., Sunshine, J. and Rutishauser, U. (1987). Neural cell adhesion molecule expression in *Xenopus* embryos. *Dev. Biol.* **119,** 540-550.

Bolce, M. E., Hemmati-Brivanlou, A., Kushner, P. D. and Harland, R. M. (1992). Ventral ectoderm of *Xenopus* forms neural tissue, including hindbrain, in response to activin. *Development* **115,** 681-688.

DeRobertis, E. M., Oliver, G. and Wright, C. V. E. (1989). Determination of axial polarity in the vertebrate embryo: Homeodomain proteins and homeogenetic induction. *Cell* **57,** 189-191.

Dixon, J. E. and Kintner, C. R. (1989). Cellular contacts required for neural induction in *Xenopus* embryos: evidence for two signals. *Development* **106,** 749-757.

Doniach, T., Phillips, C. R. and Gerhart, J. C. (1992). Planar induction of anteroposterior pattern in the developing central nervous system of *Xenopus laevis. Science,* **257,** 542-545.

Frohman, M. A., Boyle, M. and Martin, G. R. (1990). Isolation of the mouse *Hox-2.9* gene; analysis of embryonic expression suggests that positional information along the anterior-posterior axis is specified by mesoderm. *Development* **110,** 589-607.

Gerhart, J. C., Danilchik, M., Doniach, T., Roberts, S., Rowning, B. and Stewart, R. (1989). Cortical rotation of the *Xenopus* egg: consequences for the anteroposterior pattern of embryonic development. *Development* **107 Supplement,** 37-52.

Gimlich, R. L. and Braun, J. (1985). Improved fluorescent compounds for tracing cell lineage. *Dev. Biol.* **109,** 509-514.

Goodwin, B. C. and Cohen, M. H.(1969). A phase shift model for spatial and temporal organization of developing systems. *J. Theor. Biol.* **25,** 49-69.

Hamburger, V. (1988). *The Heritage of Experimental Embryology. Hans Spemann and the Organizer.* Oxford University Press, Oxford.

Harland, R. M. (1991). In situ hybridization: an improved whole mount method for *Xenopus embryos.* In *Methods in Cell Biology,* (ed. B. K. Kay and H. B. Peng), vol. 36, appendix G, pp 685-695. San Diego: Academic Press.

Hatta, K., Schilling, T. F., Bremiller, R. A. and Kimmel, C. B. (1990). Specification of jaw muscle identity in zebrafish: correlation with *engrailed*-homeoprotein expression. *Science* **250,** 802-805.

Hemmati-Brivanlou, A. and Harland, R. M. (1991). Expression of an engrailed-related protein is induced in the anterior neural ectoderm of early *Xenopus* embryos. *Development* **106,** 611-617.

Hemmati-Brivanlou, A., Stewart, R. and Harland, R. M. (1990). Region-specific neural induction of an *engrailed* protein by anterior notochord in *Xenopus. Science* **250,** 800-802.

Hemmati-Brivanlou, A., de la Torre, J. R., Holt, C. and Harland, R. M. (1991). Cephalic expression and molecular characterization of *Xenopus En-2. Development* **111,** 715 -724.

Holtfreter, J. (1933). Die totale Exogastrulation, eine Selbstablösung des Ektoderms von Entomesoderm. *Roux Arch. EntwMech. Org.* **129,** 669-793.

Holtfreter, J. and Hamburger, V. (1955). In *Analysis of Development* (eds. B. H. Willier, P. A. Weiss, and V. Hamburger), pp. 230-296. Philadelphia: W. B. Saunders..

Jacobson, M. and Rutishauser, U. (1986). Induction of neural cell adhesion molecule (N-CAM) in *Xenopus* embryos. *Dev. Biol.* **116,** 524-531.

Jacobson, A. G. and Sater, A. K. (1988). Features of embryonic induction. *Development* **104,** 341-359.

Keller, R. E. (1975). Vital dye mapping of the gastrula and neurula of *Xenopus laevis.* I. Prospective areas and morphogenetic movements of the superficial layer. *Dev. Biol.* **42,** 222-241.

Keller, R. E. (1976). Vital dye mapping of the gastrula and neurula of *Xenopus laevis.* II. Prospective areas and morphogenetic movements in the deep region. *Dev. Biol.* **51,** 118-137.

Keller, R. and Danilchik, M. (1988). Regional expression, pattern and timing of convergence and extension during gastrulation of *Xenopus laevis. Development* **103,** 193-209.

Keller, R. E., Shih, J. and Domingo, C. (1992c). The patterning and functioning of protrusive activity during convergence and extension of the *Xenopus* organizer. *Development* **1992 Supplement**

Keller, R. E., Shih, J. and Sater, A. K. (1992a). The cellular basis of the convergence and extension of the *Xenopus* neural plate. *Devel. Dynam.* **193,** 199-217.

Keller, R. E., Shih, J., Sater, A. K. and Moreno, C. (1992b). Regulation of the convergence and extension of the neural plate by the involuting marginal zone. *Devel. Dynam.* **193,** 218-234.

Kintner, C. R. and Brockes, J. P. (1984). Monoclonal antibodies identify blastemal cells derived from dedifferentiating muscle in newt limb. *Nature* **308,** 67-69.

Kintner, C. R and Melton, D. A. (1987). Expression of *Xenopus* N-CAM RNA is an early response to neural induction. *Development* **99,** 311-325.

London, C., Akers, R. and Phillips, C. (1988). Expression of Epi-1, an epidermis-specific marker in *Xenopus laevis* embryos, is specified prior to gastrulation. *Dev. Biol.* **129,** 380-389.

Mangold, O. (1933). Über die Inducktionsfähigkeit der verschiedenen Bezirke der Neurula von Urodelen. *Naturwissenschaften* **43,** 761-766.

Nieuwkoop, P. D. (1985). Inductive interactions in early amphibian development and their general nature. *J. Embryol. Exp. Morphol.* **89 Supplement** 333-347.

Nieuwkoop, P. D. and Albers, B. (1990). The role of competence in the cranio-caudal segregation of the central nervous system. *Dev. Growth and Differ.* **32**, 23-31.

Nieuwkoop, P. D. and Faber, J. (1967). *Normal Table of* Xenopus laevis *(Daudin).* North-Holland: Amsterdam.

Oliver, G., Wright, C. V. E., Hardwicke, J. and De Robertis, E. M. (1988). Differential anteroposterior expression of two proteins encoded by a homeobox gene in *Xenopus* and mouse embryos. *EMBO J.* **7,** 3199-3209.

Otte, A. P., Kramer, I. M. and Durston, A. J. (1991). Protein kinase C and regulation of the local competence of *Xenopus* ectoderm. *Science* **251,** 570-573.

Patel, N. H., Martin-Blanco, E., Coleman, K. G., Pool, S. J., Ellis, M.C., Kornberg, T. and Goodman, C. S. (1989). Expression of *engrailed* protein in arthropods, annelids and chordates. *Cell* **58,** 955-968.

Ruiz i Altaba, A. (1990). Neural expression of the *Xenopus* homeobox gene Xhox3: evidence for a patterning neural signal that spreads through the ectoderm. *Development* **108**, 595-604.

Saha, M. S. and Grainger, R. M. (1992). A labile period in the determination of the anterior-posterior axis during early neural development in Xenopus. *Neuron* **8**, 1-20.

Sala, M. (1955) . Distribution of activating and transforming influences in the archenteron roof during the induction of the nervous system in amphibians. I. Distribution in cranio-caudal direction. *Proc. Kon. Ned. Akad. Wet., Serie C,* **58,** 635-647.

Savage, R. and Phillips, C. R. (1989). Signals from the dorsal blastopore lip region during gastrulation bias the ectoderm toward a nonepidermal pathway of differentiation in *Xenopus laevis. Dev. Biol.* **133**, 157-168.

Saxen, L. (1989). Neural Induction. *International Journal of Developmental Biology* **33**, 21-48.

Servetnick, M. and Grainger, R. (1991) Changes in neural and lens competence in *Xenopus* ectoderm: evidence for an autonomous developmental timer. *Development* **112**, 177-188.

Sharpe, C. R. and Gurdon, J. B. (1990). The induction of anterior and posterior neural genes in *Xenopus laevis. Development* **109**, 765-774.

Sharpe, C. R., Fritz, A., DeRobertis, E. M. and Gurdon , J. B. (1987). A homeobox containing marker of posterior neural differentiation shows the importance of predetermination in neural induction. *Cell* **50**, 749-758.

Sive, H. L., Hattori, K. and Weintraub, H. (1989). Progressive determination during formation of the anteroposterior axis in *Xenopus laevis. Cell* **58,** 171-180.

Slack, J. M. W. and Tannahill, D. (1992). Mechanism of anteroposterior axis specification in vertebrates: lessons from the amphibians. *Development* **114**, 285-302.

Sokol, S. and Melton, D. A. (1991). Pre-existent pattern in Xenopus animal pole cells revealed by induction with activin. *Nature* **351**, 409-411.

Spemann, H. (1938). *Embryonic Development and Induction* New Haven: Yale University Press (reprinted by Garland Publishing, New York 1988).

Spemann, H. and Mangold, M. (1924). Über Induktion von Embryonalanlagen durch Implantation artfremder Organizatonen. *Roux Arch. Entwicklungsmech. Org.* **100**, 599-638. Also, translation in *Foundations of Experimental Embryology* (ed. B.H.Willier and J. M. Oppenheimer), pp. 144-184. New York:Hafner.

Ter Horst, J. (1948). Differenzierungs- und Induktionsleistungen verscheidener Abschnitte der Medullarplatte und des Urdarmdaches von Triton im Kombinat. *Wilhelm Roux's Arch. EntwMech. Org.* **143,** 275-303.

Tickle, C. (1981). The number of polarizing region cells required to specify additional digits in the developing chick wing. *Nature* **289,** 295-298.

Vincent, J-P., Oster, G. F. and Gerhart, J. C. (1986). Kinematics of grey crescent formation in *Xenopus* eggs: the displacement of subcortical cytoplasm relative to the egg surface. *Dev. Biol.* **113**, 484-500.

Wilkinson, D. G., Bhatt, S., Chavrier, P., Bravo, R. and Charnay, P. (1989). Segment-specific expression fof a zinc-finger gene in the developing nervous system of the mouse. *Science* **337,** 461-464.

Wilson, P. and Keller, R. (1991). Cell rearrangement during gastrulation of *Xenopus*: direct observation of cultured explants. *Development* **112,** 289-300.

Winklbauer, R. (1990). Mesoderm cell migration during *Xenopus* gastrulation. *Dev. Biol.* **142**, 155-168.

Winklbauer, R., Selchow, A., Nagel, M., Stoltz, C. and Angres, B. (1991). Mesoderm cell migration in the *Xenopus* gastrula. In *Gastrulation: Movements, Patterns, and Molecules* (ed. G. R. Keller, W. H. Clark, Jr. and F. Griffin) Bodega Marine Laboratory Marine Science Series, New York: Plenum Press

Wolpert, L. (1989). Positional information revisited. *Development* **107 Supplement,** 3-12.

Wright, C. V. E., Morita, E. A., Wilkin, D. J. and De Robertis, E. M. (1990). The *Xenopus* XlHbox 6 homeo protein, a marker of posterior neural induction, is expressed in proliferating neurons. *Development* **109,** 225-234.

Development 1992 Supplement, 195-202 (1992)
Printed in Great Britain © The Company of Biologists Limited 1992

Xenopus Hox-2 genes are expressed sequentially after the onset of gastrulation and are differentially inducible by retinoic acid

ERIK-JAN DEKKER[1], MARIA PANNESE[2], ERWIN HOUTZAGER[1], ANS TIMMERMANS[1], EDOARDO BONCINELLI[2,3] and ANTONY DURSTON[1]

[1]*Netherlands Institute for Developmental Biology, Uppsalalaan 8, 3584 CT Utrecht, The Netherlands*
[2]*Instituto Scientifico H San Raffaele, DIBIT, Via Olgettina, 20132 Milan, Italy*
[3]*Centro per io Studio della Farmacologia delle Infrastructure Cellulari, CNR, Via Vanvitelli, 32, 201 29 Milano, Italy*

Summary

In this paper, we review experiments to characterise the developmental expression and the responses to all-*trans* retinoic acid (RA) of six members of the Hox-2 complex of homeobox-containing genes, during the early development of *Xenopus laevis*. We showed that the six genes are expressed in a spatial sequence which is colinear with their putative 3′ to 5′ chromosomal sequence and that five of them are also expressed rapidly after the beginning of gastrulation, in a 3′ to 5′ colinear temporal sequence. The sixth gene (*Xhox2.9*) has an exceptional spatial and temporal expression pattern. The six genes all respond to RA by showing altered spatiotemporal expression patterns, and are also RA-inducible, the sequence of the magnitudes of their RA responses being colinear with their 3′ to 5′ chromosomal sequence, and with their spatial and temporal expression sequences. Our data also reveal that there is a pre-existing anteroposterior polarity in the embryo's competence for a response to RA. These results complement and extend previous findings made using murine and avian embryos and mammalian cell lines. They suggest that an endogenous retinoid could contribute to positional information in the early *Xenopus* embryo.

Key words: Hox genes, retinoic acid, *Xenopus laevis*.

Introduction

The main anteroposterior (A-P) axis of the amphibian embryo is specified via intercellular signals. Signals acting during gastrulation certainly specify A-P differences in the developing neural plate (Spemann, 1931; Mangold, 1933). There is a historical controversy (still unresolved) as to whether these signals are transacting (emitted by underlying mesoderm; Holtfreter, 1933; Mangold, 1933; Sharpe and Gurdon, 1990) or homeogenetic (emitted by the organiser and spreading from cell to cell through the neural plate; Spemann, 1931, 1938; Ruiz i Altaba and Melton, 1990) or both. It is clear, in any event, that the organiser (i.e., the dorsal lip of the blastopore) is an important signal source (Spemann and Mangold, 1924; Spemann, 1931). There is also evidence that specifying the A-P axis of the axial mesoderm requires signals during gastrulation (Nieuwkoop et al., 1985; Eyal Giladi, 1954; Kaneda and Hama, 1979), as well as possibly earlier (Ruiz i Altaba and Melton, 1989).

There is, presently, interest in the idea that an active form of vitamin A (an active retinoid) is one of the intercellular signals that specifies the vertebrate A-P axis. A first specific basis for this idea is the finding that treating early amphibian embryos with a pulse of the well-known active retinoid, all-*trans* retinoic acid (RA) can cause various A-P transformations in the main body axis (Durston et al., 1989; Sive et al., 1990; Ruiz i Altaba and Jessell, 1991a; Papalopulu et al., 1991). These effects work directly both on the developing neural plate (Durston et al., 1989; Sive et al., 1990; Ruiz i Altaba and Jessell, 1991a), as well as, on axial mesoderm (Ruiz i Altaba and Jessell, 1991b), and there is a sensitivity peak for posteriorisation during gastrulation (Durston et al., 1989; Sive et al., 1990). A second, general basis for suspecting that a retinoid is important is the general evidence, via characterisation of retinoid receptors, that retinoids are bonafide signal molecules (Dolle et al., 1989; Petkovich et al., 1987; Brand et al., 1988; Mangelsdorf et al., 1990) and specifically (from the suggestive expression patterns of transcripts for retinoid receptors and binding proteins) that retinoids are signals in early development (Ruberte et al., 1990, 1991; Vaessen et al., 1990). A third point is that studies on the chicken limb bud indicate interesting parallels with A-P specification of the main body axis as well as suggesting that a retinoid specifies A-P values in the limb bud. Here, the ZPA (zone of polarizing activity; a posterior organiser region which specifies the posterior side of the limb), can be replaced by Hensen's node; the gastrula stage organiser of the chicken embryo (Hornbruch and Wolpert, 1986). It can also be replaced, specifically by a local RA source (Tickle et al., 1982) and measurements of RA in the limb bud also show a posterior to anterior gradient of endogenous RA in an effective concentration range (Thaller and Eichele, 1987).

The endogenous signals that specify the vertebrate A-P axis are likely to work by regulating the expression of class 1 homeobox-containing genes (Hox genes). These vertebrate homologues of *Drosophila* homeotic (HOM) genes, in the *Antennpedia* and *Bithorax* complexes, are strongly suspected (from their HOM gene homologies, their expression patterns and the results of gene manipulation experiments) to provide A-P positional information in early vertebrate development (reviewed by McGinnis and Krumlauf, 1992). It is important to characterise their endogenous expression, as well as their responses to putative positional signals. An important aspect of the functioning of Hox genes is that they are organised in chromosomal complexes (four in mammals, each being a partial or total homologue of the *Drosophila Antennapedia* or *Bithorax* complexes together; Acampora et al., 1989; Boncinelli et al., 1991; Duboule and Dolle, 1989; Akam et al., 1989; Graham et al., 1989); and that these complexes act as functional units. In a recent study (Dekker et al., 1992, and unpublished observations), we have characterised the endogenous expression patterns and RA responses of six genes in one Hox gene complex (the Hox-2 complex), in the early embryo of the amphibian *Xenopus laevis*. We found that the *Xenopus* Hox-2 complex resembles mouse Hox complexes that have been studied in showing colinearity between its putative 3′ to 5′ Hox gene sequence, and the spatial and temporal sequences in which these genes are expressed in the early embryo. We found, further, that the early *Xenopus* embryo, like human and murine teratocarcinoma cell lines, shows 3′ to 5′ colinearity in the response of Hox-2 genes to RA (Simeone et al., 1990; Papalopulu et al., 1991). RA also interacts with an endogenous A-P gradient in competence to regulate Hox-2 gene expression patterns in the early embryo. These findings are discussed in relation to a potential role for an endogenous retinoid in specifying the vertebrate A-P axis, during gastrulation and neurulation.

Materials and methods

RNA isolation

Xenopus eggs were fertilized in vitro and cultured at room temperature (19-21°C) in tap water. Synchronous development allowed the collection of many embryos of defined stages (Nieuwkoop and Faber, 1967). Embryos were dejellied using 2% cysteine at pH 7.8, and then either used directly to extract RNA (Fig. 5) or else dissected, to generate embryo fragments for RNA extraction, (Figs 4, 6 and 7). The embryos were dissected in 10% Flickinger's medium (Flickinger, 1949; 58 mM NaCl, 1 mM KCl, 0.24 mM $NaHCO_3$, 1 mM Na_2HPO_4, 0.2 mM KH_2PO_4, 0.5 mM $CaCl_2$, 1 mM $MgSO_4$, pH 7.5) using tungsten needles. The embryo fragments, or whole embryos were frozen in liquid nitrogen, then kept in a −80°C freezer before being homogenised in guanidine isothiocyanate buffer, and pelleted through an 5 M CsCl cushion, to extract total RNA. 25 μg samples of this RNA were then used to perform RNase protection assays.

Probes and RNase protection assay

For our experiments, we cloned two new *Xenopus* Hox-2 genes, *Xhox2.7* and *Xhox2.9* (Dekker et al., 1992). DNA templates were made for each of the six Hox-2 genes, as well as for the *Xenopus laevis S8* (*Xom62/9*; Mariottini et al., 1988) ribosomal protein transcript. The last template was used to generate an internal standard, to quantify the amount of RNA in each sample. All of the templates were made using subcloned fragments of less than 300 bp in length (*Xhox2.9* 147 bp, *Xhox2.7* 211 bp, *Xhox-1A* 153 bp, *Xhox-1B* 150 bp, *XlHbox2* 109 bp, *XlHbox6* 227 bp, *S8* (*Xom62/9*) 89 bp). These were inserted into CsCl gradient-purified pGEM3Zf(−) (Promega) recombinants. Sp6 RNA polymerase was then used to make antisense RNA probes (labelled via [α-^{32}P] GTP), as required for the RNase protection assays. The probes were first treated with DNase-I, and then purified on a 7% urea-acrylamide gel. In the RNase protection assays, mixtures of five labelled antisense probes (Hox-2 genes, plus the internal standard) were used (at 45°C) to hybridise to RNA extracted from embryos or fragments thereby protecting their homologous transcripts from digestion by a mixture of 2 μg ml^{-1} RNase T1 (Pharmacia) and 40 μg ml^{-1} RNase A (Gibco BRL). Because all six Hox genes were assayed in the same sample together with an internal standard, it was possible to make unambiguous comparisons of the expression levels of these six genes. Autoradiograms from these gels were then analysed densitometrically, used to yield normalised values for expression of the six Hox genes (as described in a phosphorimager Molecular Dynamics, Compaq, Image Quant Desk 386/25e), and these are set out as percentages of the maximum expression. All other procedures were basically as described by Melton et al. (1984), with slight modifications as described by Simeone et al. (1990). The expression patterns were quantified by using densitometry to measure the autoradiogram exposures corresponding to each Hox gene messenger and to the internal standard (*Xom62/9*). The exposure corresponding to each messenger was then normalised (for each gel lane) to that corresponding to the internal standard, and the normalised expression was then plotted, or diagrammed.

In situ hybridisation

The procedure for whole-mount in situ hybridisation was essentially as described by Tautz and Pfiefle (1989), and Kintner and Melton (1987) and as modified by Hemmati-Brivanlou et al. (1990). A linearised template of *XlHbox6* (the same template as was used for the RNase protection experiments) was used for in vitro transcription reactions in the presence of digoxigenin-11-UTP (Boehringer Mannheim 1209 256) with SP6 RNA polymerase (as described by Melton et al., 1984). The whole-mount embryos were examined using a Zeiss Axiovert 35 microscope.

In situ hybridisation on sections was performed essentially as described by Wilkinson et al. (1987). Sections were treated with the *XlHbox6* ^{35}S-UTP-labelled antisense RNA probe in hybridisation mix at 55°C overnight. The slides were washed under stringent conditions and treated with RNase A and RNase T1 to remove unhybridised, non-specifically bound probe. Autoradiography was performed with Ilford K5 emulsion, with exposures between 2 and 4 weeks. Sections were examined under an Olympus BH-2 microscope using dark-field illumination and photographed.

Results

Spatial colinearity

We determined the spatial order in which the six Hox-2 genes were expressed along the embryo's A-P axis (Fig. 1). We dissected tailbud-stage embryos (stage 26) into seven pieces along their A-P axes (fragment one: most anterior, to seven: most posterior), and examined expression of all six Hox-2 genes in each piece. The results (Fig. 2) substantiate and extend previous studies of the expression of *Xhox-1A* (Harvey et al., 1986; Harvey and Melton, 1988) and *XlHbox6* (Sharpe et al., 1987; Wright et al., 1987; Cho and de Robertis, 1990). The spatial expression patterns of *Xhox-1B* and *XlHbox2* expression and of our two newly isolated Hox-2 genes (*Xhox2.7* and *Xhox2.9*) have not previously been described. Our results showed, for the first time, that the spatial sequence in which these members of a Hox gene complex are expressed along the main A-P axis of the *Xenopus* embryo is colinear with their putative chromosomal sequence, thus confirming that *Xenopus* resembles *Drosophila* and the mouse (Akam, 1989; Duboule and Dolle, 1989; Graham et al., 1989; Wilkinson et al., 1989), in this respect. The maximal expression of the most 3′ gene, *Xhox2.9*, was localised in fragment 2, but some expression was also observed in fragment 3. The next most 3′ located genes (*Xhox2.7*, *Xhox-1A* and *Xhox-1B* (Harvey and Melton, 1988; Harvey et al., 1986), were expressed maximally in fragment three, *Xhox2.7* expression being slightly anterior to that of *Xhox-1A* and *Xhox-1B*. *XlHbox2* (Wright et al., 1987; Muller et al., 1984) was expressed maximally in fragment 5, and showed some expression in fragments 4 and 6, and the most 5′ located gene (*XlHbox6*; Sharpe et al., 1987) was expressed in fragments 5 to 7 (Fig. 4). The A-P expression sequence for these six genes is thus *Xhox2.9* (most anterior), *Xhox2.7*, *Xhox-1A* = *Xhox-1B*, *XlHbox2*, *XlHbox6* (most posterior). Further experiments, with earlier developmental stages (Fig. 3, see below) showed that *Xhox-1A* expression is, at least transiently, anterior to *Xhox-1B* expression.

Of the six Hox-2 genes used in this study, *Xhox2.9* formed an exception in its spatial expression pattern at stage 26. Its expression is strongly localised to a relatively small region (fragment 2), in the anterior part of the embryo. The mouse homologue *Hox-2.9* and a chicken *labial* homologue *Ghox.lab*, also show a similarly restricted expression pattern at a comparable developmental stage (Wilkinson et al., 1989; Hunt et al., 1991; Sundin and Eichele, 1992).

Temporal colinearity

We also looked at the timing of Hox-2 gene expression. Fig. 3 shows that all six Hox-2 genes were expressed during gastrulation and neurulation (expression rising after the beginning of gastrulation (stage 10), maximum around stage 15-20 (neurula), decreasing to low levels by stage 35 (tadpole)). These results confirm and extend previous studies of the expression of *Xhox-1A* and *Xhox-1B* (Fritz and De Robertis, 1988; Harvey and Melton, 1988; Harvey et al., 1986; Wright et al., 1987; Sharpe et al., 1987; Fritz et al., 1989; Muller et al., 1984). We found that five of the six Hox-2 genes show temporal colinearity. They are expressed sequentially, in a temporal sequence which is strictly colinear with their 3′ to 5′ chromosomal sequence. This result parallels previous findings showing temporal colinearity in the expression of four murine Hox-4 complex genes during early development of the mouse (Izpisua-Belmonte et al., 1991b) and of the chicken limb (Dolle et al., 1989). Previous studies (eg. Gaunt, 1988; Baron et al., 1987; Kessel and Gruss, 1990), also indicate that generally more 3′ localised mouse Hox genes tend to be expressed earlier during embryogenesis than more 5′ localised Hox genes, but temporal colinearity has not been demonstrated explicitly in vivo for other Hox clusters. A biphasic expression pattern was observed for five of the six Hox-2 genes examined (*Xhox2.9* is an exception). This pattern could reflect different phases of the expression of these genes (see below).

Temporal colinearity was also found in experiments with embryo fragments. We looked at the spatiotemporal expression of five genes by harvesting (normal and RA-treated) embryos at sequential developmental stages, cutting into consecutive A-P fragments, and measuring Hox gene expression in each fragment at each stage. Embryos were dissected into three fragments: an anterior fragment (A) a middle fragment (M) and a posterior fragment (P), and this dissection was done at three stages; stage 13 (early neurula), stage 15 (neurula) and stage 20 (late neurula). RNA was isolated and RNase protections were done on the fragments using *Xhox2.7*, *Xhox-1A*, *Xhox-1B*, *XlHbox2* and *XlHbox6* as probes. Fig. 5 (solid bars) shows that *Xhox2.7*, *Xhox-1A*, *Xhox-1B* and *XlHbox2* are each initially expressed in untreated embryos at a low level, and that each then appears to show a wave of increased expression, which begins posteriorly and proceeds anteriorly. The waves are sequential (*Xhox2.7*, *Xhox-1A*, *Xhox-1B*, *XlHbox2*). *XlHbox6* is expressed last, and its expression begins and remains posterior. These results thus confirm our suspicion, already raised by the measurements on whole embryos, that Hox-2 complex gene expression occurs in the whole embryo, just as does Hox-4 complex expression in the developing limb (Dolle et al., 1989; Nohno et al., 1991; Izpisua-Belmonte et al., 1991a), genitalia (Dolle et al., 1991) and in the early mouse embryo (Izpisua-Belmonte et al., 1991b), in sequential posterior-to-anterior waves. These findings were repeated in two experiments.

The exceptional expression pattern of Xhox2.9

Xhox2.9 formed an exception to the temporal colinearity of the Hox-2 complex, since its expression is low during early development (gastrula and early neurula), and only reaches its maximum at the end of neurulation (stage 20). We can conclude from these data that, although five of the six Hox-2 genes (*Xhox2.7*, *Xhox-1A*, *Xhox-1B*, *XlHbox2* and *XlHbox6*) are expressed 3′ to 5′ sequentially during the early development of *Xenopus laevis*, *Xhox2.9*, the most 3′ located gene in the *Xenopus* Hox-2 gene complex, is an exception, which does not fit in with the 3′ to 5′ temporal

Fig. 1. The *Xenopus* Hox-2 complex. The putative *Xenopus* Hox-2 genes used in this study, their homologies with *Drosophila* HOM genes and mouse Hox-2 genes and their chromosomal linkage. Above: The *Drosophila* homeotic (HOM) genes in the *Antennapedia* and *bithorax* complexes (ANT-C and BX-C), shown, from left to right, in their 3′ to 5′ genomic sequence and their anterior to posterior (A - P) expression sequence in the early embryo. Middle: the murine Hox-2 complex, showing homologies between particular murine Hox-2 genes and particular *Drosophila* HOM genes, as deduced by Duboule and Dolle (1989), on the basis of conserved variations in the α helical subdomains of the homeodomain region of the protein products of these genes. Below: the six *Xenopus* Hox-2 genes used in this study (coloured squares and named). These are ordered (from left to right) according to their specific (vertical) DNA and amino acid sequence homologies with particular groups of *Drosophila* homeotic and murine Hox genes, as revealed by conserved variations in the homeodomain; Kappen et al., 1989; Scott et al., 1989; Fritz and De Robertis, 1988; Fritz et al., 1989; Regulski et al., 1987). *Xhox2.7* and *Xhox 2.9* are new genes, recently cloned in one of our labs (Pannese and Boncinelli, unpublished). They also each show particularly high sequence homology (both within the homeobox, and in flanking sequences) with particular mammalian Hox-2 complex genes, and have therefore tentatively been assigned to the *Xenopus* Hox-2 complex (Kappen et al.,1989; Fritz and De Robertis, 1988; Fritz et al., 1989, this study). It was already known, for *X1Hbox6* and *X1Hbox2* (Fritz et al., 1989), and for *Xhox-1A* and *Xhox-1B* (Harvey and Melton, 1988) respectively, that these two pairs of genes are closely genetically linked, as indicated in the upper copy of the complex by lines connecting these genes. We used pulsed-field gel electrophoresis to examine the linkage of these genes (Dekker et al., 1992) and thus now show that *Xhox2.9*, *Xhox2.7*, *Xhox-1A*, and *XlHbox6* are also closely linked. We can therefore conclude that all six of the putative *Xenopus* Hox-2 genes used in this study are, indeed, closely linked in the same chromosomal complex.

Fig. 2. The spatial expression patterns of the *Xenopus* Hox-2 genes. Stage-26 *Xenopus* (tailbud) embryos were dissected into seven fragments, as indicated in the diagram, and the expression of each of the six *Xenopus* Hox-2 genes was determined in each fragment, using quantitive RNase protection. RNase protection gels were made, using RNA prepared from each fragment. The gels were set up to assay expression of each of the six Hox genes, and of an internal standard (*Xom62/9*). The gels were then analysed quantitatively, using a phosphorimager (Molecular Dynamics, Compaq, Image Quant (Desk 386/25e)) to yield normalised values for expression of each of the six Hox genes. These values were calculated as percentages of the maximum expression for each gene and are diagrammed by the thickness of each appropriately coloured bar. See the materials and methods for further details. The figure shows that the spatial expression sequence of these six genes is at least approximately colinear with their putative 3′ to 5′ chromosomal sequence. Two genes (*Xhox-1A* and *Xhox-1B*) appear to be expressed at a similar A - P level.

Fig. 3. The temporal expression patterns of the *Xenopus* Hox-2 genes. RNA was extracted from *Xenopus* embryos harvested at the sequential developmental stages shown and analysed for expression of the six Hox genes, using quantitative RNase protection as in Fig. 2. The data (set out as percentages of the maximum expression of each gene) show that the temporal expression sequence of five of the six genes is colinear with their putative chromosomal sequence in the Hox-2 complex. *Xhox2.9* is an exception (see text for details).

Fig. 4. The effect of RA on the expression of the *Xenopus* Hox-2 genes. *Xenopus* embryos were treated with RA (10^{-6} M RA, continuous treatment from stage 10 onwards), and were harvested at sequential developmental stages, and analysed for Hox-2 gene expression, as in Fig. 2. The RA treatment caused overexpression of each of the six Hox-2 genes examined (see text for details). The figure shows the maximum expression level reached by each gene in RA-treated embryos (expressed as a percentage of the maximum expression for each gene in untreated embryos). It will be seen that the most 3′ Hox-2 gene (*Xhox2.9*) is the most affected by RA, and that successively more 5′ genes are successively less affected.

expression sequence demonstrated by the other five Hox-2 genes.

The effects of RA on Hox-2 gene expression

Figs 4 and 5 show that RA treatment (1×10^{-6} M RA, applied continuously from stage 10) massively enhanced the expression of all six genes. The sequence of the magnitudes with which the expression of these genes is enhanced is strictly colinear with their putative chromosomal sequence, and with the sequence of A-P locations at which they are expressed in the early embryo. RA treatment thus enhances the expression of *Xhox2.9* the most (over 10-fold; similar findings as with *Xhox.lab2*; Sive et al., 1991), *Xhox2.7* next, *Xhox-1A* next, *Xhox-1B* next, *XlHbox2* and *XlHbox6* the least. This effect was repeatable in two experiments with whole embryos (Fig. 4). The same result was also found for five of these genes (*Xhox2.7*, *Xhox-1A*, *Xhox-1B*, *XlHbox2* and *XlHbox6*), which were examined in two experiments with embryo fragments (Fig. 5, open bars). Papalopulu et al. (1991) also found, in *Xenopus* embryos treated with RA that expression of a 3′ located Hox gene (*XlHbox4*) is more enhanced by RA than that of a 5′ localised gene (*XlHbox6*). These results parallel previous findings with cell lines, which show 3′ to 5′ colinearity in the RA responses of Hox-2 genes in human and mouse teratocarcinoma cell lines (Simeone et al., 1990; Papalopulu et al., 1991).

The experiments with embryo fragments (Fig. 5) show that RA alters the normal Hox gene expression patterns, since no posterior-to-anterior waves are now observed. Expression of *Xhox2.7*, *Xhox-1A* and *Xhox-1B* now starts and remains anterior, while the *XlHbox2* and *XlHbox6* expression patterns are abnormal, but end up posterior. It is thus notable that the normal final spatially colinear expression sequence of these genes is at least approximately preserved after RA treatment. The expression sequence of these genes simply spreads out, so that 3′ genes are now expressed at more anterior positions. The more 3′ genes also tend to be induced more rapidly by RA than the more 5′ genes. This finding suggests that the early embryo contains a pre-existing, RA-insensitive, anteroposterior polarity, a point which we will take up below.

Pre-existing A-P polarity

The idea that early *Xenopus* embryos contain a pre-existing anteroposterior polarity was tested directly by exposing very anterior axial tissue to RA. We dissected out the presumptive forebrain and underlying prechordal plate mesoderm from the most anterior part of a stage-12.5 early neurula embryo (see Fig. 6), cultured these explants up to stage 20 without RA and in the presence of 10^{-8} M, 10^{-7} M and 10^{-6} M RA (continuously) and examined the expression of four Hox-2 genes. The untreated explants showed no Hox-2 gene expression, as expected, since Hox gene expression

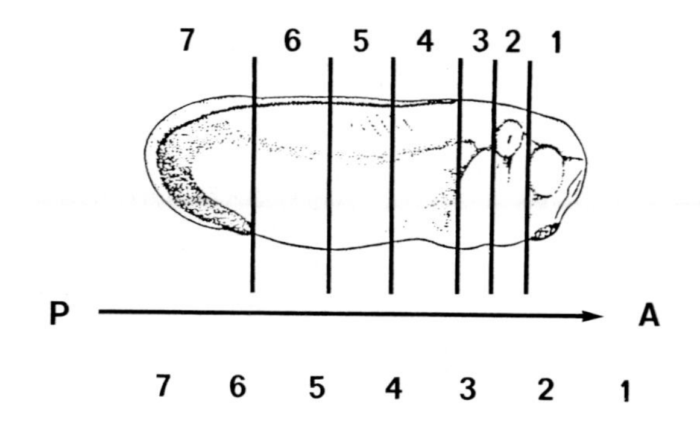
7 6 5 4 3 2 1
P
A
7 6 5 4 3 2 1

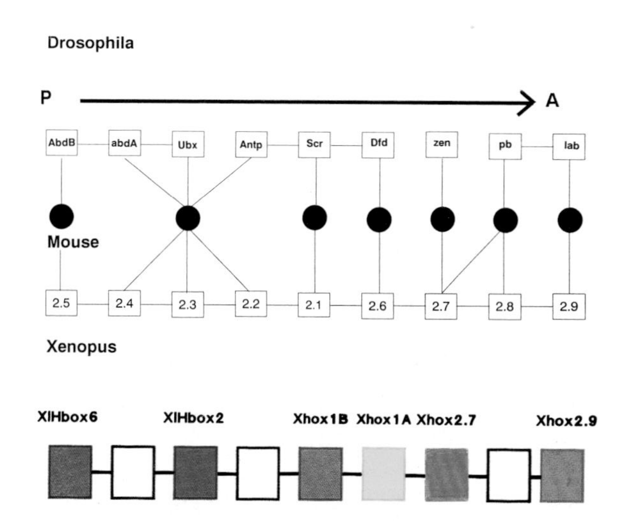
Drosophila
P
A
AbdB
abdA
Ubx
Antp
Scr
Dfd
zen
pb
lab
Mouse
2.5
2.4
2.3
2.2
2.1
2.6
2.7
2.8
2.9
Xenopus
XlHbox6
XlHbox2
Xhox1B
Xhox1A
Xhox2.7
Xhox2.9
= common ancestor

Xhox2.9
Xhox2.7
Xhox-1A
Xhox-1B
XlHbox2
XlHbox6

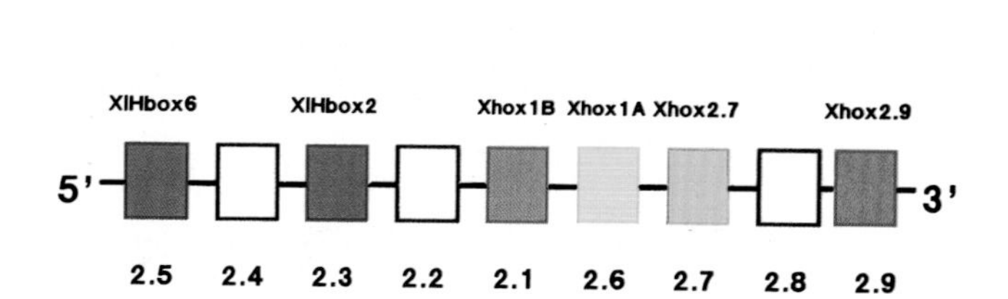
XlHbox6
XlHbox2
Xhox1B
Xhox1A
Xhox2.7
Xhox2.9
5'
3'
2.5
2.4
2.3
2.2
2.1
2.6
2.7
2.8
2.9

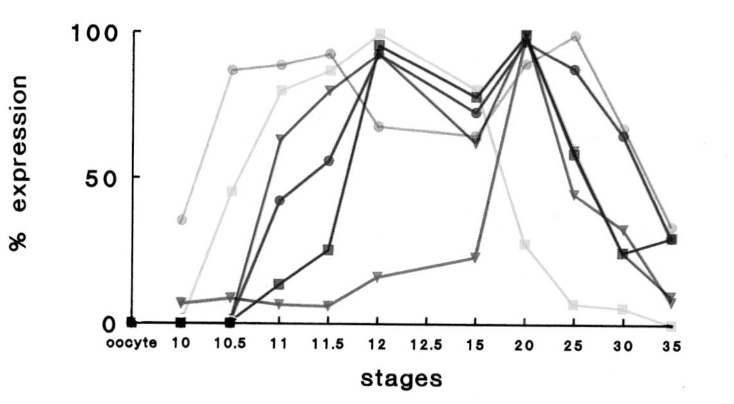
100
50
0
% expression
oocyte
10
10.5
11
11.5
12
12.5
15
20
25
30
35
stages

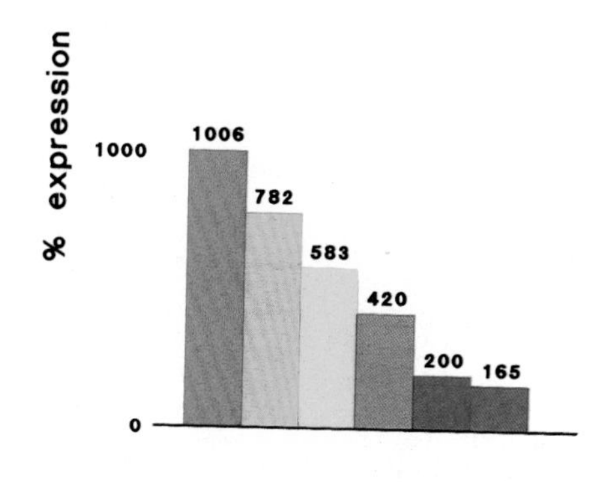
% expression
1000
0
1006
782
583
420
200
165

XlHbox6
XlHbox2
Xhox1B
Xhox1A
Xhox2.7
Xhox2.9
5'
3'
2.5
2.4
2.3
2.2
2.1
2.6
2.7
2.8
2.9

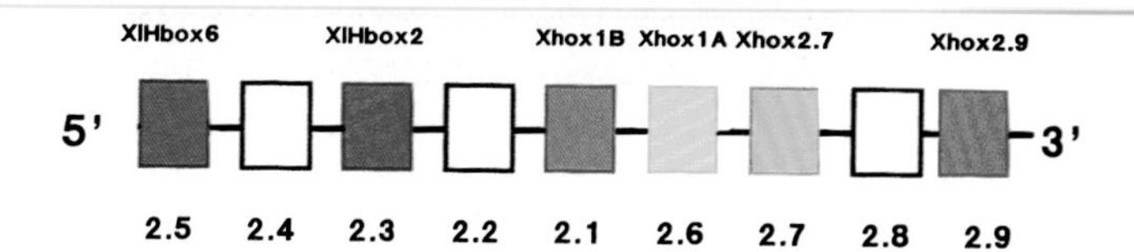
XlHbox6
XlHbox2
Xhox1B
Xhox1A
Xhox2.7
Xhox2.9
5'
3'
2.5
2.4
2.3
2.2
2.1
2.6
2.7
2.8
2.9

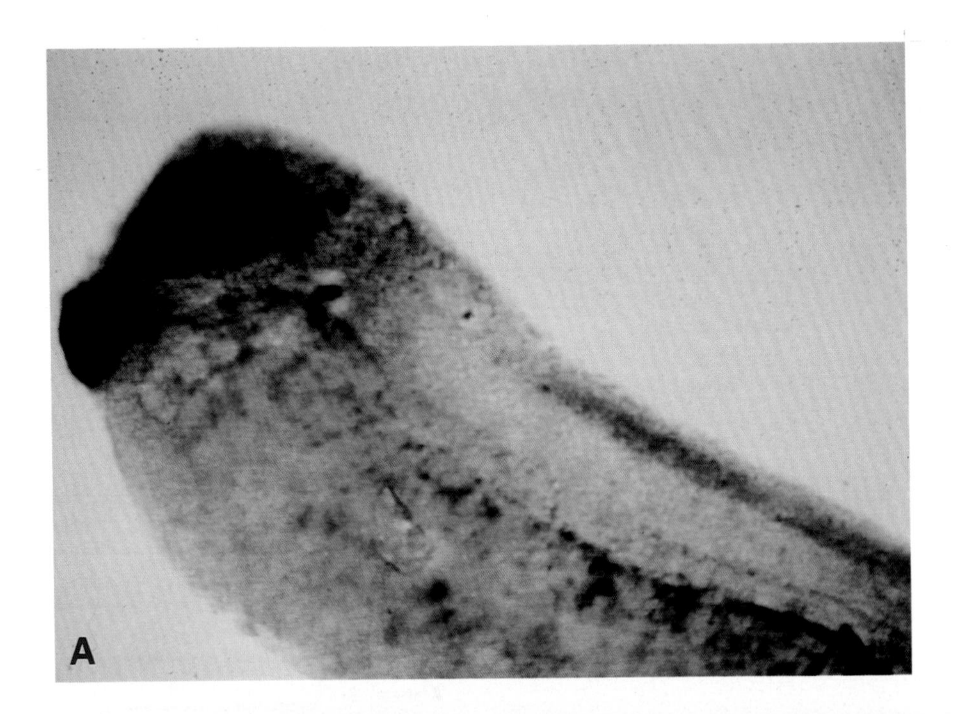

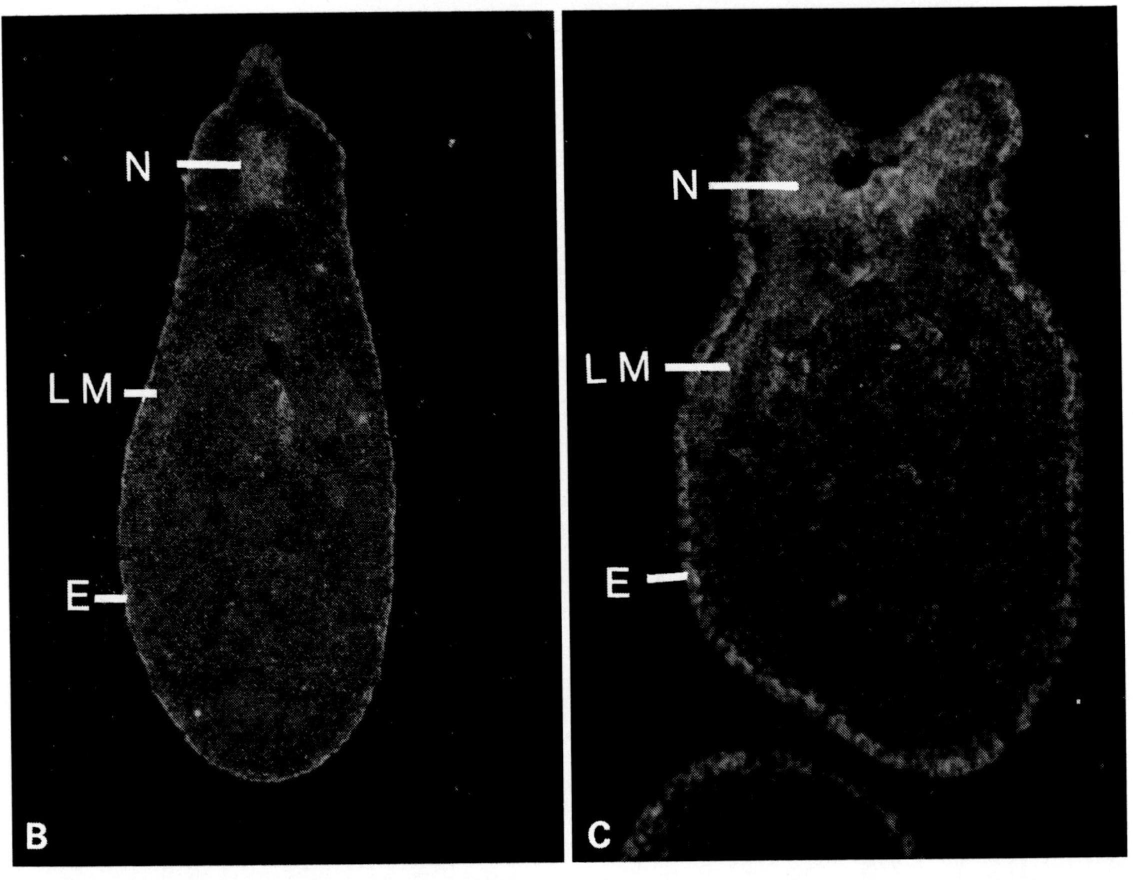

Fig. 7. Expression of *XlHbox6* in stage-28 *Xenopus laevis* embryos. (A) Detection of specific *XlHbox6* mRNAs in a normal (non-RA-treated) stage-28 embryo by whole-mount in situ hybridisation using digoxigenin-11-UTP-labelled antisense mRNA and alkaline phosphatase with nitro blue tetrazolium (NBT) as a substrate for the label. *XlHbox6* expression is localised posteriorly in the spinal cord of the embryo (blue staining; NBT). The dark color of the head is due to pigment. (B, C) Detection of *XlHbox6* mRNAs by in situ hybridisation using antisense [^{35}S]UTP-labelled mRNAs on transverse sections of the trunk region of the embryo. (B) Control embryo (non-RA treated): the mRNA is expressed abundantly in the posterior CNS (N), and also expressed at a low level in the lateral mesoderm (LM). (C) RA-treated embryo (10^{-6} M RA continuously). RA enhances *XlHbox6* expression, which now becomes very evident in the posterior mesoderm, as well as in the (open) neural tube. The anteroposterior localisation boundary of *XlHbox6* expression is not strongly affected (this remains posterior; not shown). Note: the sections show some nonspecific signal in the epidermis (E) due to pigment

has not been described in the forebrain or in anterior axial mesoderm in early embryos. Treatment with a low RA concentration (10^{-8} M) resulted in the expression of two 3′ located genes, *Xhox-1A* and *Xhox-1B*. Treatment with 10^{-7} M RA gave somewhat more expression of *Xhox-1A* and *Xhox-1B* and also induced *XlHbox2* expression. With 10^{-6} M RA, *Xhox-1A* and *Xhox-1B* were expressed still more, but *XlHbox2* was expressed less strongly than after treatment with 10^{-7} M RA. No *XlHbox6* expression was found at any of the RA concentrations tested. These results were found in two separate experiments. They suggest that RA can transform anterior embryonic axial tissue to a more posterior specification, but that a pre-existing A-P polarity also regulates Hox-2 gene expression at stage 12.5, since *XlHbox6* is not expressed after treating head tissue with RA. The expression of this gene cannot be enhanced in domains that are in this case too anterior. Similar results have been found using anterior late gastrula ectoderm cultured up to stage 28 (tailbud stage). This also showed no *XlHbox6* expression (Sharpe et al., 1987; Sharpe and Gurdon, 1990). The embryo thus appears to be polarized with respect to its competence to express Hox genes in response to RA.

Discussion

There were five main conclusions from this study.

First, we showed for the first time in a *Xenopus* embryo, that the six Hox genes studied show spatial colinearity between the sequence of A-P levels at which they are expressed along the main A-P axis of the embryo and their putative 3′ to 5′ sequence in the *Xenopus* Hox-2 chromosomal complex.

Second, we showed that five of the six Hox genes examined show temporal colinearity. They are expressed sequentially, after the beginning of gastrulation in a sequence that is colinear with their putative 3′ to 5′ sequence in the *Xenopus* Hox-2 complex. Four of the five genes (*Xhox2.7*, *Xhox-1A*, *Xhox-1B* and *XlHbox2*) seem to be expressed in sequential waves, which travel from posterior to anterior along the main A-P axis of the embryo. The fifth (*XlHbox6*) is expressed last, and its expression begins and remains posterior. The situation apparently parallels that for the Hox-4 complex in the developing mouse embryo (Izpisua-Belmonte et al., 1991b). One possible explanation would be that the opening of the Hox-2 gene complex (Gaunt and Singh, 1990) is regulated by the increasing production of a morphogen, which spreads from the posterior end of the embryo (blastopore region) during gastrulation and neurulation. We note, also, that a biphasic expression pattern was evident for five of the six Hox-2 genes. We suspect that there are two specific phases of Hox-2 gene expression. (1) A coordinated phase, characterised by sequential activation of these genes, in a temporal sequence that is colinear with their 3′ to 5′ location in the Hox-2 complex. (2) A phase in which expression is individually modulated, such that each gene in the Hox complex is finally expressed in a specific domain along the A-P axis, with a relatively sharp anterior boundary (Wilkinson et al., 1989). RA seems to affect both phases (see below).

Third, we observed that *Xhox2.9* has an exceptional tem-

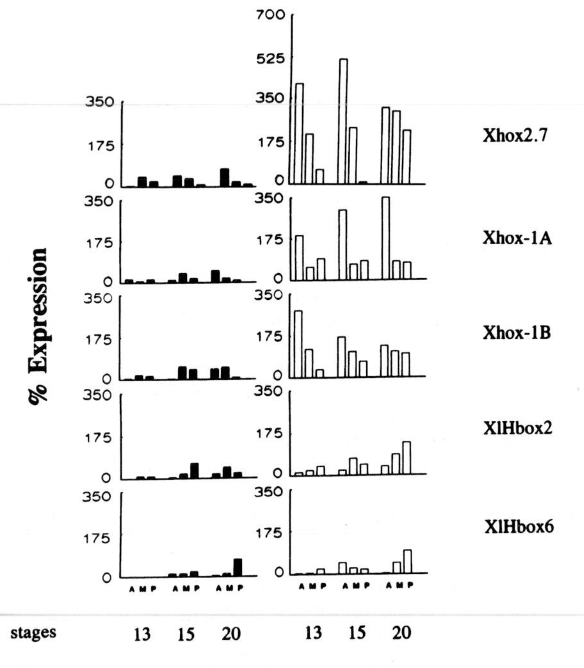

Fig. 5. Spatiotemporal expression patterns of the *Xenopus* Hox-2 genes. Solid bars: normal expression. Open bars: expression in RA treated embryos (treatment as in the Fig. 5 legend). Normal and RA-treated embryos were harvested at three sequential developmental times (stage 13 (early neurula), stage 15 (mid-neurula), stage 20 (late neurula)), and were then trisected into equal anterior (A), middle (M) and posterior (P) fragments. Expression of each of the five Hox genes examined was then measured in each fragment at each developmental stage (as in the Fig. 2 legend), and set out as a percentage of the maximum normal expression for each gene in whole embryos.

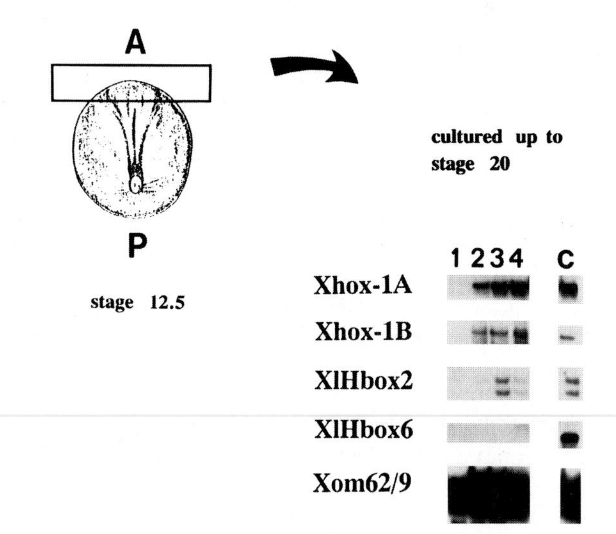

Fig. 6. Pre-existing A-P polarity. We cut out the presumptive forebrain and underlying prechordal plate mesoderm from the most anterior part of a stage-12.5 (early neurula) embryo (scheme of dissection on the left), cultured these explants up to stage 20 without RA (lane 1) and in the continuous presence of 10^{-8} M (lane 2), 10^{-7} M (lane 3) and 10^{-6} M RA (lane 4), and examined the expression of the four Hox-2 genes shown using RNAase protection assays.

poral expression pattern when compared with the other five *Xenopus* Hox-2 genes studied in this paper. This gene is expressed at a low level during gastrulation (stage 10) and neurulation (up to stage 15) and only reaches peak expression at a late stage (20; late neurula). No biphasic expression comparable to that of the other Hox-2 genes is evident, the *Xhox2.9* expression peak coincides with the second expression phase of the other Hox-2 genes. We conclude that individual modulation of Hox gene expression dominates the expression pattern of *Xhox2.9*, a conclusion that is also strongly indicated in other studies of the expression of *labial*-like Hox genes (Sundin and Eichele, 1992; Murphy et al., 1989; Wilkinson et al., 1989; Frohman et al., 1990; Hunt et al., 1991).

Fourth, we found that 10^{-6} M RA (non-localised treatment, applied from the beginning of gastrulation) destroys the wave patterns, causing Hox gene expression that is at least as high and rapid, anteriorly as posteriorly. The sequence of the magnitudes with which RA treatment affects expression of these six genes is colinear with the A-P sequence of axial positions at which they are normally expressed. It is possible that RA mimics an endogenous morphogen that contributes to regulating the expression of the Hox-2 complex during the gastrula and neurula stages of development. It could be that RA mimics the posterior morphogen postulated above and regulates opening of the Hox-2 complex, but our finding of an RA-competence gradient (below) raises the possibility that RA only regulates transcription of the complex, not its opening.

Fifth, we presented evidence for a pre-existing RA-insensitive A-P polarity in the early *Xenopus* embryo. We found that the normal spatial expression sequence of expression of the Hox-2 complex was at least partly conserved after treatment with a high RA concentration. We found further that anterior neuroectoderm and axial mesoderm, which express no Hox genes without RA treatment, expressed only the more anterior genes (and not *XlHbox6*), when treated with RA. The embryo is thus polarized with respect to its competence to express Hox genes in response to RA. These findings are compatible with the idea that an endogenous retinoid is a morphogen, but they make it unlikely that the spatial expression sequence of Hox-2 genes in the early embryo is regulated simply by a retinoid gradient. They provide an interesting puzzle concerning the nature of polarised RA competence, and the relationship of this phenomenon to the expression of retinoid receptors and to opening and transcription of the Hox-2 complex. The phenomenon of prepatterning or of a pre-existent A-P polarity in the early frog embryo has already been suggested by others (Sokol and Melton, 1991; Ruiz i Altaba and Jessell, 1991; Gerhart et al., 1989) and is confirmed by our findings.

It is clear that the findings above raise many questions that can only be answered by in situ data. We have obtained in situ hybridisation data for *XlHbox6* (Fig. 7), and these confirm that *XlHbox6* RNA, like its protein product (Wright et al., 1987) is localised posteriorly, as from stage 16, when its expression is first detectable, until the latest stage examined (stage 28). The mRNA is strongly expressed in the posterior CNS, and there is also a low level of expression in the lateral mesoderm. RA enhances the expression of *XlHbox6* mRNA such that expression now becomes very evident in the posterior mesoderm, but RA treatment does not strongly affect the anteroposterior localisation of *XlHbox6* expression (which remains posterior as was already indicated by our dissection studies).

Our findings thus revealed details of the endogenous expression, and the inducibility by RA of the Hox-2 complex of class 1 homeobox-containing genes during early *Xenopus laevis* embryogenesis. These investigations are a first step towards investigating the regulation of the expression of this Hox complex by different embryonic inducers during specification of the *Xenopus* A-P axis. They already give some clues as to the nature of the regulatory mechanism.

We would like to thank Eddy De Robertis, Doug Melton and Colin Sharpe for providing us with *Xenopus* Hox-2 cDNA clones. We would also like to thank Doug Melton for providing the *Xenopus* stage-17 cDNA library and Ali Hemmati-Brivanlou for helpful suggestions on whole-mount in situ techniques. Furthermore, we would like to thank Jacqueline Deschamps, Frits Meijlink, Olivier Destree, Anneke Koster, Antonio Simeone, Pieter Nieuwkoop, Siegfried de Laat and Pim Pijnappel for reading the manuscript; Ferdinand Vervoordeldonk for photographing the figures; and Erica Cohen and Marianne Nortier for typing the manuscript. The investigations were supported by the foundation of biological research (BION) which is subsidised by the Netherlands organisation for scientific research (NWO; Grant no. 431.122).

References

Acampora, D., D'Esposito, M., Faiella, A., Pannese, M., Migliaccio, E., Morelli, F., Stornaiuolo, A., Nigro, V., Simeone, A. and Boncinelli, E. (1989). The human Hox family. *Nucl. Acids Res.* **17**, 10385-10402.

Akam, M. (1989). Hox and HOM: Homologous gene clusters in insects and vertebrates. *Cell* **57**, 347-349.

Baron, A., Featherstone, M. S., Hill, R. E., Hall, A., Galliot, B. and Duboule, D. (1987). Hox-1.6: A mouse homeo-box-containing member of the Hox-1 complex. *EMBO J.* **6**, 2977-2986.

Boncinelli, E., Simeone, A., Acampora, D. and Mavilio, F. (1991). Hox gene activation by retinoic acid. *TIG.* **7**, 329-334.

Brand, N. J., Petkovich, M., Krust, A., Chambon, P., de The, H., Marchio, A., Tiollais, P. and Dejean, A. (1988). Identification of second human retinoic acid receptor. *Nature* **332**, 850-853.

Cho, K. W. Y. and De Robertis, E. M. (1990). Differential activation of Xenopus Homeobox genes by mesoderm inducing growth factors and retinoic acid. *Genes and Dev.* **4**, 1910-1916.

Dekker, E. J., Pannese, M., Houtzager, E., Bonicelli, E. and Durston, A. J. (1992). Colinearity in the Xenopus Hox-2 complex. *Mech. Dev.* (in Press)

Dolle, P., Izpisua-Belmonte, J. -C., Falkenstein, H., Renucci, A, and Duboule, D. (1989a). Coordinate expression of the murine Hox-5 complex homeobox-containing genes during limb bud pattern formation. *Nature* **342**, 767-772.

Dolle, P., Ruberte, E., Kastner, P., Petkovich, M., Stoner, C. M., Gudas, L. J. and Chambon, P. (1989b). Differential expression of genes encoding α, ß and γ retinoic acid receptors and CRABP in the developing limbs of the mouse. *Nature* **342**, 702-705.

Dolle, P., Izpisua-Belmonte, J. -C., Brown, J. M., Tickle, C. and Duboule, D. (1991). Hox-4 genes and the morphogenesis of mammalian genitalia. *Gènes and Dev.* **5**, 1767-1776.

Duboule, D. and Dolle, P. (1989). The structural and functional organisation of the murine Hox gene family resembles that of Drosophila homeotic genes. *EMBO J.* **8**, 1497-1505.

Durston, A. J., Timmermans, J. P. M., Hage, W. J., Hendriks, H. F. J., de Vries, N. J., Heideveld, M. and Nieuwkoop, P. D. (1989). Retinoic acid causes an anteroposterior transformation in the developing nervous system. *Nature* **340**, 140-144.

Eyal-Giladi, H. (1954). Dynamic aspects of neural induction in amphibia (experiments on *Amblystoma mexicanum* and *Pleurodeles waltii*). *Arch. Biol.* **65**, 179-259.

Flickinger, R. A. (1949). A study of the metabolism of amphibian neural crest cells during their migration and pigmentation in vitro. *J. Exp. Zool.* **112**, 165-185.

Fritz, A. F., Cho, K. W. Y., Wright, C. V. E., Jegalian, B. G. and De Robertis, E. M. (1989). Duplicated homeobox genes in Xenopus. *Dev. Biol.* **131**, 584-588.

Fritz, A. F. and De Robertis, E. M. (1988). Xenopus homeo-box-containing cDNAs expressed in early development. *Nucl. Acids Res.* **16**, 1453-1463.

Frohman, M. A., Botle, M. and Mertin, G.R. (1990). Isolation of the mouse *Hox2.9* gene; analysis of embryonic expression suggests that positional information along the anterior-posterior axis is specified by mesoderm. *Development* **110**, 589-607.

Gaunt, S. J. (1988). Mouse homeobox gene transcripts occupy different but overlapping domains in embryonic germ layers and organs: A comparison of Hox-3.1 and Hox-1.5. *Development* **103**, 135-144.

Gaunt, S. J. and Singh, P. B. (1990). Homeogene expression patterns and chromosomal imprinting. *TIG.* **6**, 209-212.

Gerhart, J., Danilchik, M., Doniach, T., Roberts, S. Browning, B. and Steward, R. (1989). Cortical rotation of the *Xenopus* egg: consequences for the anteroposterior pattern of embryonic dorsal development. *Development* **107 Supplement**, 37-51.

Graham, A., Papalopulu, N. and Krumlauf R. (1989). The murine and Drosophila homeobox gene complexes have common features of organisation and expression. *Cell* **57**, 367-378

Harvey, R. P. and Melton, D. A. (1988). Microinjection of synthetic *Xhox-1A* homeobox mRNA disrupts somite formation in developing Xenopus embryos. *Cell* **53**, 687-697.

Harvey, R. P., Tabin, C. J. and Melton, D. A. (1986). Embryonic expression and nuclear localization of Xenopus homeobox (*Xhox*) gene products. *EMBO J.* **5**, 1237-1244.

Hemmati-Brivanlou, A., Frank, D., Bolce, M. E., Brown, B. D., Sive, H. L. and Harland, R. M. (1990). Localization of specific mRNAs in *Xenopus* embryos by whole mount in situ hybridisation. *Development* **110**, 325-330.

Holtfreter, J. (1933). Der Exogastrulation, eine Selbstablosung des Ectoderms von Entomesoderm, Entwicklung und funktionelles verhalten nervenloser Organe. *Roux Arch. Etw. Mech. Org.* **129**, 669-793.

Hornbruch, A. and Wolpert, L. (1986). Positional signalling by Hensen's node when grafted to the chick limb bud. *J. Embryol. exp. Morph.* **94**, 257-265.

Hunt, P., Wilkinson, D. and Krumlauf, R. (1991). Patterning the vertebrate head: murine Hox 2 genes mark distinct subpopulations of premigratory and migrating cranial neural crest. *Development* **112**, 43-50.

Izpisua-Belmonte, J. -C., Tickle, C., Dolle, P., Wolpert, L. and Duboule, D. (1991a). Expression of the homeobox Hox-4 genes and the specification of position in chick wing development. *Nature* **350**, 855-589.

Izpisua-Belmonte, J. -C., Falkenstein, H., Dolle, P., Renucci, A. and Duboule, D. (1991b). Murine genes related to the Drosophila AbdB homeotic gene are sequentially expressed during development of the posterior part of the body. *EMBO J.* **10**, 2279-2289.

Kaneda, T. and Hama T. (1979). Studies on the formation and state of determination of the trunk organiser in the newt, *Cynops pyrhogster. Wilh. Roux Arch.. Entw. Mech.* **187**, 222-241.

Kappen, C.,Schughart, K. and Ruddle, F. H. (1989) Two steps in the evolution of Antennapedia-class vertebrate homeobox genes. *Proc. Natn. Acad. Sci., USA* **86**, 5459-5463.

Kessel, M. and Gruss, P. (1990). Murine developmental control genes. *Science* **249**, 374-379.

Kintner, C. R. and Melton, D. A. (1987). Expression of *Xenopus* N-CAM RNA in ectoderm is an early response to neural induction. *Development* **99**, 311-325.

Mangelsdorf, D. J., Ong, E. S., Dyck, J. A. and Evans, R. M. (1990). Nuclear receptor that identifies a novel retinoic acid response pathway. *Nature* **345**, 224-229.

Mangold, O. (1933). Uber die Inductionsfahigkeit der verscheidene Nezirke de Neurula von Urodelen. *Naturwiss.* **21**, 761-766.

Mariottini, P., Bagni, C., Annesi, F. and Amaldi, F. (1988). Isolation and nucleotide sequence of cDNAs for Xenopus laevis ribosomal protein *S8*: similarities in the 5′ and 3′ untranslated regions of mRNA for various r-proteins. *Gene* **67**, 69-72.

McGinnis, W. and Krumlauf, R. (1992). Homeobox genes and axial patterning. *Cell* **68**, 283-302.

Melton, D. A., Krieg, P. A., Rebagliati, M. R., Maniatis, Y. N., Zinn, K. and Green, M. R. (1984). Efficient in vitro synthesis of biologically active RNA and RNA hybridisation probes from plasmids containing bacteriophage SP6 promoter. *Nucl. Acids Res.* **12**, 7035-7056.

Muller, M., Carrasco, A. E. and De Robertis, E. M. (1984). A homeobox-containing gene expressed during oogenesis in Xenopus. *Cell* **39**, 157-162.

Murphy, P., Davidson, D. R. and Hill, R.E. (1989). Segment-specific expression of a homeobox-containing gene in the mouse hindbrain. *Nature* **341**, 156-159.

Nieuwkoop, P. D. and Faber, J. (1967). *The Normal Table Of* Xenopus laevis *(Daudin)* Second edition, Amsterdam: North Holland Publishing Co.

Nieuwkoop, P. D., Johnen, A. G. and Albers, B. (1985). *The Epigenetic Nature Of Early Chordate Development. Inductive Interaction And Competence.* Cambridge, England: Cambridge University Press.

Nohno, T., Noji, S., Koyama, E., Ohyama, K., Myokai, F., Kuroiwa, A., Salto, T. and Taniguchi, S. (1991). Involvement of Chox-4 chicken homeobox genes in determination of anteroposterior axial polarity during limb development. *Cell* **64**, 1197-1205.

Papalopulu, N., Lovell-Badge, R. and Krumlauf, R. (1991). The expression of murine Hox-2 genes is dependent on the differentiation pathway and displays a colinear sensitivity to retinoic acid in F9 cells and Xenopus embryos. *Nucl. Acids Res.* **19**, 5497-5506.

Petkovich, M., Brand, N. J., Krust, A. and Chambon, P. (1987). A human retinoic acid receptor which belongs to the family of nuclear receptors. *Nature* **330**, 444-450.

Regulski, M., McGinnis, N., Chadwick, R. and McGinnis, W. (1987). Developmental and molecular analysis of *Deformed*: a homeotic gene controlling Drosophila head development. *EMBO J.* **6**, 767-777.

Ruberte, E., Dolle, P., Krust, A., Zelent, A., Morriss-Kay, G. and Chambon, P. (1990). Specific spatial and temporal distribution of retinoic acid receptor gamma transcripts during mouse embryogenesis. *Development* **108**, 213-221.

Ruberte, E., Dolle, D., Chambon, P. and Morriss-Kay, G. (1991). Retinoic acid receptors and cellular retinoid binding proteins. II. Their differential pattern of transcription during early morphogenesis in mouse embryos. *Development* **111**, 45-60.

Ruiz i Altaba, A. and Melton, D. A. (1989). Biomodal and graded expression of the *Xenopus* homeobox gene *Xhox3* during embryonic development. *Development* **106**, 173-183.

Ruiz i Altaba, A. and Melton, D. A. (1990). Neural expression of the *Xenopus* homeobox gene *Xhox3*: Evidence for a patterning signal that spreads through the ectoderm. *Development* **108**, 595-604.

Ruiz i Altaba, A. and Jessell, T. (1991a). Retinoic acid modifies the pattern of cell differentiation in the central nervous system of neurula stage *Xenopus* embryos. *Development* **112**, 945-958.

Ruiz i Altaba, A. and Jessell, T. (1991b). Retinoic acid modifies mesodermal patterning in early Xenopus embryos. *Genes Dev.* **5**, 175-187.

Scott, M. P., Tamkun, J. W. and Hartzell E. D. (1989). The structure and function of the homeodomain. *BBA Rev. Cancer* **989**, 25-48.

Sharpe, C. R., Fritz, A., De Robertis, E. M. and Gurdon J. B. (1987). A homeobox-containing marker of positive neural differentiation shows the importance of predetermination in neural induction. *Cell* **50**, 749-758.

Sharpe, C. R. and Gurdon, J. B. (1990). The induction of anterior and posterior neural genes in *Xenopus laevis. Development* **109**, 765-774.

Simeone, A. Acampora, D., Arcioni, L., Andrews, P. W., Boncinelli, E., Mavilio, F. (1990). Sequential activation of HOX2 homeobox genes by retinoic acid in human embryonal carcinoma cells. *Nature* **346**, 763-766.

Sive, H. L., Draper, B. W., Harland, R. M. and Weintraub, H. (1991). Identification of a retinoic acid-sensitive period during primary axis formation in *Xenopus laevis. Genes Dev.* **4**, 932-942.

Sokol, S. and Melton, D. A. (1991). Pre-existent pattern in *Xenopus* animal pole cells revealed by induction with activin. *Nature* **351**, 409-411.

Spemann, H. (1931). Uber den Anteil von Implantat und Wirkskeim and der Orientierung und Beschaffenheit der induzierten Embryoanlage. *Roux Arch. Entw. Mech. Org.* **123**, 390-517.

Spemann, H. and Mangold, H. (1924). Uber die induction von Embryonanlagen durch implantation artfremder Organisatoren. *Roux Arch. Dev. Biol.* **100**, 599-638.
Spemann, H. (1938). Embryonic development and induction. New Haven: Yale University Press.
Sundin, O. H. and Eichele, G. (1992). An early marker of axial pattern in the chick embryo and its respecification by retinoic acid. *Development* **114**, 841-852
Tautz, D. and Pfiefle, C. (1989). A non-radioactive in situ hybridisation method for the localisation of specific RNAs in Drosophila embryos reveals translational control of the segmentation gene hunchback. *Chromosoma* **98**, 81-85.
Thaller, C. and Eichele, G. (1987). Identification and spatial distribution of retinoids in the developing chick limb bud. *Nature* **327**, 625-628.
Tickle, C., Alberts, B., Wolpert, L. and Lee J. (1982). Local application of retinoic acid to the limb bud mimics the action of the polarizing region. *Nature* **296**, 564-565.
Vaessen, M. -J., Meijers, H. J. C., Bootsma, D. and Geurts van Kessel, A (1990). The cellular retinoic-acid-binding protein is expressed in tissues associated with retinoic-acid-induced malformations. *Development* **110**, 371-378.
Wilkinson, D. G., Bailes, J. A. and McMahon, A. P. (1987). A molecular analysis of mouse development from 8 to 10 days post coitum detects changes only in embryonic globin expression. *Development* **99**, 493-500.
Wilkinson, D. G., Bhatt, S., Cook, M., Boncinelli, E. and Krumlauf, R. (1989). Segmental expression of Hox-2 homeobox-containing genes in the developing mouse hindbrain. *Nature* **341**, 405-409.
Wright, C. V. E., Cho, K. W. Y., Fritz, A., Burglin, T. R. and De Robertis E. M. (1987). A Xenopus laevis gene encodes both homeobox-containing and homeobox-less transcripts. *EMBO J.* **6**, 4083-4094.

Development 1992 Supplement, 203-207 (1992)
Printed in Great Britain © The Company of Biologists Limited 1992

Retinoic acid and the late phase of neural induction

C. R. SHARPE

Wellcome Trust and Cancer Research Campaign Institute of Cancer and Developmental Biology, Tennis Court Rd, Cambridge CB2 1QR, UK and *The Department of Zoology, University of Cambridge, Downing Street, Cambridge CB2 3EJ, UK*

Summary

Regional neural gene expression in *Xenopus* is the result of a number of processes that continue well beyond the end of gastrulation. By considering two of the basic features of neural induction, the duration of contact between mesoderm and ectoderm and the timing of neural competence, it has been possible to distinguish two phases in neural tissue formation. The late phase includes the period following gastrulation.

A factor in determining regional neural gene expression is the difference in inducing ability of the mesoderm that develops during gastrulation along the anterior-posterior axis. The resulting ability to express regional neural genes is subsequently refined during the late phase by a signal that progresses from the posterior part of the embryo. Using a dorsal explant system, it is shown that this progressive signal can be mimicked by the addition of retinoic acid (RA). However, the observation that regions along the anterior-posterior axis respond in different ways to the addition of RA suggests that additional factors are also important in defining regional neural gene expression. One possibilty is that the expression of retinoic acid receptors along the axis may demarcate regions that respond to RA in particular ways.

Key words: *Xenopus*, neural induction, retinoic acid.

Mechanisms of neural induction

The *Xenopus* neural tube forms as a result of cell interactions during the first 24 hours of embryonic development. The key event during this period is gastrulation when the mesoderm involutes and the anterior-posterior axis is formed. At this stage, signals from the mesoderm divert competent ectodermal cells from an epidermal to a neural pathway of development (Gurdon, 1987). It has recently become clear that at least two mechanisms generate these signals, one involves an interaction between the involuted mesoderm and the overlying ectoderm (reviewed by Hamburger, 1988; Sharpe and Gurdon, 1990) whilst the second appears to be independent of involution, occurring along a tangential interface of the two cell types (Dixon and Kintner, 1989; Ruiz i Altaba, 1990). In vitro, both mechanisms can be shown to cause the formation of a regionally distinct neural tube. In the first case, using the 'sandwich technique' of wrapping the mesoderm in competent ectoderm, regions of post-gastrulation mesoderm isolated from along the anterior-posterior axis induce the expression of different regional neural markers (Sharpe and Gurdon, 1990; Hemmati-Brivanlou et al., 1990). In the second case, planar induction is assayed in 'Keller sandwiches' (Keller and Danilchik, 1988) and the expression of regional neural markers can be followed by whole-mount in situ hybridisation (Doniach et al., 1992).

Little is known about either mechanism at the molecular level. However, they could well be manifestations of the same short-range (up to a few cell diameters) signals. In both cases, gastrulation plays an important role in generating the anterior-posterior axis; regional differences in the inducing ability of the mesoderm become more apparent following the involution and extension of the mesoderm, whilst groups of responding cells each expressing a regional neural marker become widely distributed along the axis following convergence-extension movements (Doniach et al., 1992, and this volume).

The timing of interactions between mesoderm and neurectoderm define two phases of neural induction

A feature common to all embryonic inductions is the involvement of two different cell types, one that produces an inducing signal and a second capable of responding to that signal over a window of time known as the period of competence. One of the first steps that we have taken to characterise neural induction involved analysing the timing of these interactions between mesoderm and ectoderm that give rise to neural tissue (Fig. 1; Sharpe and Gurdon, 1990).

The *duration of the inducing signal* was measured by removing the neurectoderm from the embryo during gastrulation and neurulation and culturing it in isolation to a time point equivalent to the tailbud stage then determining the levels of expression of a panel of neural markers. The neural markers included *XlHbox6*, a homeobox gene that is expressed predominantly in posterior neural tissue (Sharpe et al., 1987), *XIF6*, the *Xenopus* mid-sized neurofilament

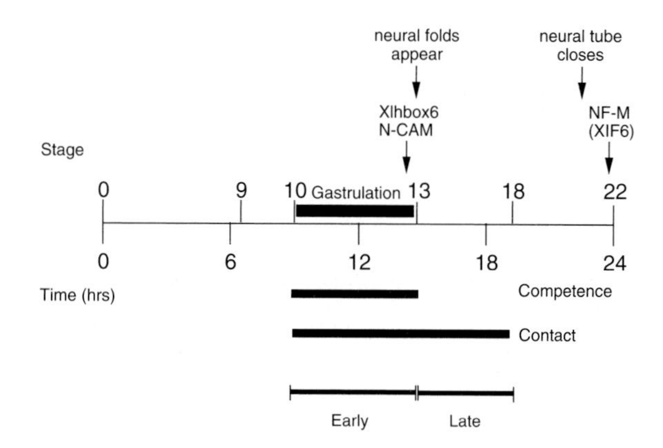

Fig. 1. Key features of neural tissue formation during the first 24 hours of development. Over the first 24 hours, equivalent to embryos reaching stage 22 of development at 23°C, all the events leading to neural tube closure have occurred. Neural induction probably starts with the onset of gastrulation though the first externally visible result is the formation of the neural folds at the end of gastrulation. Neural marker genes such as N-CAM (Kintner and Melton, 1987) and *XlHbox6* (Sharpe et al., 1987) can be detected at a slightly earlier stage. The early and late phases of neural tissue formation are defined by the duration of neural ectoderm competence and the period of contact between mesoderm and ectoderm required for the complete expression of neural markers in isolated neurectoderm.

gene (Sharpe, 1988), and *XIF3*, encoding a peripherin-like intermediate filament expressed mainly in anterior neural tissue in the embryo (Sharpe et al., 1989). Neural induction was said to be complete when the dissected neurectoderm was observed to express maximal levels of the neural markers and hence the responding tissue had received all the signals from the mesoderm that are required for independent neural differentiation. In addition to signals that convert ectoderm cells to the neural lineage, these may also include signals that maintain cells in the newly induced state. The experiment showed that the neurectoderm required signals from the embryo until the mid to late neurula stage in order to complete neural induction. Furthermore, this seemed to be the same for all the neural markers tested, there was no difference in the duration of contact required for full expression between the marker of anterior neural tissue (XIF3) and posterior neural tissue (XlHbox6).

In order to measure the *competence to respond*, ectoderm was removed from the early gastrula embryo and cultured in isolation for a period of time before recombination with a piece of inducing mesoderm. The end of competence was measured as the loss of ability of these conjugates to express a panel of neural markers when grown on to a fixed stage (equivalent to the tailbud stage). In this assay, competence is lost towards the end of gastrulation. This contrasts with the much longer period required for mesoderm-ectoderm contact. Using a transplantation technique, Servetnick and Grainger (1991) have similarly shown that competence to form neural tissue is lost at the end of gastrulation; furthermore, they observe that it is rapidly replaced by competence to form secondary neural structures such as the eye and otic vesicles.

Thus, from the comparison of the duration of contact and the time of loss of competence, it appears that neural tissue formation can be divided into two phases. The early phase is defined by the ability of the neurectoderm to respond to signals from the mesoderm and the late phase by the loss of competence and the time that neurectoderm can differentiate independently of mesodermal signals. So, what is happening in the late phase? One possibility is that this phase is required in order that the neural phenotype is maintained, suggesting that the first stages of neural induction could be reversible. Little is currently known about the stability of recently induced tissues. An additional possibility is that the late phase may be required for the patterning of regional neural gene expression along the anterior-posterior axis, and this is explored in the next section.

Regional neural gene expression in the late phase

Classical experimental embryology has outlined two possible mechanisms by which regional neural induction may occur. In the first, the anterior-posterior character of the inducing mesoderm that unfolds during gastrulation is imparted to the overlying neurectoderm (reviewed in Hamburger, 1988). This requires several inducers, which in the extreme will equal the number of regions that can be defined along the anterior-posterior axis of the neural plate. In support of this theory, it has been observed that some homeobox-containing genes are expressed in both mesoderm and neurectoderm and that each gene has its own particular boundary that is initially conserved between the two germ layers. It has been suggested that this reflects the ability of the mesoderm to impart its regional character on to the overlying neurectoderm, in a form of positional homeogenetic induction (De Robertis et al., 1989).

The second model suggests that neural induction first generates anterior neural tissue on to which a range of anterior-posterior characteristics are subsequently imposed in response to a gradient of a second signal derived from the posterior of the embryo. This 'gradient' model reduces the number of inducers to two and suggests that regional neural gene expression depends on the ability of neurectoderm cells to interpret the graded signal (reviewed by Saxen, 1989).

Recent experiments and observations have shown that neither model explains neural induction completely. For example, although anterior mesoderm taken at the end of gastrulation differs from posterior mesoderm in its ability to induce the panel of regional neural markers (Sharpe and Gurdon, 1990), these differences do not correlate with the final defined pattern of neural gene expression along the anterior-posterior axis. This is illustrated by the observation that only posterior mesoderm induces the neurofilament gene *XIF6* and yet this gene is ultimately expressed along the length of the neural tube. In another example, the ability of post-gastrula axial mesoderm (notochord) to induce expression of the *engrailed* gene has been examined. Again there is a difference in inducing ability, with the region underlying the eventual position of expression being the strongest inducer, but this ability is also found at more posterior positions along the anterior-posterior axis (Hemmati-

Brivanlou et al., 1990). In both cases, although particularly marked, the differences in the inducing ability of the mesoderm do not provide the complete solution to the problem of regional neural gene expression.

In terms of the 'gradient' model, it might be predicted that the *ability* to express neural markers is found progressively from the posterior end, whilst the ability to express anterior markers is initially found throughout the neural plate. This can be examined by isolating the dorsal tissues of the embryo that constitute the neurectoderm and underlying mesoderm (a dissection that may also contain a few dorsal endodermal cells). This explant can then be divided along the anterior-posterior axis and the ability of each piece to express neural markers determined. Repetition of this process at different time points during the late phase of neural tissue formation should make it possible to identify the progress of a signal that is propagated from the posterior end of the embryo. For example, the ability to express the neurofilament gene *XIF6* is initially confined to posterior regions but is progressively found in middle and anterior regions. *XlHbox6* shows an ability to be expressed progressively in the mid-region of the neural plate (Fig. 2; Sharpe, 1991).

By marking cells with Nile Blue sulphate, it is possible to follow their fate during subsequent development. Marked cells in the mid region at the end of gastrulation give rise predominantly to spinal cord (Fig. 3 and unpublished data) Although there is extensive distortion of the mark, this seems to be displaced both to the anterior and to the posterior. Marks placed at the lateral boundary of the mid region at the end of gastrulation converge towards the mid line and also extend both to the anterior and posterior. A much more extensive analysis of the fate map of both superficial and deep tissues in this region at these stages has been carried out by Keller (1975). Although the results are essentially the same, the observations reported here clarify the fates of the particular explants used in the above experiment. These results suggest that the observed differences in the ability of regions to express *XIF6* and *XlHbox6* are not due to sampling errors that result in different regions being examined at each stage. It seems more likely that the mid-region at the end of gastrulation is normally fated to give rise to spinal cord but does not have the ability to express *XlHbox6* because it has not received the required signal. The evidence that RA might be involved in this signal and the involvement of cell migration is considered in the next section.

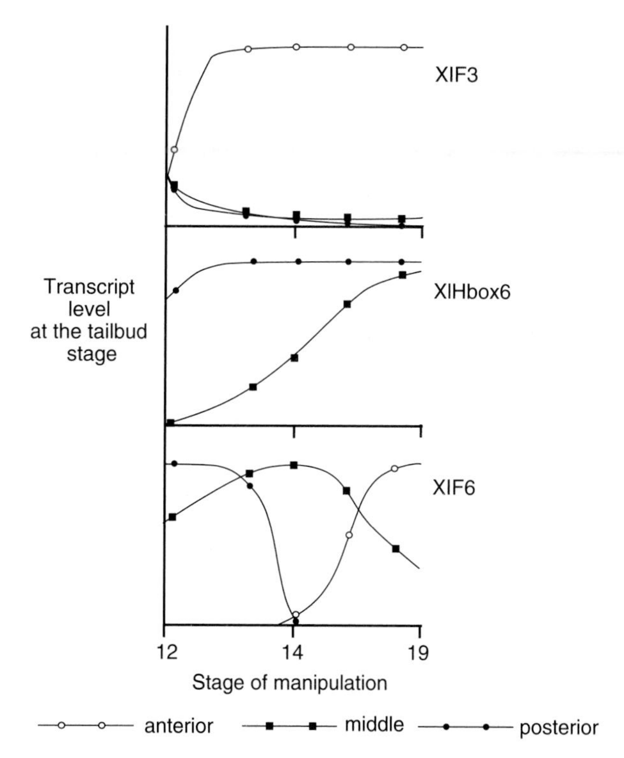

Fig. 2. The ability to express regional neural markers varies along the anterior-posterior axis in the late phase of neural tissue formation. The pattern of expression of XIF3, an anterior neural marker, remains stable in the anterior region. However, expression of *XlHbox6* in the mid region increases progressively during the late phase. Likewise there is a trend for the ability to express *XIF6* (NF-M) to be found in more anterior regions. The figure is a qualitative representation of the levels of expression of the marker genes in explants isolated at the stage shown on the x-axis and grown on to the tailbud stage determined by RNAase protection assay. A representative analysis is described in Sharpe (1991).

Retinoic acid can mimic the signals that result in regional neural gene expression in the embryo

Retinoic acid (RA) was first shown to affect *Xenopus* development by Durston and colleagues (1989) who noted that treatment of embryos resulted in altered differentiation of the neural tube. These studies have been extended (Papalopulu et al., 1991; Sive et al., 1990; Sive and Cheng, 1991; Ruiz i Altaba and Jessel, 1991a, b) to show effects both on neural patterning and mesoderm formation. Retinoic acid also affects neural gene expression and can act in combination with growth factors to alter the patterns of gene expression in isolated ectodermal tissue (Cho and De Robertis, 1990). In addition, RA has an effect on a wide range of developmental and regenerative systems (reviewed by Brockes, 1989).

It is possible to isolate the dorsal explants described above at the end of gastrulation and ask whether RA results in an altered ability to express the panel of neural markers. The observation is that physiological concentrations of RA can confer the ability to express *XIF6* onto the anterior piece and *XlHbox6* onto the mid piece (Sharpe, 1991). This shows that RA can mimic the endogenous signals that result in regional neural gene expression in the embryo. In addition, the transformation in ability takes place in isolated explants, suggesting that recruitment of cells by movement from one region to another along the anterior-posterior axis is not required. Unlike assays in which development is perturbed by RA, in this approach RA is seen to restore normal patterns of gene expression. The identification of RA and related retinoids in the *Xenopus* embryo at this stage (Durston et al., 1989) strengthens the argument that they may also be the endogenous signals . The 'gradient' model implied that the imposition of posterior character was dependent on the gradient of a signal derived from the posterior end of the embryo (reviewed by

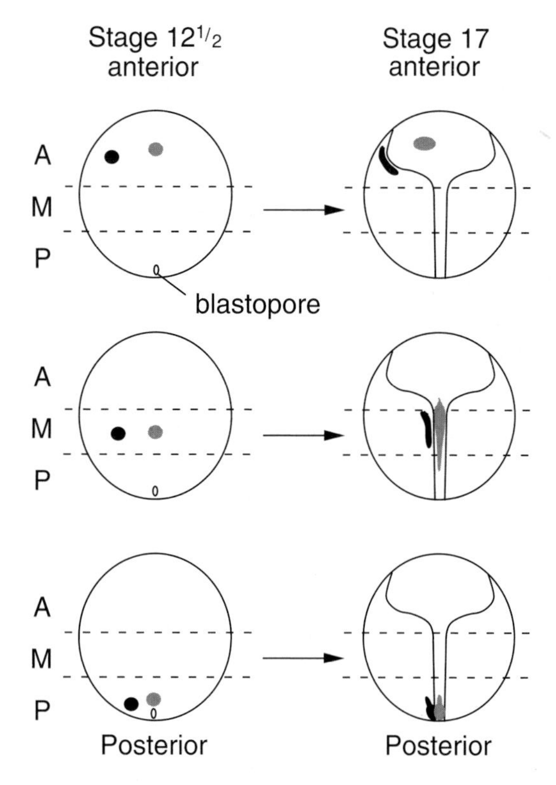

Fig. 3. Fate mapping the neural plate at the beginning of the late phase. Nile Blue sulphate marks were used to follow the fates of regions of cells along the anterior-posterior axis of the dorsal side of the embryo. One mark was applied to each embryo, the figure shows representative examples of marks placed centrally in each of the three regions that were removed as explants, with one mark on the dorsal mid-line and one displaced laterally.

Saxen, 1989). Whilst this idea remains attractive, there is no direct evidence that a gradient of any factor is involved.

Retinoic acid has been shown to activate the expression of homeobox genes expressed in an embryonal carcinoma cell line in a dose-dependent fashion (Simeone et al., 1990). Those genes at the 3′ end of the homeobox gene cluster are activated in response to low doses of RA whilst those at the 5′ end require higher concentrations (Simeone et al., 1990). This also reflects the order of expression of these genes along the axis of the embryo with the genes located at the 5′ end of the cluster being expressed at the posterior end of the embryo (Wilkinson et al., 1989; Hunt et al., 1991). It is therefore interesting to ask whether the availabilty of RA in the *Xenopus* embryo may play a part in determining the boundaries of expression of homeobox-containing genes. The explant experiments showed that RA can stimulate the expression of *XlHbox6* in the mid region suggesting an involvement in determining the anterior boundary of expression of this gene. However, when applied to the anterior explant, RA did not cause the expression of *XlHbox6* (Sharpe, 1991). This is a little surprising since cells in this region are capable of responding to RA as shown by their altered ability to express *XIF6*. This suggests that the availabilty of RA is not the sole factor that controls the boundaries of expression of genes such as *XlHbox6*. An explanation might be that the anterior-posterior axis is divided into a number of regions and that RA acts independently within each region. For example, the level of RA may affect the boundary of expression of *XlHbox6* within the mid region but be unable to affect XlHbox6 expression in the anterior region at all. One way in which these hypothetical regions may be demarcated is by the distribution of the retinoic acid receptors that provide the molecular machinery for the interpretation of endogenous signals.

Future directions

The retinoic acid receptors are ligand-dependent transcription factors and fall into two main classes. The first to be discovered were the RARs which can be subdivided into α, β and γ subclasses (Benbrook et al., 1988; Brand et al., 1989; Giguerre et al., 1987). A second group, the retinoid X receptors (RXRs) whose ligand is likely to be the 9-*cis* retinoic acid isomer can also be divided into three subgroups (Mangelsdorf et al., 1990, 1992). RARs and RXRs can act as heterodimers to modulate the expression of target genes (Kliewer et al., 1992). The proposition is that expression of the receptors will be restricted along the anterior-posterior axis and that the combination of different receptors will be used to demarcate regions that respond to RA in a particular way. In this respect, it has already been shown that *Xenopus* RARγ is restricted in its expression to particular regions along the anterior posterior axis (Ellinger-Ziegelbauer and Dreyer, 1991). Restricted patterns of expression of RAR have also been documented during the development of other species (for example, Dolle et al., 1989; Ruberte et al., 1990). Although the number of receptors available is large, and includes isoforms generated by alternative splicing (Kastner et al., 1990; Leroy et al., 1991; Zelent et al., 1991), the initial patterns of RAR expression in *Xenopus* may be quite simple if only a small number of regions need to be demarcated.

In conclusion, it is likely that differences in the inducing ability of the mesoderm and a progressive posterior signal acting during the late phase will be involved in regional neural induction. RA is able to mimic the progressive signal and this or a closely related molecule may play a role in vivo. The different abilities of regions along the anterior-posterior axis to respond to RA may reflect the domains of expression of RAR.

I would like to thank Chris Wylie, Janet Heasman, Melanie Sharpe and Tanya Whitfield for helpful comments on the manuscript.

References

Benbrook, D., Lenhardt, E. and Pfahl, M. (1988). A new retinoic acid receptor identified from a hepatocellular carcinoma. *Nature* **333**, 669-672.

Brand, N., Petkovich, M., Krust, A., Chambon, P., deThe, H., Marchio, A., Tiollis, P. and Dejean, A. (1989). Identification of a second human retinoic acid receptor. *Nature* **334**, 850-853.

Brockes, J. P. (1989). Retinoids, homeobox genes and limb morphogenesis. *Neuron* **2**, 1285-1294.

Cho, K. W. Y. and DeRobertis, E. M. (1990). Differential activation of

Xenopus homeobox genes by mesoderm inducing growth factors and retinoic acid. *Genes Dev.* **4**, 1910-1916.

DeRobertis, E. M., Oliver, G. and Wright, C. V. E. (1989). Determination of axial polarity in the vertebrate embryo: homeodomain proteins and homeogenetic induction. *Cell* **57**, 189-191.

Dixon, J. E. and Kintner, C. R. (1989). Cellular contacts required for neuralization in *Xenopus* embryos: evidence for two signals. *Development* **106**, 749-757.

Dolle, P., Ruberte, E., Kastner, P., Petkovich, M., Stoner, C. M., Gudas, L. and Chambon, P. (1989). Differential expression of genes encoding α, β and γ retinoic acid receptors and CRABP in the developing limbs of mouse. *Nature* **342**, 702-705.

Doniach, T., Phillips, C. R. and Gerhart, J. C. (1992). Planar induction of anteroposterior pattern in the developing central nervous system of *Xenopus laevis. Science* (In Press).

Durston, A. J., Timmermans, J. P. M., Hage, W. J., Hendriks, H. F. J., DeVries, N. J., Heideveld, M. and Nieuwkoop, P. D. (1989). Retinoic acid causes an anteroposterior transformation in the developing nervous system. *Nature* **340**, 140-144.

Ellinger-Ziegelbauer, H. and Dreyer, C. (1991). A retinoic acid receptor expressed in the early development of *Xenopus laevis. Genes Dev.* **5**, 94-104.

Giguerre, V., Ong, E. S., Segui, P. and Evans, R. M. (1987). Identification of a receptor for the morphogen retinoic acid. *Nature* **330**, 624-629.

Gurdon, J. B. (1987). Embryonic induction - molecular prospects. *Development* **99**, 285-306.

Hamburger, V. (1988). *The Heritage of Experimental Embryology. Hans Spemann and the Organizer.* Oxford: Oxford University Press.

Hemmati-Brivanlou, A., Stewart, R. M. and Harland, R. M. (1990). Region specific neural induction of an engrailed protein by anterior notochord in *Xenopus. Science* **250**, 800-802.

Hunt, P., Wilkinson, D. and Krumlauf, R. (1991). Patterning of the vertebrate head: murine Hox2 genes mark distinct subpopulations of premigratory and migrating neural crest. *Development* **112**, 43-51.

Kastner, P., Krust, A., Mendelsohn, C., Garnier, J. M., Zelent, A., Leroy, P., A., S. and Chambon, P. (1990). Murine isoforms of retinoic acid receptor γ with specific patterns of expression. *Proc. Natl Acad. Sci. USA* **87**, 2700-2704.

Keller, R. E. (1975). Vital dye mapping of the gastrula and neurula of *Xenopus laevis*. I. Prospective areas and morphogenetic movements of the superficial layer. *Dev. Biol.* **42**, 222-241.

Keller, R. E. and Danilchik, M. (1988). Regional expression, pattern and timing of convergence and extension during gastrulation of *Xenopus laevis. Development* **103**, 193-209.

Kliewer, S. A., Umesono, K., Mangelsdorf, D. J. and Evans, R. M. (1992). Retinoid X receptor interacts with nuclear receptors in retinoic acid, thyroid hormone and vitamin D3 signalling. *Nature* **355**, 446-449.

Leroy, P., Krust, A., Zelent, A., Mendelsohn, C., Garnier, J. M., Kastner, P., Dierich, A. and Chambon, P. (1991). Multiple isoforms of the mouse retinoic acid receptor α are generated by alternative splicing and differential induction by RA. *EMBO J.* **10**, 59-69.

Mangelsdorf, D. J., Boymeyer, U., Heyman, R. A., Zhou, J. Y., Ong, E. S., Oro, A. E., Kakizuka, A. and Evans, R. M. (1992). Characterisation of three RXR genes that mediate the action of 9 cis retinoic acid. *Genes Dev.* **6**, 329-344.

Mangelsdorf, D. J., Ong, E. S., Dyck, J. A. and Evans, R. M. (1990). Nuclear receptor that identifies a novel retinoic acid response pathway. *Nature* **345**, 224-229.

Papalopulu, N., Clarke, J. D. W., Bradley, L., Wilkinson, D., Krumlauf, R. and Holder, N. (1991). Retinoic acid causes abnormal development and segmental patterning of the anterior hindbrain in *Xenopus* embryos. *Development* **113**, 1145-1158.

Ruberte, E., Dolle, P., Krust, A., Zelent, A., Morriss-Kay, G. and P., C. (1990). Specific spatial and temporal distribution of RAR γ transcripts during mouse embryogenesis. *Development* **108**, 213-222.

Ruiz i Altaba, A. (1990). Neural expression of the *Xenopus* homeobox gene Xhox3: evidence for a patterning neural signal that spreads through the ectoderm. *Development* **108**, 595-604.

Ruiz i Altaba, A. and Jessell, T. M. (1991a). Retinoic acid modifies mesodermal patterning in early *Xenopus* embryos. *Genes Dev.* **5**, 175-188.

Ruiz i Altaba, A. and Jessell, T. M. (1991b). Retinoic acid modifies the pattern of cell differentiation in the central nervous system of neurula stage *Xenopus* embryos. *Development* **112**, 945-958.

Saxen, L. (1989). Neural Induction. *International Journal of Developmental Biology*, **33**, 21-48.

Servetnick, M. and Grainger, R. M. (1991). Changes in neural and lens competence in *Xenopus* ectoderm: Evidence for an autonomous developmental timer. *Development* **112**, 177-188.

Sharpe, C. R. (1988). Developmental expression of a neurofilament-M and two vimentin-like genes in *Xenopus laevis. Development* **103**, 269-277.

Sharpe, C. R. (1991). Retinoic acid can mimic endogenous signals involved in transformation of the *Xenopus* nervous system. *Neuron* **7**, 239-247.

Sharpe, C. R., Fritz, A., DeRobertis, E. M. and Gurdon, J. M. (1987). A homeobox containing marker of posterior neural differentiation shows the importance of predetermination in neural induction. *Cell* **50**, 749-758.

Sharpe, C. R. and Gurdon, J. B. (1990). The induction of anterior and posterior neural genes in *Xenopus . Development* **109**, 765-774.

Sharpe, C. R., Pluck, A. and Gurdon, J. B. (1989). XIF3, a Xenopus peripherin gene, requires an inductive signal for enhanced expression in anterior neural tissue. *Development* **107**, 701-714.

Simeone, A., Acampara, D., Arcioni, L., Andrews, P. W., Boncinelli, E. and Mavilio, F. (1990). Sequential activation of Hox2 homeobox genes by RA in human embryonal carcinoma cells. *Nature* **346**, 763-766.

Sive, H. L. and Cheng, P. F. (1991). Retinoic acid perturbs expression of Xhox lab genes and alters mesodermal determination in *Xenopus laevis. Genes Dev.* **5**, 1321-1332.

Sive, H. L., Draper, B. W., Harland, R. M. and Weintraub, H. (1990). Identification of a retinoic acid sensitive period during primary axis formation in *Xenopus laevis. Genes Dev.* **4**, 932-942.

Wilkinson, D., Bhatt, S., Cook, M., Boncinelli, E. and Krumlauf, R. (1989). Segmental expression of Hox2 homeobox containing genes in the developing mouse hindbrain. *Nature* **341**, 405-409.

Zelent, A., Mendelsohn, C., Kastner, P., Garnier, J. M., Ruffenach, F., Leroy, P. and Chambon, P. (1991). Differentially expressed isoforms of the mouse retinoic acid receptor b are generated by usage of two promoters and alternative splicing. *EMBO J.* **10**, 71-81.

Index

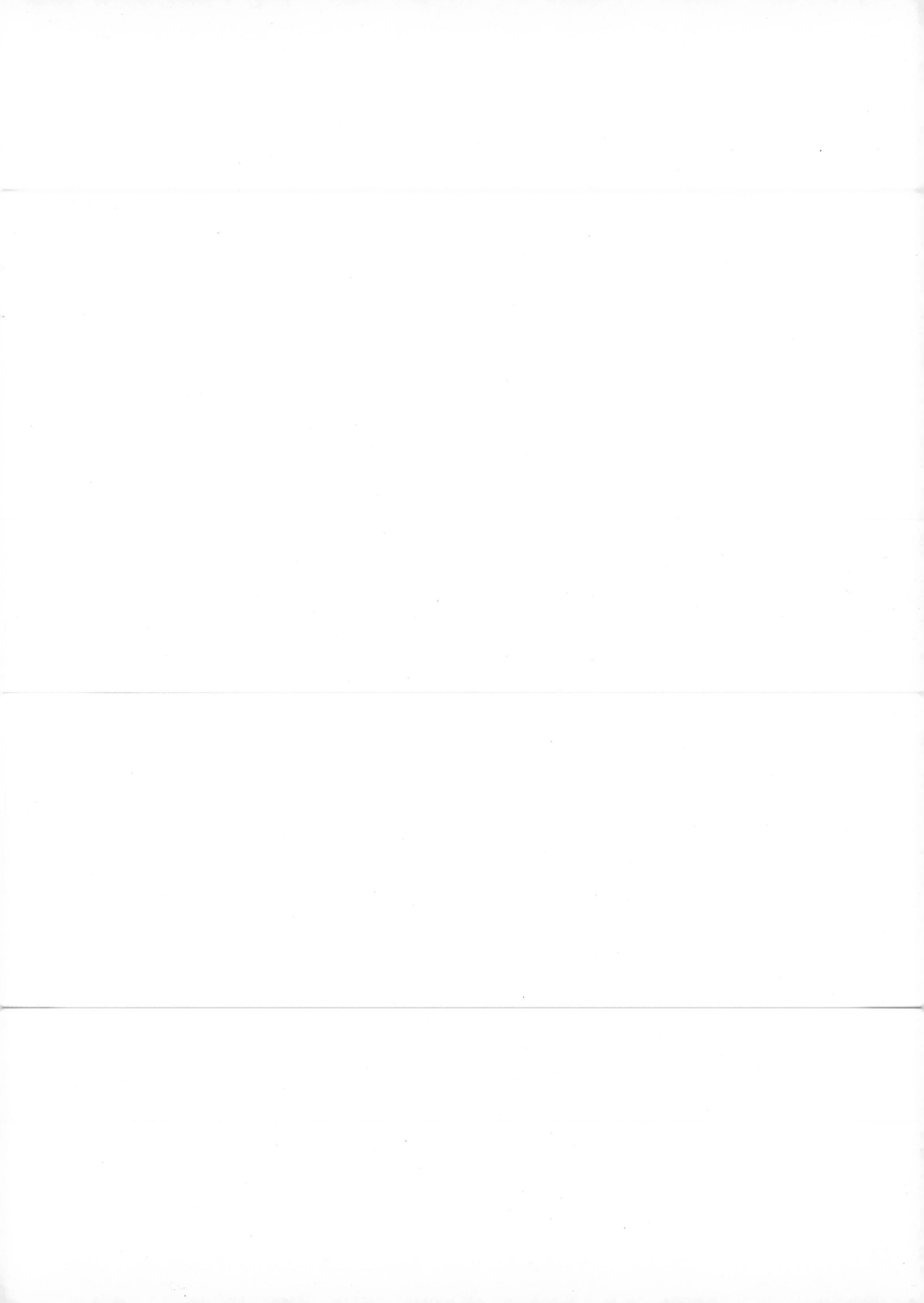